Constructions of Deviance

Social Power, Context, and Interaction

EIGHTH EDITION

PATRICIA A. ADLER
University of Colorado

PETER ADLER
University of Denver

CENGAGE
Learning·

Australia • Brazil • Japan • Korea • Mexico • Singapore • Spain • United Kingdom • United States

CENGAGE
Learning

Constructions of Deviance: Social Power, Context, and Interaction, Eighth Edition
Patricia A. Adler and Peter Adler

Product Director: Marta Lee-Perriard

Product Manager: Jennifer Harrison

Content Developer: Liana Sarkisian

Product Assistant: Julia Catalano

Media Developer: John Chell

Content Development Services Manager: Greg Albert

Marketing Manager: Kara Kindstrom

Content Project Manager: Charu Khanna, MPS Limited

Art Director: Carolyn Deacy

Manufacturing Planner: Judy Inouye

Production Service/Project Manager: Charu Khanna, MPS Limited

PMG Text Researcher: Punitha Rajamohan

Cover Image Credit: © Malcolm Tarlofsky

Compositor: MPS Limited

For product information and technology assistance, contact us at **Cengage Learning Customer & Sales Support, 1-800-354-9706**.

For permission to use material from this text or product, submit all requests online at **www.cengage.com/permissions**.

Further permissions questions can be e-mailed to **permissionrequest@cengage.com**.

Library of Congress Control Number: 2014955501

ISBN: 978-1-305-09354-6

Cengage Learning
20 Channel Center Street
Boston, MA 02210
USA

Cengage Learning is a leading provider of customized learning solutions with office locations around the globe, including Singapore, the United Kingdom, Australia, Mexico, Brazil, and Japan. Locate your local office at **www.cengage.com/global**.

Cengage Learning products are represented in Canada by Nelson Education, Ltd.

To learn more about Cengage Learning Solutions, visit **www.cengage.com**.

Purchase any of our products at your local college store or at our preferred online store **www.cengagebrain.com**.

Printed in the United States of America
Print Number: 01 Print Year: 2014

To Diane and Dana
Who remind us that the zest of life lies near the margins

and

To Chuck
Who brings out the deviance in everyone

and

To John
Who showed us the miracle of life and rebirth

and

To Jane
Who lives the ordinary as deviant

and

To Dubs and Linda
Who remind us what are friends are for

and

To Lois and David
Who allowed our nephews their independence to be deviant

and

To Marc
Who taught us that dreams can come true

and

To Asher Isaac Adler
Who made our dreams come true

Preface

Can you remember all the way back to the year 2012? It was the year that the United States doubled down and reelected its first multiracial president, a seeming impossibility just a few years ago. Kate Middleton, of British royalty, was hounded by paparazzi and photographed in the nude. Mommy porn came to legitimacy, as suburban housewives got caught up in the sexual craze of reading *Fifty Shades of Gray*. In the media, Charlie Sheen self-destructed in front of the world, streaming went viral, the most acclaimed television show was about meth cooks, and people's cell phones started talking to them. The masses occupied *Wall Street*, while people shared their state secrets with *WikiLeaks* and Big Brother spied on everyone. Flash mobs broke out all over, and so did bombs in public places and mass shootings. The weather went deviant and brought environmental disasters from storms to floods, fires, drought, tornados (*Sharknados?*), melting polar ice, and oil spills. The scandal over performance-enhancing drugs continued to grow, drawing in ever more prominent cheaters, including seven-times Tour de France cycling winner Lance Armstrong and Yankees baseball superstar A-Rod, with spillover into Major League Baseball and the National Football League. Aaron Hernandez, New England Patriots star, was imprisoned for (possibly double) murder. Dirty dealing in the ranks of sport continued, with the New Orleans Saints, 2010 Super Bowl winners, encouraging players to purposely injure opponents. In the government, Secret Service agents were caught hiring prostitutes in Colombia, and the military was exposed as a center for sexual harassment, inappropriate treatment of women, and sexual assault. Piracy continued to flourish off the Somali coasts, with men in primitive fishing vessels capturing huge cargo ships and obtaining millions of dollars in ransom. The Mexican–American border devolved into a war zone, as rival drug cartels slugged and shot it out over the multibillion-dollar illicit drug industry. On the lifestyle front, at least 14 states and the District of Columbia legalized same-sex marriage, with more expected to follow. The same can be said for medical marijuana, with dispensaries springing up legally in 20 states and with two, Colorado and

Washington, legalizing the recreational use of this drug. And while marijuana smoking gained currency, tobacco lost it, with a trend emerging among colleges to ban the smoking of cigarettes on campus. Emerging to take their place were E-cigarettes and colored and flavored cigar blunts, all targeted to a youth market.

Deviance is ubiquitous; it is all around us. Turn the pages of your local newspaper, check out your favorite current events Websites, listen to people talk in your gym locker room, and invariably they will be discussing the latest fads, the newest forms of transgressing the norms, and the ways we continue to push the boundaries of society's expectations. Contrary to what one pundit[1] claimed in the 1990s, deviance is far from dead; rather, it is flourishing like never before. In the 3 years since the sixth edition of this book was released, we were able to find more than 75 pertinent scholarly articles related to the sociology of deviant behavior. An embarrassment of riches, they made it difficult to choose among them, to select pieces that you, our readers, would find most interesting, fascinating, and relevant. It is within this context that the eighth edition of *Constructions of Deviance* was born. This book is a labor of love for us as we tweak each edition, remain in touch with the changing nuances in the field, and present to our audience what we feel is the most exciting research in the sociology of deviant behavior today.

As the shifting sands of morality in American society continue to transform our culture, deviance and its changing definitions are at the fore. More than 40 years ago, Gusfield (1967) showed us that, with continuous social change, activities that were considered nondeviant a generation ago can take on deviant characteristics (cigarette smoking, distracted driving), whereas activities that were once deviant are now acknowledged as commonplace (tattooing and piercing). For better or worse, rapid social change is occurring in front of our eyes. Against that backdrop, this is a fabulous time to study the varying definitions of deviant behavior and the subsequent consequences for individuals and society. It is our hope that this eighth edition of *Constructions of Deviance* incorporates some of this field's highlights. We want to show the liveliness of the debates, the graphic images that sociologists paint, and the wide array of activities that fall under the domain of deviance.

NEW TO THIS EDITION

In all of this book's editions, we have tried to keep pace with the transformations that have occurred in the research on deviant behavior. Despite the fact that some have decried the theoretical and empirical death of deviance, we had a surfeit of research from which to choose. Our difficulty lay not in finding new pieces, but in winnowing down our selections to the space available. We wanted to keep some basics the same, as they have come to represent the core of the book, while at the

[1]Colin. Sumner, *The Sociology of Deviance: An Obituary* (London: Open University Press, 1994).

same time infusing it with an insurgence of new material. We are pleased to offer you, then, the best of the old and a spate of exciting fresh selections.

New to this volume are chapters on the meaning of "natural law" and its relation to deviance; how, in order to gain attention, claimsmakers try to label as deviant organizations with stellar reputations; the racial profiling of young, inner-city Mexican Americans; how women on parole struggle to reclaim and manage deviant identities; how multiracial people manage their racial identities; a support group for people disabled by bowel disorders; cyber support groups for self-injurers; the pyramidal structure and international criminal activities of Hezbollah, one of the world's largest terrorist networks; the collusion and corruption between the government and the oil industry that together created the conditions leading to one of the worst environmental disasters in modern history: the Deepwater Horizon Gulf Oil spill; the rise and growth in the sales of lifelike sex dolls and the men who maintain intimate relationships with them; the hidden world of power and dominance in sadomasochistic sex play; Internet hackers' subcultural norms; the interactions between dancers and audience members in a male strip show; and the liminal social position of people who smoke cigarettes but do not see themselves as cigarette smokers: social smokers.

Relatively new pieces that have quickly become students' favorites remain, including chapters on the continuing debates (relativism vs. absolutism) over definitions of deviance; rationalizations used by shoplifters to neutralize their deviance; the negative connotations faced by male cheerleaders; the stigma management strategies of homeless children; sexual assaults and the party scene on campus; how women use drugs to maintain their eating disorders; people's decisions to commit burglary; and how people with emotional disorders relabel themselves. Some of the most popular, now classic, pieces still continue to be relevant, such as the chapters on the social construction of drug scares; the status battles over smoking; homophobia in women's sport; the mark of a criminal record; crime in the medical profession; becoming bisexual; developing a fat identity; convicted rapists' rationalizations; anorexia and bulimia; studying sex, drug trafficking, and child abuse; women in gangs; pimp-controlled prostitution; and drug dealers and smugglers' attempts to get out of the business.

We have continued to amend the part introductions and the synopses that introduce each selection. The breadth and depth of these sections enable this book to be either used as a stand-alone text–reader or easily synthesized with existing standard textbooks. As this book has gone through its various transformations, it has been our intent to convert it into more of a text in its own right: an anthology of empirical works with scholarly commentary from the framing discipline of sociology.

As it was from the beginning, the book still proudly represents the social constructionist approach, building upon our own intellectual backgrounds in symbolic interactionism and ethnographic research. As such, it retains its vibrant appeal, offering the most contemporary empirical readings that are drawn from qualitative studies rich in experiential descriptions of deviance from the everyday life perspective. At the same time, the book has increasingly incorporated more classical and mainstream theoretical and innovative methodological elements.

We offer this collection as a testimonial to the continuing vibrancy of deviance research. Our sense is that there has never been a better time to study these phenomena and to look at the changes that they represent about society. Above all else, we hope that you find the readings enjoyable, enlightening, and thought provoking.

ACKNOWLEDGMENTS

By now, literally thousands of students have been exposed to these readings and have been "christened" into the sociology of deviance. Many people provided critical feedback that has helped us in fashioning this latest edition. First and foremost are our many students, particularly in Patti's class, "Deviance in U.S. Society," at the University of Colorado, attracting more than 500 each semester, and in Peter's class, "Deviance and Society," at the University of Denver. These intrepid souls continue to brave the material and exams in these courses, despite their reputation as among the toughest on their respective campus. Extra thanks and acknowledgments go to the valiant assistant teaching assistants (ATAs) at the University of Colorado who have dedicated two semesters of their lives to this class to personalize it for other students, to keep the exams hilarious and topical as well as challenging, and to form a cohesive working group. These students have provided us with a template of what contemporary collegians desire, do, and dream about. They remind us of the diversity of sentiments—moral and immoral, normative and deviant, radical and conservative—that exist.

There have been some special people, such as Julia Cantzler, Tim Carpenter, Katherine Coroso, Marci Eads, Marc Eaton, Abby Fagan, Molly George, Joanna Gregson, Tamera Gugelmeyer, Paul Harvey, Tom Hoffman, Katy Irwin, Jennifer Lois, Adina Nack, Patrick O'Brien, Joe Settle, Katie Sirles, Jesse Smith, Jennifer (Skadi) Snook, Sarah Sutherland, Alex Thompson, and John Tribbia, who have provided much of the impetus for the changes and amendments we have made throughout. Our friends in the discipline continue to suggest studies, to supply encouragement, and to lend support for our endeavors. Whether through a quick conversation in a hallway or at a convention hotel, an email message, a lengthy letter, or a harangue over the telephone, they remind us that we should keep the edge and continue to search for the latest examples to hold their students' interest.

We would also like to thank our many colleagues in education who over the years have reviewed manuscript chapters, responded to surveys, or otherwise provided suggestions for improvement; such feedback has enabled us to improve this text edition over edition, for which we are grateful.

The stalwart staff at Thomson Wadsworth (Cengage) has provided unending support during the process of revisions and custom editions. We are fortunate to have worked with such diligent professionals as Matt Ballantyne, Tali Beesley, Paula Begley-Jenkens, Linda deStefano, Halee Dinsey, Jerilyn Emori, Peggy Francomb, Wendy Gordon, Jane Hetherington, Jennifer Jones, Bob Jucha, Bob Kauser, Ari Levenfeld, Kristin Marrs, Lin Marshall, Andrew Ogus, Reilly O'Neal, Michael Ryder, Liana Sarkisian, Erica Silverstein, Denise Simon, Steve Spangler,

Liz van der Mandele, Jennifer Walsh, Jay Whitney, Staci Wolfram, Matthew Wright, Dee Dee Zobian, and Beth Zuber. Special commendation must go to Eve Howard, our senior editor who worked hand in hand with us since the second edition until she left the company, and Serina Beauparlant, the editor who originally conceived the project. For this edition, we are pleased to welcome Seth Dobrin to our editorial team, and in the short time we have worked with him we can already see that they will have similar impacts as our previous editors and production assistants.

One of the pleasures of editing this book has been sharing it with our friends and relatives. With our first edition, we started a tradition of dedicating each volume to a different person or couple who has had a meaningful impact on our lives inside and outside the academy. We have continued this tradition throughout the subsequent editions. We respectfully dedicated the first edition to our partners in crime, Diane Duffy and Dana Larsen; the second edition to our dear and enduring friend Chuck Gallmeier; the third edition to the inimitable John Irwin; the fourth edition to our intimate friend of almost 40 years, Jane Horowitz; the fifth edition to neighbors and compadres Linda and Dubs Jacobsen; the sixth edition to Lois and David Baru, our sister and brother-in-law; and the seventh edition to Marc Taron, architect extraordinaire, brother-in-arms, and the solid foundation of our social network in our new home, Maui. Sadly, this book is also in memory of our dear friend and dedicatee of the third edition, the late John Irwin, who passed away in the first days of 2010, but whose integrity, vitality, professionalism, and honesty will live on forever through the many people he touched in his 80 years on earth. We are most pleased to be able to dedicate this book to our first grandchild, Asher, who has made our dreams come true. Finally, our children, Jori and Brye, keep us young with their irrepressible energy, enthusiasm, and zest for life. To all our readers of previous editions, thanks for the support; to the new readers of this eighth edition, welcome to the journey!

About the Editors

Patricia A. Adler (Ph.D., University of California–San Diego) is professor emerita of sociology at the University of Colorado. She has written and taught in the area of deviance, qualitative methods, and the sociology of children. A second edition of her book *Wheeling and Dealing* (Columbia University Press), a study of upper-level drug traffickers, was published in 1993. She has received many honors, including Outstanding Teacher in the Faculty of Arts and Sciences, and the Outstanding Researcher Award from the University of Colorado. In addition, she was awarded the Mentor Excellence Award in 2004 from the Society for the Study of Symbolic Interaction (SSSI).

Peter Adler (Ph.D., University of California–San Diego) is professor emeritus of sociology and criminology at the University of Denver. His research interests include social psychology, drugs and society, and sociology of work, sport, and leisure. His first book, *Momentum*, was published in 1981 by Sage. Peter has been honored with the University Lecturer Award and as the Outstanding Scholar/ Teacher at the University of Denver, as well as being named by the Society for the Study of Symbolic Interaction (SSSI) as Mentor of the Year in 2005.

Together, the Adlers served as copresidents of the Midwest Sociological Society from 2006 to 2007. They have edited the *Journal of Contemporary Ethnography* and were the founding editors of *Sociological Studies of Child Development*. In 2010, the third edition of their anthology, *Sociological Odyssey*, was published by Wadsworth Cengage, and in 2001, they released volume 1 of the *Encyclopedia of Criminology and Deviant Behavior*, coedited with Jay Corzine. In addition to publishing over one hundred journal articles, chapters in books, and book reviews, among their many books are *Membership Roles in Field Research*, a treatise on qualitative methods published by Sage in 1987; *Backboards and Blackboards*, a participant-observation study of college athletes that was published by Columbia University Press in 1991; *Peer Power*, an examination of the culture of elementary schoolchildren that was published by Rutgers University Press in 1998; and *Paradise Laborers*, a study of hotel workers in Hawai'i, published in 2004 by Cornell University Press. Their most recent project, *The Tender Cut*, focuses on self-injurers (cutters and burners) and was published by NYU Press (2011). Peter and Patti are retired and live on Maui, Hawai'i.

About the Contributors

Michael P. Arena is employed by a large state criminal justice agency, where he is an analyst and trainer. He holds an M.A. in organizational behavior and a Ph.D. in forensic psychology, with an emphasis in the administration and management of criminal justice. Along with Bruce Arrigo, he authored *The Terrorist Identity: Explaining the Terrorist Threat* (NYU Press, 2006), as well as articles in a variety of behavioral and social science journals.

Elizabeth A. Armstrong is an associate professor of sociology at the University of Michigan. Her research interests include the sociology of culture, social movements, sexuality, gender, and higher education. She is the author of *Forging Gay Identities: Organizing Sexuality in San Francisco, 1950–1994* (University of Chicago Press, 2002) and *Paying for the Party: How College Maintains Inequality* (coauthored with Laura Hamilton; Harvard University Press, 2013).

Howard S. Becker lives and works in San Francisco. He is the author of *Outsiders*, *Art Worlds*, *Writing for Social Scientists*, and *Tricks of the Trade*. He has taught at Northwestern University and the University of Washington. In 1998, the American Sociological Association bestowed upon him the Career of Distinguished Scholarship Award, the association's highest honor.

Michelle Bemiller received her Ph.D. in sociology from the University of Akron in 2005. She is now an associate professor of sociology at Walsh University in Ohio. Her interests include the sociology of deviant behavior, criminology, the sociology of gender, and the sociology of the family. Her past research has examined nontraditional mothers' (e.g., incarcerated mothers', noncustodial mothers') experiences with motherhood and occupational burnout among sexual assault and domestic violence shelter workers. Currently, Dr. Bemiller is pursuing research within the scholarship of teaching and learning, an area of study that explores the success of problem-based learning approaches in criminal justice courses.

Douglas J. Besharov, a lawyer, is the Joseph J. and Violet Jacobs Scholar in Social Welfare Studies at the American Enterprise Institute for Public Policy Research and a professor at the University of Maryland School of Public Policy. There, he directs the University's Welfare Reform Academy and teaches courses on family policy, welfare reform, evaluation, and the implementation of social policy. He was the first director of the U.S. National Center on Child Abuse and Neglect. Among his publications is *Recognizing Child Abuse: A Guide for the Concerned* (Free Press, 1990).

Joel Best is professor of sociology and criminal justice at the University of Delaware. His books include *Threatened Children* (1990), *Random Violence* (1999), *Damned Lies and Statistics* (2001), *Deviance: Career of a Concept* (2004), *The Stupidity Epidemic* (2011), and *Social Problems* (2013).

Elaine M. Blinde is emeritus professor and former chair of the Department of Kinesiology at Southern Illinois University–Carbondale before her retirement in 2010. Her research relates to the sociological analysis of sport, with a particular focus on gender issues. Recent work and publications relate to disability and sport, gender beliefs of young girls, and women's relationship to baseball.

Elizabeth Bradshaw is an assistant professor of sociology at Central Michigan University who specializes in the area of social and criminal justice, specifically the intersection of state and corporate criminality. Her dissertation examined the causes of the Deepwater Horizon explosion and the ensuing response to the 2010 Gulf of Mexico oil spill as a form of state–corporate environmental crime. She is now building on this work to develop the concept of "criminogenic industry structures" and is using this framework to study the hydraulic fracturing industry in Michigan. Beyond state–corporate crime, her additional areas of teaching and research include environmental criminology, surveillance, and social movements against corporate globalization.

William J. Chambliss was, at the time of his death, professor of sociology at George Washington University. He is the author and editor of over 20 books, including *Law, Order and Power* (with Robert Seidman); *On the Take: Petty Crooks to Presidents*; *Organizing Crime*; *Exploring Criminology*; *Boxman: A Professional Thief's Journey* (with Harry King); *Crime and the Legal Process*; *Making Law* (with Marjorie S. Zatz); *Sociology* (with Richard P. Applebaum); and *The Essence of Criminology* (with Aida Haas). He is the past president of the Society for the Study of Social Problems and the American Society of Criminology. He is currently writing a book on the political economy of piracy and smuggling.

Meda Chesney-Lind is professor and chair of Women's Studies at the University of Hawai'i at Manoa. Nationally recognized for her work on women and crime, she has authored a number of books, including *Girls, Delinquency, and Juvenile Justice*; *The Female Offender*; *Girls, Women, and Crime*; *Female Gangs in America*; *Invisible Punishment*; *Beyond Bad Girls*; *Fighting for Girls*. She has just finished an edited volume entitled *Feminist Theories of Crime* that explores the international dimensions of feminist criminology.

Terry Cluse-Tolar received her MSW and Ph.D. in social work from The Ohio State University. She is currently professor and chair of the Social Work Department at the University of Toledo. Her research interests include women and children in poverty, crisis intervention, and marginalized populations.

Donald R. Cressey was professor of sociology at the University of California, Santa Barbara, when he died in 1987. His most well known publications include *Principles of Criminology* (with Edwin H. Sutherland), *Other People's Money*, and *Theft of the Nation*. Cressey received many honors for his research and teaching, and he cherished none more than the Edwin H. Sutherland Award presented by the American Society of Criminology in 1967.

Paul Cromwell is professor emeritus of criminal justice at Wichita State University. He received his Ph.D. in criminology from Florida State University in 1986. His publications include 16 books and over 50 articles and book chapters. Recent publications include *Breaking and Entering: Burglars on Burglary* (with James Olson), *In Their Own Words: Criminals on Crime*, and *In Her Own Words: Women Offenders Perspectives on Crime and Victimization* (with Leanne Alarid). He has extensive experience in the criminal justice system, including service as parole commissioner and chairman of the Texas Board of Pardons and Paroles.

Jenny L. Davis is an assistant professor of sociology and anthropology at James Madison University. She received a Ph.D. in sociology at Texas A&M University in 2012. Her research interests include new media, social media, identity, technology and society, stigma, social psychology, and community. She has recently published in such journals as the *American Sociological Review*, *Symbolic Interaction*, *Deviant Behavior*, and the *American Behavioral Scientist*. She contributes regularly to the *Huffington Post* and *Society Pages'* Cyborgology Blog.

Scott H. Decker is Foundation Professor in, and director of, the School of Criminology and Criminal Justice at Arizona State University. He received a B.A. in social justice from DePauw University and an M.A. and a Ph.D. in criminology from Florida State University. His main research interests are in the areas of gangs, juvenile justice, criminal justice policy, and the offender's perspective. His books include *European Street Gangs and Troublesome Youth Groups* (winner of the American Society of Criminology, Division of International Criminology Outstanding Distinguished Book Award, 2006) and *The International Handbook of Juvenile Justice* (Springer-Verlag, 2006). His most recent books include *Drug Smugglers on Drug Smuggling: Lessons from the Inside* (Temple University Press, 2007) and *Criminology and Public Policy* with Hugh Barlow (Temple University Press, 2010).

Robert J. Durán is assistant professor of sociology at the University of Tennessee. He is the author of *Gang Life in Two Cities: An Insider's Journey* (Columbia

University Press, 2013). His major research interests are in urban ethnography and include forms of empowerment for marginalized groups, race and ethnic relations, social control, and violence. His current research projects focus on the states of Colorado, New Mexico, Texas, and Utah, as well as on the U.S.–Mexico border. He is the recipient of the 2011 New Scholar Award from the American Society of Criminology Division on People of Color and Crime.

Emile Durkheim (1858–1917) was a French sociologist who is generally considered to be the father of sociology. His major works are *Suicide*; *The Rules of Sociological Method*; *The Division of Labor in Society*; and *The Elementary Forms of Religious Life*.

Oskar Engdahl is an associate professor and director of the master's degree program in criminology at the Department of Sociology and Work Science, University of Gothenburg. His main research interests are in the areas of motivation, opportunity, and control of economic and white-collar crime, especially in banking and finance.

Kai T. Erikson is the William R. Kenan, Jr. Professor Emeritus of Sociology and American Studies at Yale University. He is the author of several books, including *Wayward Puritans*; *Everything in Its Path*; and *A New Species of Trouble*. He served as president of the American Sociological Association, the Society for the Study of Social Problems, and the Eastern Sociological Society.

John H. Gagnon was professor of sociology at the State University of New York, Stony Brook, until 1998. He is the author or coauthor of such books as *Sex Offenders*, *Sexual Conduct*, *Human Sexualities*, and *The Social Organization of Sexuality*. In addition, he is the coeditor of a number of books, most recently *Conceiving Sexuality* and *Encounter with AIDS: Gay Men and Lesbians Confront the AIDS Epidemic*, as well as the author of many scientific articles.

Laura Hamilton is assistant professor of sociology at the University of California, Merced. Her recent book *Paying for the Party: How College Maintains Inequality* (equally authored with Elizabeth A. Armstrong) explores how the organization of social and academic life at 4-year residential universities systematically disadvantages all but the most affluent of students.

Alex Heckert is professor of sociology at Indiana University of Pennsylvania. Most of his published research has been in the areas of family sociology, deviance, and medical sociology. His recent research in the area of domestic violence has attempted to improve the prediction of nonphysical abuse and physical reassault among batterer program participants. He has published in journals such as *Social Forces*, the *Journal of Research in Crime and Delinquency*, the *Journal of Marriage and the Family*, *Demography*, the *Journal of Family Issues*, *Rural Sociology*, *Family Relations*, *Violence and Victims*, the *Journal of Family Violence*, the *Journal of Interpersonal Violence*, and *The Sociological Quarterly*, among others.

Druann Maria Heckert received her M.A. from the University of Delaware and her Ph.D. from the University of New Hampshire. She teaches at Fayetteville (North Carolina) State University. Her research is in the areas of stigmatized appearance, positive deviance, and deviance theory, and her articles have appeared in journals such as *Deviant Behavior*, *The Sociological Quarterly*, *Symbolic Interaction*, and *Free Inquiry in Creative Sociology*.

Anne Hendershott is professor of psychology, sociology, and social work at the Franciscan University of Steubenville in Ohio. Dr. Hendershott received her Ph.D. in urban sociology from Kent State University. Her research interests focus on the role of religion and natural law as constraints on engaging in deviant behavior. Her most recent books are *The Politics of Deviance* (Encounter Books), *The Politics of Abortion* (Encounter Books), and *Status Envy* (Transaction).

Nancy J. Herman-Kinney is professor of sociology at Central Michigan University. She received her Ph.D. from McMaster University in Hamilton, Ontario, and subsequently was awarded a postdoctoral fellowship at the University of Toronto, where she conducted qualitative and quantitative research on deinstitutionalized psychiatric patients. Her primary research areas are deviance, children, youth and social problems, the sociology of mental illness, and qualitative research methods. She is the coeditor of the *Handbook of Symbolic Interactionism* (with Larry T. Reynolds), and her articles have appeared in the *Journal for the Scientific Study of Religion*, *Symbolic Interaction*, *Deviant Behavior*, and the *Journal of Contemporary Ethnography*.

Travis Hirschi received his Ph.D. in sociology from the University of California–Berkeley. He is currently Regents Professor Emeritus at the University of Arizona. He served as president of the American Society of Criminology and has received that organization's Edwin H. Sutherland Award. His books include *Delinquency Research* (with Hanan C. Selvin), *Causes of Delinquency*, *Measuring Crime* (with Michael Hindelang and Joseph Weis), and *A General Theory of Crime* (with Michael R. Gottfredson). His most recent book is a volume coedited with Michael Gottfredson: *The Generality of Deviance*.

Malcolm D. Holmes is professor of sociology at the University of Wyoming. His research primarily analyzes the relationships of race and ethnicity to criminal justice outcomes. His recent book *Race and Police Brutality: Roots of an Urban Dilemma* (with Brad W. Smith) examines the social–psychological dynamics underlying the use of excessive force by the police. Currently, he is engaged in empirical and theoretical research projects that investigate the causes of various forms of extralegal police aggression.

Thomas J. Holt is an associate professor in the School of Criminal Justice at Michigan State University. He received his Ph.D. in criminology and criminal justice from the University of Missouri–St. Louis in 2005. His research focuses on cybercrime and the ways that technology and the Internet facilitate deviance on- and off-line.

Jenna Howard received her Ph.D. from Rutgers University. She is currently a research analyst with the Department of Family and Community Health at Robert Wood Johnson Medical School. Her research interests revolve around the social psychology of individual and organizational change.

Cathryn Johnson is senior associate dean in the Laney Graduate School, and professor of sociology, at Emory University. Her work is in the areas of legitimacy, justice, and power processes in groups and organizations, in addition to identity formation and negotiation processes. Her recent research project, with Karen A. Hegtvedt, examines the relationship between collective sources of legitimacy and emotional reactions in unjust situations.

Peter Kaufman is an associate professor of sociology at the State University of New York at New Paltz. He received his B.A. from Earlham College and his Ph.D. from Stony Brook University. His teaching and research interests include education, critical pedagogy, symbolic interaction, and the sociology of sport.

Nikki Khanna received her Ph.D. in sociology from Emory University and is currently an associate professor of sociology at the University of Vermont. Her work looks at racial identity among biracial and multiracial Americans and has been published in outlets such as *Social Psychology Quarterly*, *Ethnic and Racial Studies*, *Sociological Spectrum*, *The Sociological Quarterly*, *Sociology Compass*, and *Teaching Sociology*. Her recent book, *Biracial in America: Forming and Performing Racial Identity*, looks at Black–White biracial Americans and the underlying processes shaping their racial identities.

David A. Kinney is professor of sociology at Central Michigan University. He earned his Ph.D. from Indiana University in Bloomington and then conducted qualitative research in urban schools while a postdoctoral fellow at the University of Chicago. His primary research areas are the sociology of adolescence, the sociology of education, and the sociology of identity. He has published articles and chapters on children's use of time, adolescent peer cultures, and education in venues such as *Sociology of Education*, the *Journal of Contemporary Ethnography*, *Youth and Society*, *American Behavioral Research Scientist*, and *The Praeger Handbook of American High Schools*. He was the series editor of *Sociological Studies of Children and Youth* (JAI/Elsevier/Emerald) from 1999 to 2011.

Edward O. Laumann is the George Herbert Mead Distinguished Service Professor in the Department of Sociology and the College at the University of Chicago, as well as the Director of the Ogburn Stouffer Center for Population and Social Organization. Previously, he was the editor of the *American Journal of Sociology* and dean of the Social Sciences Division and Provost at the University of Chicago. He published two volumes on sexuality in 1994: *The Social Organization of Sexuality* and *Sex in America*. Along with Robert Michael, he recently published *Sex, Love, and Health: Private Choices and Public Policy* (University of Chicago Press, 2001).

Lisa Laumann-Billings completed her doctoral work in developmental and clinical psychology at the University of Virginia. She has worked and published in the areas of divorce, family conflict, and child maltreatment for the past 10 years. She is currently working with Dr. David Olds at the Prevention Research Center in Denver, Colorado, on preventative strategies for reducing child maltreatment and family violence in high-risk families.

John Liederbach is an associate professor of criminal justice at Bowling Green State University. His primary research interests are in the areas of white-collar and professional crime, as well as police behavior and the study of community-level influences on officer activities and citizen interactions.

Kathleen Lowney is professor of sociology at Valdosta State University and is the editor of *Teaching Sociology*, an American Sociological Association journal. Her research focuses on how the mass media help create social constructions, be they about Disney, multiple personality disorder/dissociative identity disorder, or deviance. She has studied adolescent Satanism for 25 years. Currently, she is researching media accounts of "serial killer nurses," especially these accounts' constructions of gender, sexuality, and beauty and how they influence legal decisions about those nurses and their crimes. Her undergraduate degrees in sociology (honors) and comparative religions are from the University of Washington, and her Ph.D. in religion and society is from Drew University. She received the 2013 Regents' Award of Excellence in Teaching from the University System of Georgia.

Joseph Marolla is vice provost for Instruction, associate professor of sociology, and affiliated with the Sports Management Program at Virginia Commonwealth University. He is doing research in the sociology of sport and remains interested in the social construction of deviance.

Penelope A. McLorg has a Ph.D. in anthropology and was director of the Gerontology Program at Indiana University–Purdue University Fort Wayne. She specializes in the biological and sociocultural aspects of aging, with particular interests in health and aging. Dr. McLorg has conducted research in rural areas of Mexico and in the Midwest and has published on such topics as body composition, glucose metabolism, bone loss, eating disorders, and physical disability.

Robert K. Merton passed away in 2003. He was University Professor Emeritus at Columbia University at the time of his death and a member of the adjunct faculty of Rockefeller University. His books include *Social Theory and Social Structure*, *Sociology of Science*, *Sociological Ambivalence*, and *On the Shoulders of Giants*. He is the only sociologist who was awarded the National Medal of Science, the nation's highest scientific honor.

Robert T. Michael is the Eliakim Hastings Moore Distinguished Professor and dean emeritus of the Harris Graduate School of Public Policy at the University of Chicago. From 1984 to 1989, he served as director of the National Opinion Research Center (NORC).

Stuart Michaels served as project manager of the National Health and Social Life Survey (NHLS).

Ashley N. Miller graduated from Central Michigan University with a bachelor of science degree in journalism in 2012. She plans to pursue a career in elementary education.

Jody Miller is professor of criminal justice at Rutgers University-Newark. Her research focuses on gender, crime, and victimization, particularly in the contexts of urban communities and the commercial sex industry. She has published numerous articles, as well as the monographs *One of the Guys: Girls, Gangs, and Gender* (Oxford University Press, 2001) and *Getting Played: African American Girls, Urban Inequality, and Gendered* (New York University Press, 2008).

Staci Newmahr received her Ph.D. in sociology from Stony Brook University and is currently an associate professor of sociology at the State University of New York College at Buffalo. She is the author of *Playing on the Edge: Sadomasochism, Risk, and Intimacy* (Indiana University Press, 2011). Her main areas of scholarly interest are the sociology of sex and gender, deviance, and qualitative methods.

Tara Opsal is an assistant professor of sociology at Colorado State University. She received her Ph.D. from the University of Colorado–Boulder. Her research interests center on punishment, social control, and the consequences of criminal justice policy. She is the author of a number of articles that focus on parole, desistance from crime, and women's postincarceration experiences. She teaches courses in corrections, gender, and deviance.

Devah Pager is professor of sociology at Princeton University. Her research focuses on the social and economic consequences of mass incarceration, with a particular emphasis on how our crime policies contribute to enduring racial inequality. Using an experimental field methodology, Pager's research uncovers the hidden world of employment discrimination and illustrates the great barriers faced by both blacks and ex-offenders in their pursuit of economic self-sufficiency. These topics are explored in her recent book *Marked: Race, Crime, and Finding Work in an Era of Mass Incarceration*.

Douglas W. Pryor is professor of sociology and criminal justice in the Department of Sociology, Anthropology and Criminal Justice at Towson University in Maryland. He is coauthor of *Dual Attraction: Understanding Bisexuality* and author of *Unspeakable Acts: Why Men Sexually Abuse Children*. He has served as a department chair, as a cochair of his campus's Self-Study Re-accreditation Committee, and as president of his local campus AAUP Faculty Association.

Richard Quinney is professor of sociology emeritus at Northern Illinois University. His books include *The Social Reality of Crime*; *Critique of Legal Order*; *Class, State, and Crime*; *Criminology as Peacemaking* (with Harold E. Pepinsky); and *Criminal Behavior Systems* (with Marshall B. Clinard and John Wildeman). His autobiographical reflections are contained in *Journey to a Far Place* and *For the Time Being*.

Craig Reinarman is professor of sociology and legal studies at the University of California–Santa Cruz. He has been a visiting professor at Utrecht University and the University of Amsterdam. He has served on the board of directors of the College on the Problems of Drug Dependence, as a consultant to the World Health Organization's Programme on Substance Abuse, and as a principal investigator on projects for which he received research grants from the National Institute on Drug Abuse. Dr. Reinarman is the author of *States of Mind* (Yale University Press, 1987), and coauthor of *Cocaine Changes* (Temple University Press, 1991) and *Crack in America* (University of California Press, 1997).

Anne R. Roschelle is an associate professor of sociology and an affiliate in the Women's Studies Program at the State University of New York at New Paltz. She earned her Ph.D. at the State University of New York at Albany and is the author of numerous articles on the intersection of race, class, and gender, with a focus on extended kinship networks and family poverty. She is the author of *No More Kin: Exploring Race, Class, and Gender in Family Networks*, which was a recipient of *Choice Magazine*'s 1997 Outstanding Academic Book Award. Recently, she published a series of articles on domestic violence, welfare reform, and motherhood among homeless women in San Francisco, as well as an article on feminist methodology in Cuba. She is currently writing a book about homeless families in San Francisco. Dr. Roschelle is an avid hiker and plays flute (with coauthor Peter Kaufman) in a local rock band called Questionable Authorities.

Maren T. Scull teaches in the Department of Sociology at the University of Colorado–Denver. Her research interests include sexualities, deviance, social psychology, gender, and qualitative methods. She earned her Ph.D. in sociology at Indiana University–Bloomington in 2013.

Diana Scully is professor of sociology and women's studies at Virginia Commonwealth University in Richmond, Virginia, where she is also director of the Women's Studies Program. Her books include *Men Who Control Women's Health* (Teachers College Press, 1994) and *Understanding Sexual Violence* (Routledge, 1994).

Amy Shuman is professor of folklore, English, Women's Studies, and Anthropology at The Ohio State University, a core faculty member of Project Narrative and the Center for Folklore Studies, and a fellow of the Mershon Center for International Security. Her publications include *Storytelling Rights: The Uses of Oral and Written Texts Among Urban Adolescents*, *Other People's Stories: Entitlement Claims and the Critique of Empathy*, and, with Carol Bohmer, *Rejecting Refugees: Political Asylum in the 21st Century*.

Edwin H. Sutherland (1883–1950) received his Ph.D. from the University of Chicago. His major works include his enduring textbook, *Criminology*, which was first published in 1924, *The Professional Thief*, and *White Collar Crime*. He is generally regarded as the founder of differential association theory.

Brian Sweeney received his Ph.D. from Indiana University–Bloomington. He is currently an assistant professor of sociology at LIU Post in Brookville, New York. His research focuses on youth and adolescence, peer cultures, gender and sexuality, and education.

Diane E. Taub was professor of sociology and associate dean of the College of Liberal Arts at Southern Illinois University–Carbondale and, later, professor and chair of sociology at Indiana University–Purdue University Fort Wayne. She publishes primarily in the areas of eating disorders and physical disabilities, and has received many university teaching awards.

Kara L. Taylor is currently working toward her bachelor of arts in sociology, with a concentration in social and criminal justice, from Central Michigan University. She recently started an internship with the Florida Department of Juvenile Justice and plans to work with youths to prevent delinquency and offer them more opportunities for a better future.

Alex Thompson is currently pursuing a Ph.D. in sociology at the University of Colorado–Boulder. He received his B.S. degree in sociology from the University of Wisconsin–La Crosse and an M.A. in sociology from the University of Connecticut. He is interested in social psychology, the body and embodiment, crime and deviance, and substance use. He is presently conducting an ethnographic analysis of the medical and legal marijuana markets in Colorado.

Quint Thurman is provost and vice president for student and academic affairs at Sul Ross State University in Alpine, Texas. Previously, he was professor of criminal justice and department chairperson at Texas State University–San Marcos. He received a Ph.D. in sociology from the University of Massachusetts (Amherst) in 1987. His publications include seven books and more than 35 refereed articles that have appeared in such journals as the *American Behavioral Scientist, Crime and Delinquency, Criminology and Public Policy, Social Science Quarterly, Justice Quarterly, Police Quarterly*, and the *Journal of Quantitative Criminology*. His recently published books include *Controversies in Policing* (with coauthor Andrew Giacomazzi); an anthology, *Contemporary Policing: Controversies, Challenges, and Solutions* (with Jihong Zhao); and *Police Problem Solving* (with J. D. Jamieson).

Justin L. Tuggle received his B.A. from Humboldt State University and his M.A. from the University of Wyoming. He teaches third grade at Grant Elementary School in Redding, California. He is married with two children.

Katherine S. Vecitis teaches in the Department of Sociology at Tufts University. She received her Ph.D. in sociology from the University of Colorado in 2009. Her research interests include the sociology of drugs, deviant behavior, and gender.

Martin S. Weinberg received his Ph.D. from Northwestern University and is professor of sociology at Indiana University. He is coauthor of *Deviance: The Interactionist Perspective, The Study of Social Problems, Sexual Preference, Homosexualities, Male Homosexuals,* and *Dual Attraction,* and has contributed articles to such journals as the *American Sociological Review, Social Problems,* the *Journal of Contemporary Ethnography, Archives of Sexual Behavior,* and the *Journal of Sex Research.*

Jason Whitesel (Ph.D., Ohio State University) is a faculty member in Women's and Gender Studies at Pace University. His research is driven by the underlying intragroup strife among gay men that is created by their rigid body image ideal. He coauthored (with Amy Shuman) the article "Normalizing Desire: Stigma and the Carnivalesque in Gay Bigmen's Cultural Practices" in the journal *Men and Masculinities.* His book, *Fat Gay Men: Girth, Mirth, and the Politics of Stigma* (NYU Press, 2014), describes events at Girth & Mirth club gatherings and examines how gay big men use campy-queer behavior to reconfigure and reclaim their sullied images and identities.

Colin J. Williams is professor of sociology at Indiana University–Purdue University Indianapolis. He is coauthor of *Male Homosexuals, Sex and Morality in the U.S.,* and *Dual Attraction.*

Celia Williamson received her Ph.D. in social work from Indiana University. She is currently a professor in the Social Work Program at the University of Toledo, is chair of the Research and Analysis Committee for the Ohio Attorney General's Trafficking in Person Commission, and hosts the oldest annual human trafficking conference in the nation (www.prostitutionconference.com).

Richard Wright is Curators' Professor and Chair of Criminology and Criminal Justice at the University of Missouri–St. Louis and editor in chief of *Oxford Bibliographies Online [Criminology].* He has been studying active urban street criminals, especially residential burglars, armed robbers, carjackers, and drug dealers for the past two decades. His research has been funded by the National Science Foundation, National Institute of Justice, Harry Frank Guggenheim Foundation, Icelandic Research Council, National Consortium on Violence Research, and Irish Research Council for the Humanities and Social Sciences. Along with Richard Rosenfeld and Scott Jacques, he currently is conducting a major field-based study of respectability and social control among drug sellers in Amsterdam's red-light district. His most recent book, coauthored with Bruce Jacobs, is *Street Justice: Retaliation in the Criminal Underworld* (Cambridge University Press, 2006).

General Introduction

The topic of deviance has held an enduring fascination for students of sociology, gripping their interest for several reasons. Some people have career plans that include law or law enforcement and want to expand their base of practical knowledge. Others feel a special affinity for the subject of deviance on the basis of their personal experience or inclination. A third group is drawn to deviance merely because it is different, offering the promise of excitement or the exotic. Finally, some are interested in how social norms are constructed, in the ways that people and societies decide what is acceptable and what is not. The sociological study of deviance can fulfill all these goals, taking us deep into the criminal underworld, inward to the familiar, outward to the fascinating and bizarre, and finally back to the central core. In the pages that follow, we peer into the deviant realm, looking at both deviants and those who define them as such. In so doing, we look at a range of deviant behaviors, discuss why people engage in these behaviors, and analyze how the people are sociologically organized. We begin in Part I by defining deviance, in an effort to lay down the parameters of its scope.

STUDYING DEVIANCE

Reasonable theories and social policies pertaining to deviance must be based on a firm foundation of accurate knowledge. Social scientists have an array of different methodologies and sources of data at their disposal, including survey research, experimental design, historical methods, official statistics, and field research (ethnography). All of these methods have obvious strengths and weaknesses, and have been used by sociologists in studying deviance. While some generate statistical portraits about the extensiveness of deviant behaviors, we undertake, in this book, to offer a richer, more experiential understanding of what goes on in deviant worlds, showing *how* things happen and what they *mean* to participants. It is

1

the reports of field researchers—individuals who immerse themselves personally in deviant settings—that yield such depth and descriptive accounts. It is also our belief that, because of the often secretive nature of deviance, methods that objectify or distance researchers from the people being researched will be less likely to portray deviant worlds accurately. Thus, the works in this book are tied together by their descriptive richness and by the belief that researchers must study deviance as it naturally occurs in the real world. The most appropriate methodology for this task, field research, advocates that sociologists should get as close as possible to the people they are studying in order to understand their worlds (Adler and Adler, 1987). Despite the problems that arise from the secretive and hidden nature of deviant acts, sociologists have devised techniques to penetrate secluded deviant worlds. These methodological ploys often come complete with perils, so it is wise for people who are considering studying deviance to be aware of the associated problems. Part III discusses field research methods and two of the other common sources of information about deviant behavior.

CONSTRUCTING DEVIANCE

In Parts I, II, and IV we delve into the origins and definitions of deviant behavior. Most sociology courses begin by examining core definitions in the field and leave these ideas behind shortly thereafter. This is not the case with deviance. Definitions of deviance pervade all aspects of the field and are therefore addressed throughout the book. Scholars, politicians, activists, moral entrepreneurs, religious zealots, journalists, and people from all walks of life frequently discuss issues of what is deviant and what is socially acceptable. Those whose definitions come to be reflected in law and social policy gain broad moral and material resources.

There are many approaches to defining and theorizing about deviance, some of which are presented in the chapters in Parts I and II. As our title suggests, we advance a social constructionist view throughout this book. We begin, here, by discussing **three perspectives on defining deviance** and locating these perspectives in relation to social constructionism.

Proponents of the **absolutist perspective** have traditionally considered defining deviance as a simple task, implying that a widespread consensus exists about what is deviant and what is not. Emile Durkheim, a functionalist and one of sociology's founding fathers, represents this theoretical approach, arguing that the laws of any given society are objective facts. Laws, he believes, reflect the "collective consciousness" of each society and thereby reveal its true social nature. They exist before individuals enter the society, and they exist when individuals die; hence, laws represent a level of reality *sui generis* (unique unto themselves), transcending individual lives.

According to this position, there is general agreement among citizens that there is something obvious within each deviant act, belief, or condition that makes it different from the conventional norms. At its core, each such act embodies the unambiguous, objective "essence" of true or real deviance. This perspective has its roots in both religious and naturalistic assumptions; its proponents argue that certain acts are contrary to the strictures of God or to the laws of nature. Deviance is thus immoral (possibly evil), sinful, and unnatural. Contemporary religious leaders, especially those of the charismatic and evangelical persuasions, often use these arguments in advancing their moral beliefs in written and verbal oratory. Absolutist views of deviance are eternal and global: If something is judged to have been intrinsically morally wrong in the past (e.g., adultery or divorce), it should be recognized as wrong now and always in the future. Similarly, if something is considered to be morally wrong in one place, it should be judged wrong everywhere. Absolutist views on deviance flourish in homogeneous societies, in which there is a high degree of universal agreement on social values.

Deviance, then, is viewed, not as something that is determined by social norms, customs, or rules, but as something that is intrinsic to the human condition, standing apart from and existing before the creation of these socially created codes. According to absolutists, deviant attitudes, behavior, or conditions by any name would be recognized and judged similarly. People have backed up this belief system by pointing to the existence of universal taboos surrounding such acts as murder, incest, and lying. These acts, they claim, are deviant in their very essence.

Noteworthy to this perspective is its focus on the deviance itself. Proponents believe that an absolute moral order is a necessary part of reality, enabling all people to know what is right and what is wrong. Normative behavior is inherently good and deviance is inherently bad. Violations of norms, it then follows, should be met with stern reactions. People who question the norms deserve even harsher treatment, as they challenge the moral order. At its core, then, absolutism is an objectivist approach because it relies in its definition on internal, inherent features that stand apart from subjective human judgments. We see contemporary applications of the absolutist perspective in the campaign against gay marriage, with opponents arguing that homosexuality is an abomination and a sin. The morality-based conception of deviance, presented in Hendershott's Chapter 4, offers another contemporary illustration of this viewpoint.

Functionalist theories of deviance, as represented in Erikson's Chapter 1 and Durkheim's Chapter 6, incorporate elements of the absolutist perspective by suggesting that deviance is pathological (diseased) in its substance and negative in its effect. As such, deviance stands apart from, and in strong contrast to, the "normal." In all ways, the absolutist perspective persists as the foil, or antithesis, to social constructionism.

A number of theories coalesce to form the *social constructionist approach*, grounded in the interactionist theory of deviance. We can draw distinctions between theories, but they all share a focus on the norms that bound and define deviance, rather than a gaze on the deviance itself. They also share a subjectivist approach to defining deviance, guided by the belief that social meanings, values, and rules, in concrete situations, are often problematic or uncertain. Social meanings, these theories hold, arise in the situations where they occur, rather than being located within the essences of things, and are heavily influenced by people's perceptions and interpretations. Social constructionists study the ways that norms are created, the people who create them, the conditions under which those norms arise, and the consequences of such norms for different groups in society. Fundamentally, these theorists view definitions of deviance as social products and focus on those who define deviance and their definitions, rather than on the acts that generate deviant reactions.

For example, recently there has been greater social awareness about obesity and its deleterious effects. Absolutists would say that the social definition of obesity as deviant was established by doctors as a health issue and that the level of obesity in our society can be objectively measured by a scientific instrument: the scale. They would call attention to the growth in portion sizes in the United States ("supersizing"), pointing out that the average dinner plate has expanded from 10 to 12 inches. Americans, they would say, are more obese because they are eating more and exercising less. In contrast, constructionists would suggest that our collective attitudes toward obesity have changed and that levels of weight that used to raise little attention have become less tolerated now. In addition, they would show that people react to weight differently in various cultures, with citizens in some countries preferring more "full-figured" people while in other places the image of the skinny (practically anorectic) model is considered the ideal body shape. Throughout at least the developed world, clothing companies have subjectively renumbered the sizes of their garments so that size 4 is the "new" size 6 (what women should strive for) and size 8 is the "new" size 14 (you might as well forget about appearing in public). Crusaders, especially from the medical community, have waged campaigns against obesity and raised social awareness about it, making it one of the chief panics in society today. Our point is that, although being overweight can lead to numerous physical problems, we have shifted our definitions about how much weight is acceptable. As a result, when we look at people and try to assess whether they are slightly overweight, chubby, plump, heavy, fat, or outright obese, our categories have changed: What was previously considered tolerably heavy has now been redefined as obese and labeled deviant.

Falling within social constructionism is the **relativist perspective**, articulated in Becker's Chapter 3. Spurred by the rise of subcultural studies in the 1930s, deviance theorists began to note the existence of norms that differed from, and even conflicted with, those of the larger society. This awareness led them to consider the possibility that groups in society make up rules to fit the practical needs of their situations. The more the relativists examined norms in different places and times, the more they became convinced that definitions of deviance were not universal, but varied to suit the people who hold them. This finding suggested that definitions of deviance derived, not from absolute, unchanging universals such as God and nature, but from humans. Becker (1963, 9) articulated this position, noting,

> ... *social groups create deviance by making the rules whose infraction constitutes deviance*, and by applying these rules to particular people and labeling them as outsiders. From this point of view, deviance is *not* a quality of the act the person commits, but rather a consequence of the application by others of rules and sanctions to an "offender." (*Italics in original*)

Deviance, relativists argue, is thus lodged in the eye of the beholder rather than in the act itself, and it may vary by time and place in the way it is defined. Becker (1963, 147–48) further stated,

> Formal rules such as laws require the initiative or enterprise of specific individuals and groups for their reality. They do not exist independently of the actions of moral entrepreneurs to make them real....The proto-type of the rule creator is the crusading reformer. He is interested in the content of the rules. The existing rules do not satisfy him because there is some evil which profoundly disturbs him. He feels that nothing can be right in the world unless rules are made to correct it. He operates within an absolute ethic; what he sees is truly and totally evil with no qualification. Any means is justified to do away with it. The crusader is fervent and righteous, often self righteous... [He or she] typically believe[s the] mission is a holy one.

Definitions, this position suggests, are forged by crusading reformers and reforged by them in different eras and locations, leading to significant constructions and reconstructions of deviance. When different social contexts frame and give meaning to the perception and interpretation of acts, the same act may be alter-natively perceived as deviant or normative.

This idea creates the possibility of multiple definitions of acts, both deviant and nondeviant, simultaneously existing among different groups. We see that possibility clearly in cases of controversial, morally debated acts such as abortion, school prayer, gay marriage, the use of medical marijuana, and illegal immigra-tion. Major campaigns have been assembled to sway public opinion over the morality and appropriateness of these behaviors, with large segments of the country falling into oppositional camps. Groups that are fairly alike in their com-position may forge shared agreements about norms that would be received dif-ferently among a broader segment of the general public. For example, more widely acceptable behavior, such as dancing, is condemned in some thoroughly conservative environments, such as Bible-Belt religious colleges. At the same time, language that is widely used throughout the country may be condemned as politically incorrect and morally offensive (and, sometimes, as a hate crime) on extremely liberal college campuses.

Even some acts that are considered universally taboo can vary in their defi-nitions, as there are conditions under which they would be considered nondevi-ant. To kill somebody for personal gain, vengeance, or freedom, or through negligence, might be considered deviant (and, more likely, criminal), but when murder is committed by the state (e.g., by execution, in war, or through covert intelligence), in self-defense, after catching your spouse in bed with someone else (if you are a man, but not if you are a woman), by those deemed insane, or on one's own property (in states where they have "make my day" laws), it is con-sidered not only nondeviant, but also possibly heroic. Similarly, anthropologists have found cultures that condone certain forms of incest to quiet or soothe infants (Henry, 1964; Weatherford, 1986). Often, too, people differentiate between normal lies and "white" lies, accepting those designed to spare other people's feelings. Relativists argue that definitions of deviance are social products that are likely to be situationally invoked under certain circumstances. They represent the social constructionist approach because they focus on the circumstances under which social norms are differentially created and applied.

Paradoxically, extensions of functionalist theory bring it into the social constructionist realm. Despite their belief that deviance is pathological, functionalists overwhelmingly hold that all components of society contribute, somehow, to its existence; they all have positive functions in that regard. Because deviance is a universal feature of all societies, functionalists admit that it must offer benefits. Four such benefits have been outlined. First, when people react against the deviance of others, they bond together to produce cohesion and social solidarity. We saw this effect after the terrorist attacks of 9/11. Second, identifying and punishing deviance redefines and reinforces the social boundaries, which are mutable and not fixed, and reinforces the dangers of transgressing those boundaries. When, for example, business executives such as Martha Stewart, and Kenneth Lay and Jeffrey Skilling of Enron, go to prison for insider trading, "creative accounting," and income tax evasion, their punishment clarifies the limits of acceptable behavior. Similarly, politicians who get caught abusing their power represent an impetus to other politicians not only to avoid such behaviors, but to pass stronger ethics rules. Third, Durkheim noted the seeds of social change in deviance: New developments are often initially regarded skeptically or fearfully and have to go through a process of moral passage to become accepted. For instance, although Socrates was considered a political heretic in his time, he paved the way for intellectual freedom. Without deviance, Durkheim suggested, society might stagnate. Fourth, deviance promotes full employment, as the existence of all the occupations associated with it, such as the criminal justice system, the medical establishment, the media, scholars, etc., would be less robust in a society devoid of deviance. In fact, should crime and deviance disappear entirely, and we become what Durkheim called a "society of saints," new behaviors would be defined as deviant to fill this void. In sum, functionalists concluded that a certain amount of deviance is good for society. But because too much or too little is not as beneficial as just the right amount, definitions of deviance must be continually socially constructed and adjusted to ensure the smooth functioning of society.

A third position on defining deviance, presented in Quinney's Chapter 5, is the **social power perspective**, which builds on the relativist perspective by asserting that views on crime and deviance are not arbitrarily formed by just any group of "others." Closely tied to Marx's conflict theory, the social power approach focuses on the influence that powerful groups and classes have in creating and applying laws. As Quinney (1970, 43) notes, these laws are a reflection of those with the greatest social power in society:

> Criminal laws support particular interests to the neglect or negation of other interests, thus representing the concerns of only some members of society. Though some criminal laws may involve a compromise of conflicting interests, more likely than not, criminal laws mark the victory of some groups over others. The notion of a compromise of

conflicting interests is a myth perpetuated by the pluralistic model of politics. Some interests never find access to the lawmaking process. Other interests are overwhelmed in it, not compromised. But ultimately some interests succeed in becoming criminal law, and are able to control the conduct of other.

Thus, Quinney believes that laws reflect the interests and concerns of the dominant classes in a society. Dominant classes tend to formulate their definitions of deviance in accord with their own interests, defining things that threaten them as out of bounds and things that benefit them as good. All laws reflect power relations, and their existence, as well as their enforcement, illustrates which groups have the power to control and which groups are controlled. Dominant groups, then, tend to overenforce laws against the subordinate classes and underenforce them against members of their own group.

This approach incorporates social constructionism by challenging the simplistic assumption underlying the absolutist perspective that definitions of deviance are universally shared. According to the social power viewpoint, society is characterized by conflict and struggle between groups whose interests conflict with each other, with the powerful classes dominating the subordinate groups. Members of each group pursue their own needs and desires, but because of the way society is structured, what is good for one class restricts the opportunities of others. This tension pits groups against each other in active struggles. Those with the greatest social power dominate both the ability to create definitions, rules, and laws and the way those definitions, rules, and laws are negligently or aggressively enforced.

The conflicting interests of the dominant and subordinate groups can fall into the economic realm, as we see some political parties promoting the rights of big businesses to pay less money in taxes, less overtime compensation, and lower health-care benefits while other parties fight for the rights of workers to unionize, to gain job security, and to earn a decent wage. Other issues on which groups in society may conflict fall into the social realm and include environmental policies, birth control, stem cell research, school prayer, gun control, health care, and drug policies. Pervading all of these issues is a systematic differential oppression along the lines of race or ethnicity, social class, and gender, note conflict theorists.

Feminist theorists share conflict theorists' constructionist orientation, as we learn in Chesney-Lind's Chapter 10, but their focus is on the norms, policies, and laws of the patriarchal system that uphold the social, moral, economic, and political order that fosters male privilege. Their approach questions the way men forge and maintain their dominance over women, and the role of definitions of deviance in that endeavor. Feminist theorists challenge the bases and benefits of patriarchy, regarding it as a form of socially constructed social control.

Social power theorists define deviance through the social meanings that are collectively applied to people's attitudes, behavior, or conditions. These meanings are rooted in the interaction between individuals and social groups. Those who have the power to make rules and apply them to others control the normative order. The politically, socially, and economically dominant groups enforce their definitions onto the downtrodden and powerless. Deviance is thus a representation of unequal power in society.

DEVIANT IDENTITY

The second component of the social constructionist approach lies in the consequences of definitions and applications of deviance. Interactionists have argued that the two sides of constructing deviance—its articulation and its application—are each critical. Society first labels various attitudes, behaviors, and conditions as deviant and then labels specific individuals associated with those attitudes, behaviors, and conditions as deviant. Where definitions of deviance are forged but not applied, they believe, deviance does not really exist. It's as if a tree falls in the forest but we conclude that it makes no sound because no one is there to hear it. In effect, social constructionists are proposing that the essence of the sound lies, not in the impact of the tree on the ground or in the creation of the sound waves, but in the articulation of those waves against the hearer's eardrum. Specific individuals or groups of people must be labeled by society in order for deviance to be concretely envisioned. Part V takes up this change of focus away from the construction of definitions of deviance and toward looking at how the application of norms and laws affects people. Now, we move away from macrosocietal explanations to focus on the microinteraction that occurs in everyday life. Constructionists claim that deviants are people who have undergone some sort of labeling. This section looks at how deviant labels are applied and their subsequent consequences.

Social constructionists emphasize that something profound may occur when the supposed deviants and conventional others interact. In pursuing their actions, individuals may engage in deviance but not think of themselves as deviant actors. Only when they begin to apply the deviant labels "out there" in the world to themselves do they truly become deviants. This is the process of acquiring a deviant identity. People may become dislodged from their safe identity locations within the "normal" realm through their own observations, as well as through the actions and remarks of others. The greater the response of others, the stronger will be their self-conception that they are deviant and the more they may engage in further deviance.

Several aspects of this social psychological process are the most critical and will be addressed most vividly in this section. We begin by looking at how people acquire deviant identities. Many factors are influential, and we consider the ways people creatively use "accounts" or "motive talk" to explain, neutralize, or justify their actions in order to forestall being labeled as deviant. Although functionalists have suggested that labeling people as deviant has positive results for society, it has negative results for those so labeled. They acquire the stigma of being deviant, a stain or pejorative connotation associated with them or their actions. Living with known stigma makes life difficult for people, who may then be marked, disparaged, or shunned. People handle their stigma differently, and we examine these variations and their consequences in Part V.

THE SOCIAL ORGANIZATION OF DEVIANCE

We conclude this volume, in Parts VI, VII, and VIII, with a discussion of how deviants, their deviance, and their deviant careers are socially organized. Earlier sections of the book have concentrated on macro- and microlevels of addressing deviance by considering the movements and powers that shape deviant definitions at the societal level and by looking at how people's identities are shaped at the interpersonal level. Now we take a meso (midlevel) focus by looking at how deviants organize their social organization and relationships, activities and acts, and careers in connection with others. We begin with the study of deviant organization, examining the various ways members of deviant scenes organize their relations with one another. The scenarios discussed range from individuals acting on their own, outside of relationships with other deviants, to subcultures, to more tightly connected gangs, to highly committed international cartels, and, finally, to corporations and the state. We then consider the structure of deviant acts. Some forms of deviance involve cooperation between the participants, with people mutually exchanging illicit goods or services. Others are characterized by conflict, with some parties to the act taking advantage of others, often against their will. Finally, we look at the phases and contours of deviant careers, beginning with people's entry into the world of deviant behavior and associates, continuing with the way they fashion their involvement in deviance, and concluding with their often problematic, and occasionally inconclusive, retirement from the compelling world of deviance.

PART I

Defining Deviance

In order to study the topic of deviance, we must first clarify what we mean by the term. What behaviors or conditions fall into this category, and what is the relation between deviance and other categories, such as crime? When we speak of deviance, we refer to violations of social norms. Norms are behavioral codes or prescriptions that guide people into actions and self-presentations that conform to social acceptability. Norms need not be agreed upon by every member of the group doing the defining, but a clear or vocal majority must agree.

One of the founding American sociologists, William Sumner (1906), conceptualized **three types of norms:** *folkways, mores,* and *laws*. He defined folkways as simple everyday norms based on custom, tradition, or etiquette. Violations of folkway norms do not generate serious outrage, but might cause people to think of the violator as odd. Common folkway norms include standards of dress, demeanor, physical closeness to or distance from others, and eating behavior. People who come to class dressed in bathing suits, who never seem to be paying attention when they are spoken to, who sit or stand too close to others, who pick their noses in public or fail to wash their hands in a public washroom, or who eat with their hands instead of silverware (at least in the United States) would be violating a folkway norm. We would not arrest them, nor would we impugn their moral character, but we might think that there was something peculiar about them.

Mores (pronounced mor-ays) are norms based on broad societal morals whose infraction would generate more serious social condemnation. Interracial marriage, illegitimate childbearing, and drug addiction all constitute moral violations. Upholding these norms is seen as critical to the fabric of society, and their violation threatens the social order. Interracial marriage threatens racial purity and the stratification hierarchy based on race; illegitimate childbearing threatens the institution of marriage and the transference of money, status, and family responsibility from one generation to the next; drug addiction represents the triumph of hedonism over rationality, threatening the responsible behavior required to hold society together and to accomplish its necessary tasks. People who violate mores may be considered bad or wicked, and harmful to society.

Laws are the strongest norms because they are supported by codified social sanctions. People who violate them are subject to arrest and punishment ranging from fines to imprisonment (and even death). Many laws are directed toward behavior that used to be folkways or, especially, violations of mores, but that became encoded into laws. Others are regarded as necessary for maintaining social order. Although violating a traffic law such as speeding breaks society's rules, it will not usually brand the offender as deviant.

Following closely on Sumner's distinctions, Smith and Pollack (1976) suggested that deviance might be conceptualized as violations of the norms associated with *crime, sin, and poor taste*. For example, criminal acts, such as murder, rape, assault, robbery, and arson, would be violations of laws. Smith and Pollack view these acts as generally unacceptable to the large majority of the people in society. Acts of sin tend to be defined in relation to religious proscriptions and often include promiscuity, lewdness, extramarital or homosexual sex, gambling, drinking, and abortion. Although some of these may be subject to criminal sanction, the majority of them are not. Instead, societal responses tend to fall into the moral category of strong disapproval. Finally, acts of poor taste, like violations of folkways, challenge existing standards of fashion, manners, or traditions, violate social norms, and are unregulated by law.

This discussion returns us to the question about the relationship between *deviance and crime*. Are they identical terms, is one a subset of the other, or are they overlapping categories? To answer this question, we must consider one facet of it at a time. First, do some acts fall into both categories—crime and deviance? The overlap between these two is extensive, with crimes of violence, such as murder and assault, and property crimes, such as theft, arson, and vandalism, considered both deviant and illegal. But are crime and deviance always the same thing? We can see that they are not, because there is much deviance that is not criminal. Noncriminal forms of deviance include obesity, stuttering, physical handicaps, racial intermarriage, and unwed pregnancy. At the same time, some forms of crime are not considered deviant; they neither violate norms nor bring moral censure. Examples of these crimes include some white-collar crimes commonly regarded as merely aggressive business practices, such as income tax evasion; minor traffic offenses; and forms of civil disobedience, in which people break laws to protest their injustice. Crime and deviance, then, can best be seen as categories that overlap while at the same time having independent dimensions.

People can be labeled deviant as the result of the **ABCs of deviance:** their *attitudes, behaviors,* or *conditions.* First, they may be branded deviant for alternative sets of *attitudes* or belief systems. These belief systems may fall into the category of religion, with all forms of religious extremism and even moderate forms of nonnormative religious beliefs, such as those held by cults, Satanists, and fundamentalists, regarded askance. Having no religious affiliation is one of the fastest-growing forms of deviant religious belief systems in the United States (those who claimed "no religion" were the only demographic group that grew in all 50 states in the last 18 years), contravening the 85 percent of religious people who claim some religious identity and the close to 92 percent who report that they believe in the existence of a supernatural or "supreme" being. A second category of deviant people consists of those who hold extreme political attitudes (far-leftist or far-rightist terrorists). Mentally ill people also fall into this category, as do people with deviant worldviews (e.g., those who believe that the world is coming to an end and who are often considered mentally ill) and people with chemical, emotional, or psychological problems. Those in this category may also be subject to relativity, so that when a poor woman shoplifts a roast, people call her a common criminal, but if a rich woman steals a roast, her deviant status is deemed kleptomaniac, a mental illness label.

The *behavioral* category is the most familiar one, with people coming to be regarded as deviant for their outward actions. Deviant behaviors may be intentional or inadvertent and include such activities as violating dress or speech conventions, engaging in kinky sexual behavior, smoking marijuana, and committing murder. People cast into the deviant realm for their behaviors have an *achieved deviant status:* They have earned the deviant label through something they have done.

Other people regarded as deviant may have an *ascribed deviant status,* based on a *condition* they acquire from birth. This category would include those having a deviant socioeconomic status such as being either extremely poor or ultrarich; those possessing a deviant racial status such as being a person of color (in a dominantly Caucasian society); and those having a congenital physical disability. Here, there is nothing that such people have done to become deviant and little or nothing they can do to repair their deviant status. Moreover, there may be nothing inherent in these statuses that make them deviant; rather, they may become deviant through the result of a social definitional process that gives unequal weight to powerful and dominant groups in society. A conditional deviant status may be ascribed or achieved. On the one hand, people may be born with conditional deviance because of their personal, racial, or ethnic characteristics (height, weight, color). On the other hand, a conditional deviant status can be achieved, as when people burn or disfigure themselves severely, when they become too fat or too thin, or when they cover their bodies with adornments such as tattoos, piercings, or scarification. Unlike ascribed statuses, some of these achieved statuses may be changed, moving the deviant back within the norm.

Finally, it is interesting to consider the way deviance is assessed and categorized, as this varies between different eras, with different behaviors, and to different audiences. Deviance may be perceived and interpreted through the lens of the **three categories of S's:** *sin, sick, and selected.* During the Middle Ages and many earlier times when religious paradigms about the world prevailed, deviance from the norm was held to be a religious disorder and viewed as *sinful.* Nonnormative attitudes, beliefs, and conditions were attributed to satanic influences, and exorcisms were performed to cure people. Religious leaders were seen as the arbiters of official morality and were called upon to make judgments and administer sanctions. Even today, some religious zealots adhere to this perspective, not only in societies that are officially religious in their governance, but within religious pockets of secular societies. The afflicted, in the eyes of these people, are strongly condemned and often viewed as contagious.

Beginning in the first half of the twentieth century, the medicalization (*sickness*) model emerged as a strong perspective for explaining deviance. Conrad and Snyder (1980) suggested that the process of medicalization begins when a behavior or condition defined as deviant is "prospected" by people with medical interests to see if they can gain rewards from pulling it under the medical umbrella. Physicians and psychiatrists claimed that homosexuality, alcoholism, drug addiction, sexual misbehavior, mental illness, and a number of other behaviors are rooted in people's psychiatric problems, genetic abnormalities, inherited predispositions, and biochemical characteristics. Psychiatrists sought to claim ownership over the diagnosis of these forms of deviance with their *Diagnostic and Statistical Manual of Mental Disorders* (now in its current edition, the *DSM-V*) so that they could administer outpatient or inpatient therapy and be reimbursed by insurance companies. (Clinics for specialized medical conditions such as self-injury may charge as much as $1,000 a day and require a 30-day stay.) Turf wars sprang up, with psychiatrists attempting to exert domain claims over a variety of behaviors that were previously conceived as biological or sociological in order to receive research grants for further study, diagnose patients, and then treat them. For example, does your child have "dyspraxia"? Children with this new disorder ("clumsy child syndrome"), affecting 2 percent of the population (mostly male), will need the support of their families and qualified professionals to succeed in school. Are you boorish, disrespectful of authority, rude, or lacking in etiquette? Perhaps you have "impulse control disorder." Do you spend too many hours on your computer? You should consider having yourself diagnosed with "Internet addiction disorder." Are you an overly aggressive driver? Perhaps you suffer clinically from "road rage." The next edition of the *DSM* is set to include a new category on "relational disorders" that identifies sickness in groups of individuals and the relationships between them, including couples who constantly quarrel, parents and children who clash, and troubled relationships between siblings. The new *DSM* will define all of these conditions as mental illnesses, thereby encroaching into sociological behaviors, attempting to psychologize and pathologize them.

The recovery movement, with its coterie of self-help (often 12-step) programs, educators, and public health organizations, further expanded the reach of the medicalization paradigm to include addictions other than drug and alcohol addictions, as well as eating disorders, self-injury, child abuse, child hyperactivity, compulsive gambling, and interpersonal violence. Doctors and drug companies leapt into the business, manufacturing and prescribing an array of medications to "manage" or "cure" people. As a result, we are now the most legally medicated society in world history. The success of this medicalization movement, fueled by the prestige of science and the lure of the "easy fix" of a pill, drew huge areas of attitudes, behaviors, and conditions into its sphere, so that society no longer tolerates people being too sad, depressed, rebellious, rambunctious, or fidgety, eating too much or too little, or having too little or too much sex. One strong allure of the medicalization perspective has been its destigmatization of some deviance, as people who were seen as sick or acting on biological predispositions or inherited "differences" came to be seen instead as doing things "beyond their control" instead of making immoral choices. This new perspective made them more acceptable to mainstream society, albeit not quite "normal" or "healthy."

At the same time, some thought that the medicalization movement had become too pervasive, absorbing too great a swath of social life within its grasp. Some participants in deviance reached out to reclaim the rhetoric of *selection* over their attitudes, behaviors, and conditions, arguing that these were the result of voluntary choice. They, and the researchers who studied them, found groups of people who rejected the medicalized interpretation of such things as homosexuality, gambling, eating disorders, drug use, and self-injury. Participants chafed at the control that "experts" held over them, their diagnoses, their treatments, and their medicalization. Instead, they cast their behaviors as intentionally selected, as forms of recreation, lifestyle choices, and coping strategies. This demedicalization movement has in some ways empowered and destigmatized people who participated in these forms of behavior. For them, to be seen as ill was to be derogated; by contrast, to be seen as self-healing or voluntarily deviant is conventional. There is no doubt that these diverse perspectives will continue to compete for the ownership, control, and legitimacy of nonnormative forms of expression.

1

On the Sociology of Deviance

KAI T. ERIKSON

This classic selection examines the functions of deviance for society. Erikson asserts that deviance and the social reactions it evokes are key focal concerns of every community. Scrutinized by the mass media, law enforcement, and ordinary citizens, deviance leads us to continually redraw the social boundaries of acceptability. Rather than being a fixed property, norms are subject to shifts and evolution, and the interactions between deviants and agents of social control locate the margins between deviance and respectability.

Erikson notes, ironically, that the very institutions and agencies mandated to manage deviance tend to reinforce it. Once individuals have been identified as deviant, they undergo "commitment ceremonies" where they are negatively labeled, experiencing a status change that is hard to reverse. Society's expectations that deviants will not reform foster the "self-fulfilling prophesy," by which norm violators reproduce their deviance, living up to the negative images society holds of them. Erikson suggests several other valuable functions that deviance performs in a society: It fosters boundary maintenance so that people know what is acceptable and unacceptable; it bolsters cohesion, integration, and solidarity, thus preserving the stability of social life; and it promotes full employment, guaranteeing jobs for people working in the deviance- and crime-management sectors.

Can you think of any cases where people you know have been pushed into deviance by people's expectations or definitions?

H uman actors are sorted into various kinds of collectivity, ranging from relatively small units such as the nuclear family to relatively large ones such as a nation or culture. One of the most stubborn difficulties in the study of deviation is that the problem is defined differently at each one of these levels: behavior that is considered unseemly within the context of a single family may be entirely acceptable to the community in general, while behavior that attracts severe censure from the members of the community may go altogether unnoticed elsewhere in the culture. People in society, then, must learn to deal separately with deviance at each one of these levels and to distinguish among them in [their] own daily activity. A man may disinherit his son for conduct that violates old

Source: ERIKSON, KAI T., WAYWARD PURITANS: A STUDY IN THE SOCIOLOGY OF DEVIANCE, CLASSIC EDITION, 1st Edition, © 2005. Reprinted by permission of Pearson Education, Inc., Upper Saddle River, NJ.

family traditions or ostracize a neighbor for conduct that violates some local custom, but he is not expected to employ either of these standards when he serves
as a juror in a court of law. In each of the three situations he is required to use a
different set of criteria to decide whether or not the behavior in question exceeds
tolerable limits.

In the next few pages we shall be talking about deviant behavior in social
units called "communities," but the use of this term does not mean that the
argument applies only at that level of organization. In theory, at least, the argument being made here should fit all kinds of human collectivity—families as well
as whole cultures, small groups as well as nations—and the term "community" is
only being used in this context because it seems particularly convenient.[1]

The people of a community spend most of their lives in close contact with
one another, sharing a common sphere of experience which makes them feel
that they belong to a special "kind" and live in a special "place." In the formal
language of sociology, this means that communities are boundary maintaining:
each has a specific territory in the world as a whole, not only in the sense that
it occupies a defined region of geographical space but also in the sense that it
takes over a particular niche in what might be called cultural space and develops
its own "ethos" or "way" within that compass. Both of these dimensions of
group space, the geographical and the cultural, set the community apart as a special place and provide an important point of reference for its members.

When one describes any system as boundary maintaining, one is saying that it
controls the fluctuation of its consistent parts so that the whole retains a limited
range of activity, a given pattern of constancy and stability, within the larger environment. A human community can be said to maintain boundaries, then, in the
sense that its members tend to confine themselves to a particular radius of activity
and to regard any conduct which drifts outside that radius as somehow inappropriate or immoral. Thus, the group retains a kind of cultural integrity, a voluntary
restriction on its own potential for expansion, beyond that which is strictly
required for accommodation to the environment. Human behavior can vary
over an enormous range, but each community draws a symbolic set of parentheses
around a certain segment of that range and limits its own activities within that
narrower zone. These parentheses, so to speak, are the community's boundaries.

Now people who live together in communities cannot relate to one another
in any coherent way or even acquire a sense of their own stature as group members unless they learn something about the boundaries of the territory they
occupy in social space, if only because they need to sense what lies beyond the
margins of the group before they can appreciate the special quality of the experience which takes place within it. Yet how do people learn about the boundaries of their community? And how do they convey this information to the
generations which replace them?

To begin with, the only material found in a society for marking boundaries
is the behavior of its members—or rather, the networks of interaction which link
these members together in regular social relations. And the interactions which do
the most effective job of locating and publicizing the group's outer edges would
seem to be those which take place between deviant persons on the one side and
official agents of the community on the other. The deviant is a person whose

activities have moved outside the margins of the group, and when the community calls him to account for that vagrancy it is making a statement about the nature and placement of its boundaries. It is declaring how much variability and diversity can be tolerated within the group before it begins to lose its distinctive shape, its unique identity. Now there may be other moments in the life of the group which perform a similar service: wars, for instance, can publicize a group's boundaries by drawing attention to the line separating the group from an adversary, and certain kinds of religious ritual, dance ceremony, and other traditional pageantry can dramatize the difference between "we" and "they" by portraying a symbolic encounter between the two. But on the whole, members of a community inform one another about the placement of their boundaries by participating in the confrontations which occur when persons who venture out to the edges of the group are met by policing agents whose special business it is to guard the cultural integrity of the community. Whether these confrontations take the form of criminal trials, excommunication hearings, courts-martial, or even psychiatric case conferences, they act as boundary-maintaining devices in the sense that they demonstrate to whatever audience is concerned where the line is drawn between behavior that belongs in the special universe of the group and behavior that does not. In general, this kind of information is not easily relayed by the straightforward use of language. Most readers of this paragraph, for instance, have a fairly clear idea of the line separating theft from more legitimate forms of commerce, but few of them have ever seen a published statute describing these differences. More likely than not, our information on the subject has been drawn from publicized instances in which the relevant laws were applied—and for that matter, the law itself is largely a collection of past cases and decisions, a synthesis of the various confrontations which have occurred in the life of the legal order.

It may be important to note in this connection that confrontations between deviant offenders and the agents of control have always attracted a good deal of public attention. In our own past, the trial and punishment of offenders were staged in the market place and afforded the crowd a chance to participate in a direct, active way. Today, of course, we no longer parade deviants in the town square or expose them to the carnival atmosphere of a Tyburn, but it is interesting that the "reform" which brought about this change in penal practice coincided almost exactly with the development of newspapers as a medium of mass information. Perhaps this is no more than an accident of history, but it is nonetheless true that newspapers (and now radio and television) offer much the same kind of entertainment as public hangings or a Sunday visit to the local gaol. A considerable portion of what we call "news" is devoted to reports about deviant behavior and its consequences, and it is no simple matter to explain why these items should be considered newsworthy or why they should command the extraordinary attention they do. Perhaps they appeal to a number of psychological perversities among the mass audience, as commentators have suggested, but at the same time they constitute one of our main sources of information about the normative outlines of society. In a figurative sense, at least, morality and immorality meet at the public scaffold, and it is during this meeting that the line between them is drawn.

Boundaries are never a fixed property of any community. They are always shifting as the people of the group find new ways to define the outer limits of their universe, new ways to position themselves on the larger cultural map. Sometimes changes occur within the structure of the group which require its members to make a new survey of their territory—a change of leadership, a shift of mood. Sometimes changes occur in the surrounding environment, altering the background against which the people of the group have measured their own uniqueness. And always, new generations are moving in to take their turn guarding old institutions and need to be informed about the contours of the world they are inheriting. Thus, single encounters between the deviant and his community are only fragments of an ongoing social process. Like an article of common law, boundaries remain a meaningful point of reference only so long as they are repeatedly tested by persons on the fringes of the group and repeatedly defended by persons chosen to represent the group's inner morality. Each time the community moves to censure some act of deviation, then, and convenes a formal ceremony to deal with the responsible offender, it sharpens the authority of the violated norm and restates where the boundaries of the group are located.

For these reasons, deviant behavior is not a simple kind of leakage which occurs when the machinery of society is in poor working order, but may be, in controlled quantities, an important condition for preserving the stability of social life. Deviant forms of behavior, by marking the outer edges of group life, give the inner structure its special character and thus supply the framework within which the people of the group develop an orderly sense of their own cultural identity. Perhaps this is what Aldous Huxley had in mind when he wrote:

> Now tidiness is undeniably good—but a good of which it is easily
> possible to have too much and at too high a price.... The good life can
> only be lived in a society in which tidiness is preached and practised, but
> not too fanatically, and where efficiency is always haloed, as it were, by
> a tolerated margin of mess.[2]

This raises a delicate theoretical issue. If we grant that human groups often derive benefit from deviant behavior, can we then assume that they are organized in such a way as to promote this resource? Can we assume, in other words, that forces operate in the social structure to recruit offenders and to commit them to long periods of service in the deviant ranks? This is not a question which can be answered with our present store of empirical data, but one observation can be made which gives the question an interesting perspective—namely, that deviant forms of conduct often seem to derive nourishment from the very agencies devised to inhibit them. Indeed, the agencies built by society for preventing deviance are often so poorly equipped for the task that we might well ask why this is regarded as their "real" function in the first place.

It is by now a thoroughly familiar argument that many of the institutions designed to discourage deviant behavior actually operate in such a way as to perpetuate it. For one thing, prisons, hospitals, and other similar agencies provide aid and shelter to large numbers of deviant persons, sometimes giving them a certain advantage in the competition for social resources. But beyond this, such institutions gather marginal people into tightly segregated groups, give them an opportunity to teach one another the skills and attitudes of a deviant career, and even provoke them into using these skills by reinforcing their sense of alienation from the rest of society.[3] Nor is this observation a modern one:

> The misery suffered in gaols is not half their evil; they are filled with every sort of corruption that poverty and wickedness can generate; with all the shameless and profligate enormities that can be produced by the impudence of ignominy, the range of want, and the malignity of dispair. In a prison the check of the public eye is removed; and the power of the law is spent. There are few fears, there are no blushes. The lewd inflame the more modest; the audacious harden the timid. Everyone fortifies himself as he can against his own remaining sensibility; endeavoring to practise on others the arts that are practised on himself; and to gain the applause of his worst associates by imitating their manners.[4]

These lines, written almost two centuries ago, are a harsh indictment of prisons, but many of the conditions they describe continue to be reported in even the most modern studies of prison life. Looking at the matter from a long-range historical perspective, it is fair to conclude that prisons have done a conspicuously poor job of reforming the convicts placed in their custody; but the very consistency of this failure may have a peculiar logic of its own. Perhaps we find it difficult to change the worst of our penal practices because we *expect* the prison to harden the inmate's commitment to deviant forms of behavior and draw him more deeply into the deviant ranks. On the whole, we are a people who do not really expect deviants to change very much as they are processed through the control agencies we provide for them, and we are often reluctant to devote much of the community's resources to the job of rehabilitation. In this sense, the prison which graduates long rows of accomplished criminals (or, for that matter, the state asylum which stores its most severe cases away in some back ward) may do serious violence to the aims of its founders; but it does very little violence to the expectations of the population it serves.

These expectations, moreover, are found in every corner of society and constitute an important part of the climate in which we deal with deviant forms of behavior.

To begin with, the community's decision to bring deviant sanctions against one of its members is not a simple act of censure. It is an intricate rite of transition, at once moving the individual out of his ordinary place in society and transferring him into a special deviant position.[5] The ceremonies which mark this change of status, generally, have a number of related phases. They supply a formal stage on which the deviant and his community can confront one another (as in the criminal trial); they make an announcement about the nature of his deviancy (e.g., a verdict or diagnosis); and they place him in a particular role which is thought to neutralize the harmful effects of his misconduct (like the role of prisoner or patient). These commitment ceremonies tend to be occasions of wide public interest and ordinarily take place in a highly dramatic setting.[6] Perhaps the most obvious example of a commitment ceremony is the criminal trial, with its elaborate formality and exaggerated ritual, but more modest equivalents can be found wherever procedures are set up to judge whether or not someone is legitimately deviant.

Now an important feature of these ceremonies in our own culture is that they are almost irreversible. Most provisional roles conferred by society—those of the student or conscripted soldier, for example—include some kind of terminal ceremony to mark the individual's movement back out of the role once its temporary advantages have been exhausted. But the roles allotted the deviant seldom make allowance for this type of passage. He is ushered into the deviant position by a decisive and often dramatic ceremony, yet is retired from it with scarcely a word of public notice. And as a result, the deviant often returns home with no proper license to resume a normal life in the community. Nothing has happened to cancel out the stigmas imposed upon him by earlier commitment ceremonies; nothing has happened to revoke the verdict or diagnosis pronounced upon him at that time. It should not be surprising, then, that the people of the community are apt to greet the returning deviant with a considerable degree of apprehension and distrust, for in a very real sense they are not at all sure who he is.

A circularity is thus set into motion which has all the earmarks of a "self-fulfilling prophecy," to use Merton's fine phrase. On the one hand, it seems quite obvious that the community's apprehensions help reduce whatever chances the deviant might otherwise have had for a successful return home. Yet at the same time, everyday experience seems to show that these suspicions are wholly reasonable, for it is a well-known and highly publicized fact that many if not most ex-convicts return to crime after leaving prison and that large numbers of mental patients require further treatment after an initial hospitalization. The common feeling that deviant persons never really change, then, may derive from a faulty premise; but the feeling is expressed so frequently and with such conviction that it eventually creates the facts which later "prove" it to be correct. If the returning deviant encounters this circularity often enough, it is quite understandable that he, too, may begin to wonder whether he has fully graduated from the deviant role, and he may respond to the uncertainty by resuming some kind of deviant activity. In many respects, this may be the only way for the individual and his community to agree what kind of person he is.

Moreover, this prophecy is found in the official policies of even the most responsible agencies of control. Police departments could not operate with any real effectiveness if they did not regard ex-convicts as a ready pool of suspects to be tapped in the event of trouble, and psychiatric clinics could not do a successful job in the community if they were not always alert to the possibility of former patients suffering relapses. Thus, the prophecy gains currency at many levels within the social order, not only in the poorly informed attitudes of the community at large, but in the best informed theories of most control agencies as well.

In one form or another this problem has been recognized in the West for many hundreds of years, and this simple fact has a curious implication. For if our culture has supported a steady flow of deviation throughout long periods of historical change, the rules which apply to any kind of evolutionary thinking would suggest that strong forces must be at work to keep the flow intact—and this because it contributes in some important way to the survival of the culture as a whole. This does not furnish us with sufficient warrant to declare that deviance is "functional" (in any of the many senses of that term), but it should certainly make us wary of the assumption so often made in sociological circles that any well-structured society is somehow designed to prevent deviant behavior from occurring.[7]

It might be then argued that we need new metaphors to carry our thinking about deviance onto a different plane. On the whole, American sociologists have devoted most of their attention to those forces in society which seem to assert a centralizing influence on human behavior, gathering people together into tight clusters called "groups" and bringing them under the jurisdiction of governing principles called "norms" or "standards." The questions which sociologists have traditionally asked of their data, then, are addressed to the uniformities rather than the divergencies of social life: how is it that people learn to think in similar ways, to accept the same group moralities, to move by the same rhythms of behavior, to see life with the same eyes? How is it, in short, that cultures accomplish the incredible alchemy of making unity out of diversity, harmony out of conflict, order out of confusion? Somehow we often act as if the differences between people can be taken for granted, being too natural to require comment, but that the symmetry which human groups manage to achieve must be explained by referring to the molding influence of the social structure.

But variety, too, is a product of the social structure. It is certainly remarkable that members of a culture come to look so much alike; but it is also remarkable that out of all this sameness a people can develop a complex division of labor, move off into diverging career lines, scatter across the surface of the territory they share in common, and create so many differences of temper, ideology, fashion, and mood. Perhaps we can conclude, then, that two separate yet often competing currents are found in any society: those forces which promote a high degree of conformity among the people of the community so that they know what to expect from one another, and those forces which encourage a certain degree of diversity so that people can be deployed across the range of group space to survey its potential, measure its capacity, and, in the case of those we call deviants, patrol its boundaries. In such a scheme, the deviant would appear as a natural product of group differentiation. He is not a bit of debris spun out by faulty social machinery, but a relevant figure in the community's overall division of labor.

NOTES

1. In fact, the first statement of the general notion presented here was concerned with the study of small groups. See Robert A. Dentier and Kai T. Erikson, "The Functions of Deviance in Groups," *Social Problems,* VII (Fall 1959), pp. 98–107.

2. Aldous Huxley, *Prisons: The "Carceri" Etchings by Piranesi* (London: The Trianon Press, 1949), p. 13.

3. For a good description of this process in the modern prison, see Gresham Sykes, *The Society of Captives* (Princeton, N.J.: Princeton University Press, 1958). For discussions of similar problems in two different kinds of mental hospital, see Erving Goffman, *Asylums* (New York: Bobbs-Merrill, 1962) and Kai T. Erikson, "Patient Role and Social Uncertainty: A Dilemma of the Mentally Ill," *Psychiatry*, XX (August 1957), pp. 263–274.

4. Written by "a celebrated" but not otherwise identified author (perhaps Henry Fielding) and quoted in John Howard, *The State of the Prisons*, London, 1777 (London: J. M. Dent and Sons, 1929), p. 10.

5. The classic description of this process as it applies to the medical patient is found in Talcott Parsons, *The Social System* (Glencoe, Ill.: The Free Press, 1951).

6. See Harold Garfinkel, "Successful Degradation Ceremonies," *American Journal of Sociology*, LXI (January 1956), pp. 420–424.

7. Albert K. Cohen, for example, speaking for a dominant strain in sociological thinking, takes the question quite for granted: "It would seem that the control of deviant behavior is, by definition, a culture goal." See "The Study of Social Disorganization and Deviant Behavior" in Merton et al., *Sociology Today* (New York: Basic Books, 1959), p. 465.

2

Applying an Integrated Typology of Deviance to Middle-Class Norms

ALEX HECKERT AND DRUANN MARIA HECKERT

Deviance has traditionally been regarded as behavior that underconforms to society's norms of acceptability and is negatively received. Deviants were considered antisocial misfits, either through their failure to live up to societal standards or through their intentional defiance of norms. In previous works, Heckert and Heckert challenged this view, exploring both of its dimensions: the excessive enactment of prosocial behavior and people's positive reactions to deviance. They offered a controversial fourfold table interrelating the axis of under/non- and overconformity (how well the behavior fits in with normative expectations) with the suggestion that people do not always react to deviance negatively, but sometimes admire it (by their social reactions and collective evaluations). By cross-tabulating the dimension of underconformity and overconformity with that of social evaluation, positive and negative, they analytically came up with four types of deviance: negative deviance, rate busting, deviance admiration, and positive deviance.

In this chapter, new to this edition, they further articulate their typology by applying it to a list of 10 middle-class norms developed by Tittle and Paternoster (2000). Addressing such issues as lying, disloyalty, hedonism, irresponsibility, and invasion of privacy, Heckert and Heckert evaluate these forms of "negative deviance" and suggest how such acts might translate into their deviance admiration, positive deviance, and rate-busting counterparts. This creative analysis strengthens their previous typology by fleshing it out with concrete examples that further challenge traditional thinking about definitions of deviance.

What do you think of Heckert and Heckert's typologies? Should we stick with the more conventional definition of deviance as norm violations that are negatively received, or do you buy their assertions about positive deviance? Is too much of a good thing a good thing, or is it really deviant?

Arguably, the subfield of the sociology of deviance began in the late 1800s with the publication of Durkheim's *The Rules of Sociological Method*. Despite its rich history, some scholars have recently claimed that the field of deviance is in a declining state (Best, 2004). Best asserts that a key problem is definitional; he contends that analytical problems are created when too many attitudes, behaviors, and conditions are defined as deviant. He conjectures that, as the field of criminology began to ascend after the mid-1970s, the field of deviance began to descend. Although crimes are defined as violations of criminal statutes, the boundaries regarding what defines deviance are less clear and more fluid. We have argued, however, that, although criminology has a unifying definition, like deviance, it is not homogeneous and covers widely diverging behaviors. For example, the following behaviors all violate criminal statutes and are therefore criminal: homicide, theft, rape, underage drinking, jaywalking, speeding, and littering. Yet, these behaviors are quite diverse. With regard to studying their etiology, what does a murderer or a thief have in common with a person who listened to popular music under the Taliban or an American adolescent who drinks illegally, jaywalks, and then downloads music illegally? Criminologists need to study a wide range of behaviors, as do sociologists who specialize in deviance. Compared with criminologists, what sociologists of deviance have lacked is a unifying definition.

We recently proposed an integrated typology of deviance that synthesizes normative and reactivist definitions of deviance in an attempt to ameliorate the definitional problems that have afflicted the field (Heckert and Heckert, 2002). We conclude that deviance is still an integral area of sociology. Our typology proposed the following four categories: negative deviance, deviance admiration, rate busting, and positive deviance. Negative deviance deals with nonconformity or underconformity that is negatively evaluated. Deviance admiration is nonconformity or underconformity that is positively evaluated. Rate busting involves overconformity or hyperconformity that is negatively evaluated. Finally, positive deviance has to do with overconformity or hyperconformity that is positively evaluated.

In this article, we review the various ways that deviance has been defined. We next discuss a contested concept in deviance called positive deviance and articulate how it has been defined. We then discuss our integrated typology of deviance and show how it conceptually situates positive deviance and resolves the definitional challenges in the field. Finally, we apply our typology to the 10 key middle-class norms proposed by Tittle and Paternoster (2000).

DEFINITIONS OF DEVIANCE

Deviance has been defined in four main ways: the absolutist approach, the statistical approach, the normative/objectivist approach, and the reactivist/subjectivist approach. The absolutist and statistical approaches have been thoroughly rejected; the predominant bifurcation within the discipline is between the normative and reactivist perspectives. The absolutist approach suggests that there are absolute standards of behavior that are moral and good, and any deviation from these standards constitutes deviance. The empirical relativity across social groups and across time regarding what constitutes good behavior leads to the rejection of the absolutist approach to defining deviance. The statistical approach defines deviance as attitudes, behaviors, or conditions that are statistically rare.

This definition is easily rejected because of the number of behaviors and conditions that are considered deviant even though they are not rare, such as drinking to excess, lying, and adultery, among many others. There are also conditions and behaviors that are rare but not considered to be deviant, such as certain kinds of cancer, having blonde hair, and running marathons. Normative, or objectivist, definitions focus on the violation of norms. Some scholars simply defined deviance as behavior that violates rules or normative expectations. Reactivist or subjectivists definitions, by contrast, focus on the dynamics of the reactions and evaluations of a social audience. As Becker (1963:11) wrote in his seminal book *The Outsiders*, "social groups constitute deviance by making rules whose infractions constitute deviance, and by applying those rules to particular people and labeling them as outsiders." Each definition constitutes a major paradigmatic understanding of deviance. All four definitions tend to emphasize behaviors; nonetheless, most definitions of deviance include what Adler and Adler call in this book the ABCs of deviance: attitudes, behaviors, and conditions. Although the examples we provide in this paper typically involve behaviors, it is also the case that attitudes (or beliefs) and conditions are essential to deviance because normative expectations and social reactions pertain to both of them as well.

Positive Deviance

Positive deviance is a contested term in the sociology of deviance. While some theorists have rejected it as oxymoronic or as an unviable concept, other theorists have argued that a substantial amount of deviant behavior is actually defined in a *positive* fashion. Unfortunately, the same bifurcation that characterizes the field of deviance as a whole also exists with regard to defining positive deviance. Scholars who take a normative perspective define deviance as attitudes, behaviors, and conditions that exceed or overconform to the norms. Scholars who take a reactivist perspective define positive deviance as any attitude, behavior, or condition that is positively evaluated.

INTEGRATED TYPOLOGY OF DEVIANCE

In an attempt to resolve the definitional tensions in the field and to conceptually locate the concept of positive deviance, we have proposed a typology that integrates the normative and reactivist definitions of deviance (Heckert and Heckert, 2002). This typology treats the distinction as a false dichotomy and acknowledges that both norms (or rules) of behavior (attitudes and conditions) and social reactions and evaluations exist. Behavior and conditions can underconform (or fail to conform) or overconform to normative expectations, as well as lead to negative or positive reactions. Accordingly, we have cross-classified the significance of both norms and reactions, producing four possible scenarios. As Figure 2.1 demonstrates, negative deviance involves underconformity or nonconformity that results in negative reactions. Deviance admiration has to do with underconformity or nonconformity that is positively evaluated. Rate busting involves overconformity

Normative Expectations

Social Reactions and Collective Evaluations	Underconformity or Nonconformity	Overconformity
Negative Evaluations	Negative Deviance	Rate Busting
Positive Evaluations	Deviance Admiration	Positive Deviance

FIGURE 2.1 Deviance Typology

or hyperconformity to norms that is negatively evaluated, and positive deviance deals with overconformity or hyperconformity that is positively evaluated. We will briefly describe each type of deviance.

Historically, the substantive area of deviance has focused on negative deviance, which refers to nonconformity or underconformity that is also negatively evaluated. We have metaphorically labeled it the "Jeffrey Dahmer phenomenon" after the serial murderer who was universally reviled for his behavior. Negative deviance can range from the behavior of most criminals, to that of the mentally ill, to the behavior of substance abusers, to that of individuals with unpopular religious or political stances. The study of deviance has traditionally been confined to the study of negative deviance.

Deviance admiration focuses on nonconformity or underconformity that is *positively* evaluated. We have labeled such underconformity the "John Gotti" phenomenon. John Gotti was an organized crime leader who was admired by many conformists in the United States. Kooistra (1989) concluded that a surprising number of criminals, even brutal ones, have been transformed by the collective imagination from thugs into icons. In other words, the mythological Robin Hood has bona fide real-life counterparts, such as Billy the Kid, D. B. Cooper, Butch Cassidy, and the Sundance Kid. As another example, in a society that often rejects individuals who do not meet culturally dominant and socially constructed aesthetic images of appearance, there are also many who still admire the stigmatized attribute (e.g., being tall and having red hair) and individuals with that attribute.

Rate busting refers to overconformity or hyperconformity that is negatively evaluated. The "geek phenomenon" serves as the quintessential example of this cell of the typology. Various social scientists have noted that norms operate at two levels: the idealized (i.e., that which is believed sublimely better, but improbable, for most people) and the realistic (i.e., that which is viewed as achievable by typical people). Regardless, this overconformity, even if idealized—or perhaps because it is idealized—is often subjected to negative reactions. Rate busting has occurred in various contexts in social life. For example, several studies have found that gifted students are often rejected by their peers (Huryn, 1986; Shoenberger, Heckert, and Heckert, 2012). Krebs and Adinolfi (1975) concluded that attractive individuals are often slighted by members of their same sex, and Heckert (2003) found that blonde women (defined as overconforming to traditional European standards of beauty) are subjected to epic stereotyping, especially with regard to their (assumed lack of) intellectual capacity.

The last cell of our typology highlights positive deviance, which refers to overconformity or hyperconformity that is responded to in a positive or esteemed fashion. We have labeled this category the "Mother Teresa phenomenon," after the universally admired nun who worked with the poor. As with negative deviance, positive deviance has been defined in various ways, including from a normative perspective, from a reactivist perspective, and from the perspective that positive deviance refers to overconformity which is positively evaluated (Dodge, 1985; Heckert and Heckert, 2002). Some examples of positive deviance include saints and good neighbors (Sorokin, 1950), winners of the Congressional Medal of Honor (Steffensmeier and Terry, 1975), and the physically attractive. Exceeding normative expectations of conformity can, at times, create situations in which positive reactions and consequences are abundant, often producing additional advantages that can transcend even the norm at stake.

Overall, our typology recognizes that the traditional distinction between normative expectations and social reactions constitutes a false dichotomy. Also, our typology seeks to integrate normative definitions with reactivist definitions. Doing so allows attitudes, behaviors, and conditions to be conceptualized as negative deviance, deviance admiration, rate busting, and positive deviance. Of course, contexts do have to be considered because audiences can react differentially to the same behaviors or conditions. For example, gifted and overachieving students are rate busters to their peers, but positive deviants to their teachers (Shoenberger, Heckert, and Heckert, 2012). Similarly, elite tattoo collectors are simultaneously positive deviants in their subculture and negative deviants to the dominant culture in society (Irwin, 2003). As another example, some individuals are castigated in their own era for not adhering to the normative expectations of that time, only to be admired as extraordinary in a later era (e.g., Martin Luther King, Galileo, Socrates, and the French Impressionists). For different reasons, nonconformists and underconformers are sometimes positively evaluated while overconformers are sometimes negatively evaluated.

This integrated typology allows deviance scholars to continue to acknowledge that deviance is relative and context is critical. It is important, moreover, to analyze why nonconformity or underconformity can result in positive evaluations or negative evaluations, depending on the era, place, or social group involved. The same is true for overconformity or hyperconformity. Power and peoples' self-interest are critical factors. With regard to both underconformity and overconformity, negative reactions and evaluations are likely to occur when the attitudes, behaviors, or conditions threaten the interests of the social group. As Heckert and Heckert (2002:468–469) asserted,

> Whether or not the negative reactions have an impact (or become "sticky" to use a labeling term) is primarily determined by the relative power of the (potential) deviant(s) and the social audience. The relative power is determined by a number of factors such as the numbers in each group (deviant and reactors), the amount of wealth and income (property) of each group, the relative prestige of each group, the level of organization of each group (from individualized to subcultural to organized), and the relative quality of their discourse or claims (ability to persuade and manipulate symbols). These various sources of power interact in determining the ability of a given social group or audience to apply negative or positive labels to a type of behavior, condition, or particular group of (potential) deviants. Viewed in this light, deviant behavior is, in essence, a test of power relationships and serves as a potential threat to the power of the dominant group(s).

The same processes apply to overconformity. Other factors are also important, such as the degree of violation of the norm (minor to extreme) and the definition of the situation. High-consensus norms (e.g., first-degree homicide) will generally generate more consistent evaluations than will low-consensus norms (Thio, 2001). Future scholarship should investigate these factors, as well as consider others.

On the basis of the insights provided by our integrated typology of deviance, we propose the following definition of deviance that accommodates all four cells: Deviance consists of behaviors, attitudes, or conditions that violate norms, under-conform to norms, or overconform to norms and that are either negatively or positively evaluated and/or negatively or positively sanctioned (or would be if detected).

THE PREDOMINANT MIDDLE-CLASS NORMS
IN THE UNITED STATES

Norms are integral components of any social system; they guide expected behavior and are based on values viewed as important. For example, on the basis of his review of 14 common definitions of *norm*, Gibbs (1981:7) stated that the most common definition of norm is "a belief shared to some extent by members of a social unit as to what conduct *ought to be* in particular situations or circumstances." Gusfield (1963:65) suggested, moreover, that cultural systems produce "regularities of action."

Some sociologists have suggested that norms can be dually conceptualized as encompassing the idealized and the more realistic or operative. In essence, the more idealized norms are really values. Williams (1965) created the seminal and most comprehensive description of the value system of the United States. His description outlined the following dominant American values: achievement and success, individualism, activity and work, efficiency and practicality, science and technology, progress, material comfort, humanitarianism, freedom, democracy, equality, and racism and group superiority. Henslin (1975) added the following values to that list: education, religiosity, romantic love, and monogamy. Overall, the American value system has been sufficiently addressed.

More recently, Tittle and Paternoster (2000) expanded sociological understanding by outlining the normative system itself, as opposed to the idealized scheme of values. They restrict their outline to the primary middle-class values in the United States and highlight the following dominant norms: group loyalty, privacy, prudence, conventionality, responsibility, participation, moderation, honesty, peacefulness, and courtesy. Loyalty is valued because of the group's need to survive over individual concerns. Privacy suggests that individuals need to be in command of certain spaces and places. Prudence places pleasure into perspective and encourages practicing pleasure in moderation. Conventionality involves people choosing habits and life scenarios that are similar to those of other people. Responsibility entails dependability, especially when others must count on an individual. Participation implies personal involvement in both social and economic spheres, with alienation constituting the nonconforming counterpart. Moderation is the avoidance of the extremes in life. Honesty has to do with veracity and candor. Peacefulness involves a sedate and calm lifestyle. Finally, courtesy pertains to refined social etiquette in human interaction. Clearly, Tittle and Paternoster have thoroughly delineated one set of consequential norms.

APPLYING THE INTEGRATED TYPOLOGY
OF DEVIANCE TO MIDDLE-CLASS NORMS

To illustrate the utility of our integrated typology of deviance, as shown in Table 2.1, we apply it to the normative system of U.S. middle-class norms

T A B L E   2.1   A Classification of U.S. Middle-Class Deviance

Norm	Negative Deviance	Deviance Admiration	Rate Busting	Positive Deviance
Group loyalty	Apostasy	Rebellion	Fanaticism	Altruism
Privacy	Intrusion	Investigation	Seclusion	Circumspection
Prudence	Indiscretion	Exhibitionism	Puritanism	Discretion
Conventionality	Bizarreness	Faddishness	Provincialism	Properness
Responsibility	Irresponsibility	Adventuresome	Priggishness	Hyperresponsibility
Participation	Alienation	Independence	Dependence	Cooperation
Moderation	Hedonism	Roguishness	Asceticism	Temperance
Honesty	Deceitfulness	Tactfulness	Tactlessness	Forthrightness
Peacefulness	Disruption	Revelry	Wimpishness	Pacifism
Courtesy	Uncouthness	Irreverence	Obsequiousness	Gentility

Note: Types of negative deviance taken from Tittle & Paternoster, 2000:35.

outlined by Tittle and Paternoster (2000). In addition to documenting the dominant middle-class norms in contemporary United States, they highlighted the types of (negative) deviance associated with each norm. In addition to reviewing the forms of negative deviance they featured, Table 2.1 shows the forms of deviance admiration, rate busting, and positive deviance associated with each norm.

Negative Deviance

As noted earlier, negative deviance refers to nonconformity or underconformity that is negatively evaluated. Tittle and Paternoster (2000) emphasized negative deviance and provided examples of each category, as shown in Table 2.2. These examples are quintessentially representative of the traditional nucleus of the substantive area of the sociological study of deviance. We will highlight just a few types of negative deviance in this section.

TABLE 2.2   Negative Deviance

Norm	Negative Deviance	Examples
Group loyalty	Apostasy	Revolution; betraying national secrets; treason; draft dodging; defiling the flag; giving up citizenship; advocating contrary government philosophy
Privacy	Intrusion	Theft; burglary; rape; homicide; voyeurism; forgery; spying on records
Prudence	Indiscretion	Prostitution; homosexual behavior; incest; bestiality; adultery; swinging; gambling; substance abuse
Conventionality	Bizarreness	"Mentally ill behavior" (handling excrement, talking nonsense, eating human flesh, fetishes); separatist lifestyles
Responsibility	Irresponsibility	Deserting the family desertion; reneging on debts; unprofessional conduct; improper role performance; violations of trust; pollution; conducting fraudulent business
Participation	Alienation	Nonparticipatory lifestyles (being a hermit, street living); perpetual unemployment; receiving public assistance; suicide
Moderation	Hedonism	Chiseling; atheism; alcoholism; total deceit; wasting; ignoring children
Honesty	Deceitfulness	Selfish lying; price fixing; exploitation of the weak and helpless; bigamy; welfare cheating
Peacefulness	Disruption	Noisy disorganizing behavior; boisterous reveling; quarreling; fighting; contentiousness
Courtesy	Uncouthness	Private behavior in public places (picking nose, burping); rudeness (smoking in prohibited places, breaking in line); uncleanliness

Note: Table from Tittle & Paternoster, 2000:35.

Tittle and Paternoster (2000) contend that, because survival of the group is viewed as paramount, apostasy becomes deviantized. Accordingly, actions such as revolution and treason are viewed as deviant. Given that prudence is valued, indiscretions such as adultery, gambling, and substance abuse are negatively deviantized. Because responsibility is highly valued, irresponsible behaviors such as family desertion and fraudulent business practices are treated as deviant. In light of the fact that honesty is valued, deceitful behaviors such as bigamy and selfish lying are considered deviant. Because peacefulness constitutes an important American norm, disruptive behaviors, including quarreling and boisterous reveling, are negatively valued. In sum, Tittle and Paternoster (2000) provided a comprehensive conceptualization of various deviantized behaviors within the middle class in the United States. Their framework highlights a number of forms of negative deviance and provides illustrative examples, all of which are comfortably ensconced within the long-established core of the field of deviance.

Deviance Admiration

As noted earlier, deviance admiration involves nonconformity or underconformity that is positively evaluated. Although people do not necessarily view norm violators as doing right, in certain cases they nevertheless respond with positive appraisals. Table 2.3 reveals the types of deviance admiration associated with each of the 10 middle-class norms articulated by Tittle and Paternoster (2000) and also provides examples of each type. We will highlight a few examples to illustrate.

The deviance admiration form of deviance deriving from the norm of group loyalty is rebellion. As the quintessential example, the American revolutionaries, such as George Washington, Thomas Jefferson, Nathan Hale, and Paul Revere, are some of the most revered people in American history. Their rebellion is now emblematic of the highest level of patriotism. The type of deviance admiration associated with the norm of conventionality is faddishness. Examples of faddishness include body piercing and subcultures such as Wiccan and certain religious sects. Roguishness is illustrative of the deviance admiration component of moderation. The dominant middle-class culture has often venerated the charming rogue, as opposed to the earnest, but dull, individual. Some of the charismatic charm of former President Bill Clinton was attributable to his roguish violations of the norm of moderation. Nonconformity to the norm of honesty is often appreciated as tactfulness. The individual who tells "white lies" to spare a friend's feelings, as well as the cop who plays loose with the truth to extract a confession from a criminal, are often admired for their violation of the norm of honesty. Finally, irreverence is the deviance admiration behavior associated with the norm of courtesy. Mischievous individuals such as the class clown and characters on shows like *The Simpsons* and *South Park* are often appreciated for their underconformity to the norm of courtesy.

Obviously, there is a corresponding form of deviance admiration associated with each of the 10 middle-class norms outlined by Tittle and Paternoster (2000). Various attitudes, behaviors, and conditions exemplify each category.

TABLE 2.3   Deviance Admiration

Norm	Deviance Admiration	Examples
Group loyalty	Rebellion	James Dean (*Rebel Without a Cause*); American Revolution; Pentagon papers; admiration for men who went to Canada to avoid draft during Vietnam War
Privacy	Investigation	Investigative journalism; private investigators; revelations on Jerry Springer show
Prudence	Exhibitionism	Famous strippers; gay rights movement; famous gamblers; flamboyant entertainers (e.g., Liberace, RuPaul; transvestites, female impersonators); extreme sports; reality television with people like the Osbournes and Anna Nicole Smith
Conventionality	Faddishness	The Osbournes' show; tattoos; bizarre subcultures; body piercing; Wiccan and similar subcultures
Responsibility	Adventuresome	Extreme sports (e.g., hang gliding, mountain climbing, free diving); *Jackass: the Movie* humor; explorers
Participation	Independence	Romantic loners; mysterious strangers; drifters; hobos; explorers
Moderation	Roguishness	Charming rogue; the lovable drunk (e.g., Otis on *The Andy Griffith Show*); Bill Clinton
Honesty	Tactfulness	Honest humor on *The Man Show*; telling "tall tales"; the liar who is admired; telling "white lies"; the person who lies to achieve a greater goal (e.g., national security, confession from serial killer using "good cop," "bad cop" strategies; an operative who lies to be able to capture an "enemy" such as a terrorist, and the like)
Peacefulness	Revelry	Mardi Gras; fraternity partying; loud entertainer at a party; spring break phenomenon
Courtesy	Irreverence	Class clown; humor on *The Man Show*; rude humor on shows such as *The Simpsons, South Park, Beavis and Butthead*, and the like

Rate Busting

Rate busting is the negative reaction to overconformity, a reaction we have met-aphorically referred to as the "geek phenomenon." In Table 2.4, we highlight a rate-busting form of deviance associated with each of the 10 middle-class norms outlined by Tittle and Paternoster (2000). We also provide a nonexhaustive list of examples for each category of rate busting and discuss a few of them to illustrate several types of rate busting.

With regard to the norm of group loyalty, overconforming to the group can result in negative evaluations, which we refer to as fanaticism. As examples, hate

TABLE 2.4   Rate Busting

Norm	Rate Busting	Examples
Group loyalty	Fanaticism	Various religious cults; Ku Klux Klan (which, of course, engages in acts of negative deviance); Aryan Nation; National Rifle Association; religious fanatics; superpatriots; people who are "holier than thou"
Privacy	Seclusion	Hermits; loners; Amish and other reclusive religious sects; secretive or reclusive behavior; Howard Hughes
Prudence	Puritanism	Negative attitudes and behavior toward Amish and other conservative religious sects
Conventionality	Provincialism	Stepford wives; Martha Stewart and her followers; keeping up with the Joneses in following every little rule so as to be accepted; individuals who are negatively evaluated for ritualistically following convention
Responsibility	Priggishness	Negative attitudes regarding people who are self-Righteous; jokes about, and meanness toward Martha Stewart and her followers; some workaholics; straight-A student
Participation	Dependence	Brownnoser; people trying so hard to be accepted by every group that they lose their individuality; the concepts of codependency and enabling behavior
Moderation	Asceticism	Negative reactions to not drinking; People so meek that they never take a stand on anything so as not to offend anyone
Honesty	Tactlessness	Being too honest with friends (e.g., about ugly clothing or an unattractive haircut); People so honest that they won't tell a "white lie" or who will say something mean to a child because it is honest rather than protect the child's feelings
Peacefulness	Wimpishness	People who never party or "let their hair down"; a person who will never take a stand or stand up to people; a "yes man"
Courtesy	Obsequiousness	The person who is overly polite; making fun of Miss Manners and people who are overly courteous

group members and religious fanatics, who may be accepted by their subculture, are too responsive to their groups and can be deviantized by the dominant middle-class culture. Individuals who are "holier than thou" within dominant religious traditions often alienate members of their own group. The rate-busting form of deviance affiliated with the norm of prudence is puritanism. For example, very conservative religious sects, such as the Amish, Hutterites, and Jehovah's Witnesses, that seem to eschew what dominant Americans view as necessary pleasures are often shunned or not accepted by the dominant culture. Overconformity to the norm of responsibility can be negatively evaluated as a deviant form of priggishness. Some workaholics who are motivated by an intense sense of responsibility have been subjected to negative treatment by their coworkers, even if praised by their supervisors. Similarly, peers have long taunted overachieving students, labeling them as bookworms or geeks, while their teachers view them as praiseworthy.

With regard to the norm of participation, dependence is the rate-busting form of deviance. Among those who are negatively evaluated for their overparticipation in social life are the brownnoser, the codependent and enabling relative of a substance abuser, and the individual who wants to be accepted so badly by a group that she or he sacrifices individuality. With regard to honesty, tactlessness is the rate-busting form of deviance. For example, some individuals are so honest that they won't tell a white lie, even to preserve the feelings of a child. Imagine how the typical middle-class person would respond to an adult who told children that they were ugly! The rate-busting form of deviance associated with the norm of peacefulness is wimpishness. Meek individuals who never "let their hair down" or who never stand up to anyone are often negatively appraised as "mousy," "wimpy," or a "yes man."

Clearly, a rate-busting form of deviance is affiliated with overconformity or hyperconformity to each of the 10 norms listed by Tittle and Paternoster (2000). We have provided examples for a few types of rate-busting deviance.

Positive Deviance

As suggested by our typology, positive deviance consists of overconformity or hyperconformity to the norms that is positively evaluated. We have symbolically labeled this type of deviance as the "Mother Teresa phenomenon." Mother Teresa overconformed to the norm of altruism, and she was almost universally praised as a saintly woman. In Table 2.5, we list the types of positive deviance associated with each middle-class norm and provide corresponding examples.

Overconformity to the norm of altruism that is positively evaluated is called loyalty. Examples of individuals who have exhibited loyalty include religious and political martyrs, many of whom have been viewed as heroes. Political leaders who were assassinated, such as Abraham Lincoln, John F. Kennedy, and Martin Luther King, are greatly admired. Every soldier is honored on Veteran's Day, and soldiers who have died for the United States are venerated on Memorial Day. With regard to the norm of prudence, discretion constitutes its positive deviance counterpart. People, such as Olympic athletes and novelists, who forgo unnecessary and frivolous pleasures to pursue valued goals, are often admired.

Temperance illustrates the positive deviance form of moderation. When ascetics, or those who overconform to the moderation norm for a more transcendent purpose, are positively evaluated, their behavior is a form positive deviance. One example is individuals who choose the religious life as the calling of their life, such as priests, monks, nuns, preachers, and rabbis. Individuals who overconform to the norm of honesty, but who are positively appraised,

TABLE 2.5   **Positive Deviance**

Norm	Positive Deviance	Examples
Group loyalty	Altruism	Kamikaze pilots; martyrs; sharing food or other resources; daring rescues (e.g., at sea, in the mountains); patriot
Privacy	Circumspection	CIA operatives; FBI/Justice Department agents; loyal company employees (e.g., Oliver North)
Prudence	Discretion	The good friend who practices discretion; people who deny themselves pleasures to achieve a goal (e.g., Olympians and other athletes)
Conventionality	Properness	Junior League members; Martha Stewart's followers; Boy Scouts, Girl Scouts, and similar groups
Responsibility	Hyperresponsibility	Overachiever; straight-A student (as viewed by parents and teachers); workaholic (as viewed by management); overzealous athlete (e.g., Tiger Woods; Michael Jordan)
Participation	Cooperation	Athletic team on which individual talents are de-emphasized so that the team can win; employees who are positively viewed as team players
Moderation	Temperance	Monks; nuns; Women's Temperance Movement; Mothers Against Drunk Drivers movement
Honesty	Forthrightness	Honest Abe Lincoln; story of George Washington cutting down the cherry tree
Peacefulness	Pacifism	Gandhi; Martin Luther King; Jimmy Carter
Courtesy	Gentility	Miss Manners; the old-fashioned practitioners of southern hospitality who are admired; the gentleman; the "Southern Belle"

are engaging in forthrightness. Virtually every schoolchild in this country is successfully socialized to revere the importance of honesty through the story of George Washington and the cherry tree. The type of positive deviance associated with the norm of peacefulness is pacifism. The Nobel Prize for Peace is given annually to venerate individuals for their contributions to peace. Among individuals who are nearly universally admired for pacifism are Gandhi, Martin Luther King, and even Jimmy Carter, who attempted peaceful approaches to accomplishing social change. The final form of positive deviance listed in Table 2.5 is gentility, which is associated with the norm of courtesy. As examples, positive appraisals of overconformity to the norm of courtesy occur with individuals who give up their seat on a crowded bus or who give up their place in line in a crowded grocery store.

Clearly, there is a type of positive deviance associated with each of the 10 middle-class norms developed by Tittle and Paternoster (2000). We have provided various examples to illustrate some of those types.

DISCUSSION

In this article, we have demonstrated how our integrated typology of deviance resolves the definitional disputes that have plagued the field of sociology of deviance. In that vein, we offered the following definition of deviance: deviance is defined as behaviors, attitudes, or conditions that violate norms, underconform to norms, or overconform to norms and that are either negatively or positively evaluated and/or negatively or positively sanctioned (or would be if detected). By applying our integrated typology to 10 middle-class American norms, we have illuminated the efficacy of the integrated typology as a heuristic device and conceptually located the contested concept of positive deviance.

Long ago, Durkheim (1982) discussed the inevitability or normalcy of deviance. Moreover, Durkheim (1964; 1982) illuminated the positive functions of deviance. A variety of social psychological experiments, such as those conducted by Sherif (1936), Asch (1952), and Schachter (1951), suggested the fundamental existence of the processes of norms production, conformity, deviance production, and individual and social reactions to (potentially) deviant behaviors, attitudes, and conditions. The importance of power and social context in influencing norms, social reactions, and deviance outcomes should not be ignored. The conditions and behaviors of more powerful actors are less likely to be deviantized than those of less powerful actors. Similarly, the reactions of powerful actors and social groups are more important in determining what and who are successfully labeled as deviant. As we have argued, deviant behavior actually serves as a test of power relationships (Heckert and Heckert, 2002). When children lie, for example, they are testing the parent's ability to discern and react to the lie. When they "get away" with the lie, a shift in the relative power between child and parent has occurred. The same is true for other social relationships and forms of deviance.

The central place of norms, social reactions, efforts at social control, and deviance processes, therefore, suggests the importance of an integrated framework for understanding these factors and their interrelationships. Our integrated typology constitutes an initial attempt to eliminate the false dichotomy imposed by previous conceptualizations and definitions of deviance.

REFERENCES

Asch, Solomon. (1952). *Social Psychology.* Englewood Cliffs, New Jersey: Prentice Hall.

Becker, Howard. (1963). *Outsiders.* New York: The Free Press of Glencoe.

Best, Joel. (2004). *Deviance: Career of a Concept.* Belmont, California: Thomson/Wadsworth.

Dodge, David L. (1985). "The Over-Negativized Conceptualization of Deviance: A Programmatic Exploration." *Deviant Behavior* 6:17–37.

Durkheim, Emile. (1895/1964). *The Division of Labor in Society.* Trs. J. Solovay and J. Mueller. New York: The Free Press.

Durkheim, Emile. (1893/1982). *The Rules of Sociological Method*. Tr. S. Lukes. London: Macmillan.

Gibbs, Jack P. (1981). *Norms, Deviance and Social Control*. New York: Elsevier.

Gusfield, Joseph R. (1963). *Symbolic Crusade*. Urbana, Illinois: University of Illinois Press.

Heckert, Druann. (2003). "Mixed Blessings: Blonde Women as Positive Deviants and as Rate-Busters." *Free Inquiry in Creative Sociology* 31:47–72.

Heckert, Alex, and Druann Maria Heckert. (2002). "A New Typology of Deviance: Integrating Normative and Reactivist Definitions of Deviance." *Deviant Behavior* 23:449–479.

Henslin, James M. (1975). *Introducing Sociology: Toward Understanding Life in Society*. New York: Free Press.

Huryn, Jean Scherz. (1986). "Giftedness as Deviance: A Test of Interaction Theories." *Deviant Behavior* 7:175–186.

Irwin, Katherine. (2003). "Saints and Sinners: Elite Tattoo Collectors and Tattooists as Positive and Negative Deviants." *Sociological Spectrum* 23:27–58.

Kooistra, Paul. (1989). *Criminals as Heroes: Structure, Power and Identity*. Bowling Green, Ohio: Bowling Green State University Popular Press.

Krebs, Dennis, and Allen A. Adinolfi. (1975). "Physical Attractiveness, Social Relations, and Personality Style." *Journal of Personality and Social Psychology* 31:245–253.

Schachter, Stanley. (1951). "Deviance, Rejection, and Communication." *Journal of Abnormal Social Psychology* 46:190–207.

Sherif, Muzifer. (1936). *The Psychology of Social Norms*. New York: Harper.

Shoenberger, Nicole, Heckert, Alex, and Druann M. Heckert. (2012). "Techniques of Neutralization and Positive Deviance." *Deviant Behavior* 33:774–791.

Sorokin, Pitirim A. (1950). *Altruistic Love*. Boston: The Beacon Press.

Steffensmeier, Darrell J., and Robert M. Terry. (1975). *Examining Deviance Experimentally*. Port Washington, New York: Alfred Publishing.

Thio, Alex. (2001). *Deviant Behavior*, 6th ed. Boston: Allyn and Bacon.

Tittle, Charles R., and Raymond Paternoster. (2000). *Social Deviance and Crime*. Los Angeles, California: Roxbury Publishing Company.

Williams, Robin M., Jr. (1965). *American Society: A Sociological Interpretation*, 2nd ed. New York: Knopf.

3

Relativism: Labeling Theory

HOWARD S. BECKER

Becker's classic statement setting forth labeling theory advances the relativistic perspective on defining deviance. Here, he argues that the essence of deviance is contained, not within individuals' behaviors, but in the response others have to those behaviors. Deviance, he claims, is a social construction forged by diverse audiences. Becker assesses the level of deviance attached to a behavior by the social reactions to it. He supports this idea by pointing out that the same behaviors may be received very differently under varying conditions. He notes that variations in the degree of deviance attached to an act may arise because of the temporal or historical contexts framing the act, the social position and power of those who committed or were harmed by the act, and the consequences that arise from the act. These framing elements, which are sometimes unrelated to the behavior itself, may lead one act to be designated as heinous and relegate another, similar one, to obscurity. Becker thus locates the root of deviance in the response of people rather than the act itself, and in the chain of events that is unleashed once people have labeled acts and their perpetrators as deviant.

The interactionist perspective ... defines deviance as the infraction of some agreed-upon rule. It then goes on to ask who breaks rules, and to search for the factors in their personalities and life situations that might account for the infractions. This assumes that those who have broken a rule constitute a homogeneous category, because they have committed the same deviant act.

Such an assumption seems to me to ignore the central fact about deviance: it is created by society. I do not mean this in the way it is ordinarily understood, in which the causes of deviance are located in the social situation of the deviant or in "social factors" which prompt his action. I mean, rather, that *social groups create deviance by making the rules whose infraction constitutes deviance*, and by applying those rules to particular people and labeling them as outsiders. From this point of view, deviance is *not* a quality of the act the person commits, but rather a consequence of the application by others of rules and sanctions to an "offender." The deviant is one to whom the label has successfully been applied; deviant behavior is behavior that people so label.[1]

Since deviance is, among other things, a consequence of the responses of others to a person's act, students of deviance cannot assume that they are dealing with a homogeneous category when they study people who have been labeled deviant. That is, they cannot assume that those people have actually committed a deviant act or broken some rule, because the process of labeling may not be infallible; some people may be labeled deviant who, in fact, have not broken a rule. Furthermore, they cannot assume that the category of those labeled deviant will contain all those who actually have broken a rule, for many offenders may escape apprehension and thus fail to be included in the population of "deviants" they study. Insofar as the category lacks homogeneity and fails to include all the cases that belong in it, one cannot reasonably expect to find common factors of personality or life situation that will account for the supposed deviance. What, then, do people who have been labeled deviant have in common? At the least, they share the label and the experience of being labeled as outsiders. I will begin my analysis with this basic similarity and view deviance as the product of a transaction that takes place between some social group and one who is viewed by that group as a rule-breaker. I will be less concerned with the personal and social characteristics of deviants than with the process by which they come to be thought of as outsiders and their reactions to that judgment....

The point is that the response of other people has to be regarded as problematic. Just because one has committed an infraction of a rule does not mean that others will respond as though this had happened. (Conversely, just because one has not violated a rule does not mean that he may not be treated, in some circumstances, as though he had.)

The degree to which other people will respond to a given act as deviant varies greatly. Several kinds of variation seem worth noting. First of all, there is variation over time. A person believed to have committed a given "deviant" act may at one time be responded to much more leniently than he would be at some other time. The occurrence of "drives" against various kinds of deviance illustrates this clearly. At various times, enforcement officials may decide to make an all-out attack on some particular kind of deviance, such as gambling, drug addiction, or homosexuality. It is obviously much more dangerous to engage in one of these activities when a drive is on than at any other time. (In a very interesting study of crime news in Colorado newspapers, Davis found that the amount of crime reported in Colorado newspapers showed very little association with actual changes in the amount of crime taking place in Colorado. And, further, that people's estimate of how much increase there had been in crime in Colorado was associated with the increase in the amount of crime news but not with any increase in the amount of crime.)[2]

The degree to which an act will be treated as deviant depends also on who commits the act and who feels he has been harmed by it. Rules tend to be applied more to some persons than others. Studies of juvenile delinquency make the point clearly. Boys from middle-class areas do not get as far in the legal process when they are apprehended as do boys from slum areas. The middle-class boy is less likely, when picked up by the police, to be taken to the station; less likely when taken to the station to be booked; and it is extremely unlikely that he will be convicted and sentenced.[3] This variation occurs even though the original infraction of the rule is the same in the two cases. Similarly, the law is differentially applied to Negroes and whites. It is well known that a Negro believed to have attacked a white woman is much more likely to be punished than a white man who commits the same offense; it is only slightly less well known that a Negro who murders another Negro is much less likely to be punished than a white man who commits murder.[4] This, of course, is one of the main points of Sutherland's analysis of white-collar crime: crimes committed by corporations are almost always prosecuted as civil cases, but the same crime committed by an individual is ordinarily treated as a criminal offense.[5]

Some rules are enforced only when they result in certain consequences. The unmarried mother furnishes a clear example. Vincent[6] points out that illicit sexual relations seldom result in severe punishment or social censure for the offenders. If, however, a girl becomes pregnant as a result of such activities, the reaction of others is likely to be severe. (The illicit pregnancy is also an interesting example of the differential enforcement of rules on different categories of people. Vincent notes that unmarried fathers escape the severe censure visited on the mother.)

Why repeat these commonplace observations? Because, taken together, they support the proposition that deviance is not a simple quality, present in some kinds of behavior and absent in others. Rather, it is the product of a process which involves responses of other people to the behavior. The same behavior may be an infraction of the rules at one time and not at another or may be an infraction when committed by one person, but not when committed by another; some rules are broken with impunity, others are not. In short, whether a given act is deviant or not depends in part on the nature of the act (that is, whether or not it violates some rule) and in part on what other people do about it.

Some people may object that this is merely a terminological quibble, that one can, after all, define terms any way he wants to and that if some people want to speak of *rule-breaking behavior* as *deviant* without reference to the reactions of others they are free to do so. This, of course, is true. Yet it might be worthwhile to refer to such behavior as rule-breaking behavior and reserve the term deviant for those labeled as deviant by some segment of society. I do not insist that this usage be followed. But it should be clear that insofar as a scientist uses "deviant" to refer to any rule-breaking behavior and takes as his subject of study only those who have been *labeled* deviant, he would be hampered by the disparities between the two categories.

If we take as the object of our attention behavior which comes to be labeled as deviant, we must recognize that we cannot know whether a given act will be categorized as deviant until the response of others has occurred. Deviance is not a quality that lies in behavior itself, but in the interaction between the person who commits an act and those who respond to it....

In any case, being branded as deviant has important consequences for one's further social participation and self-image. The most important consequence is a drastic change in the individual's public identity. Committing the improper act and being publicly caught at it place him in a new status. He has been revealed as a different kind of person from the kind he was supposed to be. He is labeled a "fairy," "dope fiend," "nut," or "lunatic," and treated accordingly.

In analyzing the consequences of assuming a deviant identity let us make use of Hughes' distinction between master and auxiliary status traits.[7] Hughes notes that most statuses have one key trait which serves to distinguish those who belong from those who do not. Thus the doctor, whatever else he may be, is a person who has a certificate stating that he has fulfilled certain requirements and is licensed to practice medicine; this is the master trait. As Hughes points out, in our society a doctor is also informally expected to have a number of auxiliary traits: most people expect him to be upper middle-class, white, male, and Protestant. When he is not, there is a sense that he has in some way failed to fill the bill. Similarly, though skin color is the master status trait determining who is Negro and who is white, Negroes are informally expected to have certain status traits and not to have others; people are surprised and find it anomalous if a Negro turns out to be a doctor or a college professor. People often have the master status trait but lack some of the auxiliary, informally expected characteristics; for example, one may be a doctor but be a female or a Negro.

Hughes deals with this phenomenon in regard to statuses that are well thought of, desired, and desirable (noting that one may have the formal qualifications for entry into a status but be denied full entry because of lack of the proper auxiliary traits), but the same process occurs in the case of deviant statuses. Possession of one deviant trait may have a generalized symbolic value, so that people automatically assume that its bearer possesses other undesirable traits allegedly associated with it.

To be labeled a criminal one need only commit a single criminal offense, and this is all the term formally refers to. Yet the word carries a number of connotations specifying auxiliary traits characteristic of anyone bearing the label. A man who has been convicted of housebreaking and thereby labeled criminal is presumed to be a person likely to break into other houses; the police, in rounding up known offenders for investigation after a crime has been committed, operate on this premise. Further, he is considered likely to commit other kinds of crimes as well, because he has shown himself to be a person without "respect for the law." Thus, apprehension for one deviant act exposes a person to the likelihood that he will be regarded as deviant or undesirable in other respects.

There is one other element in Hughes' analysis we can borrow with profit: the distinction between master and subordinate statuses.[8] Some statuses, in our society as in others, override all other statuses and have a certain priority. Race is one of them. Membership in the Negro race, as socially defined, will override most other status considerations in most other situations; the fact that one is a physician or middle-class or female will not protect one from being treated as a Negro first and any of these other things second. The status of deviant (depending on the kind of deviance) is this kind of master status. One receives the status as a result of breaking a rule, and the identification proves to be more important than most others. One will be identified as a deviant first, before other identifications are made....

NOTES

1. The most important earlier statements of this view can be found in Frank Tannenbaum, *Crime and the Community* (New York: Columbia University Press, 1938), and E. M. Lemert, *Social Pathology* (New York: McGraw-Hill Book Co., 1951). A recent article stating a position very similar to mine is John Kitsuse, "Societal Reaction to Deviance: Problems of Theory and Method," *Social Problems* 9 (Winter 1962): 247–256.

2. F. James Davis, "Crime News in Colorado Newspapers," *American Journal of Sociology* LVII (January 1952): 325–330.

3. See Albert K. Cohen and James F. Short, Jr., "Juvenile Delinquency," p. 87 in Robert K. Merton and Robert A. Nisbet, eds., *Contemporary Social Problems* (New York: Harcourt, Brace and World, 1961).

4. See Harold Garfinkel, "Research Notes on Inter- and Intra-Racial Homicides," *Social Forces* 27 (May 1949): 369–381.

5. Edwin Sutherland, "White Collar Criminality," *American Sociological Review* V (February 1940): 1–12.

6. Clark Vincent, *Unmarried Mothers* (New York: The Free Press of Glencoe, 1961), pp. 3–5.

7. Everett C. Hughes, "Dilemmas and Contradictions of Status," *American Journal of Sociology* L (March 1945): 353–359.

8. *Ibid.*

4

Natural Law and the Sociology of Deviance

ANNE HENDERSHOTT

Hendershott takes the absolutist perspective on defining deviance, proposing a morally based view of what is right and wrong. She outlines the "Natural Law" doctrine, rooting it in religious precepts. Societal values that define deviance, she suggests, are morally based unwritten laws that stretch from community to community over history. These laws can be found, she argues, in the founding precepts of sociology, a discipline that arose to articulate and explain the collective consciousness that bound people together by internalizing a set of social controls.

Hendershott suggests that society is founded on consensus, with most people agreeing about right and wrong. Urging us to adopt a more sin-based model, she criticizes contemporary sociologists and politicians for failing to speak in this language of right and wrong and failing to condemn the immorality of deviance. Clear boundaries, she argues, protect us from the instability of moral panics. She further argues that we should construct the hard principles of the moral order and enforce them. Like many strains of Marxism and feminism, Hendershott's morality-based approach to defining deviance is inherently absolutist because it advances a situationally and temporally consistent universal yardstick for assessing the meaning of behavior—a yardstick that is lodged in the essence of the behavior itself.

What does Hendershott have to say about relativist definitions of deviance, about how definitions of deviance have changed over time, and about those who have proposed these kinds of changes? How does she feel about the medical model of deviance? About the "experts" who support it? About the difference between "psychological" and "Christian" man? Finally, what view does she propose about how social definitions are formed?

In 1993, Pope John Paul II cautioned, "No one can escape from the fundamental questions: What must I do? How do I distinguish good from evil?" (*Veritatis Splendor*, Introduction). Although Pope John Paul II (a sociologist by training) was speaking of revealed truth, natural law, and moral theology, the founding sociologists shared many of these same concerns about the moral

Reprinted with permission from the author, Anne Hendershott.

choices we make and the moral order we create. In some important ways, the earliest sociologists were asking some of the same questions.

In fact, from the earliest days of the fledgling discipline of sociology, the founders were concerned about social order and the common good. They warned of the threat to the social order of the community that comes with the breakdown of traditional moral boundaries. And, from this time onward, sociologists continued to assert that social stability is founded on *moral order*—a *common worldview* that binds people to their families, to their communities, and to the larger economic and political institutions. Integral to this moral order is a shared concept of what constitutes deviant behavior—behavior that is defined as "outside the norm"—and a willingness to identify the boundaries of appropriate behavior.

Today, many sociologists have become reluctant to acknowledge that there are *moral judgments* to be made when discussing a subject such as deviance. Globalization has created societies based less on shared culture than on narrow calculations of individual self-interest. A commitment to a common moral order is much more difficult within a culture of such strong individualism.

DEFINING DEVIANCE DOWN AND UP

Émile Durkheim, the "founding father" of the sociology of deviance, wrote that deviance is an integral part of all societies because it affirms cultural norms and values. Durkheim acknowledged that all societies require moral definition; some behaviors and attitudes must be identified as more salutary than others. As a sociologist, he saw that moral unity could be ensured only if all members of a society were anchored to common assumptions about the world around them; without these assumptions, a society was bound to degenerate and decay. Yet contemporary sociologists have often embraced the dangerous principle of shifting moral boundaries.

In 1993, Daniel Patrick Moynihan, a sociologist and four-term U.S. senator from New York, wrote a seminal piece on deviance in *The American Scholar*. In his clever alliteration *"defining deviancy down"* (the title of the piece), Moynihan captured the essence of a disturbing trend in the United States: the decline of our quality of life through our unqualified acceptance of too many activities formerly considered unacceptable. Out-of-wedlock births, teenage pregnancy, promiscuity, abortion, drug abuse, welfare dependency, and homelessness all seemed to be increasing, even in a climate of prosperity. Worse, these behaviors appeared to be nominally condoned.

At the same time, there was a parallel, but opposite, development that the senator did not touch on in his speech: a movement of *"defining deviancy up."* Powerful advocacy groups were successfully stigmatizing behaviors that had formerly been regarded as "normal" and even benign. Some of these redefinitions have had positive consequences for society. For example, the efforts of advocacy groups such as Mothers Against Drunk Drivers to stigmatize drunk driving has achieved success and saved lives, and the civil rights movement has succeeded in stigmatizing racism such that, if anyone dares utter a racist joke or attempts to make a negative racial comment, that person will be immediately stigmatized as a racist deserving of punitive sanctions.

But more importantly, these kinds of shifts in the definitions of deviance work for the convenience of society and ignore the kind of deeply seated moral imperatives on which a consensually based society with shared boundaries and a common moral order rests.

CULTURAL RELATIVISM AND EXPERTS

The process of redefining deviance is a subtle one, and changes in language are so incremental and seemingly innocuous that the new meanings appear almost invisibly. Social philosopher Alasdair MacIntyre says that ours is a culture dominated by experts, **experts** who profess to assist the rest of us, but who often instead make us their **victims**. MacIntyre says that we must be able to identify the particular set of precepts that will help us achieve that which contributes to the common good. Most of us know that there are unwritten, morally based "laws" that tell us what kind of behavior is deviant. These laws have not necessarily been codified within the legal system, but are nonetheless binding.

In an age of technology and expertise, we have been convinced that we should listen to the "experts" rather than **common sense** in determining the norms, values, and attitudes of our families, our churches, and other trusted institutions. Cultural relativists urge us to adapt to the changes of our times—changes that they define as "progress" rather than mere change whose inevitability is not assured.

The continued attempts to psychologize and "understand" deviance—even in the face of evil such as that which appeared in America on September 11—show the distance some will go to avoid applying moral categories of judgment. Sociologists Peter Conrad and Joseph Schneider (1980, 6) cautioned us more than three decades ago that the **medicalization of deviance** would eventually "shroud conditions, events and people and prevent them from being confronted as evil." Although medicalizing deviance does not automatically render evil consequences good, the assumption that behavior is the product of a "sick" mind or body gives it a status similar to that of "accidents." It infers that removing intent or motive relieves us from the human element in the decisions we make, the actions we take, and the social structures we create.

The reluctance of sociologists to acknowledge that there are moral judgments to be made when discussing a subject like deviance shows how far the discipline has strayed from its origins. From the earliest days of sociology, scholars were concerned about the question of social order and the common good. Yet, in a secular society, most—including most sociologists—believe that there are no objective properties which all deviant acts can be said to share in common—even within the confines of a given group.

THE ABSOLUTIST PERSPECTIVE ON DEVIANCE

The United States Conference of Catholic Bishops' Committee on Doctrine is so concerned about the dismissal of natural law because, from Catholic social teaching, **natural law** includes an acknowledgment that we are not the ultimate creators of the moral order—that there is a moral order prior to all human creation. Pope John Paul II (1993: 40) provides the foundation for the bishops' argument on natural law when he writes, "The moral law has its origin in God and always finds it source in him…Indeed, as we have seen, the natural law is nothing other than the light of understanding infused in us by God, whereby we understand what must be done and what must be avoided. God gave this light and this law to man at creation."

From a sociological perspective, drawing upon natural law is described as an "absolutist perspective" of deviance. Yet Catholics and other conservatives are not the only ones who use an absolutist perspective today. Any point of view which asserts that certain behaviors or conditions are intrinsically good or bad falls within this approach. For example, Marxists believe that the oppression of subordinate groups by the dominant group is wrong; workers of the world, like teachers, policemen, and factory workers, should throw off the chains of their economic, capitalist suppression and revolt against the corporate ownership of their governments by means of which the rich get richer and the middle class progressively loses ground. Feminists can also be recognized as absolutists, because they maintain that an absolute moral standard must be applied to some behaviors; women should be liberated from the systems of patriarchal oppression by men. Included among such absolutist feminists are proabortion feminists who regard any restrictions on a woman's right to choose, even in the last few months of pregnancy, as deviant because they deprive women of equal rights. Some radical gay and lesbian sociologists are also absolutist, because they define "social justice" so broadly that the term includes the right to marriage by same-sex couples. For these sociologists, any behavior that results in the exploitation of one person or a category of persons for the benefit of another or that threatens the dignity and quality of life for specific people humanity as a whole is inherently evil, and thereby deviant. These are all absolutist perspectives of deviance— yet because they support causes that many sociologists support, few define the Marxist or the radical gay and feminist perspectives in that way.

SOCIAL ORDER AND SACRED ORDER

More than 50 years ago, the poet T. S. Eliot wrote about the sense of alienation that occurs when social regulators begin to splinter and the controlling moral authority of a society is no longer effective. In his play *The Cocktail Party*, a troubled young protagonist visits a psychiatrist and confides that she feels a *"sense of sin"* because of her relationship with a married man. She is distressed not so much by the illicit relationship, but rather by the strange feeling of sinfulness. Eliot (1950: 156) writes, "Having a sense of sin seems abnormal to her—she had never noticed before that such behavior might be seen in those terms. She believed that she had become ill."

When Eliot writes of his protagonist's feeling unease or uncertainty about her behavior, he is really speaking of the sense of normlessness that has traditionally been a focus of sociology. In many ways, Eliot's play is about *anomie*—the state of *normlessness* that sociologists identify as resulting when one is caught between the loosening moral norms regulating behavior and one's own moral misgivings. Eliot's play echoes the scholarship of Durkheim. Both men saw that the identification and stigmatization of deviant behavior is functional for society because it can produce certainty for individuals and solidarity for the group. Both recognized that dramatic social change through the rapid redefinition of deviance can be dysfunctional for society. Strong cultural values and clear concepts of good and evil integrate members into the group and provide meaning. When traditional cultural attachments are disrupted, or when behavior is no longer regulated by these common norms and values, individuals are left without a moral compass.

Durkheim knew that social facts, like crime statistics, abortion rates, and poll data on support for gay marriage, can be explained only by analyzing the unique social conditions that evolve when norms break down. The resulting anomic state leads to deviant behavior as the individual's attachment to social bonds is weakened. According to this view, people care what others think of them and attempt to conform to expectations because they accept what others expect. However, when these same people are unsure about the norms, or when the norms are changing rapidly—as they have for the past few decades—there is a growing unwillingness to make moral judgments about behavior.

This value-free ideology was predicted over 40 years ago by sociologist Philip Rieff, who warned, in his now classic book *The Triumph of the Therapeutic*, that "*psychological man*" was beginning to replace "*Christian man*" as the dominant character type in our society. Unlike traditional Christianity, which made moral demands on believers, the secular world of "psychological man" rejected both the idea of sin and the need for salvation. Replacing the concept of sin with the concept of sickness was documented by Rieff, who wrote in 1966 that the authority that had been vested in Christian culture had been all but shattered. Nothing had succeeded it. What worried him was that the institutions of morality—especially the Church—lacked authority and could no longer persuade others to follow them. Further, Rieff (1966: 205) believed that this failure of authority was no accident, but rather the program of "the modern cultural revolution," which was conducted "not in the name of any new order of communal purpose, but rather for the permanent disestablishment of any deeply internalized moral demands."

The faithful know that teachings on life issues and on marriage and the family cannot be changed. For Catholics, there remain the enduring truths, those which Philip Rieff (1966: 59) calls "*commanding truths*," that cannot be changed: "Commanding truths will not be mocked, except to the destruction of everything sacred." Of the family as a commanding truth, Rieff (1966: 107) wrote, "the destruction of the family is the key regimen of technological innovation and moral deviance." And of life itself, Rieff (1966: 42) wrote, "We must stand against the re-creation of life in the laboratory and the taking of life in the abortion clinic." Rieff knew, as the sociologists of the past knew, that culture survives by faith in the highest absolute authority and its interdicts. For Catholics, there can be no Catholic culture and no true Catholic Church without such commanding truths.

The sociology of deviance provides a useful framework to help us understand the success that the gay and lesbian community has achieved in defining down what had long been viewed as the deviance of homosexuality. Beyond redefining homosexuality, the continued refinement of the theory of deviance is probably one of the greatest contributions that sociology can make to understanding social change. Yet, in an effort to avoid alienating those with diverse lifestyles and values, sociologists have become hesitant to make judgments about the behavior of others.

In the aftermath of September 11, 2001, President George W. Bush repeatedly called the terrorist acts "evil" and those who perpetrated them "evildoers." This language drew only a minor protest from those who, on September 10, would have excoriated the president for such inflammatory language. Reassessing the politics and culture after the terrorists declared war on the United States, we were reminded again that evil exists. We again realized that there are those who are capable of doing monstrous acts. And, to achieve social order, we must be willing to identify and defend ourselves against those who want to do us harm. Perhaps the "remoralization" of our public discourse that occurred after September 11 was the only good to come out of the terrorist attacks. Most of us were reminded again that evil exists and that good people must recognize this fact. Perhaps, in time, sociologists will again be willing to recognize that a society which continues to define down the deviant acts our common sense tells us are destructive is a society that has lost the capacity to confront evil.

REFERENCES

Conrad, Peter, and Schneider, Joseph. (1980). *Deviance and Medicalization.* St. Louis: Mosby.

Eliot, T. S. (1950). *The Cocktail Party.* Orlando, Florida: Harcourt Brace Jovanovich.

Moynihan, Daniel Patrick. (1993). "Defining Deviancy Down," *The American Scholar* 62:1 (Winter), pp. 17–30.

Pope John Paul II. (1993). *Veritatis Splendor.* [The Splendor of Truth]. *Encyclical published by the Vatican.*

Rieff, Philip. (1966). *The Triumph of the Therapeutic: The Uses of Faith after Freud.* New York: Harper & Row.

5

Social Power: Conflict Theory of Crime

RICHARD QUINNEY

Quinney's conflict theory of crime represents the social power perspective on defining deviance. He builds on Becker's relativist approach by asserting that definitions of deviance are social constructions and not absolute "givens." He, too, rejects the essentialist view of deviance as inherent in specific acts. But while Becker is not specific about the identity of those who formulate the definitions of deviance, Quinney locates the decision makers in the dominant class. Definitions of deviance, then, stem from the views of those who have the power to make and enforce them. Members of this group formulate the definitions of deviance with the express purpose of advancing themselves by labeling behaviors that threaten their class interests as criminal. They then enforce the definitions unequally: more harshly against their opponents and more leniently against members of their own group. Yet they disguise the self-serving basis of their rules, creating and enforcing them through justifications that legitimate their actions and that cast the rules as rational or beneficial to others. In turn, those who are defined as criminal become more likely to engage in future behavior that will be defined as criminal.

Quinney thus goes beyond Becker to offer a view of society that radically departs from previous perspectives. The conflict perspective envisions two groups in society: the dominant class and those they dominate. Criminals are conceptualized as powerless and oppressed people who threaten the interests of the ruling class. Defining and enforcing crime becomes a means of reproducing the power and socioeconomic inequalities between these groups. Quinney's conflict theory suggests that definitions of deviance represent one of the coercive means through which the elite maintain their dominance over the masses.

How does Quinney's conflict view of society compare or contrast with Erikson's?

A theory that helps us begin to examine the legal order critically is the one I call the *social reality of crime*. Applying this theory, we think of crime as it is affected by the dynamics that mold the society's social, economic, and political

From Richard Quinney, *Criminology* (Boston: Little, Brown, 1975). Reprinted by permission of the author.

structure. First, we recognize how criminal law fits into capitalist society. The legal order gives reality to the crime problem in the United States. Everything that makes up crime's social reality, including the application of criminal law, the behavior patterns of those who are defined as criminal, and the construction of an ideology of crime, is related to the established legal order. The social reality of crime is constructed on conflict in our society.

The theory of the social reality of crime is formulated as follows.

I.   **The Official Definition of Crime:** *Crime as a legal definition of human conduct is created by agents of the dominant class in a politically organized society.*

The essential starting point is a definition of crime that itself is based on the legal definition. Crime, as *officially* determined, is a *definition* of behavior that is conferred on some people by those in power. Agents of the law (such as legislators, police, prosecutors, and judges) are responsible for formulating and administering criminal law. Upon *formulation and application* of these definitions of crime, persons and behaviors become criminal.

Crime, according to this first proposition, is not inherent in behavior, but is a judgment made by some about the actions and characteristics of others. This proposition allows us to focus on the formulation and administration of the criminal law as it applies to the behaviors that become defined as criminal. Crime is seen as a result of the class-dynamic process that culminates in defining persons and behaviors as criminal. It follows, then, that the greater the number of definitions of crime that are formulated and applied, the greater the amount of crime.

II. **Formulating Definitions of Crime:** *Definitions of crime are composed of behaviors that conflict with the interests of the dominant class.*

Definitions of crime are formulated according to the interests of those who have the power to translate their interests into public policy. Those definitions are ultimately incorporated into the criminal law. Furthermore, definitions of crime in a society change as the interests of the dominant class change. In other words, those who are able to have their interests represented in public policy regulate the formulation of definitions of crime.

The powerful interests are reflected not only in the definitions of crime and the kinds of penal sanctions attached to them, but also in the *legal policies* on handling those defined as criminals. Procedural rules are created for enforcing and administering the criminal law. Policies are also established on programs for treating and punishing the criminally defined and programs for controlling and preventing crime. From the initial definitions of crime to the subsequent procedures, correctional and penal programs, and policies for controlling and preventing crime, those who have the power regulate the behavior of those without power.

III. **Applying Definitions of Crime:** *Definitions of crime are applied by the class that has the power to shape the enforcement and administration of criminal law.*

The dominant interests intervene in all the stages at which definitions of crime are created. Because class interests cannot be effectively protected merely by formulating criminal law, the law must be enforced and administered. The interests of the powerful, therefore, also operate where the definitions of crime reach the *application* stage. As Vold (1958, 163) has argued, crime is "political behavior and the criminal becomes in fact a member of a 'minority group' without sufficient public support to dominate the control of the police power of the state." Those whose interests conflict with the ones represented in the law must either change their behavior or possibly find it defined as criminal.

The probability that definitions of crime will be applied varies according to how much the behaviors of the powerless conflict with the interests of those in power. Law enforcement efforts and judicial activity are likely to increase when the interests of the dominant class are threatened. Fluctuations and variations in applying definitions of crime reflect shifts in class relations.

Obviously, the criminal law is not applied directly by those in power; its enforcement and administration are delegated to authorized *legal agents*. Because the groups responsible for creating the definitions of crime are physically separated from the groups that have the authority to enforce and administer law, local conditions determine how the definitions will be applied. In particular, communities vary in their expectations of law enforcement and the administration of justice. The application of definitions is also influenced by the visibility of offenses in a community and by the public's norms about reporting possible violations. And especially important in enforcing and administering the criminal law are the legal agents' occupational organization and ideology.

The probability that these definitions will be applied depends on the actions of the legal agents who have the authority to enforce and administer the law. A definition of crime is applied depending on their evaluation. Turk (1969) has argued that during "criminalization," a criminal label may be affixed to people because of real or fancied attributes: "Indeed, a person is evaluated, either favorably or unfavorably, not because he *does* something, or even because he *is* something, but because others react to their perceptions of him as offensive or inoffensive." Evaluation by the definers is affected by the way in which the suspect handles the situation, but ultimately the legal agents' evaluations and subsequent decisions are the crucial factors in determining the criminality of human acts. As legal agents evaluate more behaviors and persons as worthy of being defined as crimes [and criminals], the probability that definitions of crime will be applied grows.

IV. How Behavior Patterns Develop in Relation to Definitions of Crime: *Behavior patterns are structured in relation to definitions of crime, and within this context people engage in actions that have relative probabilities of being defined as criminal.*

Although behavior varies, all behaviors are similar in the way they represent patterns within society. All persons—whether they create definitions of crime or are the objects of these definitions—act in reference to *normative systems* learned in relative social and cultural settings. Because it is not the quality of the behavior but the action taken against the behavior that gives it the character of criminality, which is defined as criminal is relative to the behavior patterns of the class that formulates and applies definitions. Consequently, people whose behavior patterns are not represented when the definitions of crime are formulated and applied are more likely to act in ways that will be defined as criminal than those who formulate and apply the definitions.

Once behavior patterns become established with some regularity within the segments of society, individuals have a framework for creating *personal action patterns*. These continually develop for each person as he moves from one experience to another. Specific action patterns give behavior an individual substance in relation to the definitions of crime.

People construct their own patterns of action in participating with others. It follows, then, that the probability that persons will develop action patterns with a high potential for being defined as criminal depends on (1) structured opportunities, (2) learning experiences, (3) interpersonal associations and identifications, and (4) self-conceptions. Throughout the experiences, each person creates a conception of self as a human social being. Thus prepared, he behaves according to the anticipated consequences of his actions.

In the experiences shared by the definers of crime and the criminally defined, personal-action patterns develop among the latter because they are so defined. After they have had continued experience in being defined as criminal, they learn to manipulate the application of criminal definitions.

Furthermore, those who have been defined as criminal begin to conceive of themselves as criminal. As they adjust to the definitions imposed on them, they learn to play the criminal role. As a result of others' reactions, therefore, people may develop personal-action patterns that increase the likelihood of their being defined as criminal in the future. That is, increased experience with definitions of crime increases the probability of their developing actions that may be subsequently defined as criminal.

Thus, both the definers of crime and the criminally defined are involved in reciprocal action patterns. The personal-action patterns of both the definers and the defined are shaped by their common, continued, and related experiences. The fate of each is bound to that of the other.

V. Constructing an Ideology of Crime: *An ideology of crime is constructed and diffused by the dominant class to secure its hegemony.*

This ideology is created in the kinds of ideas people are exposed to, the manner in which they select information to fit the world they are shaping, and their way of interpreting this information. People behave in reference to the *social meanings* they attach to their experiences.

Among the conceptions that develop in a society are those relating to what people regard as crime. The concept of crime must of course be accompanied by ideas about the nature of crime. Images develop about the relevance of crime, the offender's characteristics, the appropriate reaction to crime, and the relation of crime to the social order. These conceptions are constructed by communication, and, in fact, an ideology of crime depends on the portrayal of crime in all personal and mass communication. This ideology is thus diffused throughout the society.

One of the most concrete ways by which an ideology of crime is formed and transmitted is the official investigation of crime. The President's Commission on Law Enforcement and Administration of Justice is the best contemporary example of the state's role in shaping an ideology of crime. Not only are we as citizens more aware of crime today because of the President's Commission, but official policy on crime has also been established in a crime bill, the Omnibus Crime Control and Safe Streets Act of 1968. The crime bill, itself a reaction to the growing fears of class conflict in American society, creates an image of a severe crime problem and, in so doing, threatens to negate some of our basic constitutional guarantees in the name of controlling crime.

Consequently, the conceptions that are most critical in actually formulating and applying the definitions of crime are those held by the dominant class. These conceptions are certain to be incorporated into the social reality of crime. The more the government acts in reference to crime, the more probable it is that definitions of crime will be created and that behavior patterns will develop in opposition to those definitions. The formulation of definitions of crime, their application, and the development of behavior patterns in relation to the definitions are thus joined in full circle by the construction of an ideological hegemony toward crime.

VI. Constructing the Social Reality of Crime: *The social reality of crime is constructed by the formulation and application of definitions of crime, the development of behavior patterns in relation to these definitions, and the construction of an ideology of crime.*

The first five propositions are collected here into a final composition proposition. The theory of the social reality of crime, accordingly, postulates creating a series of phenomena that increase the probability of crime. The result, holistically, is the social reality of crime.

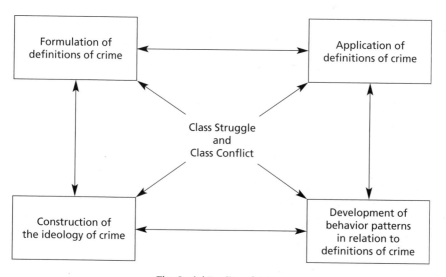

The Social Reality of Crime

Because the first proposition of the theory is a definition and the sixth is a composite, the body of the theory consists of the four middle propositions. These form a model of crime's social reality. The model, as diagrammed, relates the proposition units into a theoretical system (see figure on p. 55). Each unit is related to the others. The theory is thus a system of interacting developmental propositions. The phenomena denoted in the propositions and their relationships culminate in what is regarded as the amount and character of crime at any time—that is, in the social reality of crime.

The theory of the social reality of crime as I have formulated it is inspired by a change that is occurring in our view of the world. This change, pervading all levels of society, pertains to the world that we all construct and from which, at the same time, we pretend to separate ourselves in our human experiences. For the study of crime, a revision in thought has directed attention to the criminal process: All relevant phenomena contribute to creating definitions of crime, [developing] behaviors by those involved in criminal-defining situations, and constructing an ideology of crime. The result is the social reality of crime that is constantly being constructed in society.

REFERENCES

Turk, Austin. (1969). *Criminality and the Legal Order*. Chicago: Rand McNally.

Vold, George B. (1958). *Theoretical Criminology*. New York: Oxford University Press.

PART II

Theories of Deviance

Deviance holds a special intrigue for scholars of theory. Given its pervasive nature in society, its enigmatic conditions, and its generic appeal, even the earliest sociologists attempted to explain how and why deviance occurs. Especially considering people's inclination to conform, the pressing question for scholars has been *why* individuals engage in norm-violating behavior. Explanations for deviant behavior are as divergent as the acts they explain, which range from acts of delinquency to professional theft, acts of integrity to a search for kicks, and acts of desperation to those of bravado and daring. We next outline some of the major attempts at understanding deviant behavior.

BIOLOGICAL AND PSYCHOLOGICAL THEORIES

The language of deviance used in everyday life tends to follow biological or psychological assumptions about the causes of behavior. Sociology students often have difficulty thinking in other terms about why people would violate norms. Reinarman, in Chapter 15, calls everyday language having to do with deviance a "vocabulary of attribution," suggesting that the cultural language of talking about how and why people do things is individualistic, rather than collective or sociological. This book aims to remedy that problem.

In earlier times, scholars of crime approached deviant behavior as rooted in people's *biological* abnormalities or predispositions. They tried to find links between incarcerated criminals and genetic deficiencies. In the 1800s, Cesare Lombroso and his followers suggested that criminals were more like primitive human beings, resembling their apelike ancestors (Lombroso, 1876).

Women, he believed, were evolutionarily inferior to men (Lombroso, 1920). This approach viewed criminals as born, not made, and therefore unresponsive to rehabilitation or treatment. People's "defective" criminal tendencies could be classified into distinctive criminal types and inherited from one generation to the next. Other researchers of this era, such as Charles Goring (1913) and Earnest Hooton (1939), connected people with physical inferiorities, such as being shorter and lighter, to born criminal types.

At around the same time, body-type theorists correlated criminality with three "somatotypes": body builds thought to be related to certain personality characteristics or temperaments. William Sheldon (1949) suggested that neither the slow, soft, and comfort-loving endomorphs nor the lean, fragile ectomorphs were as criminally inclined as the muscular and active mesomorphs, who were more aggressive. Contemporary studies of body type focus on things such as one's body mass index (BMI: a person's weight divided by his or her height squared) and large early muscular development, suggesting that strongly built or well-muscled people are more prone toward violence.

More recent biological approaches have pursued genetic explanations. Some looked for chromosomal patterns in deviants, suggesting that variations on the typical male (XY) and female (XX) configurations could foster abnormal behavior. In particular, criminologists proposed an XYY syndrome that creates "double male" or "supermale" individuals who were unusually tall and predisposed to aggressive and violent behavior. An accused murderer in Australia was acquitted on the basis of this genetic defense as recently as the 1960s, although it did not work in the trial of Richard Speck, the man charged with killing eight Chicago nurses. Studies of twins have been very popular in the nineteenth century as well. The strongest connection between criminally convicted twins, people who would likely be similarly inclined because of their biological imprinting, was found to be only a 35 percent rate of offense for their identical sibling, with an even smaller 12 percent for fraternal siblings.

Brain studies, popular in the 1930s, suggested that some people might cease their deviant ways if their brains were surgically altered. It was believed that chronic deviants, unresponsive to other treatments, might benefit from lobotomies that destroyed the frontal lobe of the brain (as portrayed in the book and film *One Flew Over the Cuckoo's Nest*). From the 1930s to 1950s, approximately 50,000 such operations were performed. Lobotomies finally gave way to other forms of psychosurgery involving inserting electric needles into the skull and searing part of the cingulum, the emotional center of the brain. A host of other sociobiological explanations related to diet, learning disorders, endocrine or hormonal imbalance, and allergies thought to influence behavior have since arisen. One example of this kind of explanation can been seen in the "twinkie defense" from the trial of Dan White, charged with killing San Francisco Mayor George Moscone (portrayed in the movie *Milk*). Defense attorneys brought in psychiatrists who testified that individuals might become depressed and driven to uncontrollable violence as the result of excessively consuming junk food and sugary soft drinks. They argued that, because of the chemicals these foods produced in his body, White exploded, went onto "autopilot," and diminished his capacity for rational thought. To some degree, jurors bought this argument, rejecting the charge of premeditated murder and instead convicting White of the lesser crime of manslaughter. Brain theories continue to be popular with the public and are being applied to attribute deviant behaviors such as rape, spousal and date assault, child abuse and neglect, and cheating to hormonal functioning (the PMS syndrome defense), cerebral and neuro-allergies to food substances, EEG abnormalities, and a wide range of evolutionary theories.

Although early biological theories focused more on body types, more recent thinking examines how people's inherited characteristics translate into predispositions toward traits such as thrill seeking, risk taking, or substance abuse. But biological theories fail to explain why people who share common biological characteristics differ, with some turning to deviance while others remain conventional. Biological factors may indeed exert some influence on people's behavior, but these theories tend to be limited and offer less convincing explanations than explanations involving social and cultural factors. Many are now denounced as racist or for condoning genetic research on deviant behavior. They are most commonly used in contemporary society to provide a rationale for conservative ideological principles and support a political climate that blames social ills such as crime and poverty on individuals rather than on social policies.

Psychological theories have their roots in the late eighteenth century, drawing on psychiatric, psychoanalytical, and psychological explanations of how individuals' minds and personalities affect their deviance. Many have tried to articulate a deviant or criminal personality, dating back to Sigmund Freud's (1925) model of the id, ego, and superego. Freud believed that people with too little ability to resist their impulses had Oedipus or Electra complexes, death wishes, inferiority complexes, or fears of castration, or suffered from frustration–aggression syndromes or penis envy, leading them to commit hostile acts. Freud's early work was succeeded by more contemporary theories about a range of impulse control disorders. Psychologists have also linked personality traits, especially defiance, hostility, ambivalence toward authority, and emotional psychopathologies, to crime and deviance. We see the popularity of this approach in criminal profiling, which attempts to construct typical characterizations of certain offenders, although this technique may be more successful in books, movies, and television than in real life.

Other psychological approaches have addressed "operant conditioning," examining the way behavior modification (B. F. Skinner, 1953) or human conditioning (Hans Eysenck, 1977) can lead individuals to commit crime. Albert Bandura (1973) proposed a social learning theory which suggested that exposure to aggressive or aversive behaviors could reinforce people's tendencies to become aggressive.

Many people have tried to link distinct personality types to deviant behavior. Succeeding in this endeavor is difficult because deviance comes in so many forms; thus, identity thieves are likely to be rather different from robbers, gamblers, and drug users. Samuel Yochelson and Stanton Samenow (1976) proposed that criminals have distinctive personalities and thinking patterns, such as unrelenting optimism, manipulativeness, intense anger, fear of being injured or insulted, an inclination toward chronic lying, and an inflexibly high self-image. Although scholars have yet to find evidence of a "criminal mind," Yochelson and Samenow believed that criminals were victimizers rather than victims of society.

Finally, intelligence theorists have put forth a range of IQ theories, suggesting that people whose "mental age" lags behind their chronological age might be predisposed toward criminality, with one scholar (Henry Goddard, 1979) even suggesting that criminals were feebleminded morons. Although the connection between IQ and deviance was convincingly refuted by Edwin Sutherland in 1934, people still believe that low intelligence is linked to deviance.

The problem with many of these psychological theories is that they focus almost exclusively on individuals' personalities, ignoring their social conditions or life situations. They tend to blame individuals, rather than the social structure or environmental factors that may have fostered their deviance—a form of victim blaming. Many psychological theorists have approached deviance as consisting of intrinsically defined acts, rather than behavior that is regarded differently by different groups in different times and places. Consequently, they distinctly overlook the element of social power and its role in defining deviance and in implementing such definitions. It is left up to sociology courses to expand students' horizons so that they can better understand the invisible forces of culture and social structure.

THE STRUCTURAL PERSPECTIVE

The dominant theory in sociology for the first half of the twentieth century, **structural functionalism**, also commanded the greatest amount of sway in explaining deviant behavior. Durkheim advanced the theory that society is a moral phenomenon. He believed that, at its root, the morals (norms, values, and laws) that individuals are taught constrain their behavior. Youngsters are taught the "rights" and "wrongs" of society early in life, with most people conforming to these expectations throughout adulthood. These moral beliefs determine, in large measure, how people behave, what they want, and who they are. Durkheim suggested that societies with high degrees of social integration (bonding, cohesion, community involvement) would increase the conformity of its members. However—and this is what concerned Durkheim—in the modern French society in which he was living, more and more people were becoming distanced from each other, people were losing some of their sense of belonging to their communities, and the norms and expectations of their groups were becoming less clearly defined. He believed that this condition, which he referred to as *anomie*, was producing a gradual social disintegration, leading to greater degrees of deviance. Thus, for Durkheim, although norms still existed on the societal level, the lack of social integration created a situation in which they were no longer becoming as significant a part of each individual.

Despite his concerns about the increasing rates of deviance that society would produce, Durkheim also subscribed to the idea that deviance was functional for society. As we noted earlier, Durkheim felt that, despite its obvious negative effects, deviance produces some benefits as well. At a time when people are worrying about the moral breakdown and social disintegration of society, deviance serves to remind us of the moral boundaries in society. Each time a deviant act is committed and publicly announced, society is united in indignation against the perpetrator. This unity serves to bring people together, rather than tearing them apart. At the same time, society is reminded about what is "right" and "wrong" and, for those who conform, greater social integration ensues. These ideas were perhaps best illustrated by Yale sociologist Kai Erikson, who, in his 1966 book *Wayward Puritans* (excerpted in Chapter 1), demonstrated the role of deviance in defining morality and bringing people together. Erikson examined Puritan patterns of isolating and treating offenders. He believed that deviance serves as a means of promoting a contrast with the rest of the community, thus giving members of the larger society more strength in their moral convictions. Erikson's analysis focused on the transformation of seventeenth-century Bay Colony as a group of revolutionaries tried to establish a new community in New England. These deviants, the revolutionaries, played an important role in the transformation of norms and values: Their behavior elicited societal reaction, which served to define the new community's norms and values clearly. In addition, punishing some people for violations of norms reminded others of the rewards for conformity.

Chapter 6, "Functionalism: The Normal and the Pathological," by Durkheim, lays out his theory of the inevitably of deviance in all societies. In his ironic twist, Durkheim argues that deviance is normal rather than pathological, serving a positive function in society. To achieve the maximum benefit, however, a society needs a manageable amount of deviance. When the number of people declared deviant by current moral standards rises or falls too much, society alters its moral criteria, to maintain the level of deviance in the optimal range. At different times, society may "define deviancy down," as Moynihan (1993) has suggested in looking at the way the bar defining acceptable behavior has been lowered. When society lowers the bar of acceptability, fewer acts are viewed as deviant and more become recast as tolerable. Over the years, we have normalized violence, divorce, smoking marijuana, homosexuality, premarital sex, tattooing and piercing, and unwed pregnancy. Or society may "define deviancy up," as Krauthammer (1993) has suggested, *raising* the bar defining normality. When the bar is raised, behavior formerly considered acceptable becomes redefined as deviant. Thus, things now considered deviant (or even criminal) that used to be regarded as tolerable, even if not exactly embraced, include cigarette smoking, spanking, date rape, sexual harassment, panhandling, talking in movie theaters, and hate speech. What deviance does for society is define the moral boundaries for everyone. The violation of norms serves to remind the masses what is acceptable and what is not; in Durkheim's words, it enforces the "collective conscience" of the group. We saw this behavior in the public outcry over the inadequate governmental response to the victims of Hurricane Katrina. Perhaps it is difficult to imagine that behaviors that disgust, revile, or even nauseate you are not the acts of immoral, sick, or evil people, but are typical, and even beneficial, parts of all societies. Durkheim's theory suggests that structural needs of the society as a whole, beyond the scope of its individual members, foster the continuing recurrence of deviance.

The structural perspective locates the root cause of crime and deviance outside of individuals, in the invisible social structures that make up any society. Structural explanations for deviance look at features of society that seem to generate higher rates of crime or deviance among some societies or groups within them. In looking for explanations for why some societies are likely to have higher crime rates than others, sociologists have suggested that those with greater degrees of inequality are likely to show more crime than those in which people have roughly similar amounts of what the society values. Looking within each society, structuralists locate the cause of crime in two main factors: the differential opportunity structure of society, and prejudice and discrimination toward certain groups. In a society with inequality, some groups will clearly have greater structural access to certain opportunities than others. Groups with access to greater power and to more political and economic opportunity may use these factors to define their acts as legitimate and the acts of others as deviant, at the same time that they corruptly use their power to their own advantage. Not everyone has equal opportunity to dispense political favors, to manipulate stock prices, or to conduct covert operations. In the same society, groups with less access to the legitimate opportunity structure through reduced educational opportunity, diminished access to health, a lower class background, and disadvantaged legitimate networks and connections do not have the same opportunity to succeed normatively. Members of these groups may be propelled into alternative pathways by their position in the social structure.

It was Robert Merton, a mid-twentieth-century sociologist from Columbia University, who actually extended Durkheim's ideas and built them into a specific structural **strain theory** of deviant behavior. Merton claimed that contradictions are implicit in a stratified system in which the culture dictates success goals for all citizens while institutional access is limited to just the middle and upper strata. In other words, despite the American Dream of rags-to-riches opportunities, some people, most often lower class individuals, are systematically excluded from the competition. Instead of merely going through the motions while knowing that their legitimate path to success (measured in American society by financial wealth) is blocked, some members of the lower class retaliate by choosing a deviant alternative. Merton believed that these people have accepted society's goals (to be comfortable, to get rich) but have insufficient access to the approved means of attaining those goals (deferred gratification, education, hard work). The problem lies in the social structure of society, whereby, even if people follow the approved means, there are "roadblocks" prohibiting them from rising through the stratification system. Deviant behavior occurs when socially sanctioned means are not available for the realization of highly desirable goals. In that case, the only way to achieve these goals is to "detour" around them, to bypass the approved means in order to get at the approved goals. For example, the road to "success" for young men raised in urban ghettos with poor housing facilities, dilapidated schools, and inconsistent family lives is more likely to be through dealing drugs, pimping, or robbing than it is through the normative route of attending school and engaging in hard work. According to Merton, then, *anomie* results from the lack of access to culturally prescribed goals and the unavailability of legitimate means for attaining those goals. Deviance (or, more specifically, crime) is the obvious alternative. Once again, the lack of structural opportunities, rather than some psychological or individual pathology, is seen as the root cause of deviance. You can find an articulation of Merton's theory in Chapter 7.

Richard Cloward and Lloyd Ohlin, in *Delinquency and Opportunity* (1960), thought that Merton was correct in directing us toward the notion that members of disadvantaged socioeconomic groups have less opportunity for achieving success in a legitimate manner, but they thought that Merton wrongly assumed that those groups would automatically choose deviance and crime when confronted with the problem of differential opportunity. In their **differential opportunity theory**, Cloward and Ohlin suggested that all disadvantaged people have some lack of opportunity for pursuing legitimate societal goals but they do not have the same opportunity for participating in illegitimate practices. What Cloward and Ohlin believed was that deviant behavior depends on people's access to illegitimate opportunities. They found that three types of deviant opportunities are present: *criminal, conflict,* and *retreatist.* Criminal opportunities, similar to the type Merton described, arise from access to deviant subcultures, although not all disadvantaged youths enjoy these avenues. Conflict opportunities attract people who have a propensity for violence and fighting. Retreatist opportunities attract people, such as drug users, who are not inclined toward illegitimate means or violent actions but who want to withdraw from society. According to Cloward and Ohlin, groups of people may have greater or lesser opportunity to climb the illicit opportunity ladder by virtue of several factors: (1) Some neighborhoods are rife with more criminal opportunities, networks, and enterprises than others, and people reared in these neighborhoods grow up amidst these increased opportunities, (2) some forms of illicit enterprise are dominated by people of particular racial or ethnic groups, so members of these groups have an easier time rising to the top of those businesses or organizations, and (3) the upper echelons of crime display a distinct glass ceiling for women, with men dominating the positions of decision making, earning, and power. Thus, Cloward and Ohlin extended Merton's theory by specifying the existence of differential illegitimate opportunities available to members of disadvantaged groups. It is in this illegitimate-opportunity structure, rather than individual motivation, they argue, that the explanation for deviance can be found.

Chapter 5, "Social Power: Conflict Theory of Crime," by Richard Quinney (discussed in connection with the social power perspective in the general introduction), offers a **conflict theory** explanation that is not functionalist but is still structural. Conflict theorists view society as pluralistic, heterogeneous, and conflictual, rather than unified and consensual as the functionalist sees it. Social conflict arises out of the incompatible interests of diverse groups in society, such as businesses versus their workers, conservatives versus liberals, Whites versus people of color, and the rich versus the poor. All of these groups have a structural conflict of interest with each other that stands above and beyond the individual members and that frames the way those members come to recognize their interests and act in the world. Not only is conflict a natural outcome of this arrangement, but crime is as well. In a succinct summation of conflict theory's major tenets, Quinney tells us that crime exists because some behaviors conflict with the interests of the dominant class. These powerful members of society create legal definitions of human conduct, casting those behaviors that threaten its interests as criminal. Then, the dominant class enforces those laws onto the less powerful groups in society, through the police, the legal, and the criminal justice system, ensuring that the interests of the dominant class are protected. Members of subordinate classes are compelled to commit those actions which have been defined as crimes because their poverty presses them to do so. The dominant group can then create and disseminate its ideology of crime, which is that the most dangerous criminal elements in society can be found in the subordinate classes and that these groups deserve arrest, prosecution, and imprisonment. Through class struggle and class conflict, crime is constructed, formulated, and applied so that less powerful groups are subdued and more powerful groups are strengthened. These processes are illustrated by the diagram in Quinney's chapter. This approach shows how larger social forces, such as group and class interests, shape the behavior of individual members, leading some to use their advantage to dominate over others while the others react to their structural subordination by engaging in those behaviors already defined as deviant and deserving of punishment. All of these structural theories place the cause of deviance on the structures of society, rather than on individuals and their problems.

Feminist theory, the subject of Chapter 10, takes a structural approach as well, locating the pervasive discrimination and oppression of women in society in the overarching patriarchal system. Through the intertwined effects of major institutions and social structures, such as our legal codes, the economy, our political system, social and cultural practices, religion, the family, the educational system, and the media, women are systematically disadvantaged. Women, feminists argue, are unprotected against verbal, physical, and sexual abuse, and their individual attempts to rise up and protect themselves often subject them to being labeled as offenders. When they flee abusive situations, the patriarchy of the streets oppresses them further, funneling them into acts of survival defined as deviant by the male hegemony. Feminists maintain that theories of deviance are male centered when they impose stereotyped gender role requirements onto teenage girls and when they problematize women's attempts to survive under oppressive conditions in a system that systematically deprives them of resources.

THE CULTURAL PERSPECTIVE

Although structuralist theories had an enormous impact on sociologists' thinking about deviance, other authors arose who felt that the structuralist explanations were not all-encompassing. These theorists believed that deviance was a collective act, driven and carried out by groups of people. Building on conflict theory's view that multiple groups with different interests exist in society, the subcultural theorists examined the implications of membership in these groups. Groups with conflicting interests include not only dominant and subordinate groups, but also a variety of social, religious, political, ethnic, and economic factions. Membership in each of these groups places people in distinct subcultures, each of which contains its own set of distinct norms and values. A pluralistic nation that was once thought of as the world's "melting pot," we have become, in part, a nation of many different groups, each with its own distinct subculture.

Thorsten Sellin, writing about "The Conflict of Conduct Norms" (1938) suggested in his **culture conflict theory** that, to some extent, the norms and values of these subcultures incorporated and meshed with the norms and values of the overarching American culture, but, to some extent, they were different and in conflict. The disparities between, and different cultural codes of, subcultural groups may become apparent in three situations. First, when people from one culture "migrate," or cross over into the territory of another culture, they may experience a disconnect. For example, when people from a rural area move to the city, they find that their country ways do not mesh with modern urbanity. In these cases, they may find themselves subject to the urban norms and values. Second, cultural conflict may occur during a "takeover" situation, when one group moves into and takes over the territory of another. In that case, the laws of the cultural group that moves in are extended to apply to the group that is taken over. A prime example is when middle-class people gentrify a run-down neighborhood and the latitude that used to be enjoyed by the former occupants to congregate outdoors, be homeless, do or deal drugs, or solicit prostitution becomes lost. In these cases, the norms of the group that has taken over may apply. Third, cultural codes may clash on the "border" of contiguous cultural areas, as when people from different cultural groups find themselves in contact. They may be on a national border or a neighborhood border, or they may simply be people encountering members of another subculture. In these cases, no clear set of norms and values necessarily dominates, but individuals have to negotiate their cultural understandings delicately, trying to understand each other's norms and values. Such a situation could come about when new first-year students find themselves rooming with someone from a different cultural background and neither of their ways predominates. Sometimes this even happens when college students go home for the holidays, only to find that the norms under which they were living at college are rather different from those prevailing in their parents' houses.

In each of these three cases, people may find themselves torn between the norms and values of different group memberships. Following the norms and values of their subcultural group may produce behavior that becomes defined as deviant by the standards of the broader culture. Yet, from their subcultural perspective, their behavior may be viewed as representing the acts of good people working to uphold the behaviors they honor. In his writing, Sellin was thinking particularly about the deviance of children who belong to immigrant ethnic or racial groups moving into the United States, caught in the struggle between two cultures, but his theory applies equally well to the large number of diverse subcultural groups in our country. He extended his model to apply to all conflicts between cultural groups that share a close geographic area, especially when one culture dominates another normatively and imposes its values on the other culture.

Building on this idea, Albert Cohen, in *Delinquent Boys* (1955), posited a **reaction theory**, according to which working-class adolescent males develop a subculture with a different value system from the dominant American culture. These boys, Cohen asserted, have the greatest degree of difficulty in achieving success, because the establishment's standards are so different from their own. They try, at first, to fit in with the cultural expectations, but find that they are unsuccessful. Exposed to middle-class aspirations and judgments they cannot reasonably fulfill, they develop a blockage (or strain) that leads them to experience "status frustration." What results from this frustration is the reactive formation of an oppositional subculture that allows them to achieve status based on nonutilitarian, malicious, and negativistic behavior. These boys, reacting against what they perceive as society's unfairness toward them, substitute norms that are the reverse of those of the larger society. Cohen claimed that because of their rejection by society, delinquent boys turn society's norms "upside down," rejecting middle-class standards and adopting values in direct opposition to those of the majority.

Walter Miller (1958), writing just after Cohen, further delineated the importance of subcultural values for the development of deviant behavior. Miller believed that the values of the lower class culture produce deviance for its members because those values are "naturally" in discord with middle-class values. Young people who conform to the lower class culture in which they were born almost automatically become deviant. That culture, by which members attain status in the eyes of their peers, is characterized by several "focal concerns": getting into trouble, showing toughness, maintaining autonomy, demonstrating street-smartness, searching for excitement, and being tied in their lot to the capricious whims of fate. **Lower class culture theory** suggests that, when these individuals follow the norms of their subculture, they become deviant according to the predominantly middle-class societal norms and values.

The lasting impact of subcultural theories has been to suggest that conflicting values may exist in society. When one part of society can impose its definitions on other parts, the dominant group has the ability to label the minority group's norms and values, and the behavior that results from these, as deviant. Thus, any act can be considered deviant if it is so defined. Subcultural theories are suited to illustrating the motivations of people from minority, youth, alternative, or disadvantaged subcultures that are not well aligned with the dominant culture. They locate the explanation for deviance not in the structures that shape society, but in the flesh of the norms and values that compose different subcultural groups. Through cultural transmission, groups pass their norms and values down from one generation to the next, ensuring the survival and social placement of those norms and values, as well as the continuation of cultural conflict.

THE INTERACTIONIST PERSPECTIVE

Although the theories discussed up to now produced insight into some explanations of deviance, there are interactional forces that inevitably intervene between the larger causes that the sociologists who espouse those theories propose and the way deviant behavior takes shape. Many people are exposed to the same structural conditions and the same cultural conflicts and pressures that have been theorized as accounting for deviance, but still resist engaging in deviant behaviors. Left unaddressed is how people from the same structural groups and same subcultures can turn out so different, how members of some families turn to deviance while others do not, and how members of the same family turn out so different from one another. Interactionist theories fill this void by looking in a more microlevel fashion at people's everyday life behavior to try to understand why some people engage in deviance and become so labeled while others do not. Interactionist theories deal with real flesh-and-blood people in specific times and places. They look at how people actually encounter specific others, and they look at the influence of these others. They seek to understand not only why deviance occurs, but how it happens. Many of these theories look at specific social-psychological and interactional dynamics, such as family dynamics, the influence of role models, and the role of peer groups. When people confront the problems, pressures, excitements, and allures of the world, they most often do so in conjunction with their peer groups. It is within peer groups that people make decisions about what they will do and how they will do it. Their core feelings about themselves develop and become rooted in such groups. People's actions and reactions are thus guided by the collective perceptions, interpretations, and actions of their peer groups.

Edwin Sutherland and Donald Cressey recognized this point when they proposed their **differential association theory** of deviance, the subject of Chapter 8. The key feature of their view is the belief that deviant behavior is socially learned, and not from just anyone, but from people's most intimate friends and family members. People may be exposed to a variety of deviant and nondeviant ideas and contacts without that exposure necessarily leading them to engage in deviance. But as their circle of contacts shifts from being composed primarily of people who hold nondeviant ideas to having greater numbers holding deviant ideas and favorable definitions of deviant acts, they become more likely to engage in deviance. The more their friends hold deviant attitudes and engage in deviant behavior, the more likely they are to follow suit.

Sutherland and Cressey further suggested that people learn a variety of elements critical to deviance from their associates: the norms and values of the deviant subculture, the rationalizations for legitimizing deviant behavior, the techniques necessary to commit deviant acts, and the status system of the subculture, by which members evaluate themselves and others. People thus do not decide, at a fixed point in time, to become deviant, but move toward deviant attitudes and behavior as they shift their circle of associates from more normative friends to more deviant friends. Sutherland argued that people rarely stumble onto deviance through their own devices or by seeing acts of deviance in the mass media (as many would suggest), but do so rather by having the knowledge, skills, attitudes, values, traditions, and motives passed down to them through interpersonal (not impersonal) means. Influencing this tendency toward deviance is the age at which they encounter the deviance (earlier in life is likely to be more significant) and the intimacy of the deviant relationships (closer friends and relatives will have greater sway).

Also looking at the interactional level, David Matza (1964) proposed **drift theory**, noting that the movement into deviant subcultures occurs through a process of drift, as people gradually leave their old crowd and become enmeshed in a circle of deviant associates. In proposing his idea, Matza suggested that, rather than just jumping immediately into deviance, people may drift between deviance and legitimacy, keeping one foot in each world. By simultaneously participating in both deviant and "legitimate" worlds, people can learn about and experience the nuances of the deviant world without having to abandon the advantages of their status within the "legitimate" world. In fact, they may drift indefinitely, without having to make a commitment to either for quite some time.

For example, college students may experiment with a different sexual orientation without revealing this behavior to everyone (or, indeed, anyone) and without necessarily giving up their claim to heterosexuality. At some point, they may decide to align more firmly with one side, or a commitment may be forced on them by outside events (getting caught, moving away, becoming sick). Being confronted by someone who discovers the deviance may force people to make a choice and get off the fence, or perhaps just leaving college and having to choose which lifestyle and social network to align themselves with may force a choice. An alternative is that, after a time, individuals choose one path and decide to follow it. But Matza suggests that the dual-membership condition, in which individuals try out both alternatives for a time without making a commitment, may precede such a decision. Thus, Matza proposes that it is rare for people to turn to deviance overnight; more commonly, they take smaller steps, gradually moving to making deviant acquaintances, becoming familiar with deviant ideology, thinking about engaging in deviant acts, trying some out, and then expanding their frequency and range of deviance. Quitting deviance may be a similarly gradual and difficult process, requiring the abandonment of the group of deviant friends and reintegration into conventional circles, perhaps with one foot in both worlds again, before normative behavior becomes the mode that is finally chosen.

A third theory under the interactionist perspective is **labeling theory**, the subject of Chapter 3. This approach suggests that many people dabble to greater or lesser degrees in various forms of deviance. Studies of juvenile delinquency suggest that high rates of youthful participation are extremely widespread, nearly universal. How many people can claim to have reached adulthood without experimenting in illicit drinking, drug use, cheating, stealing, or vandalism? Yet, do all of these people consider themselves deviants? Most do not. Many people retire out of deviance as they mature, avoiding developing the deviant identity altogether. Others go on to engage in what Becker (1963) has called "secret deviance," violating norms without ever seriously encountering the deviant label. Still others, many of them no more experienced in the ways of deviance than the youthful delinquent or the secret deviant, become identified, and identify themselves, as deviants. What causes this difference? One critical difference, labeling theory suggests, lies in who gets caught. Getting caught sets off a chain reaction of events that leads to profound social and self-conceptual consequences. Frank Tannenbaum (1938) has described how individuals are publicly identified as norm violators and branded with that tag. They may go through official or unofficial social sanctioning by which people identify and treat them as deviant. Becker (1963, 9) noted that "the deviant is one to whom that label has successfully been applied; deviant behavior is behavior that people so label." Deviance exists at the societal macrolevel of social norms and definitions through the collective attitudes we assign to certain acts and conditions. But it also comes into being at the everyday-life microlevel when the deviant label is applied to someone. The thrust of labeling theory is twofold, focusing on diverse levels and forces. As Edwin Schur (1979,160) summarized its complexity,

> The twin emphases in such an approach are on *definition* and *process* at all the levels that are involved in the production of deviant situations and outcomes. Thus, the perspective is concerned not only with what happens to specific individuals when they are branded with deviantness ("labeling," in the narrow sense) but also with the wider domains and processes of social definitions and collective rule-making that frequently lie behind such concrete applications of negative labels. (*italics in original*)

In the chapter on labeling theory, Becker emphasizes that deviance lies in the eye of the beholder. There is nothing inherently deviant in any particular act, he claims, until some powerful group defines the act as deviant. Taking the onus off of the individual, Becker emphasizes the importance of looking at the process by which people are labeled deviant and of understanding that deviance is a consequence of others' reactions. This approach forces us to look, then, at how people are defined as deviant, why some acts are labeled and others ignored, and the circumstances that surround the commission of the act. Thus, deviance exists only when it is created by society. The key emphasis of the labeling theory approach to deviance, then, lies in the importance of human peer interaction in understanding the cause of human behavior.

Rooted at the microlevel, but looking less at the specific dynamics of interaction and more at the relationship between individuals and society, is Travis Hirschi's **control theory of delinquency**, the subject of Chapter 9. Like labeling theory, which took as a given that people readily engage in acts of deviance but focused its explanation on the process of identity change that occurs when individuals are caught and labeled, control theory finds it unnecessary to look for the causes of deviant behavior. These are obvious, Hirschi asserts, as deviance and crime not only may be fun, but offer shortcuts and yield immediate, tangible benefits (albeit with a risk). What we should be seeking to understand instead is what holds people back from committing these acts—what forces constrain and control deviance. Hirschi's answer is that social control lies in the extent to which people develop a stake in conformity—a *bond to society*. People who have a greater investment in society will be less likely to risk losing that investment through violations of norms or laws and will follow the rules more willingly. They may have such a stake through their job, through relationships to friends and family, or by virtue of their reputation in the community. Their stake may be fostered by any of the four components discussed in this chapter: attachment to conventional others; commitment to conventional institutions; involvement in conventional activities, and deep beliefs in conventional norms. The extent to which a society is able to foster greater bonds between itself and its potentially deviant members, by giving them a greater stake in achieving success, will affect the constraint or spread of its deviance, particularly, Hirschi notes, of the delinquent variety. It is these ties, forged and maintained through interactions, that influence individuals in their choice between deviant and nondeviant pathways.

Although these overarching perspectives and the theories nested within them differ in the level at which they place their explanations, they all locate them squarely in the social domain. In this aspect, they renounce the prevailing tendency toward unidimensional psychological explanations that lodge causation in pathology, compulsion, neurosis, or maladjustment—simplistic explanations whose inadequacy in a modern, complex world cannot be overstated.

The most recent perspective on deviance theory has been advanced by social constructionists, and Joel Best, one of its leading proponents, describes its history and views in Chapter 11, "The Constructionist Stance." Initially coming from a microinteractionist approach, constructionist theorists sought to revitalize labeling theory by bridging the gap between the way labels are applied to individuals, who then internalize them in their everyday life context, and a larger, more macroawareness of the power structure in society that influences the way these labels are defined and enforced. When some citizens have access to greater degrees of social power that enables the dominant groups to rationalize their ideologies and behavior as legitimate, they are simultaneously defining the actions of less powerful groups as deviant. In so doing, they use the vehicle of deviance and its enforcement to boost their own power while disempowering those they construct as illegitimate. The definitions they forge are then applied to individuals who lack the power to repel them, and the agents of social control act to carry out their moral edicts. The powerless segments of society become the recipients of the activities of moral entrepreneurs (rule creators and rule enforcers) who lay claims that certain behaviors are menacing and dangerous. Individuals connected to these activities may be defined as deviant and may be isolated, sought out, labeled, and stigmatized. Social constructionism thus builds on the basic labeling theory foundation of identity construction by integrating conflict theory's sensitivity to inequality, looking at how the power struggle between dominant and subordinate groups is directly tied to interactional and identity consequences.

6

Functionalism: The Normal and the Pathological

ÉMILE DURKHEIM

One of the founders of the functionalist approach to sociology, Durkheim integrates deviance into his overarching view of society. If, as he believes, society can be compared to a living organism, then all of its social institutions, like the parts of a human body, must contribute to its continuing existence. Under this view, deviance, because of its pervasive presence cross-culturally and over the entire course of history, is not an illness or pathology of the system, but rather something that contributes to society's positive functioning. In fact, like Erikson, Durkheim notes several positive functions that deviance provides, including serving as a means of introducing social change, as new behaviors move through criminal and deviant status into respectability. Durkheim argues that deviance is so critical to society that, if people stopped engaging in it immediately (if we lived in a "society of saints"), we would have to redefine acts now considered acceptable as deviant. Punishing and curing criminals cannot simply, then, be regarded as the objective of society.

Crime is present not only in the majority of societies of one particular species but also in all societies of all types. There is no society that is not confronted with the problem of criminality. Its form changes; the acts thus characterized are not the same everywhere; but, everywhere and always, there have been men who have behaved in such a way as to draw upon themselves penal repression. If, in proportion as societies pass from the lower to the higher types, the rate of criminality, that is the relation between the yearly number of crimes and the population, tended to decline, it might be believed that crime, while still normal, is tending to lose this character of normality. But we have no reason to believe that such a regression is substantiated. Many facts would seem rather to indicate a movement in the opposite direction. From the beginning of the [nineteenth] century, statistics enable us to follow the course of criminality. It has everywhere increased. In France, the increase is nearly 300 percent. There is, then, no phenomenon that presents more indisputably all the symptoms of normality, since it appears closely connected with the conditions of all collective life. To make of crime a form of social morbidity would be to admit that morbidity is not something accidental, but, on the contrary, that in certain cases it grows out of the fundamental constitution of the living organism; it would result in wiping out all distinction between the physiological and the pathological. No doubt it is possible that crime itself will have abnormal forms, as, for example, when its rate is unusually high. This excess is, indeed, undoubtedly morbid in nature. What is normal, simply, is the existence of criminality, provided that it attains and does not exceed, for each social type, a certain level, which it is perhaps not impossible to fix in conformity with the preceding rules.[1]

Here we are, then, in the presence of a conclusion in appearance quite paradoxical. Let us make no mistake. To classify crime among the phenomena of normal sociology is not to say merely that it is an inevitable, although regrettable phenomenon, due to the incorrigible wickedness of men; it is to affirm that it is a factor in public health, an integral part of all healthy societies. This result is, at first glance, surprising enough to have puzzled even ourselves for a long time. Once this first surprise has been overcome, however, it is not difficult to find reasons explaining this normality and at the same time confirming it.

In the first place crime is normal because a society exempt from it is utterly impossible. Crime, we have shown elsewhere, consists of an act that offends certain very strong collective sentiments. In a society in which criminal acts are no longer committed, the sentiments they offend would have to be found without exception in all individual consciousnesses, and they must be found to exist with the same degree as sentiments contrary to them. Assuming that this condition could actually be realized, crime would not thereby disappear; it would only change its form, for the very cause which would thus dry up the sources of criminality would immediately open up new ones.

Indeed, for the collective sentiments which are protected by the penal law of a people at a specified moment of its history to take possession of the public conscience or for them to acquire a stronger hold where they have an insufficient grip, they must acquire an intensity greater than that which they had hitherto had. The community as a whole must experience them more vividly, for it can acquire from no other source the greater force necessary to control these individuals who formerly were the most refractory. For murderers to disappear, the horror of bloodshed must become greater in those social strata from which murderers are recruited; but, first it must become greater throughout the entire society. Moreover, the very absence of crime would directly contribute to produce this horror, because any sentiment seems much more respectable when it is always and uniformly respected.

One easily overlooks the consideration that these strong states of the common consciousness cannot be thus reinforced without reinforcing at the same time the more feeble states, whose violation previously gave birth to mere infraction of convention—since the weaker ones are only the prolongation, the attenuated form, of the stronger. Thus, robbery and simple bad taste injure the same single altruistic sentiment, the respect for that which is another's. However, this same sentiment is less grievously offended by bad taste than by robbery; and since, in addition, the average consciousness has not sufficient intensity to react keenly to the bad taste, it is treated with greater tolerance. That is why the person guilty of bad taste is merely blamed, whereas the thief is punished. But, if this sentiment grows stronger, to the point of silencing in all consciousnesses the inclination which disposes man to steal, he will become more sensitive to the offenses which, until then, touched him but lightly. He will react against them, then, with more energy; they will be the object of greater opprobrium, which will transform certain of them from the simple moral faults that they were and give them the quality of crimes. For example, improper contracts, or contracts improperly executed, which only incur public blame or civil damages, will become offenses in law.

Imagine a society of saints, a perfect cloister of exemplary individuals. Crimes, properly so called, will there be unknown; but faults which appear venial to the layman will create there the same scandal that the ordinary offense does in ordinary consciousness. If, then, this society has the power to judge and punish, it will define these acts as criminal and will treat them as such. For the same reason, the perfect and upright man judges his smallest failings with a severity that the majority reserve for acts more truly in the nature of an offense. Formerly, acts of violence against persons were more frequent than they are today, because respect for individual dignity was less strong. As this has increased, these crimes have become more rare; and also, many acts violating this sentiment have been introduced into the penal law which were not included there in primitive times.[2]

In order to exhaust all the hypotheses logically possible, it will perhaps be asked why this unanimity does not extend to all collective sentiments without exception. Why should not even the most feeble sentiment gather enough energy to prevent all dissent? The moral consciousness of the society would be present in its entirety in all the individuals, with a vitality sufficient to prevent all acts offending it—the purely conventional faults as well as the crimes. But a uniformity so universal and absolute is utterly impossible; for the immediate physical milieu in which each one of us is placed, the hereditary antecedents, and the social influences vary from one individual to the next, and consequently diversify consciousnesses. It is impossible for all to be alike, if only because each one has his own organism and that these organisms occupy different areas in space. That is why, even among the lower peoples, where individual originality is very little developed, it nevertheless does exist.

Thus, since there cannot be a society in which the individuals do not differ more or less from the collective type, it is also inevitable that, among these divergences, there are some with a criminal character. What confers this character upon them is not the intrinsic quality of a given act but that definition which the collective conscience lends them. If the collective conscience is stronger, if it has enough authority practically to suppress these divergences, it will also be more sensitive, more exacting; and, reacting against the slightest deviations with the energy it otherwise displays only against more considerable infractions, it will attribute to them the same gravity as formerly to crimes. In other words, it will designate them as criminal.

Crime is, then, necessary; it is bound up with fundamental conditions of all social life, and by that very fact it is useful because these conditions of which it is part are themselves indispensable to the normal evolution of morality and law.

Indeed, it is no longer possible today to dispute the fact that law and morality vary from one social type to the next, nor that they change within the same type if the conditions of life are modified. But, in order that these transformations may be possible, the collective sentiments at the basis of morality must not be hostile to change, and consequently must have but moderate energy. If they were too strong, they would no longer be plastic. Every pattern is an obstacle to new patterns, to the extent that the first pattern is inflexible. The better a structure is articulated, the more it offers a healthy resistance to all modification; and this is equally true of functional, as of anatomical, organization. If there were no crimes, this condition could not have been fulfilled; for such a hypothesis presupposes that collective sentiments have arrived at a degree of intensity unexampled in history. Nothing is good indefinitely and to an unlimited extent. The authority which the moral conscience enjoys must not be excessive; otherwise no one would dare criticize it, and it would too easily congeal into an immutable form. To make progress, individual originality must be able to express itself. In order that the originality of the idealist whose dreams transcend his century may find expression, it is necessary that the originality of the criminal, who is below the level of his time, shall also be possible. One does not occur without the other.

Nor is this all. Aside from this indirect utility, it happens that crime itself plays a useful role in this evolution. Crime implies not only that the way remains open to necessary changes but that in certain cases it directly prepares these changes. Where crime exists, collective sentiments are sufficiently flexible to take on a new form, and crime sometimes helps to determine the form they will take. How many times, indeed, it is only an anticipation of future morality—a step toward what will be! According to Athenian law, Socrates was a criminal, and his condemnation was no more than just. However, his crime, namely, the independence of his thought, rendered a service not only to humanity but also to his country. It served to prepare a new morality and faith which the Athenians needed, since the traditions by which they had lived until then were no longer in harmony with the current conditions of life. Nor is the case of Socrates unique; it is reproduced periodically in history. It would never have been possible to establish the freedom of thought we now enjoy if the regulations prohibiting it had not been violated before being solemnly abrogated. At that time, however, the violation was a crime, since it was an offense against sentiments still very keen in the average conscience. And yet this crime was useful as a prelude to reforms which daily became more necessary. Liberal philosophy had as its precursors the heretics of all kinds who were justly punished by secular authorities during the entire course of the Middle Ages and until the eve of modern times.

From this point of view the fundamental facts of criminality present themselves to us in an entirely new light. Contrary to current ideas, the criminal no longer seems a totally unsociable being, a sort of parasitic element, a strange and unassimilable body, introduced into the midst of society.[3] On the contrary, he plays a definite role in social life. Crime, for its part, must no longer be conceived as an evil that cannot be too much suppressed. There is no occasion for self-congratulation when the crime rate drops noticeably below the average level, for we may be certain that this apparent progress is associated with some social disorder. Thus, the number of assault cases never falls so low as in times of want.[4] With the drop in the crime rate, and as a reaction to it, comes a revision, or the need of a revision in the theory of punishment. If, indeed, crime is a disease, its punishment is its remedy and cannot be otherwise conceived; thus, all the discussions it arouses bear on the point of determining what the punishment must be in order to fulfill this role of remedy. If crime is not pathological at all, the object of punishment cannot be to cure it, and its true function must be sought elsewhere.

NOTES

1. From the fact that crime is a phenomenon of normal sociology, it does not follow that the criminal is an individual normally constituted from the biological and psychological points of view. The two questions are independent of each other. This independence will be better understood when we have shown, later on, the difference between psychological and sociological facts.

2. Calumny, insults, slander, fraud, etc.

3. We have ourselves committed the error of speaking thus of the criminal, because of a failure to apply our rule (*Division du travail social,* pp. 395–96).

4. Although crime is a fact of normal sociology, it does not follow that we must not abhor it. Pain itself has nothing desirable about it; the individual dislikes it as a society does crime, and yet it is a function of normal physiology. Not only is it necessarily derived from the very constitution of every living organism, but it also plays a useful role in life, for which reason it cannot be replaced. It would, then, be a singular distortion of our thought to present it as an apology for crime. We would not even think of protesting against such an interpretation, did we not know to what strange accusations and misunderstandings one exposes oneself when one undertakes to study moral facts objectively and to speak of them in a different language from that of the layman.

7

Social Structure and Anomie

ROBERT K. MERTON

Merton's notable contribution to deviance theory locates the root causes of crime in the structure of society, not in the individuals involved in criminal actions. Broader patterns of crime characterize certain groups of people because of the social "strain" they face between the good things society offers and their inability to legitimately attain them. Merton identifies a disjuncture for people between what he calls cultural goals, or the ends toward which we are all socialized to strive (largely conceived by him in a material sense) and institutional norms, or the acceptable means that we use to reach these goals. He refers to this condition, in which people cannot attain their goals legitimately, as a "blocked opportunity structure." Societies in which goals are more strongly emphasized than the legitimate means of achieving them are likely to see people pursue the innovation adaptation, as when schools make grades more critically important than the learning required to attain them and when some groups lack the resources to legitimately earn good grades. Merton sketches other ways that groups adapt to their balance of means and ends (conformity, ritualism, retreatism, rebellion) as alternative adaptations to their structural conditions in society.

How can Merton's theory of deviance be seen as the pull of opposing forces in the social structure rather than the decisions of specific individuals?

The framework set out in this essay is designed to provide one systematic approach to the analysis of social and cultural sources of deviant behavior. Our primary aim is to discover how some *social structures exert a definite pressure upon certain persons in the society to engage in nonconforming rather than conforming conduct.* If we can locate groups peculiarly subject to such pressures, we should expect to find fairly high rates of deviant behavior in these groups, not because the human beings comprising them are compounded of distinctive biological tendencies but because they are responding normally to the social situation in which they find themselves. Our perspective is sociological. We look at variations in the *rates* of deviant behavior, not at its incidence. Should our quest be at all successful, some forms of deviant behavior will be found to be as psychologically normal as conformist behavior, and the equation of deviation and psychological abnormality will be put in question.

PATTERNS OF CULTURAL GOALS
AND INSTITUTIONAL NORMS

Among the several elements of social and cultural structures, two are of immediate importance. These elements are analytically separable although they merge in concrete situations. The first consists of culturally defined goals, purposes, and interests, held out as legitimate objectives for all or for diversely located members of the society. The goals are more or less integrated—the degree is a question of empirical fact—and roughly ordered in some hierarchy of value. Involving various degrees of sentiment and significance, the prevailing goals comprise a frame of aspirational reference. They are the things "worth striving for:" They are a basic, though not the exclusive, component of what Linton has called "designs for group living." And though some, not all, of these cultural goals are directly related to the biological drives of man, they are not determined by them.

A second element of the cultural structure defines, regulates, and controls the acceptable modes of reaching out for these goals. Every social group invariably couples its cultural objectives with regulations, rooted in the mores or institutions of allowable procedures for moving toward these objectives. These regulatory norms are not necessarily identical with technical or efficiency norms. Many procedures which from the standpoint of particular individuals would be most efficient in securing desired values—the exercise of force, fraud, power—are ruled out of the institutional area of permitted conduct. At times, the disallowed procedures include some which would be efficient for the group itself—for example, historic taboos on vivisection, on medical experimentation, on the sociological analysis of "sacred" norms—since the criterion of acceptability is not technical efficiency but value-laden sentiments (supported by most members of the group or by those able to promote these sentiments through the composite use of power and propaganda). In all instances, the choice of expedients for striving toward cultural goals is limited by institutionalized norms.

We shall be primarily concerned with the first—a society in which there is an exceptionally strong emphasis upon specific goals without a corresponding emphasis upon institutional procedures. If it is not to be misunderstood, this statement must be elaborated. No society lacks norms governing conduct. But societies do differ in the degree to which the folkways, mores, and institutional controls are effectively integrated with the goals which stand high in the hierarchy of cultural values. The culture may be such as to lead individuals to center their emotional convictions upon the complex of culturally acclaimed ends, with far less emotional support for prescribed methods of reaching out for these ends. With such differential emphases upon goals and institutional procedures, the latter may be so vitiated by the stress on goals as to have the behavior of many individuals limited only by considerations of technical expediency. In this context, the sole significant question becomes: Which of the available procedures is most efficient in netting the culturally approved value?

The technically most effective procedure, whether culturally legitimate or not, becomes typically preferred to institutionally prescribed conduct. As this process of attenuation continues, the society becomes unstable and there develops what Durkheim called "anomie" (or normlessness).

The working of this process eventuating in anomie can be easily glimpsed in a series of familiar and instructive, though perhaps trivial, episodes. Thus, in competitive athletics, when the aim of victory is shorn of its institutional trappings and success becomes construed as "winning the game" rather than "winning under the rules of the game," a premium is implicitly set upon the use of illegitimate but technically efficient means. The star of the opposing football team is surreptitiously slugged; the wrestler incapacitates his opponent through ingenious but illicit techniques; university alumni covertly subsidize "students" whose talents are confined to the athletic field. The emphasis on the goal has so attenuated the satisfactions deriving from sheer participation in the competitive activity that only a successful outcome provides gratification. Through the same process, tension generated by the desire to win in a poker game is relieved by successfully dealing one's self four aces, or when the cult of success has truly flowered, by sagaciously shuffling the cards in a game of solitaire. The faint twinge of uneasiness in the last instance and the surreptitious nature of public delicts indicate clearly that the institutional rules of the game are *known* to those who evade them. But cultural (or idiosyncratic) exaggeration of the success–goal leads men to withdraw emotional support from the rules.

This process is of course not restricted to the realm of competitive sport, which has simply provided us with microcosmic images of the social macrocosm. The process whereby exaltation of the end generates a literal *demoralization,* that is, a deinstitutionalization, of the means occurs in many groups where the two components of the social structure are not highly integrated.

Contemporary American culture appears to approximate the polar type in which great emphasis upon certain success–goals occurs without equivalent emphasis upon institutional means. It would of course be fanciful to assert that accumulated wealth stands alone as a symbol of success just as it would be fanciful to deny that Americans assign it a place high in their scale of values. In some large measure, money has been consecrated as a value in itself, over and above its expenditure for articles of consumption or its use for the enhancement of power. "Money" is peculiarly well adapted to become a symbol of prestige. As Simmel emphasized, money is highly abstract and impersonal. However acquired, fraudulently or institutionally, it can be used to purchase the same goods and services. The anonymity of an urban society, in conjunction with these peculiarities of money, permits wealth, the sources of which may be unknown to the community in which the plutocrat lives or, if known, to become purified in the course of time, to serve as a symbol of high status. Moreover, in the American Dream there is no final stopping point. The measure of "monetary success" is conveniently indefinite and relative. At each income level, as H. F. Clark found,

Americans want just about 25 percent more (but of course this "just a bit more" continues to operate once it is obtained).

In this flux of shifting standards, there is no stable resting point, or rather, it is the point which manages always to be "just ahead." An observer of a community in which annual salaries in six figures are not uncommon reports the anguished words of one victim of the American Dream: "In this town, I'm snubbed socially because I only get a thousand a week. That hurts."

To say that the goal of monetary success is entrenched in American culture is only to say that Americans are bombarded on every side by precepts which affirm the right or, often, the duty of retaining the goal even in the face of repeated frustration. Prestigeful representatives of the society reinforce the cultural emphasis. The family, the school, and the workplace—the major agencies shaping the personality structure and goal formation of Americans—join to provide the intensive disciplining required if an individual is to retain intact a goal that remains elusively beyond reach, if he is to be motivated by the promise of a gratification which is not redeemed. As we shall presently see, parents serve as a transmission belt for the values and goals of the groups of which they are a part—above all, of their social class or of the class with which they identify themselves. And the schools are, of course, the official agency for the passing on of the prevailing values, with a large proportion of the textbooks used in city schools implying or stating explicitly "that education leads to intelligence and consequently to job and money success." Central to this process of disciplining people to maintain their unfulfilled aspirations are the cultural prototypes of success, the living documents testifying that the American Dream can be realized if one but has the requisite abilities.

Coupled with this positive emphasis upon the obligation to maintain lofty goals is a correlative emphasis upon the penalizing of those who draw in their ambitions. Americans are admonished "not to be a quitter" for in the dictionary of American culture, as in the lexicon of youth, "there is no such word as 'fail.'" The cultural manifesto is clear: one must not quit, must not cease striving, must not lessen his goals, for "not failure, but low aim, is crime."

Thus, the culture enjoins the acceptance of three cultural axioms: First, all should strive for the same lofty goals because these are open to all; second, present seeming failure is but a way–station to ultimate success; and third, genuine failure consists only in the lessening or withdrawal of ambition.

In rough psychological paraphrase, these axioms represent, first a symbolic secondary reinforcement of incentive; second, curbing the threatened extinction of a response through an associated stimulus; third, increasing the motive–strength to evoke continued responses despite the continued absence of reward.

In sociological paraphrase, these axioms represent, first, the deflection of criticism of the social structure onto one's self among those so situated in the society that they do not have full and equal access to opportunity; second, the preservation of a structure of social power by having individuals in the lower social strata identify themselves, not with their compeers, but with those at the top (whom they will ultimately join); and third, providing pressures for conformity with the cultural dictates of unslackened ambition by the threat of less than full membership in the society for those who fail to conform.

It is in these terms and through these processes that contemporary American culture continues to be characterized by a heavy emphasis on wealth as a basic symbol of success, without a corresponding emphasis upon the legitimate avenues on which to march toward this goal. How do individuals living in this cultural context respond? And how do our observations bear upon the doctrine that deviant behavior typically derives from biological impulses breaking through the restraints imposed by culture? What, in short, are the consequences for the behavior of people variously situated in a social structure of a culture in which the emphasis on dominant success–goals has become increasingly separated from an equivalent emphasis on institutionalized procedures for seeking these goals?

TYPES OF INDIVIDUAL ADAPTATION

Turning from these culture patterns, we now examine types of adaptation by individuals within the culture-bearing society. Though our focus is still the cultural and social genesis of varying rates and types of deviant behavior, our perspective shifts from the plane of patterns of cultural values to the plane of types of adaptation to these values among those occupying different portions in the social structure.

We here consider five types of adaptation, as these are schematically set out in the following table, where (+) signifies "acceptance," (−) signifies "rejection," and (±) signifies "rejection of prevailing values and substitution of new values."

I. Conformity

To the extent that a society is stable, adaptation type I—conformity to both cultural goals and institutionalized means—is the most common and widely diffused. Were this not so, the stability and continuity of the society could not be maintained ....

II. Innovation

Great cultural emphasis upon the success–goal invites this mode of adaptation through the use of institutionally proscribed but often effective means of attaining

TABLE 7.1   A Typology of Modes of Individual Adaptation

Modes of Adaptation	Culture Goals	Institutionalized Means
I. Conformity	+	+
II. Innovation	+	−
III. Ritualism	−	+
IV. Retreatism	−	−
V. Rebellion	±	±

at least the simulacrum of success—wealth and power. This response occurs when the individual has assimilated the cultural emphasis upon the goal without equally internalizing the institutional norms governing ways and means for its attainment.

It appears from our analysis that the greatest pressures toward deviation are exerted upon the lower strata. Cases in point permit us to detect the sociological mechanisms involved in producing these pressures. Several researchers have shown that specialized areas of vice and crime constitute a "normal" response to a situation where the cultural emphasis upon pecuniary success has been absorbed, but where there is little access to conventional and legitimate means for becoming successful. The occupational opportunities of people in these areas are largely confined to manual labor and the lesser white-collar jobs. Given the American stigmatization of manual labor *which has been found* to hold *rather uniformly in all social classes,* and the absence of realistic opportunities for advancement beyond this level, the result is a marked tendency toward deviant behavior. The status of unskilled labor and the consequent low income cannot readily compete *in terms of established standards of worth* with the promises of power and high income from organized vice, rackets, and crime.

For our purposes, these situations exhibit two salient features. First, incentives for success are provided by the established values of the culture *and* second, the avenues available for moving toward this goal are largely limited by the class structure to those of deviant behavior. It is the *combination* of the cultural emphasis and the social structure which produces intense pressure for deviation ....

III. Ritualism

The ritualistic type of adaptation can be readily identified. It involves the abandoning or scaling down of the lofty cultural goals of great pecuniary success and rapid social mobility to the point where one's aspirations can be satisfied. But though one rejects the cultural obligation to attempt "to get ahead in the world," though one draws in one's horizons, one continues to abide almost compulsively by institutional norms ....

We should expect this type of adaptation to be fairly frequent in a society which makes one's social status largely dependent upon one's achievements. For, as has so often been observed, this ceaseless competitive struggle produces acute status anxiety. One device for allaying these anxieties is to lower one's level of aspiration—permanently. Fear produces inaction, or, more accurately, routinized action.

The syndrome of the social ritualist is both familiar and instructive. His implicit life-philosophy finds expression in a series of cultural clichés: "I'm not sticking *my* neck out," "I'm playing safe," "I'm satisfied with what I've got," "Don't aim high and you won't be disappointed." The theme threaded through these attitudes is that high ambitions invite frustration and danger whereas lower aspirations produce satisfaction and security. It is the perspective of the frightened employee, the zealously conformist bureaucrat in the teller's cage of the private banking enterprise or in the front office of the public works enterprise.

IV. Retreatism

Just as Adaptation I (conformity) remains the most frequent, Adaptation IV (the rejection of cultural goals and institutional means) is probably the least common. People who adapt (or maladapt) in this fashion are, strictly speaking, *in* the society but not *of* it. Sociologically these constitute the true aliens. Not sharing the common frame of values, they can be included as members of the *society* (in distinction from the *population*) only in a fictional sense.

In this category fall some of the adaptive activities of psychotics, autists, pariahs, outcasts, vagrants, vagabonds, tramps, chronic drunkards, and, drug addicts. They have relinquished culturally prescribed goals and their behavior does not accord with institutional norms. The competitive order is maintained but the frustrated and handicapped individual who cannot cope with this order drops out. Defeatism, quietism, and resignation are manifested in escape mechanisms which ultimately lead him to "escape" from the requirements of the society. It is thus an expedient which arises from continued failure to near the goal by legitimate measures and from an inability to use the illegitimate route because of internalized prohibitions.

V. Rebellion

This adaptation leads men outside the environing social structure to envisage and seek to bring into being a new, that is to say, a greatly modified social structure. It presupposes alienation from reigning goals and standards. These come to be regarded as purely arbitrary. And the arbitrary is precisely that which can neither exact allegiance nor possess legitimacy, for it might as well be otherwise. In our society, organized movements for rebellion apparently aim to introduce a social structure in which the cultural standards of success would be sharply modified and provision would be made for a closer correspondence between merit, effort, and reward.

THE STRAIN TOWARD ANOMIE

The social structure we have examined produces a strain toward anomie and deviant behavior. The pressure of such a social order is upon outdoing one's competitors. So long as the sentiments supporting this competitive system are distributed throughout the entire range of activities and are not confined to the final result of "success," the choice of means will remain largely within the ambit of institutional control. When, however, the cultural emphasis shifts from the satisfactions deriving from competition itself to almost exclusive concern with the outcome, the resultant stress makes for the breakdown of the regulatory structure.

8

Differential Association

EDWIN H. SUTHERLAND AND DONALD R. CRESSEY

Like Durkheim, Sutherland and Cressey do not regard crime and deviance as the result of either pathology in society or pathological behavior patterns. Crime, they argue, is learned in much the same way as all ordinary behavior and represents the expression of the same behavioral needs and values as other behavior. Crime is less likely to be learned, in fact, from frightening or suspicious outsiders than from people's own intimate associates. In this way, Sutherland and Cressey cast the learning of crime and deviance as a normal process and note that it is a likely occurrence as people become surrounded with increasing numbers of deviant friends. Sutherland and Cressey place the learning of deviance within people's most intense and personal relations: families and peer groups. By repeatedly watching others modeling crime, by learning from them how to do it effectively, and by becoming convinced by their rationalizations and neutralizations that such behaviors are acceptable, individuals move into criminal or deviant behavior patterns.

How does Sutherland and Cressey's perspective on the causes of deviance separate the behavior of individuals from that of members of their subcultural group or broader society?

The following statements refer to the process by which a particular person comes to engage in criminal behavior.

1. *Criminal behavior is learned.* Negatively, this means that criminal behavior is not inherited, as such; the person who is not already trained in crime does not also invent criminal behavior, just as a person does not make mechanical inventions unless he has had training in mechanics.

2. *Criminal behavior is learned in interaction with other persons in a process of communication.* This communication is verbal in many respects and includes also "the communication of gestures."

3. *The principal part of the learning of criminal behavior occurs within intimate personal groups.* Negatively, this means that the impersonal agencies of communication, such as movies and newspapers, play a relatively unimportant part in the genesis of criminal behavior.

Principles of Criminology by Edwin H. Sutherland, Donald R. Cressey, and David F. Luckenbill. Reproduced with permission of General Hall via Copyright Clearance Center.

4. *When criminal behavior is learned, the learning includes (a) techniques of committing the crime, which are sometimes very complicated, sometimes very simple; (b) the specific direction of motive, drives, rationalizations, and attitudes.*

5. *The specific direction of motives and drives is learned from definitions of the legal codes as favorable or unfavorable.* In some societies an individual is surrounded by persons who invariably define the legal codes as rules to be observed, while in others he is surrounded by persons whose definitions are favorable to the violation of the legal codes. In our American society these definitions are almost always mixed, with the consequence that we have culture conflict in relation to the legal codes.

6. *A person becomes delinquent because of an excess of definitions favorable to violation of law over definitions unfavorable to violation of law.* This is the principle of differential association. It refers to both criminal and anticriminal associations and has to do with counteracting forces. When persons become criminal, they do so because of contacts with criminal patterns and also because of isolation from anticriminal patterns. Any person inevitably assimilates the surrounding culture unless other patterns are in conflict; a southerner does not pronounce *r* because other southerners do not pronounce *r*. Negatively, this proposition of differential association means that associations which are neutral so far as crime is concerned have little or no effect on the genesis of criminal behavior. Much of the experience of a person is neutral in this sense, for example, learning to brush one's teeth. This behavior has no negative or positive effect on criminal behavior except as it may be related to associations which are concerned with the legal codes. This neutral behavior is important especially as an occupier of the time of a child so that he is not in contact with criminal behavior during the time he is so engaged in the neutral behavior.

7. *Differential associations may vary in frequency, duration, priority, and intensity.* This means that associations with criminal behavior and also associations with anticriminal behavior vary in those respects. "Frequency" and "duration" as modalities of associations are obvious and need no explanation. "Priority" is assumed to be important in the sense that lawful behavior developed in early childhood may persist throughout life, and also that delinquent behavior developed in early childhood may persist throughout life. This tendency, however, has not been adequately demonstrated, and priority seems to be important principally through its selective influence. "Intensity" is not precisely defined, but it has to do with such things as the prestige of the source of a criminal or anticriminal pattern and with emotional reactions related to the associations. In a precise description of the criminal behavior of a person, these modalities would be rated in quantitative form and a mathematical ratio reached. A formula in this sense has not been developed, and the development of such a formula would be extremely difficult.

8. *The process of learning criminal behavior by association with criminal and anticriminal patterns involves all of the mechanisms that are involved in any other learning.*

Negatively, this means that the learning of criminal behavior is not restricted to the process of imitation. A person who is seduced, for instance, learns criminal behavior by association, but this process would not ordinarily be described as imitation.

9. *While criminal behavior is an expression of general needs and values, it is not explained by those general needs and values, since noncriminal behavior is an expression of the same needs and values.* Thieves generally steal in order to secure money, but likewise honest laborers work in order to secure money. The attempts by many scholars to explain criminal behavior by general drives and values, such as the happiness principle, striving for social status, the money motive, or frustration, have been, and must continue to be, futile, since they explain lawful behavior as completely as they explain criminal behavior. They are similar to respiration, which is necessary for any behavior, but which does not differentiate criminal from noncriminal behavior.

It is not necessary, at this level of explanation, to explain why a person has the associations he has; this certainly involves a complex of many things. In an area where the delinquency rate is high, a boy who is sociable, gregarious, active, and athletic is very likely to come in contact with other boys in the neighborhood, learn delinquent behavior patterns from them, and become a criminal; in the same neighborhood the psychopathic boy who is isolated, introverted, and inert may remain at home, not become acquainted with the other boys in the neighborhood, and not become delinquent. In another situation, the sociable, athletic, aggressive boy may become a member of a scout troop and not become involved in delinquent behavior. The person's associations are determined in a general context of social organization. A child is ordinarily reared in a family; the place of residence of the family is determined largely by family income; and the delinquency rate is in many respects related to the rental value of the houses. Many other aspects of social organization affect the kinds of associations a person has.

The preceding explanation of criminal behavior purports to explain the criminal and noncriminal behavior of individual persons. It is possible to state sociological theories of criminal behavior which explain the criminality of a community, nation, or other group. The problem, when thus stated, is to account for variations in crime rates and involves a comparison of the crime rates of various groups or the crime rates of a particular group at different times. The explanation of a crime rate must be consistent with the explanation of the criminal behavior of the person, since the crime rate is a summary statement of the number of persons in the group who commit crimes and the frequency with which they commit crimes. One of the best explanations of crime rates from this point of view is that a high crime rate is due to social disorganization. The term *social disorganization* is not entirely satisfactory, and it seems preferable to substitute for it the term *differential social organization*. The postulate on which this theory is based, regardless of the name, is that crime is rooted in the social organization and is an expression of that social organization. A group may be organized for criminal behavior or organized against criminal behavior. Most communities are organized for both criminal and anticriminal behavior, and in that sense the crime rate is an expression of the differential group organization. Differential group organization as an explanation of variations in crime rates is consistent with the differential association theory of the processes by which persons become criminals.

9

Control Theory

TRAVIS HIRSCHI

Hirschi focuses on youthful delinquency, the age at which most deviation from the norms generally occurs. In contrast to previous perspectives, Hirschi's model is a more individualistic one, looking at the bond between each person and society. Conforming behavior is reinforced by individuals' attachment to norm-abiding members of society; by their commitment to, and investment in, a legitimate life and identity (e.g., earning educational credentials and a respectable reputation); by their level of involvement in legitimate activities and organizations; and by their subscription to the commonly held beliefs and values characterizing normative society. People who violate norms have a flaw in one or more of these bonds to society and can be brought back into the normative ranks by strengthening and reinforcing those weak bonds. Hirschi's perspective is thus more social psychological than structural.

How can Hirschi's theory suggest specific types of social policies that could hold people back from entry into deviance?

Control theories assume that delinquent acts result when an individual's bond to society is weak or broken. Since these theories embrace two highly complex concepts, the *bond* of the individual to *society*, it is not surprising that they have at one time or another formed the basis of explanations of most forms of aberrant or unusual behavior. It is also not surprising that control theories have described the elements of the bond to society in many ways, and that they have focused on a variety of units as the point of control ....

ELEMENTS OF THE BOND

Attachment

In explaining conforming behavior, sociologists justly emphasize sensitivity to the opinion of others.[1] Unfortunately, ... they tend to suggest that man *is* sensitive to the opinion of others and thus exclude sensitivity from their explanations of deviant behavior. In explaining deviant behavior, psychologists, in contrast,

From Travis Hirschi, *Causes of Delinquency*, pp. 16–26, Berkeley: University of California Press, 1969. Reprinted by permission of the author.

emphasize insensitivity to the opinion of others.[2] Unfortunately, they too tend to ignore variation, and, in addition, they tend to tie sensitivity inextricably to other variables, to make it part of a syndrome or "type," and thus seriously to reduce its value as an explanatory concept. The psychopath is characterized only in part by "deficient attachment to or affection for others, a failure to respond to the ordinary motivations founded in respect or regard for one's fellow";[3] he is also characterized by such things as "excessive aggressiveness," "lack of superego control," and "an infantile level of response."[4] Unfortunately, too, the behavior that psychopathy is used to explain often becomes part of the *definition* of psychopath. As a result, in Barbara Wootton's words: "[The psychopath] is ... *par excellence*, and without shame or qualification, the model of the circular process by which mental abnormality is inferred from antisocial behavior while antisocial behavior is explained by mental abnormality."[5]

The problems of diagnosis, tautology, and name-calling are avoided if the dimensions of psychopathy are treated as causally and therefore problematically interrelated, rather than as logically and therefore necessarily bound to each other. In fact, it can be argued that all of the characteristics attributed to the psychopath follow from, are effects of, his lack of attachment to others. To say that to lack attachment to others is to be free from moral restraints is to use lack of attachment to explain the guiltlessness of the psychopath, the fact that he apparently has no conscience or superego. In this view, lack of attachment to others is not merely a symptom of psychopathy, it *is* psychopathy; lack of conscience is just another way of saying the same thing; and the violation of norms is (or may be) a consequence.

For that matter, given that man is an animal, "impulsivity" and "aggressiveness" can also be seen as natural consequences of freedom from moral restraints. However, since the view of man as endowed with natural propensities and capacities like other animals is peculiarly unpalatable to sociologists, we need not fall back on such a view to explain the amoral man's aggressiveness.[6] The process of becoming alienated from others often involves or is based on active interpersonal conflict. Such conflict could easily supply a reservoir of *socially derived* hostility sufficient to account for the aggressiveness of those whose attachments to others have been weakened.

Durkheim said it many years ago: "We are moral beings to the extent that we are social beings."[7] This may be interpreted to mean that we are moral beings to the extent that we have "internalized the norms" of society. But what does it mean to say that a person has internalized the norms of society? To violate a norm is, therefore, to act contrary to the wishes and expectations of other people. If a person does not care about the wishes and expectations of other people—that is, if he is insensitive to the opinion of others—then he is to that extent not bound by the norms. He is free to deviate.

The essence of internalization of norms, conscience, or superego thus lies in the attachment of the individual to others.[8] This view has several advantages over the concept of internalization. For one, explanations of deviant behavior based on attachment do not beg the question, since the extent to which a person is attached to others can be measured independently of his deviant behavior.

Furthermore, change or variation in behavior is explainable in a way that it is not when notions of internalization or superego are used. For example, the divorced man is more likely after divorce to commit a number of deviant acts, such as suicide or forgery. If we explain these acts by reference to the superego (or internal control), we are forced to say that the man "lost his conscience" when he got a divorce; and, of course, if he remarries, we have to conclude that he gets his conscience back.

This dimension of the bond to conventional society is encountered in most social control–oriented research and theory. F. Ivan Nye's "internal control" and "indirect control" refer to the same element, although we avoid the problem of explaining changes over time by locating the "conscience" in the bond to others rather than making it part of the personality.[9] Attachment to others is just one aspect of Albert J. Reiss's "personal controls"; we avoid his problems of tautological empirical *observations* by making the relationship between attachment and delinquency problematic rather than definitional.[10] Finally, Scott Briar and Irving Piliavin's "commitment" or "stake in conformity" subsumes attachment, as their discussion illustrates, although the terms they use are more closely associated with the next element to be discussed.[11]

Commitment

"Of all passions, that which inclineth men least to break the laws, is fear. Nay, excepting some generous natures, it is the only thing, when there is the appearance of profit or pleasure by breaking the laws, that makes men keep them."[12] Few would deny that men on occasion obey the rules simply from fear of the consequences. This rational component in conformity we label commitment. What does it mean to say that a person is committed to conformity? In Howard S. Becker's formulation it means the following:

> First, the individual is in a position in which his decision with regard to some particular line of action has consequences for other interests and activities not necessarily [directly] related to it. Second, he has placed himself in that position by his own prior actions. A third element is present though so obvious as not to be apparent; the committed person must be aware [of these other interests] and must recognize that his decision in this case will have ramifications beyond it.[13]

The idea, then, is that the person invests time, energy, himself, in a certain line of activity—say, getting an education, building up a business, acquiring a reputation for virtue. When or whenever he considers deviant behavior, he must consider the costs of this deviant behavior, the risk he runs of losing the investment he has made in conventional behavior.

If attachment to others is the sociological counterpart of the superego or conscience, commitment is the counterpart of the ego or common sense. To the person committed to conventional lines of action, risking one to ten years in prison for a ten-dollar holdup is stupidity, because to the committed person the costs and risks obviously exceed ten dollars in value. (To the psychoanalyst, such an act exhibits failure to be governed by the "reality-principle.") In the sociological control theory, it can be and is generally assumed that the decision to commit a criminal act may well be rationally determined—that the actor's decision was not irrational given the risks and costs he faces. Of course, as Becker points out, if the actor is capable of in some sense calculating the costs of a line of action, he is also capable of calculational errors: ignorance and error return, in the control theory, as possible explanations of deviant behavior.

The concept of commitment assumes that the organization of society is such that the interest of most persons would be endangered if they were to engage in criminal acts. Most people, simply by the process of living in an organized society, acquire goods, reputations, prospects that they do not want to risk losing. These accumulations are society's insurance that they will abide by the rules. Many hypotheses about the antecedents of delinquent behavior are based on this premise. For example, Arthur L. Stinchcombe's hypothesis that "high school rebellion ... occurs when future status is not clearly related to present performance"[14] suggests that one is committed to conformity not only by what one has but also by what one hoped to obtain. Thus, "ambition" and/or "aspiration" play an important role in producing conformity. The person becomes committed to a conventional line of action, and he is therefore committed to conformity.

Most lines of action in a society are of course conventional. The clearest examples are educational and occupational careers. Actions thought to jeopardize one's chances in these areas are presumably avoided. Interestingly enough, even nonconventional commitments may operate to produce conventional conformity. We are told, at least, that boys aspiring to careers in the rackets or professional thievery are judged by their "honesty" and "reliability"—traits traditionally in demand among seekers of office boys.[15]

Involvement

Many people undoubtedly owe a life of virtue to a lack of opportunity to do otherwise. Time and energy are inherently limited: "Not that I would not, if I could, be both handsome and fat and well dressed, and a great athlete, and make a million a year, be a wit, a bon vivant, and a lady killer, as well as a philosopher, a philanthropist, a statesman, warrior, and African explorer, as well as a 'tone-poet' and saint. But the thing is simply impossible."[16] The things that William James here says he would like to be or do are all, I suppose, within the realm of conventionality, but if he were to include illicit actions he would still have to eliminate some of them as simply impossible.

Involvement or engrossment in conventional activities is thus often part of a control theory. The assumption, widely shared, is that a person may be simply too busy doing conventional things to find time to engage in deviant behavior. The person involved in conventional activities is tied to appointments, deadlines, working hours, plans, and the like, so the opportunity to commit deviant acts rarely arises. To the extent that he is engrossed in conventional activities, he cannot even think about deviant acts, let alone act out his inclinations.[17]

This line of reasoning is responsible for the stress placed on recreational facilities in many programs to reduce delinquency, for much of the concern with the high school dropout, and for the idea that boys should be drafted into the army to keep them out of trouble. So obvious and persuasive is the idea that involvement in conventional activities is a major deterrent to delinquency that it was accepted even by Sutherland: "In the general area of juvenile delinquency it is probable that the most significant difference between juveniles who engage in delinquency and those who do not is that the latter are provided abundant opportunities of a conventional type for satisfying their recreational interests, while the former lack those opportunities or facilities."[18]

The view that "idle hands are the devil's workshop" has received more sophisticated treatment in recent sociological writings on delinquency. David Matza and Gresham M. Sykes, for example, suggest that delinquents have the values of a leisure class, the same values ascribed by Veblen to *the* leisure class: a search for kicks, disdain of work, a desire for the big score, and acceptance of aggressive toughness as proof of masculinity.[19] Matza and Sykes explain delinquency by reference to this system of values, but they note that adolescents at all class levels are "to some extent" members of a leisure class, that they "move in a limbo between earlier parental domination and future integration with the social structure through the bonds of work and marriage."[20] In the end, then, the leisure of the adolescent produces a set of values, which, in turn, leads to delinquency.

Belief

Unlike the cultural deviance theory, the control theory assumes the existence of a common value system within the society or group whose norms are being violated. If the deviant is committed to a value system different from that of conventional society, there is, within the context of the theory, nothing to explain. The question is, "Why does a man violate the rules in which he believes?" It is not, "Why do men differ in their beliefs about what constitutes good and desirable conduct?" The person is assumed to have been socialized (perhaps imperfectly) into the group whose rules he is violating; deviance is not a question of one group imposing its rules on the members of another group. In other words, we not only assume the deviant *has* believed the rules, but we also assume he believes the rules even as he violates them.

How can a person believe it is wrong to steal at the same time he is stealing? In the strain theory, this is not a difficult problem. (In fact, ... the strain theory was devised specifically to deal with this question.) The motivation to deviance adduced by the strain theorist is so strong that we can well understand the deviant act even assuming the deviator believes strongly that it is wrong.[21] However, given the control theory's assumptions about motivation, if both the deviant and the nondeviant believe the deviant act is wrong, how do we account for the fact that one commits it and the other does not?

Control theories have taken two approaches to this problem. In one approach, beliefs are treated as mere words that mean little or nothing if the other forms of control are missing. "Semantic dementia," the dissociation between rational faculties and emotional control which is said to be characteristic of the psychopath, illustrates this way of handling the problem.[22] In short, beliefs, at least insofar as they are expressed in words, drop out of the picture; since they do not differentiate between deviants and nondeviants, they are in the same class as "language" or any other characteristic common to all members of the group. Since they represent no real obstacle to the commission of delinquent acts, nothing need be said about how they are handled by those committing such acts. The control theories that do not mention beliefs (or values), and many do not, may be assumed to take this approach to the problem.

The second approach argues that the deviant rationalizes his behavior so that he can at once violate the rule and maintain his belief in it. Donald R. Cressey had advanced this argument with respect to embezzlement,[23] and Sykes and Matza have advanced it with respect to delinquency.[24] In both Cressey's and Sykes and Matza's treatments, these rationalizations (Cressey calls them "verbalizations," Sykes and Matza term them "techniques of neutralization") occur prior to the commission of the deviant act. If the neutralization is successful, the person is free to commit the act(s) in question. Both in Cressey and in Sykes and Matza, the strain that prompts the effort at neutralization also provides the motive force that results in the subsequent deviant act. Their theories are thus, in this sense, strain theories. Neutralization is difficult to handle within the context of a theory that adheres closely to control theory assumptions because in the control theory there is no special motivational force to account for the neutralization. This difficulty is especially noticeable in Matza's later treatment of this topic, where the motivational component, the "will to delinquency," appears *after* the moral vacuum has been created by the techniques of neutralization.[25] The question thus becomes: Why neutralize?

In attempting to solve a strain-theory problem with control-theory tools, the control theorist is thus led into a trap. He cannot answer the crucial question. The concept of neutralization assumes the existence of moral obstacles to the commission of deviant acts. In order plausibly to account for a deviant act, it is necessary to generate motivation to deviance that is at least equivalent in force to the resistance provided by these moral obstacles. However, if the moral obstacles are removed, neutralization and special motivation are no longer required. We therefore follow the implicit logic of control theory and remove these moral obstacles by hypothesis. Many persons do not have an attitude of respect toward the rules of society; many persons feel no moral obligation to conform regardless of personal advantage. Insofar as the values and beliefs of these persons are consistent with their feelings, and there should be tendency toward consistency, neutralization is unnecessary; it has already occurred.

Does this merely push the question back a step and at the same time produce conflict with the assumption of a common value system? I think not. In the first place, we do not assume, as does Cressey, that neutralization occurs in order to make a specific criminal act possible.[26] We do not assume, as do Sykes and Matza, that neutralization occurs to make many delinquent acts possible. We do not assume, in other words, that the person constructs a system of rationalizations in order to justify commission of acts he *wants* to commit. We assume, in contrast, that the beliefs that free a man to commit deviant acts are *unmotivated* in the sense that he does not construct or adopt them in order to facilitate the attainment of illicit ends. In the second place, we do not assume, as does Matza, that "delinquents concur in the conventional assessment of delinquency."[27] We assume, in contrast, that there is *variation* in the extent to which people believe they should obey the rules of society, and, furthermore, that the less a person believes he should obey the rules, the more likely he is to violate them.[28]

In chronological order, then, a person's beliefs in the moral validity of norms are, for no teleological reason, weakened. The probability that he will commit delinquent acts is therefore increased. When and if he commits a delinquent act, we may justifiably use the weakness of his beliefs in explaining it, but no special motivation is required to explain either the weakness of his beliefs or, perhaps, his delinquent act.

The keystone of this argument is of course the assumption that there is variation in belief in the moral validity of social rules. This assumption is amenable to direct empirical test and can thus survive at least until its first confrontation with data. For the present, we must return to the idea of a common value system with which this section was begun.

The idea of a common (or perhaps better, a single) value system is consistent with the fact, or presumption, of variation in the strength of moral beliefs. We have not suggested that delinquency is based on beliefs counter to conventional morality; we have not suggested that delinquents do not believe delinquent acts are wrong. They may well believe these acts are wrong, but the meaning and efficacy of such beliefs are contingent on other beliefs and, indeed, on the strength of other ties to the conventional order.[29]

NOTES

1. Books have been written on the increasing importance of interpersonal sensitivity in modern life. According to this view, controls from within have become less important than controls from without in *producing* conformity. Whether or not this observation is true as a description of historical trends, it is true that interpersonal sensitivity has become more important in *explaining* conformity. Although logically it should also have become more important in explaining nonconformity, the opposite has been the case, once again showing that Cohen's observation that an explanation of conformity should be an explanation of deviance cannot be translated as "an explanation of conformity has to be an explanation of deviance." For the view that interpersonal sensitivity currently plays a greater role than formerly in producing conformity, see William J. Goode, "Norm Commitment and Conformity to Role-Status Obligations," *American Journal of Sociology* LXVI (1960): 246–258. And, of course, also see David Riesman, Nathan Glazer, and Reuel Denney, *The Lonely Crowd* (Garden City, New

York: Doubleday, 1950), especially Part I.

2. The literature on psychopathy is voluminous. See William McCord and Joan McCord, *The Psychopath* (Princeton: D. Van Nostrand, 1964).

3. John M. Martin and Joseph P. Fitzpatrick, *Delinquent Behavior* (New York: Random House, 1964), p. 130.

4. *Ibid.* For additional properties of the psychopath, see McCord and McCord, *The Psychopath,* pp. 1–22.

5. Barbara Wootton, *Social Science and Social Pathology* (New York: Macmillan, 1959), p. 250.

6. "The logical untenability [of the position that there are forces in man 'resistant to socialization'] was ably demonstrated by Parsons over 30 years ago, and it is widely recognized that the position is empirically unsound because it assumes [!] some universal biological drive system distinctly separate from socialization and social context—a basic and intransigent human nature" (Judith Blake and Kingsley Davis, "Norms, Values, and Sanctions," *Handbook of Modem Sociology,* ed. Robert E. L. Pans [Chicago: Rand McNally, 1964], p. 471).

7. Emile Durkheim, *Moral Education,* trans. Everett K. Wilson and Herman Schnurer (New York: The Free Press, 1961), p. 64.

8. Although attachment alone does not exhaust the meaning of internalization, attachments and beliefs combined would appear to leave only a small residue of "internal control" not susceptible in principle to direct measurement.

9. F. Ivan Nye, *Family Relationships and Delinquent Behavior* (New York: Wiley, 1958), pp. 5–7.

10. Albert J. Reiss, Jr., "Delinquency as the Failure of Personal and Social Controls," *American Sociological Review* XVI (1951): 196–207. For example,

"Our observations show ... that delinquent recidivists are less often persons with mature ego ideals or nondelinquent social roles" (p. 204).

11. Scott Briar and Irving Piliavin, "Delinquency, Situational Inducements, and Commitment to Conformity," *Social Problems* XIII (1965): 41–42. The concept "stake in conformity" was introduced by Jackson Toby in his "Social Disorganization and Stake in Conformity: Complementary Factors in the Predatory Behavior of Hoodlums," *Journal of Criminal Law, Criminology and Police Science* XLVIII (1957): 12–17. See also his "Hoodlum or Business Man: An American Dilemma," in *The Jews,* ed. Marshall Sklare (New York: The Free Press, 1958), pp. 542–550. Throughout the text, I occasionally use "stake in conformity" in speaking in general of the strength of the bond to conventional society. So used, the concept is somewhat broader than is true for either Toby or Briar and Piliavin, where the concept is roughly equivalent to what is here called "commitment."

12. Thomas Hobbes, *Leviathan* (Oxford: Basil Blackwell, 1957), p. 195.

13. Howard S. Becker, "Notes on the Concept of Commitment," *American Journal of Sociology* LXVI (1960): 35–36.

14. Arthur L. Stinchcombe, *Rebellion in a High School* (Chicago: Quadrangle, 1964), p. 5.

15. Richard A. Cloward and Lloyd E. Ohlin, *Delinquency and Opportunity* (New York: The Free Press, 1960), p. 147, quoting Edwin H. Sutherland, ed., *The Professional Thief* (Chicago: University of Chicago Press, 1937), pp. 211–213.

16. William James, *Psychology* (Cleveland: World Publishing Co., 1948), p. 186.

17. Few activities appear to be so engrossing that they rule out

contemplation of alternative lines of behavior, at least if estimates of the amount of time men spend plotting sexual deviations have any validity.

18. *The Sutherland Papers,* ed. Albert K. Cohen et al. (Bloomington: Indiana University Press, 1956), p. 37.

19. David Matza and Gresham M. Sykes, "Juvenile Delinquency and Subterranean Values," *American Sociological Review* XXVI (1961): 712–719.

20. *Ibid.,* p. 718.

21. The starving man stealing the loaf of bread is the image evoked by most strain theories. In this image, the starving man's belief in the wrongness of his act is clearly not something that must be explained away. It can be assumed to be present without causing embarrassment to the explanation.

22. McCord and McCord, *The Psychopath,* pp. 12–15.

23. Donald R. Cressey, *Other People's Money* (New York: The Free Press, 1953).

24. Gresham M. Sykes and David Matza, "Techniques of Neutralization: A Theory of Delinquency," *American Sociological Review* XXII (1957), 664–670.

25. David Matza, *Delinquency and Drift* (New York: Wiley, 1964), pp. 181–191.

26. In asserting that Cressey's assumption is invalid with respect to delinquency, I do not wish to suggest that it is invalid for the question of embezzlement, where the problem faced by the deviator is fairly specific and he can reasonably be assumed to be an upstanding citizen. (Although even here the fact that the embezzler's non-sharable financial problem often results from some sort of hanky-panky suggests that "verbalizations" may be less necessary than might otherwise be assumed.)

27. *Delinquency and Drift,* p. 43.

28. This assumption is not, I think, contradicted by the evidence presented by Matza against the existence of a delinquent subculture. In comparing the attitudes and actions of delinquents with the picture painted by delinquent subculture theorists, Matza emphasizes—and perhaps exaggerates—the extent to which delinquents are tied to the conventional order. In implicitly comparing delinquents with a supermoral man, I emphasize—and perhaps exaggerate—the extent to which they are not tied to the conventional order.

29. The position taken here is therefore somewhere between the "semantic dementia" and the "neutralization" positions. Assuming variation, the delinquent is, at the extremes, freer than the neutralization argument assumes. Although the possibility of wide discrepancy between what the delinquent professes and what he practices still exists, it is presumably much rarer than is suggested by studies of articulate "psychopaths."

10

Feminist Theory

MEDA CHESNEY-LIND

Chesney-Lind speaks strongly for the feminist perspective in pointing out that most theories of crime and deviance have focused overly on a male model of offending. She points out that girls and women hold a very different structural position in society and that the experience they are likely to encounter along with the opportunities (or lack thereof) they face are often markedly at odds with those of boys and men. Girls grow up in the United States facing vastly different sets of pressures, encounters, and forms of social control than their male counterparts.

Gender, notes Chesney-Lind, is a master status, which means that most of girls' experiences in society are filtered through this lens. From early youth, their role in society is affected by the structure of male domination. When we consider the effect of patriarchy on women, the special pathways that girls take into crime and deviance are illuminated. Girls are more vulnerable to physical and sexual abuse in the home than boys are, and girls lack the resources to rebuff or escape such abuse, not only interpersonally, within the family, but because of the double standard of sexuality and sexual control embedded within the juvenile justice system. Runaways who escape from abusive families are systematically returned right back into them. Girls who stay and remain subjected to further abuse often marry at a young age to get out, but frequently find themselves domestically abused by their partners. Girls who are strong enough to fight for survival escape abusive family contexts to the street, where their opportunities (both legally and within the world of crime) are extremely limited. Once again, then, they find themselves subject to the domination and abuse of male predators. In these situations, girls' background experiences and lack of other survival techniques lead them into adult situations of abuse to which they voluntarily and involuntarily subject themselves in order to stay fed, sheltered, and clothed.

This cycle of victimization to criminalization characterizes the gender-related pathway of girls into deviance and crime, and illustrates the structural disadvantage faced by younger and older women in society.

Where should we locate Chesney-Lind's feminist theory among the broader perspectives on deviance? Where is she placing the root causes of women's deviance: in society, subcultures, or individuals?

From *Crime & Delinquency*, 35(1), pp. 10–11, pp. 19–27. Copyright © 1989 by Sage Publications. Reprinted by permission of Sage Publications, Inc.

There is considerable question as to whether existing theories that were admittedly developed to explain male delinquency can adequately explain female delinquency. Clearly, these theories were much influenced by the notion that class and protest masculinity were at the core of delinquency. Will the "add women and stir approach" be sufficient? Are these really theories of delinquent behavior as some have argued?

This article will suggest that they are not. The extensive focus on male delinquency and the inattention to the role played by patriarchal arrangements in the generation of adolescent delinquency and conformity has rendered the major delinquency theories fundamentally inadequate to the task of explaining female behavior. There is, in short, an urgent need to rethink current models in light of girls' situation in patriarchal society.

… This discussion will also establish that the proposed overhaul of delinquency theory is not, as some might think, solely an academic exercise. Specifically, it is incorrect to assume that because girls are charged with less serious offenses, they actually have few problems and are treated gently when they are drawn into the juvenile justice system. Indeed, the extensive focus on disadvantaged males in public settings has meant that girls' victimization and the relationship between that experience and girls' crime has been systematically ignored. Also missed has been the central role played by the juvenile justice system in the sexualization of girls' delinquency and the criminalization of girls' survival strategies. Finally, it will be suggested that the official actions of the juvenile justice system should be understood as major forces in girls' oppression as they have historically served to reinforce the obedience of all young women to demands of patriarchal authority no matter how abusive and arbitrary….

TOWARD A FEMINIST THEORY OF DELINQUENCY

To sketch out completely a feminist theory of delinquency is a task beyond the scope of this article. It may be sufficient, at this point, simply to identify a few of the most obvious problems with attempts to adapt male-oriented theory to explain female conformity and deviance. Most significant of these is the fact that all existing theories were developed with no concern about gender stratification.

Note that this is not simply an observation about the power of gender roles (though this power is undeniable). It is increasingly clear that gender stratification in patriarchal society is as powerful system as is class. A feminist approach to delinquency means construction of explanations of female behavior that are sensitive to its patriarchal context. Feminist analysis of delinquency would also examine ways in which agencies of social control—the police, the courts, and the persons—act in ways to reinforce woman's place in male society. Efforts to construct a feminist model of delinquency must first and foremost be sensitive to the situations of girls. Failure to consider the existing empirical evidence on girls' lives and behavior can quickly lead to stereotypical thinking and theoretical dead ends.

An example of this sort of flawed theory building was the early fascination with the notion that the women's movement was causing an increase in women's crime, a notion that is now more or less discredited. A more recent example of the same sort of thinking can be found in recent work on the "power-control" model of delinquency. Here, the authors speculate that girls commit less delinquency in part because their behavior is more closely controlled by the patriarchal family. The authors' promising beginning quickly gets bogged down in a very limited definition of patriarchal control (focusing on parental supervision and variations in power within the family). Ultimately, the authors' narrow formulation of patriarchal control results in their arguing that mother's work force participation (particularly in high status occupations) leads to increases in daughters' delinquency since these girls find themselves in more "egalitarian families."

This is essentially a not-too-subtle variation on the earlier "liberation" hypothesis. Now, mother's liberation causes daughter's crime. Aside from the methodological problems with the study (e.g., the authors argue that female-headed households are equivalent to upper-class "egalitarian" families where both parents work, and they measure delinquency using a six-item scale that contains no status offense items), there is a more fundamental problem with the hypothesis. There is no evidence to suggest that as women's labor force participation has increased, girls' delinquency has increased. Indeed, during the last decade when both women's labor force participation accelerated and the number of female-headed households soared, aggregate female delinquency measured both by self-report and official statistics either declined or remained stable.

By contrast, a feminist model of delinquency would focus more extensively on the few pieces of information about girls' actual lives and the role played by girls' problems, including those caused by racism and poverty, in their delinquency behavior. Fortunately, a considerable literature is now developing on girls' lives and much of it bears directly on girls' crime.

CRIMINALIZING GIRLS' SURVIVAL

It has long been understood that a major reason for girls' presence in juvenile courts was the fact that their parents insisted on their arrest. In the early years, conflicts with parents were by far the most significant referral source; in Honolulu 44 percent of the girls who appeared in court in 1929 and 1930 were referred by parents.

Recent national data, while slightly less explicit, also show that girls are more likely to be referred to court by "sources other than law enforcement agencies" (which would include parents). In 1983, nearly a quarter (23 percent) of all girls but only 16 percent of boys charged with delinquent offenses were referred to court by non–law enforcement agencies. The pattern among youth referred for status offenses (for which girls are overrepresented) was even more pronounced. Well over half (56 percent) of the girls charged with these offenses and 45 percent of the boys were referred by sources other than law enforcement.

The fact that parents are often committed to two standards of adolescent behavior is one explanation for such a disparity—and one that should not be discounted as a major source of tension even in modern families. Despite expectations to the contrary, gender-specific socialization patterns have not changed very much and this is especially true for parents' relationships with their daughters. It appears that even parents who oppose sexism in general feel "uncomfortable tampering with existing traditions" and "do not want to risk their children becoming misfits." Clearly, parental attempts to adhere to and enforce these traditional notions will continue to be a source of conflict between girls and their elders. Another important explanation for girls' problems with their parents, which has received attention only in more recent years, is the problem of physical and sexual abuse. Looking specifically at the problem of childhood sexual abuse, it is increasingly clear that this form of abuse is a particular problem for girls.

Girls are, for example, much more likely to be the victims of child sexual abuse than are boys. Finkelhor and Baron estimate from a review of community studies that roughly 70 percent of the victims of sexual abuse are female, they are more likely than boys to be assaulted by a family member (often a stepfather), and, as a consequence, their abuse tends to last longer than male sexual abuse. All of these factors are associated with more severe trauma—causing dramatic short- and long-term effects in victims. The effects noted by researchers in this area move from the more well known "fear, anxiety, depression, anger and hostility, and inappropriate sexual behavior" to behaviors of greater familiarity to criminologists, including running away from home, difficulties in school, truancy, and early marriage. Herman's study of incest survivors in therapy found that they were more likely to have run away from home than a matched sample of women whose fathers were "seductive" (33 percent compared to 5 percent). Another study of women patients found that 50 percent of the victims of child sexual abuse, but only 20 percent of the nonvictim group, had left home before the age of 18.

Not surprisingly, then, studies of girls on the streets or in court populations are showing high rates of both physical and sexual abuse. Silbert and Pines (1981, p. 409) found, for example, that 60 percent of the street prostitutes they interviewed had been sexually abused as juveniles. Girls at an Arkansas diagnostic unit and school who had been adjudicated for either status or delinquent offenses reported similarly high levels of physical abuse; 53 percent indicated they had been sexually abused, 25 percent recalled scars, 38 percent recalled bleeding from abuse, and 51 percent recalled bruises. A sample survey of girls in the juvenile justice system in Wisconsin revealed that 79 percent had been subjected to physical abuse that resulted in some form of injury, and 32 percent had been sexually abused by parents or other persons who were closely connected to their families. Moreover, 50 percent had been sexually assaulted ("raped" or forced to participate in sexual acts). Even higher figures were reported by McCormack and her associates in their study of youth in a runaway shelter in Toronto. They found that 73 percent of the females and 38 percent of the males had been sexually abused. Finally, a study of youth charged with running away, truancy, or listed as missing persons in Arizona found that 55 percent were incest victims.

Many young women, then, are running away from profound sexual victimization at home, and once on the streets they are forced further into crime in order to survive. Interviews with girls who have run away from home show, very clearly, that they do not have a lot of attachment to their delinquent activities. In fact, they are angry about being labeled as delinquent, yet all engaged in illegal acts. The Wisconsin study found that 54 percent of the girls who ran away found it necessary to steal money, food, and clothing in order to survive. A few exchanged sexual contact for money, food, and/or shelter. In their study of runaway youth, McCormack, Janus, and Burgess found that sexually abused female runaways were significantly more likely than their nonabused counterparts to engage in delinquent or criminal activities such as substance abuse, petty theft, and prostitution. No such pattern was found among male runaways.

Research on the backgrounds of adult women in prison underscores the important links between women's childhood victimizations and their later criminal careers. The interviews revealed that virtually all of this sample were the victims of physical and/or sexual abuse as youngsters; over 60 percent had been sexually abused and about half had been raped as young women. This situation promoted these women to run away from home (three-quarters had been arrested for status offenses) where once on the streets they began engaging in prostitution and other forms of petty property crime. They also begin what becomes a lifetime problem with drugs. As adults, the women continue in these activities since they possess truncated educational backgrounds and virtually no marketable occupational skills.

Confirmation of the consequences of childhood sexual and physical abuse on adult female criminal behavior has also recently come from a large quantitative study of 908 individuals with substantiated and validated histories of these victimizations. Widom (1988) found that abused or neglected females were twice as likely as a matched group of controls to have an adult record (16 percent compared to 7.5). The difference was also found among men, but it was not as dramatic (42 percent compared to 33 percent). Men with abuse backgrounds were also more likely to contribute to the "cycle of violence" with more arrests for violent offenses as adult offenders than the control group. In contrast, when women with abuse backgrounds did become involved with the criminal justice system, their arrests tended to involve property and order offenses (such as disorderly conduct, curfew, and loitering violations).

Given this information, a brief example of how a feminist perspective on the causes of female delinquency might look seems appropriate. First, like young men, girls are frequently the recipients of violence and sexual abuse. But unlike boys, girls' victimization and their response to that victimization is specifically shaped by their status as young women. Perhaps because of the gender and sexual scripts found in patriarchal families, girls are much more likely than boys to be victims of family-related sexual abuse. Men, particularly men with traditional attitudes toward women, are likely to define their daughters or stepdaughters as their sexual property. In a society that idealizes inequality in male/female relationships and venerates youth in women, girls are easily defined as sexually attractive by older men. In addition, girls' vulnerability to both physical and sexual abuse is heightened by norms that require that they stay at home where their victimizers have access to them.

Moreover, their victimizers (usually males) have the ability to invoke official agencies of social control in their efforts to keep young women at home and vulnerable. That is to say, abusers have traditionally been able to utilize the uncritical commitment of the juvenile justice system toward parental authority to force girls to obey them. Girls' complaints about abuse were, until recently, routinely ignored. For this reason, statutes that were originally placed in law to "protect" young people have, in the case of girls' delinquency, criminalized their survival strategies. As they run away from abusive homes, parents have been able to employ agencies to enforce their return. If they persisted in their refusal to stay in that home, however intolerable, they were incarcerated.

Young women, a large number of whom are on the run from homes characterized by sexual abuse and parental neglect, are forced by the very statutes designed to protect them into the lives of escaped convicts. Unable to enroll in school or take a job to support themselves because they fear detection, young female runaways are forced into the streets. Here they engage in panhandling, petty theft, and occasional prostitution in order to survive. Young women in conflict with their parents (often for very legitimate reasons) may actually be forced by present laws into petty criminal activity, prostitution, and drug use.

In addition, the fact that young girls (but not necessarily young boys) are defined as sexually desirable and, in fact, more desirable than their older sisters due to the double standard of aging means that their lives on the streets (and their survival strategies) take on unique shape—once again shaped by patriarchal values. It is no accident that girls on the run from abusive homes, or on the streets because of profound poverty, get involved in criminal activities that exploit their sexual object status. The U.S. society has defined as desirable youthful, physically perfect women. This means that girls on the streets, who have little else of value to trade, are encouraged to utilize this "resource." It also means that the criminal subculture views them from this perspective.

FEMALE DELINQUENCY, PATRIARCHAL AUTHORITY, AND FAMILY COURTS

The early insights into male delinquency were largely gleaned by intensive field observation of delinquent boys. Very little of this sort of work has been done in the case of girls' delinquency, though it is vital to an understanding of girls' definitions of their own situations, choices, and behavior. Time must be spent listening to girls. Fuller research on the settings, such as families and schools, that girls find themselves in and the impact of variations in those settings should also be undertaken. A more complete understanding of how poverty and racism shape girls' lives is also vital.

Finally, current qualitative research on the reaction of official agencies to girls' delinquency must be conducted. This latter task, admittedly more difficult, is particularly critical to the development of delinquency theory that is as sensitive to gender as it is to race and class.

It is clear that throughout most of the court's history, virtually all female delinquency has been placed within the larger context of girls' sexual behavior. One explanation for this pattern is that familial control over girls' sexual capital has historically been central to the maintenance of patriarchy. The fact that young women have relatively more of this capital has been one reason for the excessive concern that both families and official agencies of social control have expressed about youthful female defiance (otherwise much of the behavior of criminal justice personnel makes virtually no sense). Only if one considers the role of women's control over their sexuality at the point in their lives that their value to patriarchal society is so pronounced, does the historic pattern of jailing of huge numbers of girls guilty of minor misconduct make sense.

This framework also explains the enormous resistance that the movement to curb the juvenile justice system's authority over status offenders encountered. Supporters of this change were not really prepared for the political significance of giving youth the freedom to run. Horror stories told by the opponents of deinstitutionalization about victimized youth, youthful prostitution, and youthful involvement in pornography all neglected the unpleasant reality that most of these behaviors were often in direct response to earlier victimization, frequently by parents, that officials had, for years, routinely ignored. What may be at stake in efforts to roll back deinstitutionalization efforts is not so much "protection" of youth as it is curbing the right of young women to defy patriarchy.

In sum, research in both the dynamics of girls' delinquency and official reactions to that behavior is essential to the development of theories of delinquency that are sensitive to its patriarchal as well as class and racial context.

REFERENCES

Silbert, Mimi. H., and Ayah Pines. (1981). "Sexual Child Abuse as an Antecedent to Prostitution." *International Journal of Child Abuse ana Neglect* 5:407–411.

Widom, Cathy Spatz. (1988). "Sampling Biases and Implications for Child Abuse Research." *American Journal of Orthopsychiatry* 58(2):260–270.

11

The Constructionist Stance

JOEL BEST

Best, one of the leading practitioners of the social constructionist approach, offers a historical analysis of the way theories of deviance have evolved. He notes the rise and decline in significance of several approaches to deviance theory, showing how social constructionism arose from its early roots in the sociology of deviance and moved into explaining social problems. The constructionist perspective represents a wedding of the views of labeling and conflict theories, as we noted in the general introduction to this text. By looking at how individuals encounter societal reactions and become labeled as deviants, the constructionist approach joins the microanalysis of labeling theory with the broader, more structural, social power contribution of conflict theory, isolating certain groups as more likely to have social reactions and definitions formed and applied by and against them. It is this social constructionist stance that will frame the organization and selections making up the remainder of the book. We begin our investigation therein by examining the process by which groups vie for power in society and try to legislate their views into morality. Then we focus on how people develop deviant identities and manage their stigma as a result of the definitions and enforcements which come out of that legislation.

What parts of Best's analysis of the social construction of social problems can be applied to defining deviance?

What does it mean to say that deviance is "socially constructed?" Some people assume that social construction is the opposite of real, but this is a mistake. Reality, that is everything we understand about the world, is socially constructed. The term calls attention to the processes by which people make sense of the world: we create—or construct—meaning. When we define some behavior as deviant, we are socially constructing deviance. The constructionist approach recognizes that people can only understand the world in terms of words and categories that they create and share with one another.

THE EMERGENCE OF CONSTRUCTIONISM

The constructionist stance had its roots in two developments. The first was the publication of Peter L. Berger and Thomas Luckmann's (1966) *The Social Construction of Reality*. Berger and Luckmann were writing about the sociology of

Reprinted by permission of the author.

knowledge—how social life shapes everything that people know. Their book introduced the term "social construction" to a wide sociological audience, and soon other sociologists were writing about the construction of science, news, and other sorts of knowledge, including what we think about deviance.

Second, labeling theory, which had become the leading approach to studying deviance during the 1960s, came under attack from several different directions by the mid-1970s. Conflict theorists charged that labeling theory ignored how elites shaped definitions of deviance and social control policies. Feminists complained that labeling ignored the victimization of women at the hands of both male offenders and male-dominated social control agencies. Activists for gay rights and disability rights insisted that homosexuals and the disabled should be viewed as political minorities, rather than deviants. At the same time, mainstream sociologists began challenging labeling theory's claims about the ways social control operated and affected deviants' identities.

THE CONSTRUCTIONIST RESPONSE

In response to these attacks, some sociologists sympathetic to the labeling approach moved away from studying deviance. Led by John I. Kitsuse, a sociologist whose work had helped shape labeling theory, these sociologists of deviance turned to studying the sociology of social problems. With Malcolm Spector, Kitsuse published *Constructing Social Problems* (1977)—a book that would inspire many sociologists to begin studying how and why particular social problems emerged as topics of public concern. They argued that sociologists ought to redefine social problems as claims that various conditions constituted social problems; therefore, the constructionist approach involved studying claims and those who made them—the claimsmakers. In this view, sociologists ought to study how and why particular issues such as date rape or binge drinking on college campuses suddenly became the focus of attention and concern. How were these problems constructed?

There were several advantages to studying social problems. First, constructionists had the field virtually to themselves. Although many sociology departments taught social problems courses, there were no rival well-established, coherent theories of social problems. In contrast, labeling had to struggle against functionalism, conflict theory, and other influential approaches to studying deviance.

The constructionist approach was also flexible. Analysts of social problems construction might concentrate on various actors: some examined the power of political and economic elites in shaping definitions of social problems; others focused on the role of activists in bringing attention to problems; and still others concentrated on how media coverage shaped the public's and policymakers' understandings of problems. This flexibility meant that constructionists might criticize some claims as exaggerated, distorted, or unfounded (the sort of critique found in several studies of claims about the menace of Satanism), but they might also celebrate the efforts of claimsmakers to draw attention to neglected problems (e.g., researchers tended to treat claims about domestic violence sympathetically).

Again, it is important to appreciate that "socially constructed" is not a synonym for erroneous or mistaken. All knowledge is socially constructed; to say that a social problem is socially constructed is not to imply that it does not exist, but rather that it is through social interaction that the problem is assigned particular meanings.

THE RETURN TO DEVIANCE

Although constructionists studied the emergence and evolution of many different social problems, ranging from global warming to homelessness, much of their work remained focused on deviance. They studied the construction of rape, child abduction, illicit drugs, family violence, and other forms of deviance.

Closely related to the rise of constructionism were studies of medicalization (Conrad and Schneider, 1980). Medicalization—defining deviance as a form of illness requiring medical treatment—was one popular, contemporary way of constructing deviance. By the end of the twentieth century, medical language— "disease," "symptom," "therapy," and so on—was used, not only by medical authorities, but even by amateurs (e.g., in the many Twelve-Step programs of the recovery movement).

A large share of constructionist studies traced the rise of social problems to national attention; for example, the construction of the federal War on Drugs was studied by several constructionist researchers. However, other sociologists began studying how deviance was constructed in smaller settings, through interpersonal interaction. In particular, they examined social problems work (Holstein and Miller, 1993). Even after claimsmakers have managed to draw attention to some social problem and shape the creation of social policies to deal with it, those claims must be translated into action. Police officers, social workers, and other social problems workers must apply broad constructions to particular cases. Thus, after wife abuse is defined as a social problem, it is still necessary for the police officer investigating a domestic disturbance call to define—or construct—these particular events as an instance of wife abuse (Loseke, 1992). Studies of this sort of social problems work are a continuation of earlier research on the labeling process.

CONSTRUCTIONISM'S DOMAIN

Social constructionism, then, has become an influential stance for thinking about deviance, particularly for understanding how concerns about particular forms of deviance emerge and evolve, and for studying how social control agents construct particular acts as deviance and individuals as deviants. Constructionism emphasizes the role of interpretation, of people assigning meaning, or making sense of the behaviors they classify as deviant. This can occur at a societal level, as when the mass media draw attention to a new form of deviance and legislators pass laws against it, and it can also occur in face-to-face interaction, when one individual expresses disapproval of another's rule breaking. Deviance, like all reality, is constantly being constructed.

REFERENCES

Berger, Peter L., and Thomas Luckmann. (1966). *The Social Construction of Reality*. New York: Doubleday.

Conrad, Peter, and Joseph W. Schneider. (1980). *Deviance and Medicalization*. St. Louis: Mosby.

Holstein, James A., and Gale Miller. (1993). "Social Constructionist and Social Problems Work." Pp. 151–72. *Reconsidering Social Constructionism*, edited by James Holstein and Gale Miller. Hawthorne, NY: Aldine de Gruyter.

Loseke, Donileen R. (1992). *The Battered Woman and Shelters*. Albany: State University of New York Press.

Spector, Malcolm, and John I. Kitsuse. (1977). *Constructing Social Problems*. Menlo Park, California: Cummings.

PART III

Studying Deviance

Accurate and reliable knowledge about deviance is critically important to many groups in society. First, policy makers are concerned with deviant groups such as the homeless and transient, the chronically mentally ill, high school dropouts, criminal offenders, prostitutes, juvenile delinquents, gang members, runaways, and other members of disadvantaged and disenfranchised populations. These people pose social problems that lawmakers and social welfare agencies want to alleviate. Second, sociologists and other researchers have an interest in deviance based on their goal of understanding human nature, human behavior, and human society. Deviants are a critical group to this enterprise because they reside near the margins of social definition: They help define the boundaries of what is acceptable and what is unacceptable by given groups.

In this part, we will discuss information about deviance coming from three primary data sources: official statistics, survey research, and field research. **Official statistics** are numerical tabulations compiled by government officials and employees of social service agencies (often, those receiving financial grants or assistance from the government) in the course of doing their jobs. These people routinely collect information about their clients as they process them and combine them into statistics about deviance. This information includes arrest data that are compiled by the police and published by the FBI (the *Uniform Crime Reports*), census data on various shifting populations (such as the homeless), victim data from helping agencies (such as shelters for battered women), medical data from emergency rooms (such as DAWN, the Drug Abuse Warning Network) or from state public health agencies (such as the offices of the coroner and medical examiner), and prosecution data on cases that are tried in the courts. These official statistics are then compiled by the various government organizations responsible for collecting them and are made available to the public.

Official statistics are connected to the absolutist perspective on defining deviance because they are considered an objective source of measurement.

Another source of statistical data about deviance is **survey research**. Sometimes sociologists want information about behavior that is not routinely collected by the government or official agencies. For instance, when we wanted to find out how emergency room admissions for self-injury (e.g., cutting, burning, and branding oneself) varied over time, we were unable to do so, because the only category emergency rooms use to document such cases is "self-inflicted injury," which also includes (and is made up mostly of) suicide attempts. Emergency rooms simply do not differentiate between these very different types of acts, and therefore there are no official statistics on hospital admissions for self-injury. Accordingly, rather than being limited to official sources of data or to categories that officials use, many sociologists choose to gather their own data. One very popular way is to use large-scale questionnaire surveys. Prominent ones include the National Youth Survey, a self-report questionnaire about delinquent behavior; "Monitoring the Future," an annual survey from the Institute for Social Research at the University of Michigan on the drug use of high school seniors (a survey in which many of you reading this book have probably participated); and some of the Kinsey surveys about sexual behavior. Survey research can inquire into instances of various behaviors, but it can also collect information about people's attitudes. Survey research is particularly prevalent in the political domain, where we see opinion polls conducted all the time.

A third kind of information, richly descriptive and analytical rather than numerical, comes from sociologists who conduct **field research** (also called participant observation) on deviance. Much like anthropologists who go out to live among native peoples, sociological fieldworkers live with members of deviant groups and become intimately familiar with their lives. They hang around with people, befriend them, become part of their lives, and learn how and why they do things. They learn what their subjects think about things, how their attitudes and behaviors change over time, and how they resolve the many contradictions people find in life. This type of research yields information more deeply based in the subjects' own perspectives, detailing how they see the world, the allure of deviance for them, the problems they encounter, the ways they resolve those problems, the significant individuals and groups in their lives, and their role among these others. Unlike other types of research, participant observation is generally a longitudinal method, entailing years of involvement with subjects. Researchers must gain acceptance by group members, develop meaningful relationships with them, and learn about their deepest thoughts and emotions.

There are many differences among these types of data and among the methods used to gather them. Each has its advantages and disadvantages, and each may do a better job than the others of answering certain research questions. Thus, depending on people's particular needs, they may turn to one type of data or use a mixed-method approach. Official statistics have the advantage of being inexpensive to gather and quick to access, because they are already collected and published. Moreover, they aim to include the entire population of those they address—for example, *all* criminals, *all* victims, and *all* emergency room admittees—not just some subsample thereof. In addition, records about these occurrences can be accessed as far back as the official statistics have been collected, potentially a rather long time. Yet, official statistics have certain validity problems and tend to be inaccurate in patterned and systematic ways.

Official police statistics, the *Uniform Crime Reports*, for example, fail to include a host of crimes for several reasons: Crimes may be unrecognized by victims who do not notice their occurrence or who lack the power to define them as deviant; crimes may be unreported by individuals who see no gain by, or who fear embarrassment, censure, or retaliation from, calling police attention to their victimization; or crimes may be unrecorded by police officers who use their broad discretion to handle problems informally. Official statistics on suicide, determined and collected by coroners and medical examiners, also fluctuate (Whitt, 2006). They may be unreliable in rural areas because officials know families, making them loath to render a verdict on a cause of death that would stigmatize a family or impede the family's collecting an insurance settlement. In urban areas, the collection of official statistics may rise and fall because of political pressures, turnovers in personnel, fluctuating resources, or policy changes and the statistics themselves may be suppressed by placing them under other labels (as when statistics on drug overdoses are placed into the category of "indeterminate causes") or may be "found" (as when statistics on single-car accidents are ruled as suicides).

Other types of official statistics vastly underrepresent criminal activity for similar and other reasons, and may be problematic because the categories used to conceive of them and the way they are assembled and reported change over time, making comparisons over the years frustrating. In sum, although official statistics yield information about a broad spectrum of people, they may be fairly shallow and unreliable in nature. Besharov and Laumann-Billings's chapter on child abuse statistics (Chapter 12) discusses the dramatic rise in the number of reported cases of child abuse and some of the sociological factors that have accounted for this wild swing in the official statistics. Here, the authors examine factors that artificially both inflate and deflate our official estimates of child abuse and consider the consequences of those factors for the protection of abused children.

Survey research lets social scientists collect data on topics of their choice, but it is an expensive and time-consuming enterprise. Still, through careful sampling procedures, researchers can gather data about a smaller population and generalize from it to a much larger group to a high degree of accuracy (external reliability). Strict controls over the standardization of procedures and the detachment of data gatherers make this method a relatively objective one. Correlations (although not causal relations) between social factors can be established carefully. But survey research, like official statistics, has internal validity problems: Chiefly, it may not yield an accurate portrait of the sample group it is studying, especially with topics as sensitive as deviance. To elaborate, first, it is unlikely that people—especially deviants—will fill out a questionnaire and readily disclose information about the hidden aspects of their lives. Second, in responding to the questionnaire, subjects may not define their behavior the same way or use the same terms as the researchers who are writing the questions (e.g., prostitutes' conceptions of a "date" may be different from those of survey researchers, and runaways may mean different things when they refer to their "home" than researchers intend). Researchers are then likely to misinterpret the nature and extent of behavior from the answers they receive. Third, sometimes the correlations produced by survey research—that is, what trends occur together (such as deviance and divorce, or violence in the media and violence in everyday life)—are mistakenly assumed to be causal connections. But survey research cannot tell us *why* or *how* people act; it can only tell us *what* people are doing, even if these trends range across a broad spread of the population. A much-heralded study of sexual behavior is featured in Chapter 13, Laumann and colleagues' description of their survey, a major, highly professionally designed and conducted study that gives us a glimpse into the problems and creative adaptations that can arise when a comprehensive effort is made to conduct large-scale survey research into Americans' sexual practices.

Field research, in contrast, cannot reach as many people, but yields deeper and more accurate information about the people studied, information that is backed by researchers' own direct observations, to enhance its internal validity. Participant observers spend long amounts of time in the field, becoming close to the people they study and learning how their subjects perceive, interpret, and act upon the complex and often contradictory nature of their social worlds. In contrast to the detached and objective relationships between survey researchers and their subjects, field researchers rely on the subjectivity and strength of the close personal relationships they forge with people they study in order to get behind false fronts and to find out what is really going on. An in-depth understanding is especially important in studying a topic such as deviance because so much behavior is hidden due to its stigma and illicit status. Also critically important is the ability of field research to study deviance as it occurs *in situ*, in its natural setting, not via the structural constraints of police reporting or the interpretation and recollection of questionnaire research (Polsky, 1967). Although often less costly than survey research to conduct, field research is very time consuming, as building a rapport and trust with subjects takes a long time to develop. Field research also lacks the generalizability of careful survey research, as subjects tend to be gathered through a referral ("snowball") technique or because they are members of a common "scene." The assurance of randomization and objectivity, then, is not as strong as it is in other methodologies. In Chapter 14, on field research, we share with readers our own experiences with participant observation, talking about what it is like to carry out such research with a criminal, and potentially dangerous, group.

Figure 1 presents a comparison of the strengths and weakness of the preceding three methods or sources of data.

FIGURE 1   Strengths and Weakness of the Three Sources of Data

Category	Survey Research	Field Research	Official Statistics
Cost	High	Low	Free
Time	Medium	Long	Short
Approach	Objective	Subjective	Clerical
Generalizability	High	Low	High
Accuracy	Medium	High	Low

The empirical chapters that fill the remainder of this book are based primarily on participant observation studies of deviance, for two main reasons. First, as Becker (1973) remarked, participant observation is the method of the interactionist perspective; it offers direct access to the way definitions and laws are socially constructed, to the way people's actions are influenced by their associates, and to the way people's identities are affected by the deviant labels cast on them. Second, these types of studies offer a deeper view of people's feelings, experiences, motivations, and social-psychological states—a richer and more vivid portrayal of deviance than any charts or numbers can reveal.

12

Child Abuse Reporting

DOUGLAS J. BESHAROV WITH LISA A. LAUMANN-BILLINGS

Besharov and Laumann-Billings discuss official statistics in our first chapter on the varieties of ways that deviance is studied. They note the spectacular rise in our official rates of child abuse, with figures increasing by 300 percent over a recent 30-year period. Such a dramatic change cannot be attributable solely to changes in deviant behavior, but must also involve a measurement artifact. The authors root the increase in child abuse statistics in three factors: mandatory reporting laws, media campaigns surrounding child abuse, and the changed social definition of what constitutes abuse. Besharov and Laumann-Billings discuss two ironically opposing problems associated with child abuse statistics: unreported cases and unsubstantiated cases. On the one hand, they claim, we are still unaware of many cases of child abuse because the abuse tends to be hidden, defined as a private family matter, and regarded as "normal" child-rearing practice.

On the other hand, the way we, as a society, tumultuously attacked this "discovered" social problem and deputized numerous social groups to document it resulted in cases that could not be substantiated. Some of these cases were unfounded because they were investigated and found to be lacking in substance; others, however, were unprovable because the families could not be located, the child abuse was able to remain hidden even in the face of investigation, or the huge increase in the number of cases requiring investigation overburdened the dockets of social service agencies and diminished their ability to resolve all allegations. Some desperate situations are being attended to, but others are slipping through the cracks because of overreporting problems. All of these cases, both underreported and overreported, signal continued ambiguity over definitions of child abuse.

Together, the problems encountered shed light on some of the enormous inaccuracies that are inherent in the official statistics gathered by social welfare agents. Official statistics have a number of advantages: They are inexpensive and easy to collect (they are collected by people in the course of doing their jobs); they go backward in time over long periods; and, unlike sampling, they contain information about the entire population of interest. But official statistics are notoriously inaccurate, relying for their credibility on the impartiality of disinterested data gatherers, always subject to personal and political pressures aimed at skewing official reports, and beseeching public agencies to stretch their limited resources to collect the relevant data.

What do Besharov and Laumann-Billings suggest about the relative strengths and weaknesses of these kinds of official statistics?

For 30 years, advocates, program administrators, and politicians have joined to encourage even more reports of suspected child abuse and neglect. Their efforts have been spectacularly successful, with about three million cases of suspected child abuse having been reported in 1993. Large numbers of endangered children still go unreported, but an equally serious problem has developed: Upon investigation, as many as 65 percent of the reports now being made are determined to be "unsubstantiated," raising serious civil liberties concerns and placing a heavy burden on already overwhelmed investigative staffs.

These two problems—nonreporting and inappropriate reporting—are linked and must be addressed together before further progress can be made in combating child abuse and neglect. To lessen both problems, there must be a shift in priorities—away from simply seeking more reports and toward encouraging better reports.

REPORTING LAWS

Since the early 1960s, all states have passed laws that require designated professionals to report specified types of child maltreatment. Over the years, both the range of designated professionals and the scope of reportable conditions have been steadily expanded.

Initially, mandatory reporting laws applied only to physicians, who were required to report only "serious physical injuries" and "nonaccidental injuries." In the ensuing years, however, increased public and professional attention, sparked in part by the number of abused children revealed by these initial reporting laws, led many states to expand their reporting requirements. Now almost all states have laws that require the reporting of all forms of suspected child maltreatment, including physical abuse, physical neglect, emotional maltreatment, and of course, sexual abuse and exploitation.

Under threat of civil and criminal penalties, these laws require most professionals who serve children to report suspected child abuse and neglect. About twenty states require all citizens to report, but in every state, any citizen is permitted to report.

These reporting laws, associated public awareness campaigns, and professional education programs have been strikingly successful. In 1993, there were about three million reports of children suspected of being abused or neglected. This is a twenty-fold increase since 1963, when about 150,000 cases were reported to the authorities. (As we will see, however, [the three million] figure is bloated by reports that later turn out to be unfounded.)

Many people ask whether this vast increase in reporting signals a rise in the incidence of child maltreatment. Recent increases in social problems such as out-of-wedlock births, inner-city poverty, and drug abuse have probably raised the underlying rates of child maltreatment, at least somewhat. Unfortunately, so many maltreated children previously went unreported that earlier reporting statistics do not provide a reliable baseline against which to make comparisons. One thing is clear, however: The great bulk of reports now received by child protective agencies would not be made but for the passage of mandatory reporting laws and the media campaigns that accompanied them.

This increase in reporting was accompanied by a substantial expansion of prevention and treatment programs. Every community, for example, is now served by specialized child protective agencies that receive and investigate reports. Federal and state expenditures for child protective programs and associated foster care services now exceed $6 billion a year. (Federal expenditures for foster care, child welfare, and related services make up less than 50 percent of total state and federal expenditures for these services; in 1992, they amounted to a total of $2,773.7 million. In addition, states may use a portion of the $2.8 billion federal Social Services Block Grant for such services, though detailed data on these expenditures are not available. Beginning in 1994, additional federal appropriations funded family preservation and support services.)

As a result, many thousands of children have been saved from serious injury and even death. The best estimate is that over the past twenty years, child abuse and neglect deaths have fallen from over 3,000 a year—and perhaps as many as 5,000—to about 1,000 a year. In New York State, for example, within five years of the passage of a comprehensive reporting law, which also created specialized investigative staffs, there was a 50 percent reduction in child fatalities, from about two hundred a year to less than one hundred. (This is not meant to minimize the remaining problem. Even at this level, maltreatment is the sixth largest cause of death for children under fourteen.)

UNREPORTED CASES

Most experts agree that reports have increased over the past 30 years because professionals and laypersons have become more likely to report apparently abusive and neglectful situations. But the question remains: How many more cases still go unreported?

Two studies performed for the National Center on Child Abuse and Neglect by Westat, Inc., provide a partial answer. In 1980 and then again in 1986, Westat conducted national studies of the incidence of child abuse and neglect. (A third Westat incidence study is now underway.) Each study used essentially the same methodology: In a stratified sample of counties, a broadly representative sample of professionals who serve children was asked whether, during the study period, the children they had seen in their professional capacities appeared to have been abused or neglected. (Actually, the professionals were not asked the ultimate question of whether the children appeared to be "abused" or "neglected." Instead, they were asked to identify children with certain specified harms or conditions, which were then decoded into a count of various types of child abuse and neglect.)

Because the information these selected professionals provided could be matched against pending cases in the local child protective agency, Westat was able to estimate rates of nonreporting among the surveyed professionals. It could not, of course, estimate the level of unintentional nonreporting, since there is no way to know of the situations in which professionals did not recognize signs of possible maltreatment. There is also no way to know how many children the professionals recognized as being maltreated but chose not to report to the study. Obviously, since the study methodology involved asking professionals about children they had seen in their professional capacities, it also did not allow Westat to estimate the number of children seen by nonprofessionals, let alone their nonreporting rate.

Westat found that professionals failed to report many of the children they saw who had observable signs of child abuse and neglect. Specifically, it found that in 1986, 56 percent of apparently abused or neglected children, or about 500,000 children, were not reported to the authorities. This figure, however, seems more alarming than it is: Basically, the more serious the case, the more likely the report. For example, the surveyed professionals reported over 85 percent of the fatal or serious physical abuse cases they saw, 72 percent of the sexual abuse cases, and 60 percent of the moderate physical abuse cases. In contrast, they only reported 15 percent of the educational neglect cases they saw, 24 percent of the emotional neglect cases, and 25 percent of the moderate physical neglect cases.

Nevertheless, there is no reason for complacency. Translating these raw percentages into actual cases means that in 1986, about 2,000 children with observable physical injuries severe enough to require hospitalization were not reported and that more than 100,000 children with moderate physical injuries went unreported, as did more than 30,000 apparently sexually abused children. And these are the rates of nonreporting among relatively well-trained professionals. One assumes that nonreporting is higher among less-well-trained professionals and higher still among laypersons.

Obtaining and maintaining a high level of reporting requires a continuation of the public education and professional training begun 30 years ago. But, now, such efforts must also address a problem as serious as nonreporting: inappropriate reporting.

At the same time that many seriously abused children go unreported, an equally serious problem further undercuts efforts to prevent child maltreatment: The nation's child protective agencies are being inundated by inappropriate reports. Although rules, procedures, and even terminology vary—some states use the phrase "unfounded," others "unsubstantiated" or "not indicated"—an "unfounded" report, in essence, is one that is dismissed after an investigation finds insufficient evidence upon which to proceed.

UNSUBSTANTIATED REPORTS

Nationwide, between 60 and 65 percent of all reports are closed after an initial investigation determines that they are "unfounded" or "unprovable." This is in sharp contrast to 1974, when only about 45 percent of all reports were unfounded. Unfounded cases are those where investigation occurs and it is determined that the child abuse did not occur. Unprovable cases are those that quite possibly did occur, but where definitive proof was not able to be obtained.

A few advocates, in a misguided effort to shield child protective programs from criticism, have sought to quarrel with estimates that I and others have made that the national unfounded rate is between 60 and 65 percent. They have grasped at various inconsistencies in the data collected by different organizations to claim either that the problem is not so bad or that it has always been this bad.

To help settle this dispute, the American Public Welfare Association (APWA) conducted a special survey of child welfare agencies in 1989. The APWA researchers found that between fiscal years 1986 and 1988, the weighted average for the substantiation rates in 31 states declined 6.7 percent—from 41.8 percent in fiscal year 1986 to 39 percent in fiscal year 1988.

Most recently, the existence of this high unsubstantiated rate was reconfirmed by the annual Fifty State Survey of the National Committee to Prevent Child Abuse (NCPCA), which found that in 1993 only about 34 percent of the reports received by child protective agencies were substantiated.

The experience of New York City indicates what these statistics mean in practice. Between 1989 and 1993, as the number of reports received by the city's child welfare agency increased by over 30 percent (from 40,217 to 52,472), the percentage of substantiated reports fell by about 47 percent (from 45 percent to 24 percent). In fact, the number of substantiated cases—a number of families were reported more than once—actually fell by about 41 percent, from 14,026 to 8,326. Thus, 12,255 additional families were investigated, while 5,700 fewer families received child protective help.

The determination that a report is unfounded can only be made after an unavoidably traumatic investigation that is inherently a breach of parental and family privacy. To determine whether a particular child is in danger, caseworkers must inquire into the most intimate personal and family matters. Often it is necessary to question friends, relatives, and neighbors, as well as school teachers, day-care personnel, doctors, clergy, and others who know the family.

Laws against child abuse are an implicit recognition that family privacy must give way to the need to protect helpless children. But in seeking to protect children, it is all too easy to ignore the legitimate rights of parents. Each year, about 700,000 families are put through investigations of unfounded reports. This is a massive and unjustified violation of parental rights.

Few unsubstantiated reports are made maliciously. Studies of sexual abuse reports, for example, suggest that, at most, from 4 to 10 percent of these reports are knowingly false. Many involve situations in which the person reporting, in a well-intentioned effort to protect a child, overreacts to a vague and often misleading possibility that the child may be maltreated. Others involve situations of poor child care that, though of legitimate concern, simply do not amount to child abuse or neglect. In fact, a substantial proportion of unfounded cases are referred to other agencies for them to provide needed services for the family.

Moreover, an unsubstantiated report does not necessarily mean that the child was not actually abused or neglected. Evidence of child maltreatment is hard to obtain and might not be uncovered when agencies lack the time and resources to complete a thorough investigation or when inaccurate information is given to the investigator. Other cases are labeled unfounded when no services are available to help the family. Some cases must be closed because the child or family cannot be located.

A certain proportion of unsubstantiated reports, therefore, is an inherent—and legitimate—aspect of reporting *suspected* child maltreatment and is necessary to ensure adequate child protection. Hundreds of thousands of strangers report their suspicions; they cannot all be right. But unfounded rates of the current magnitude go beyond anything reasonably needed. Worse, they endanger children who are really abused.

The current flood of unsubstantiated reports is overwhelming the limited resources of child protective agencies. For fear of missing even one abused child, workers perform extensive investigations of vague and apparently unsupported reports. Even when a home visit based on an anonymous report turns up no evidence of maltreatment, they usually interview neighbors, school teachers, and day-care personnel to make sure that the child is not abused. And even repeated anonymous and unsubstantiated reports do not prevent a further investigation. But all this takes time.

As a result, children in real danger are getting lost in the press of inappropriate cases. Forced to allocate a substantial portion of their limited resources to unfounded reports, child protective agencies are less able to respond promptly and effectively when children are in serious danger. Some reports are left uninvestigated for a week and even two weeks after they are received. Investigations often miss key facts, as workers rush to clear cases, and dangerous home situations receive inadequate supervision, as workers must ignore pending cases as they investigate the new reports that arrive daily on their desks. Decision making also suffers. With so many cases of unsubstantiated or unproven risk to children, caseworkers are desensitized to the obvious warning signals of immediate and serious danger.

These nationwide conditions help explain why from 25 to 50 percent of child abuse deaths involve children previously known to the authorities. In 1993, the NCPCA reported that of the 1,149 child maltreatment deaths, 42 percent had already been reported to the authorities. Tens of thousands of other children suffer serious injuries short of death while under child protective agency supervision.

In a 1992 New York City case, for example, five-month-old Jeffrey Harden died from burns caused by scalding water and three broken ribs while under the supervision of New York City's Child Welfare Administration. Jeffrey Harden's family had been known to the administration for more than a year and a half.

Over this period, the case had been handled by four separate caseworkers, each conducting only partial investigations before resigning or being reassigned to new cases. It is unclear whether Jeffrey's death was caused by his mother or her boyfriend, but because of insufficient time and overburdened caseloads, all four workers failed to pay attention to a whole host of obvious warning signals: Jeffrey's mother had broken her parole for an earlier conviction of child sexual abuse, she had a past record of beating Jeffrey's older sister, and she had a history of crack addiction and past involvement with violent boyfriends.

Here is how two of the Hardens' caseworkers explained what happened: Their first caseworker could not find Ms. Harden at the address she had listed in her files. She commented, "It was an easy case. We couldn't find the mother so we closed it." Their second caseworker stated that he was unable to spend a sufficient amount of time investigating the case, let alone make the minimum monthly visits because he was tied down with an overabundance of cases and paperwork. He stated, "It's impossible to visit these people within a month. They're all over New York City." Just before Jeffrey's death every worker who had been on the case had left the department. Ironically, by weakening the system's ability to respond, unfounded reports actually discourage appropriate ones. The sad fact is that many responsible individuals are not reporting endangered children because they feel that the system's response will be so weak that reporting will do no good or may even make things worse....

13

Survey of Sexual Behavior
of Americans

EDWARD O. LAUMANN, JOHN H. GAGNON, ROBERT T. MICHAEL,
AND STUART MICHAELS

*A much-heralded study of sexual behavior is featured in this selection by Laumann
et al. This major study, which is highly professionally designed and conducted,
gives us a glimpse into the problems and creative adaptations that arise when a
comprehensive effort is made to conduct large-scale survey research into Americans'
sexual practices. The selection outlines the procedures for conceiving and carrying
out the study, from initial conceptualization to sampling, administration,
interviewer training, questionnaire design, and issues of privacy and confidentiality.
Readers can get some feel for the generic features of survey research, the specific
decisions as to how this project was implemented, and the strengths and weaknesses
associated with this mode of gathering data about deviance. Survey research is an
objective methodology, scientifically controlled through the standardization of the
interview questionnaire and the careful training of interviewers so that they do not
lead respondents into or away from particular answers. When careful probability
sampling is used, such as that which we see here, survey research holds the greatest
potential for generalizability from the sample population to the larger population of
interest. Political, voting, and public opinion polls, for example, use this kind of
methodology and (with careful sampling) have a high degree of accurately predicting
the attitudes and behavior of large swaths of people. Yet surveys have their
disadvantages as well, especially in their problems of internal validity, or
accuracy. People regularly misrepresent themselves on surveys because they can't
understanding the meaning of the questions, because they misremember their past
attitudes or behavior, or just plain intentionally. When we look at the cost of survey
research, we see that this approach can be rather expensive, making a solid, social-
scientific study unaffordable without grant funding. Survey research represents one
of the major sources of information about deviance and is preferred by public policy
analysts because of its quantitative results and its use of the rhetoric of science. Its
strength lies in gathering broad (though not deep) levels of information about less
sensitive subjects from a wide spectrum of people.*

From Edward O. Laumann, et al., *The Social Organization of Sexuality*, © 1994. Reprinted
by permission of the University of Chicago Press and the author.

What kinds of sociological questions can best be answered with the survey research approach? How is it better in some ways than using official statistics or field research? How is it worse in other ways?

Most people with whom we talked when we first broached the idea of a national survey of sexual behavior were skeptical that it could be done. Scientists and laypeople alike had similar reactions: "Nobody will agree to participate in such a study." "Nobody will answer questions like these, and, even if they do, they won't tell the truth." "People don't know enough about sexual practices as they relate to disease transmission or even to pleasure or physical and emotional satisfaction to be able to answer questions accurately." It would be dishonest to say that we did not share these and other concerns. But our experiences over the past seven years, rooted in extensive pilot work, focus-group discussions, and the fielding of the survey itself, resolved these doubts, fully vindicating our growing conviction that a national survey could be conducted according to high standards of scientific rigor and replicability....

The society in which we live treats sex and everything related to sex in a most ambiguous and ambivalent fashion. Sex is at once highly fascinating, attractive, and, for many at certain stages in their lives, preoccupying, but it can also be frightening, disturbing, or guilt inducing. For many, sex is considered to be an extremely private matter, to be discussed only with one's closest friends or intimates, if at all. And, certainly for most if not all of us, there are elements of our sexual lives never acknowledged to others, reserved for our own personal fantasies and self-contemplation. It is thus hardly surprising that the proposal to study sex scientifically, or any other way for that matter, elicits confounding and confusing reactions. Mass advertising, for example, unremittingly inundates the public with explicit and implicit sexual messages, eroticizing products and using sex to sell. At the same time, participants in political discourse are incredibly squeamish when handling sexual themes, as exemplified in the curious combination of horror and fascination displayed in the public discourse about Long Dong Silver and pubic hairs on pop cans during the Senate hearings in September 1991 on the appointment of Clarence Thomas to the Supreme Court. We suspect, in fact, that with respect to discourse on sexuality there is a major discontinuity between the sensibilities of politicians and other self-appointed guardians of the moral order and those of the public at large, who, on the whole, display few hang-ups in discussing sexual issues in appropriately structured circumstances. This work is a testament to that proposition.

The fact remains that, until quite recently, scientific research on sexuality has been taboo and therefore to be avoided or at best marginalized. While there is a visible tradition of (in)famous sex research, what is, in fact, most striking is how little prior research exists on sexuality in the general population. Aside from the research on adolescence, premarital sex, and problems attendant to sex such as fertility, most research attention seems to have been directed toward those believed to be abnormal, deviant, criminal, perverted, rare, or unusual, toward sexual pathology, dysfunction, and sexually transmitted disease—the label used typically reflecting the way in which the behavior or condition in question is to be regarded. "Normal sex" was somehow off limits, perhaps because it was considered too ordinary, trivial, and self-evident to deserve attention. To be fair, then, we cannot blame the public and the politicians entirely for the lack of sustained work on sexuality at large—it also reflects the prejudices and understandings of researchers about what are "interesting" scientific questions. There has simply been a dearth of mainstream scientific thinking and speculation about sexual issues. We have repeatedly encountered this relative lack of systematic thinking about sexuality to guide us in interpreting and understanding the many findings reported in this book.

... In order to understand the results of our survey, the National Health and Social Life Survey (NHSLS), one must understand how these results were generated. To construct a questionnaire and field a large-scale survey, many research design decisions must be made. To understand the decisions made, one needs to understand the multiple purposes that underlie this research project. Research design is never just a theoretical exercise. It is a set of practical solutions to a multitude of problems and considerations that are chosen under the constraints of limited resources of money, time, and prior knowledge.

SAMPLE DESIGN

The sample design for the NHSLS is the most straightforward element of our methodology because nothing about probability sampling is specific to or changes in a survey of sexual behavior....

Probability sampling, that is, sampling where every member of a clearly specified population has a known probability of selection—what lay commentators often somewhat inaccurately call random sampling—is the sine qua non of modern survey research (see Kish, 1965, the classic text on the subject). There is no other scientifically acceptable way to construct a representative sample and thereby to be able to generalize from the actual sample on which data are collected to the population that that sample is designed to represent. Probability sampling as practiced in survey research is a highly developed practical application of statistical theory to the problem of selecting a sample. Not only does this type of sampling avoid the problems of bias introduced by the researcher or by subject self-selection bias that come from more casual techniques, but it also allows one to quantify the variability in the estimates derived from the sample.

In order to determine how large a sample size for a given study should be, one must first decide how precise the estimates to be derived need to be. To illustrate this reasoning process, let us take one of the simplest and most commonly used statistics in survey research, the proportion. Many of the most important results reported are proportions. For example, what proportion of the population had more than five sex partners in the last year? What proportion engaged in anal intercourse? With condoms? Estimates based on our sample will differ from the true proportion in the population because of sampling error (i.e., the random fluctuations in our estimates that are due to the fact that they are based on samples rather than on complete enumerations or censuses). If one drew repeated samples using the same methodology, each would produce a slightly different estimate. If one looks at the distribution of these *estimates*, it turns out that they will be normally distributed (i.e., will follow the famous bell-shaped curve known as the Gaussian or normal distribution) and centered around the true proportion in the population. The larger the sample size, the tighter the distribution of estimates will be.

This analysis applies to an estimate of a single proportion based on the whole sample. In deciding the sample size needed for a study, one must consider the subpopulations for which one will want to construct estimates. For example, one almost always wants to know not just a single parameter for the whole population but parameters for subpopulations such as men and women, whites, blacks, and Hispanics, and younger people and older people. Furthermore, one is usually interested in the intersections of these various breakdowns of the population, for example, young black women. The size of the interval estimate for a proportion based on a subpopulation depends on the size of that group in the sample (sometimes called the *base "N,"* i.e., the number in the sample on which the estimate is based). It is actually this kind of number that one needs to consider in determining the sample size for a study.

When we were designing the national survey of sexual behavior in the United States for the NICHD (National Institute of Child Health and Development), we applied just these sorts of considerations to come to the conclusion that we needed a sample size of about 20,000 people....

GAINING COOPERATION: THE RESPONSE RATE

First, let us consider the cooperation or response rate. No survey of any size and complexity is able to get every sampling-designated respondent to complete an interview. Individuals can have many perfectly valid reasons why they cannot participate in the survey: being too ill, too busy, or always absent when an effort to schedule an interview is made or simply being unwilling to grant an interview. While the face-to-face or in-person survey is considerably more expensive than other techniques, such as mail or telephone surveys, it usually gets the highest response rate. Even so, a face-to-face, household-based survey such as the General Social Survey successfully interviews, on the average, only about 75 percent of the target sample (Davis and Smith, 1991). The missing 25 percent pose a serious problem for the reliability and validity of a survey: is there some systematic (i.e., nonrandom) process at work that distinguishes respondents from nonrespondents? That is, if the people who refuse to participate or who can never be reached to be interviewed differ systematically in terms of the issues being researched from those who are interviewed, then one will not have a representative sample of the population from which the sample was drawn. If the respondents and nonrespondents do not differ systematically, then the results will not be affected. Unfortunately, one usually has no (or only minimal) information about nonrespondents. It is thus a challenge to devise ways of evaluating the extent of bias in the selection of respondents and nonrespondents. Experience tells us that, in most well-studied fields in which survey research has been applied, such moderately high response rates as 75 percent do not lead to biased results. And it is difficult and expensive to push response rates much higher than that. Experience suggests that a response rate close to 90 percent may well represent a kind of upper limit.

Because of our subject matter and the widespread skepticism that survey methods would be effective, we set a completion rate of 75 percent as the survey organization's goal. In fact, we did much better than this; our final completion rate was close to 80 percent. We have extensively investigated whether there are detectable participation biases in the final sample.... To summarize these investigations, we have compared our sample and our results with other surveys of various sorts and have been unable to detect systematic biases of any substantive significance that would lead us to qualify our findings at least with respect to bias due to sampling.

One might well ask what the secret was of our remarkably high response rate, by far the highest of any national sexual behavior survey conducted so far. There is no secret. Working closely with the NORC (National Opinion Research Center) senior survey and field management team, we proceeded in the same way as one would in any other national area probability survey. We did not scrimp on interviewer training or on securing a highly mobilized field staff that was determined to get respondent participation in a professional and respectful manner. It was an expensive operation: the average cost of a completed interview was approximately $450.

We began with an area probability sample, which is a sample of households, that is, of addresses, not names. Rather than approach a household by knocking on the door without advance warning, we followed NORC's standard practice of sending an advance letter, hand addressed by the interviewer, about a week before the interviewer expected to visit the address. In this case, the letter was signed by the principal investigator, Robert Michael, who was identified as the dean of the Irving B. Harris Graduate School of Public Policy Studies of the University of Chicago. The letter briefly explained the purpose of the survey as helping "doctors, teachers, and counselors better understand and prevent the spread of diseases like AIDS and better understand the nature and extent of harmful and of healthy sexual behavior in our country." The intent was to convince the potential respondent that this was a legitimate scientific study addressing personal and potentially sensitive topics for a socially useful purpose. AIDS was the original impetus for the research, and it certainly seemed to provide a timely justification for the study. But any general purpose approach has drawbacks. One problem that the interviewers frequently encountered was potential respondents who did not think that AIDS affected them and therefore that information about their sex lives would be of little use.

Mode of Administration: Face-to-Face, Telephone, or Self-Administered

Perhaps the most fundamental design decision, one that distinguishes this study from many others, concerned how the interview itself was to be conducted.

In survey research, this is usually called the *mode* of interviewing or of question-naire administration. We chose face-to-face interviewing, the most costly mode, as the primary vehicle for data collection in the NHSLS. What follows is the reasoning behind this decision.

A number of recent sex surveys have been conducted over the telephone,... The principal advantage of the telephone survey is its much lower cost. Its major disadvantages are the length and complexity of a questionnaire that can be real-istically administered over the telephone and problems of sampling and sample control.... The NHSLS, cut to its absolute minimum length, averaged about ninety minutes. Extensive field experience suggests an upper limit of about forty-five minutes for phone interviews of a cross-sectional survey of the popu-lation at large. Another disadvantage of phone surveys is that it is more difficult to find people at home by phone and, even once contact has been made, to get them to participate.... One further consideration in evaluating the phone as a mode of interviewing is its unknown effect on the quality of responses. Are peo-ple more likely to answer questions honestly and candidly or to dissemble on the telephone as opposed to face-to-face? Nobody knows for sure.

The other major mode of interviewing is through self-administered forms distributed either face-to-face or through the mail.[1] When the survey is con-ducted by mail, the questions must be self-explanatory, and much prodding is typically required to obtain an acceptable response rate.... This procedure has been shown to produce somewhat higher rates of reporting socially undesirable behaviors, such as engaging in criminal acts and substance abuse. We adopted the mixed-mode strategy to a limited extent by using four short, self-administered forms, totaling nine pages altogether, as part of our interview. When filled out, these forms were placed in a "privacy envelope" by the respondent so that the interviewer never saw the answers that were given to these questions....

The fundamental disadvantage of self-administered forms is that the ques-tions must be much simpler in form and language than those that an interviewer can ask. Complex skip patterns must be avoided. Even the simplest skip patterns are usually incorrectly filled out by some respondents on self-administered forms. One has much less control over whether (and therefore much less confidence that) respondents have read and understood the questions on a self-administered form. The NHSLS questionnaire (discussed below) was based on the idea that questions about sexual behavior must be framed as much as possible in the spe-cific contexts of particular patterns and occasions. We found that it is impossible to do this using self-administered questions that are easily and fully comprehen-sible to people of modest educational attainments.

To summarize, we decided to use face-to-face interviewing as our primary mode of administration of the NHSLS for two principal reasons: it was most likely to yield a substantially higher response rate for a more inclusive cross sec-tion of the population at large, and it would permit more complex and detailed questions to be asked. While by far the most expensive approach, such a strategy provides a solid benchmark against which other modes of interviewing can and should be judged. The main unresolved question is whether another mode has an edge over face-to-face interviewing when highly sensitive questions likely to be upsetting or threatening to the respondent are being asked. As a partial control and test of this question, we have asked a number of sensitive questions in both formats so that an individual's responses can be systematically compared.... Suffice it to say at this point that there is a stunning consistency in the responses secured by the different modes of administration.

Recruiting and Training Interviewers

Gaining respondents' cooperation requires mastery of a broad spectrum of techniques that successful interviewers develop with experience, guidance from the research team, and careful field supervision. This project required extensive training before entering the field. While interviewers are generally trained to be neutral toward topics covered in the interview, this was especially important when discussing sex, a topic that seems particularly likely to elicit emotionally freighted sensitivities both in the respondents and in the interviewers. Interviewers needed to be fully persuaded about the legitimacy and importance of the research. Toward this end, almost a full day of training was devoted to presentations and discussions with the principal investigators in addition to the extensive advance study materials to read and comprehend. Sample answers to frequently asked questions by skeptical respondents and brainstorming about strategies to convert reluctant respondents were part of the training exercises. A set of endorsement letters from prominent local and national notables and refusal conversion letters were also provided to interviewers. A hotline to the research office at the University of Chicago was set up to allow potential respondents to call in with their concerns. Concerns ranged from those about the legitimacy of the survey, most fearing that it was a commercial ploy to sell them something, to fears that the interviewers were interested in robbing them. Ironically, the fact that the interviewer initially did not know the name of the respondent (all he or she knew was the address) often led to behavior by the interviewer that appeared suspicious to the respondent. For example, asking neighbors for the name of the family in the selected household and/or questions about when the potential respondent was likely to be home induced worries that had to be assuaged. Another major concern was confidentiality—respondents wanted to know how they had come to be selected and how their answers were going to be kept anonymous.

THE QUESTIONNAIRE

The questionnaire itself is probably the most important element of the study design. It determines the content and quality of the information gathered for analysis. Unlike issues related to sample design, the construction of a questionnaire is driven less by technical precepts and more by the concepts and ideas motivating the research. It demands even more art than applied sampling design requires.

Before turning to the specific forms that this took in the NHSLS, we should first discuss several general problems that any survey questionnaire must address.

The essence of survey research is to ask a large sample of people from a defined population the *same set of questions*. To do this in a relatively short period of time, many interviewers are needed. In our case, about 220 interviewers from all over the country collected the NHSLS data. The field period, beginning on 14 February 1992 and ending in September, was a time in which over 7,800 households were contacted (many of which turned out to be ineligible for the study) and 3,432 interviews were completed. Central to this effort was gathering comparable information on the same attributes from each and every one of these respondents. The attributes measured by the questionnaire become the variables used in the data analysis. They range from demographic characteristics (e.g., gender, age, and race/ethnicity) to sexual experience measures (e.g., numbers of sex partners in given time periods, frequency of particular practices, and timing of various sexual events) to measures of mental states (e.g., attitudes toward premarital sex, the appeal of particular techniques like oral sex, and levels of satisfaction with particular sexual relationships).

Very early in the design of a national sexual behavior survey, in line with our goal of not reducing this research to a simple behavioral risk inventory, we faced the issue of where to draw the boundaries in defining the behavioral domain that would be encompassed by the concept of sex. This was particularly crucial in defining sexual activity that would lead to the enumeration of a set of sex partners. There are a number of activities that commonly serve as markers for sex and the status of [one's] sex partner, especially intercourse and orgasm. While we certainly wanted to include these events and their extent in given relationships and events, we also felt that using them to define and ask about sexual activity might exclude transactions or partners that should be included. Since the common meaning and uses of the term *intercourse* involve the idea of the intromission of a penis, intercourse in that sense as a defining act would at the very least exclude a sexual relationship between two women. There are also many events that we would call sexual that may not involve orgasm on the part of either or both partners.

Another major issue is what sort of language is appropriate in asking questions about sex. It seemed obvious that one should avoid highly technical language because it is unlikely to be understood by many people. One tempting alternative is to use colloquial language and even slang because that is the only language that some people ever use in discussing sexual matters. There is even some evidence that one can improve reporting somewhat by allowing respondents to select their own preferred terminology (Blair et al., 1977; Bradburn, Sudman, et al., 1979; Bradburn and Sudman, 1983). Slang and other forms of colloquial speech, however, are likely to be problematic in several ways. First, the use of slang can produce a tone in the interview that is counterproductive because it downplays the distinctiveness of the interviewing situation itself. An essential goal in survey interviewing, especially on sensitive topics like sex, is to create a neutral, nonjudgmental and confiding atmosphere and to maintain a certain professional distance between the interviewer and the respondent. A key advantage that the interviewer has in initiating a topic for discussion is being a stranger or an outsider who is highly unlikely to come in contact with the respondent again. It is not intended that a longer-term bond between the interviewer and the respondent be formed, whether as an advice giver or a counselor or as a potential sex partner.[2]

The second major shortcoming of slang is that it is highly variable across class and education levels, ages, regions, and other social groupings. It changes meanings rapidly and is often imprecise. Our solution was to seek the simplest possible language—standard English—that was neither colloquial nor highly technical. For example, we chose to use the term *oral sex* rather than the slang *blow job* and *eating pussy* or the precise technical but unfamiliar *terras, fellatio,* and *cunnilingus.* Whenever possible, we provided definitions when terms were first introduced in a questionnaire—that is, we tried to train our respondents to speak about sex in our terms. Many terms that seemed clear to us may not, of course, be universally understood; for example, terms like *vaginal* or *heterosexual* are not understood very well by substantial portions of the population. Coming up with simple and direct speech was quite a challenge because most of the people working on the questionnaire were highly educated, with strong inclinations toward the circumlocutions and indirections of middle-class discourse on sexual themes. The detailed reactions from field interviewers and managers and extensive pilot testing with a broad cross section of recruited subjects helped minimize these language problems.

ON PRIVACY, CONFIDENTIALITY, AND SECURITY

Issues of respondent confidentiality are at the very heart of survey research. The willingness of respondents to report their views and experiences fully and honestly depends on the rationale offered for why the study is important and on the assurance that the information provided will be treated as confidential. We offered respondents a strong rationale for the study, our interviewers made great efforts to conduct the interview in a manner that protected respondents' privacy, and we went to great lengths to honor the assurances that the information would be treated confidentially. The subject matter of the NHSLS makes the issues of confidentiality especially salient and problematic because there are so many easily imagined ways in which information voluntarily disclosed in an interview might be useful to interested parties in civil and criminal cases involving wrongful harm, divorce proceedings, criminal behavior, or similar matters.

NOTES

1. We ruled out the idea of a mail survey because its response rate is likely to be very much lower than any other mode of interviewing (see Bradburn, Sudman, et al., 1979).

2. Interviewers are not there to give information or to correct misinformation. But such information is often requested in the course of an interview. Interviewers are given training

in how to avoid answering such questions (other than clarification of the meaning of particular questions). They are not themselves experts on the topics raised and often do not know the correct answers to questions. For this reason, and also in case emotionally freighted issues for the respondent were raised during the interview process, we provided interviewers with a list of toll-free phone numbers for a variety of professional sex- and health-related referral services (e.g., the National AIDS Hotline, an STD hotline, the National Child Abuse Hotline, a domestic violence hotline, and the phone number of a national rape and sexual assault organization able to provide local referrals).

REFERENCES

Blair, Ellen, Seymour Sudman, Norman M. Bradburn, and Carol Stacking. (1977). "How to Ask Questions About Drinking and Sex: Response Effects in Measuring Consumer Behavior." *Journal of Marketing Research* 14: 316–321.

Bradburn, Norman M., and Seymour Sudman. (1983). *Asking Questions: A Practical Guide to Questionnaire Design*. San Francisco: Jossey-Bass.

Bradburn, Norman M., Seymour Sudman, Ed Blair, and Carol Stacking. (1979). *Improving Interview Method and Design*. San Francisco: Jossey-Bass.

Davis, James Allan, and Tom W. Smith. (1991). *General Social Surveys, 1972–1991: Cumulative Codebook*. Chicago: National Opinion Research Center.

Kish, Leslie. (1965). *Survey Sampling*. New York: Wiley.

14

Researching Dealers and Smugglers

PATRICIA A. ADLER

Adler offers us a glimpse of what it is like to carry out participant observation research with a deviant group in this description of her study of upper-level drug traffickers. This natural history carefully explains the process used in field research, the relationships formed with setting members, and the feelings researchers experience. A stage-by-stage analysis of the activities, pitfalls, mishaps, intimacies, and relationships Adler encountered, the piece shows the connection and overlap between the development of research ties and those found in natural, everyday life. Field research, we learn, cannot be carried out by just anyone in every setting, but is dependent on researchers' ability to build a bridge of understanding, rapport, and trust between themselves and their subjects. Adler's experiences vividly show us the dangers posed to fieldworkers in criminal and deviant settings and the intimacy of the connections forged there. These kinds of ties, which form the basis of data gathering, make participant observation a subjective, rather than an objective, type of research. The strength of the data rests on the real-world bonds forged in the field as well as on researchers' ability not only to hear what their subjects have to say, but also to see them in action, and to cross-check their self-presentations against hard facts, the accounts of others, and common sense.

Field research, while costing considerably less than survey research, takes much longer to conduct, requiring years to find deviants, develop trust and relationships, and obtain deeply meaningful information. Nonetheless, it not only offers us a better idea of the sequential development of causal forces in the field, but also gives us the best insight into what is really going on in a deviant scene inhabited by a hidden population. Thus, although field research may not yield the ability to generalize with as much scientific precision as survey research, it is the preferred approach for those who want to deeply and accurately understand the perceptions, interpretations, analyses, life worlds, and unfolding careers of secretive deviants.

What kinds of research questions can best be answered by field research? Do its strengths outweigh its potential ethical and interpersonal dilemmas?

From Patricia A. Adler, *Wheeling and Dealing* (New York: Columbia University Press, 1985). Reprinted by permission of the publisher.

I strongly believe that investigative field research (Douglas, 1976), with emphasis on direct personal observation, interaction, and experience, is the only way to acquire accurate knowledge about deviant behavior. Investigative techniques are especially necessary for studying groups such as drug dealers and smugglers because the highly illegal nature of their occupation makes them secretive, deceitful, mistrustful, and paranoid. To insulate themselves from the straight world, they construct multiple false fronts, offer lies and misinformation, and withdraw into their group. In fact, detailed, scientific information about upper-level drug dealers and smugglers is lacking precisely because of the difficulty sociological researchers have had in penetrating into their midst. As a result, the only way I could possibly get close enough to these individuals to discover what they were doing and to understand their world from their perspectives (Blumer, 1969) was to take a membership role in the setting. While my different values and goals precluded my becoming converted to complete membership in the subculture, and my fears prevented my ever becoming "actively" involved in their trafficking activities, I was able to assume a "peripheral" membership role (Adler and Adler, 1987). I became a member of the dealers' and smugglers' social world and participated in their daily activities on that basis. In this chapter, I discuss how I gained access to this group, established research relations with members, and how personally involved I became in their activities.

GETTING IN

When I moved to Southwest County [not the real name] in the summer of 1974, I had no idea that I would soon be swept up in a subculture of vast drug trafficking and unending partying, mixed with occasional cloak-and-dagger subterfuge. I had moved to California with my husband, Peter, to attend graduate school in sociology. We rented a condominium town house near the beach and started taking classes in the fall. We had always felt that socializing exclusively with academicians left us nowhere to escape from our work, so we tried to meet people in the nearby community. One of the first friends we made was our closest neighbor, a fellow in his late twenties with a tall, hulking frame and gentle expression. Dave, as he introduced himself, was always dressed rather casually, if not sloppily, in T-shirts and jeans. He spent most of his time hanging out or walking on the beach with a variety of friends who visited his house, and taking care of his two young boys, who lived alternately with him and his estranged wife. He also went out of town a lot. We started spending much of our free time over at his house, talking, playing board games late into the night, and smoking marijuana together. We were glad to find someone from whom we could buy marijuana in this new place, since we did not know too many people. He also began treating us to a fairly regular supply of cocaine, which was a thrill because this was a drug we could rarely afford on our student budgets. We noticed right away, however, that there was something unusual about his use and knowledge of drugs: while he always had a plentiful supply and was fairly expert about marijuana and cocaine, when we tried to buy a small bag of marijuana from him he had little idea of the going price. This incongruity piqued our curiosity and raised suspicion. We wondered if he might be dealing in larger quantities. Keeping our suspicions to ourselves, we began observing Dave's activities a little more closely. Most of his friends were in their late twenties and early thirties and, judging by their lifestyles and automobiles, rather wealthy. They came and left his house at all hours, occasionally extending their parties through the night and the next day into the following night. Yet throughout this time we never saw Dave or any of his friends engage in any activity that resembled a legitimate job. In most places this might have evoked community suspicion, but few of the people we encountered in Southwest County seemed to hold traditionally structured jobs. Dave, in fact, had no visible means of financial support. When we asked him what he did for a living, he said something vague about being a real estate speculator, and we let it go at that. We never voiced our suspicions directly since he chose not to broach the subject with us.

We did discuss the subject with our mentor, Jack Douglas, however. He was excited by the prospect that we might be living among a group of big dealers, and urged us to follow our instincts and develop leads into the group. He knew that the local area was rife with drug trafficking, since he had begun a life history case study of two drug dealers with another graduate student several years previously. That earlier study was aborted when the graduate student quit school, but Jack still had many hours of taped interviews he had conducted with them, as well as an interview that he had done with an undergraduate student who had known the two dealers independently, to serve as a cross-check on their accounts. He therefore encouraged us to become friendlier with Dave and his friends. We decided that if anything did develop out of our observations of Dave, it might make a nice paper for a field methods class or independent study.

Our interests and background made us well suited to study drug dealing. First, we had already done research in the field of drugs. As undergraduates at Washington University we had participated in a nationally funded project on urban heroin use (see Cummins et al., 1972). Our role in the study involved using fieldwork techniques to investigate the extent of heroin use and distribution in St. Louis. In talking with heroin users, dealers, and rehabilitation personnel, we acquired a base of knowledge about the drug world and the subculture of drug trafficking. Second, we had a generally open view toward soft drug use, considering moderate consumption of marijuana and cocaine to be generally nondeviant. This outlook was partially etched by our 1960s-formed attitudes, as we had first been introduced to drug use in an environment of communal friendship, sharing, and counterculture ideology. It also partially reflected the widespread acceptance accorded to marijuana and cocaine use in the surrounding local culture. Third, our age (mid-twenties at the start of the study) and general appearance gave us compatibility with most of the people we were observing.

We thus watched Dave and continued to develop our friendship with him. We also watched his friends and got to know a few of his more regular visitors. We continued to build friendly relations by doing, quite naturally, what Becker (1963), Polsky (1969), and Douglas (1972) had advocated for the early stages of field research: we gave them a chance to know us and form judgments about our trustworthiness by jointly pursuing those interests and activities which we had in common.

Then one day something happened which forced a breakthrough in the research. Dave had two guys visiting him from out of town and, after snorting quite a bit of cocaine, they turned their conversation to a trip they had just made from Mexico, where they piloted a load of marijuana back across the border in a small plane. Dave made a few efforts to shift the conversation to another subject, telling them to "button their lips," but they apparently thought that he was joking. They thought that anybody as close to Dave as we seemed to be undoubtedly knew the nature of his business. They made further allusions to his involvement in the operation and discussed the outcome of the sale. We could feel the wave of tension and awkwardness from Dave when this conversation began, as he looked toward us to see if we understood the implications of what was being said, but then he just shrugged it off as done. Later, after the two guys left, he discussed with us what happened. He admitted to us that he was a member of a smuggling crew and a major marijuana dealer on the side. He said that he knew he could trust us, but that it was his practice to say as little as possible to outsiders about his activities. This inadvertent slip, and Dave's subsequent opening up, were highly significant in forging our entry into Southwest County's drug world. From then on he was open in discussing the nature of his dealing and smuggling activities with us.

He was, it turned out, a member of a smuggling crew that was importing a ton of marijuana weekly and 40 kilos of cocaine every few months. During that first winter and spring, we observed Dave at work and also got to know the other members of his crew, including Ben, the smuggler himself. Ben was also very tall and broad shouldered, but his long black hair, now flecked with gray, bespoke his earlier membership in the hippie subculture. A large physical stature, we observed, was common to most of the male participants involved in this drug community. The women also had a unifying physical trait: they were extremely attractive and stylishly dressed. This included Dave's ex-wife, Jean, with whom he reconciled during the spring. We therefore became friendly with Jean and through her met a number of women ("dope chicks") who hung around the dealers and smugglers. As we continued to gain the friendship of Dave and Jean's associates, we were progressively admitted into their inner circle and apprised of each person's dealing or smuggling role.

Once we realized the scope of Ben's and his associates' activities, we saw the enormous research potential in studying them. This scene was different from any analysis of drug trafficking that we had read in the sociological literature because of the amounts they were dealing and the fact that they were importing it themselves. We decided that, if it was at all possible, we would capitalize on this situation, to "opportunistically" (Riemer, 1977) take advantage of our prior expertise and of the knowledge, entrée, and rapport we had already developed with several key people in this setting. We therefore discussed the idea of doing a study of the general subculture with Dave and several of his closest friends (now becoming our friends). We assured them of the anonymity, confidentiality, and innocuousness of our work. They were happy to reciprocate our friendship by being of help to our professional careers. In fact, they basked in the subsequent attention we gave their lives.

We began by turning first Dave, then others, into key informants and collecting their life histories in detail. We conducted a series of taped, in-depth interviews with an unstructured, open-ended format. We questioned them about such topics as their backgrounds, their recruitment into the occupation, the stages of their dealing careers, their relations with others, their motivations, their lifestyle, and their general impressions about the community as a whole.

We continued to do taped interviews with key informants for the next six years until 1980, when we moved away from the area. After that, we occasionally did follow-up interviews when we returned for vacation visits. These later interviews focused on recording the continuing unfolding of events and included detailed probing into specific conceptual areas, such as dealing networks, types of dealers, secrecy, trust, paranoia, reputation, the law, occupational mobility, and occupational stratification. The number of taped interviews we did with each key informant varied, ranging between 10 and 30 hours of discussion.

Our relationship with Dave and the others thus took on an added dimension—the research relationship. As Douglas (1976), Henslin (1972), and Wax (1952) have noted, research relationships involve some form of mutual exchange. In our case, we offered everything that friendship could entail. We did routine favors for them in the course of our everyday lives, offered them insights and advice about their lives from the perspective of our more respectable position, wrote letters on their behalf to the authorities when they got in trouble, testified as character witnesses at their non-drug-related trials, and loaned them money when they were down and out. When Dave was arrested and brought to trial for check-kiting, we helped Jean organize his defense and raise the money to pay his fines. We spelled her in taking care of the children so that she could work on his behalf. When he was eventually sent to the state prison we maintained close ties with her and discussed our mutual efforts to buoy Dave up and secure his release. We also visited him in jail. During Dave's incarceration, however, Jean was courted by an old boyfriend and gave up her reconciliation with Dave. This proved to be another significant turning point in our research because, desperate for money, Jean looked up Dave's old dealing connections and went into the business herself. She did not stay with these marijuana dealers and smugglers for long, but soon moved into the cocaine business. Over the next several years her experiences in the world of cocaine dealing brought us into contact with a different group of people. While these people knew Dave and his associates (this was very common in the Southwest County dealing and smuggling community), they did not deal with them directly. We were thus able to gain access to a much wider and more diverse range of subjects than we would have had she not branched out on her own.

Dave's eventual release from prison three months later brought our involvement in the research to an even deeper level. He was broke and had nowhere to go. When he showed up on our doorstep, we took him in. We offered to let him stay with us until he was back on his feet again and could afford a place of his own. He lived with us for seven months, intimately sharing his daily experiences with us. During this time we witnessed, firsthand, his transformation from a scared ex-con who would never break the law again to a hard-working legitimate employee who only dealt to get money for his children's Christmas presents, to a full-time dealer with no pretensions at legitimate work. Both his process of changing attitudes and the community's gradual reacceptance of him proved very revealing.

We socialized with Dave, Jean, and other members of Southwest County's dealing and smuggling community on a near-daily basis, especially during the first four years of the research (before we had a child). We worked in their legitimate businesses, vacationed together, attended their weddings, and cared for their children. Throughout their relationship with us, several participants became co-opted to the researcher's perspective[1] and actively sought out instances of behavior which filled holes in the conceptualizations we were developing. Dave, for one, became so intrigued by our conceptual dilemmas that he undertook a "natural experiment" entirely on his own, offering an unlimited supply of drugs to a lower-level dealer to see if he could work up to higher levels of dealing, and what factors would enhance or impinge upon his upward mobility.

In addition to helping us directly through their own experiences, our key informants aided us in widening our circle of contacts. For instance, they let us know when someone in whom we might be interested was planning on dropping by, vouching for our trustworthiness and reliability as friends who could be included in business conversations. Several times we were even awakened in the night by phone calls informing us that someone had dropped by for a visit, should we want to "casually" drop over too. We rubbed the sleep from our eyes, dressed, and walked or drove over, feeling like sleuths out of a television series. We thus were able to snowball, through the active efforts of our key informants,[2] into an expanded study population. This was supplemented by our own efforts to cast a research net and befriend other dealers, moving from contact to contact slowly and carefully through the domino effect.

THE COVERT ROLE

The highly illegal nature of dealing in illicit drugs and dealers' and smugglers' general level of suspicion made the adoption of an overt research role highly sensitive and problematic. In discussing this issue with our key informants, they all agreed that we should be extremely discreet (for both our sakes and theirs). We carefully approached new individuals before we admitted that we were studying them. With many of these people, then, we took a covert posture in the research setting. As nonparticipants in the business activities which bound members together into the group, it was difficult to become fully accepted as peers. We therefore tried to establish some sort of peripheral, social membership in the general crowd, where we could be accepted as "wise" (Goffman, 1963) individuals and granted a courtesy membership. This seemed an attainable goal, since we had begun our involvement by forming such relationships with our key informants. By being introduced to others in this wise rather than overt role, we were able to interact with people who would otherwise have shied away from us. Adopting a courtesy membership caused us to bear a courtesy stigma,[3] however, and we suffered since we, at times, had to disguise the nature of our research from both lay outsiders and academicians.

In our overt posture we showed interest in dealers' and smugglers' activities, encouraged them to talk about themselves (within limits, so as to avoid acting like narcs), and ran home to write field notes. This role offered us the advantage of gaining access to unapproachable people while avoiding researcher effects, but it prevented us from asking some necessary, probing questions and from tape recording conversations.[4] We therefore sought, at all times, to build toward a conversion to the overt role. We did this by working to develop their trust.

DEVELOPING TRUST

Like achieving entrée, the process of developing trust with members of unorganized deviant groups can be slow and difficult. In the absence of a formal structure separating members from outsiders, each individual must form his or her own judgment about whether new persons can be admitted to their confidence. No gatekeeper existed to smooth our path to being trusted, although our key informants acted in this role whenever they could by providing introductions and references. In addition, the unorganized nature of this group meant that we met people at different times and were constantly at different levels in our developing relationships with them. We were thus trusted more by some people than by others, in part because of their greater familiarity with us. But as Douglas (1976) has noted, just because someone knew us or even liked us did not automatically guarantee that they would trust us.

We actively tried to cultivate the trust of our respondents by tying them to us with favors. Small things, like offering the use of our phone, were followed with bigger favors, like offering the use of our car, and finally really meaningful favors, like offering the use of our home. Here we often trod a thin line, trying to ensure our personal safety while putting ourselves in enough of a risk position, along with our research subjects, so that they would trust us. While we were able to build a "web of trust" (Douglas, 1976) with some members, we found that trust, in large part, was not a simple status to attain in the drug world. Johnson (1975) has pointed out that trust is not a one-time phenomenon, but an ongoing developmental process. From my experiences in this research I would add that it cannot be simply assumed to be a one-way process either, for it can be diminished, withdrawn, reinstated to varying degrees, and requestioned at any point. Carey (1972) and Douglas (1972) have remarked on this waxing and waning process, but it was especially pronounced for us because our subjects used large amounts of cocaine over an extended period of time. This tended to make them alternately warm and cold to us. We thus lived through a series of ups and downs with the people we were trying to cultivate as research informants.

THE OVERT ROLE

After this initial covert phase, we began to feel that some new people trusted us. We tried to intuitively feel when the time was right to approach them and go overt. We used two means of approaching people to inform them that we were involved in a study of dealing and smuggling: direct and indirect. In some cases our key informants approached their friends or connections and, after vouching for our absolute trustworthiness, convinced these associates to talk to us. In other instances, we approached people directly, asking for their help with our project. We worked our way through a progression with these secondary contacts, first discussing the dealing scene overtly and later moving to taped life history interviews. Some people reacted well to us, but others responded skittishly, making appointments to do taped interviews only to break them as the day drew near, and going through fluctuating stages of being honest with us or putting up fronts about their dealing activities. This varied, for some, with their degree of active involvement in the business. During the times when they had quit dealing, they would tell us about their present and past activities, but when they became actively involved again, they would hide it from us.

This progression of covert to overt roles generated a number of tactical difficulties. The first was the problem of *coming on too fast* and blowing it. Early in the research we had a dealer's old lady (we thought) all set up for the direct approach. We knew many dealers in common and had discussed many things tangential to dealing with her without actually mentioning the subject. When we asked her to do a taped interview of her bohemian lifestyle, she agreed without hesitation. When the interview began, though, and she found out why we were interested in her, she balked, gave us a lot of incoherent jumble, and ended the session as quickly as possible. Even though she lived only three houses away we never saw her again. We tried to move more slowly after that.

A second problem involved simultaneously *juggling our overt and covert roles* with different people. This created the danger of getting our cover blown with people who did not know about our research (Henslin, 1972). It was very confusing to separate the people who knew about our study from those who did not, especially in the minds of our informants. They would make occasional veiled references in front of people, especially when loosened by intoxicants, that made us extremely uncomfortable. We also frequently worried that our snooping would someday be mistaken for police tactics. Fortunately, this never happened.

CROSS-CHECKING

The hidden and conflictual nature of the drug-dealing world made me feel the need for extreme certainty about the reliability of my data. I therefore based all my conclusions on independent sources and accounts that we carefully verified. First, we tested information against our own common sense and general knowledge of the scene. We adopted a hard-nosed attitude of suspicion, assuming people were up to more than they would originally admit. We kept our attention especially riveted on "reformed" dealers and smugglers who were living better than they could outwardly afford, and [we] were thereby able to penetrate their public fronts.

Second, we checked out information against a variety of reliable sources. Our own observations of the scene formed a primary reliable source, since we were involved with many of the principals on a daily basis and knew exactly what they were doing. Having Dave live with us was a particular advantage because we could contrast his statements to us with what we could clearly see was happening. Even after he moved out, we knew him so well that we could generally tell when he was lying to us or, more commonly, fooling himself with optimistic dreams. We also observed other dealers' and smugglers' evasions and misperceptions about themselves and their activities. These usually occurred when they broke their own rules by selling to people they did not know, or when they commingled other people's money with their own. We also cross-checked our data against independent, alternative accounts. We were lucky, for this purpose, that Jean got reinvolved in the drug world. By interviewing her, we gained additional insight into Dave's past, his early dealing and smuggling activities, and his ongoing involvement from another person's perspective. Jean (and her connections) also talked to us about Dave's associates, thereby helping us to validate or disprove their statements. We even used this pincer effect to verify information about people we had never directly interviewed. This occurred, for instance, with the tapes that Jack Douglas gave us from his earlier study. After doing our first round of taped interviews with Dave, we discovered that he knew the dealers Jack had interviewed. We were excited by the prospect of finding out what had happened to these people and if their earlier stories checked out. We therefore sent Dave to do some investigative work. Through some mutual friends he got back in touch with them and found out what they had been doing for the past several years.

Finally, wherever possible, we checked out accounts against hard facts: newspaper and magazine reports; arrest records; material possessions; and visible evidence. Throughout the research, we used all these cross-checking measures to evaluate the veracity of new information and to prod our respondents to be more accurate (by abandoning both their lies and their self-deceptions).[5]

After about four years of near-daily participant observation, we began to diminish our involvement in the research. This occurred gradually, as first pregnancy and then a child hindered our ability to follow the scene as intensely and spontaneously as we had before. In addition, after having a child, we were less willing to incur as many risks as we had before; we no longer felt free to make decisions based solely on our own welfare. We thus pulled back from what many have referred to as the "difficult hours and dangerous situations" inevitably present in field research on deviants (see Becker, 1963; Carey, 1972; Douglas, 1972). We did, however, actively maintain close ties with research informants (those with whom we had gone overt), seeing them regularly and periodically doing follow-up interviews.

PROBLEMS AND ISSUES

Reflecting on the research process, I have isolated a number of issues which I believe merit additional discussion. These are rooted in experiences which have the potential for greater generic applicability.

The first is the *effect of drugs on the data-gathering process*. Carey (1972) has elaborated on some of the problems he encountered when trying to interview respondents who used amphetamines, while Wax (1952, 1957) has mentioned the difficulty of trying to record field notes while drinking sake. I found that marijuana and cocaine had nearly opposite effects from each other. The latter helped the interview process, while the former hindered it. Our attempts to interview respondents who were stoned on marijuana were unproductive for a number of reasons. The primary obstacle was the effects of the drug. Often, people became confused, sleepy, or involved in eating to varying degrees. This distracted them from our purpose. At times, people even simulated overreactions to marijuana to hide behind the drug's supposed disorienting influence and thereby avoid divulging information. Cocaine, in contrast, proved to be a research aid. The drug's warming and sociable influence opened people up, diminished their inhibitions, and generally increased their enthusiasm for both the interview experience and us.

A second problem I encountered involved *assuming risks while doing research*. As I noted earlier, dangerous situations are often generic to research on deviant behavior. We were most afraid of the people we studied. As Carey (1972), Henslin (1972), and Whyte (1955) have stated, members of deviant groups can become hostile toward a researcher if they think that they are being treated wrongfully. This could have happened at any time from a simple occurrence, such as a misunderstanding, or from something more serious, such as our covert posture being exposed. Because of the inordinate amount of drugs they consumed, drug dealers and smugglers were particularly volatile, capable of becoming malicious toward each other or us with little warning. They were also likely to behave erratically owing to the great risks they faced from the police and other dealers. These factors made them moody, and they vacillated between trusting us and being suspicious of us.

At various times we also had to protect our research tapes. We encountered several threats to our collection of taped interviews from people who had granted us these interviews. This made us anxious, since we had taken great pains to acquire these tapes and felt strongly about maintaining confidences entrusted to us by our informants. When threatened, we became extremely frightened and shifted the tapes between different hiding places. We even ventured forth one rainy night with our tapes packed in a suitcase to meet a person who was uninvolved in the research at a secret rendezvous so that he could guard the tapes for us.

We were fearful, lastly, of the police. We often worried about local police or drug agents discovering the nature of our study and confiscating or subpoenaing our tapes and field notes. Sociologists have no privileged relationship with their subjects that would enable us legally to withhold evidence from the authorities should they subpoena it.[6] For this reason we studiously avoided any publicity about the research, even holding back on publishing articles in scholarly journals until we were nearly ready to move out of the setting. The closest we came to being publicly exposed as drug researchers came when a former sociology graduate student (turned dealer, we had heard from inside sources) was arrested at the scene of a cocaine deal. His lawyer wanted us to testify about the dangers of doing drug-related research, since he was using his research status as his defense. Fortunately, the crisis was averted when his lawyer succeeded in suppressing evidence and had the case dismissed before the trial was to have begun. Had we been exposed, however, our respondents would have acquired guilt by association through their friendship with us.

Our fear of the police went beyond our concern for protecting our research subjects, however. We risked the danger of arrest ourselves through our own violations of the law. Many sociologists (Becker, 1963; Carey, 1972; Polsky, 1969; Whyte, 1955) have remarked that field researchers studying deviance must inevitably break the law in order to acquire valid participant observation data. This occurs in its most innocuous form from having "guilty knowledge": information about crimes that are committed. Being aware of major dealing and smuggling operations made us an accessory to their commission, since we failed to notify the police. We broke the law, secondly, through our "guilty observations," by being present at the scene of a crime and witnessing its occurrence (see also Carey, 1972). We knew it was possible to get caught in a bust involving others, yet buying and selling was so pervasive that to leave every time it occurred would have been unnatural and highly suspicious. Sometimes drug transactions even occurred in our home, especially when Dave was living there, but we finally had to put a stop to that because we could not handle the anxiety. Lastly, we broke the law through our "guilty actions," by taking part in illegal behavior ourselves. Although we never dealt drugs (we were too scared to be seriously tempted), we consumed drugs and possessed them in small quantities. Quite frankly, it would have been impossible for a nonuser to have gained access to this group to gather the data presented here. This was the minimum involvement necessary to obtain even the courtesy membership we achieved. Some kind of illegal action was also found to be a necessary or helpful component of the research by Becker (1963), Carey (1972), Johnson (1975), Polsky (1969), and Whyte (1955).

Another methodological issue arose from the *cultural clash between our research subjects and ourselves*. While other sociologists have alluded to these kinds of differences (Humphreys, 1970; Whyte, 1955), few have discussed how the research relationships affected them. Relationships with research subjects are unique because they involve a bond of intimacy between persons who might not ordinarily associate together, or who might otherwise be no more than casual friends. When field-workers undertake a major project, they commit themselves to maintaining a long-term relationship with the people they study. However, as researchers try to get depth involvement, they are apt to come across fundamental differences in character, values, and attitudes between their subjects and themselves. In our case, we were most strongly confronted by differences in present versus future orientations, a desire for risk versus security, and feelings of spontaneity versus self-discipline. These differences often caused us great frustration. We repeatedly saw dealers act irrationally, setting themselves up for failure. We wrestled with our desire to point out their patterns of foolhardy behavior and offer advice, feeling competing pulls between our detached, observer role which advised us not to influence the natural setting, and our involved, participant role which called for us to offer friendly help whenever possible.[7]

Each time these differences struck us anew, we gained deeper insights into our core, existential selves. We suspended our own taken-for-granted feelings and were able to reflect on our culturally formed attitudes, character, and life choices from the perspective of the other. When comparing how we might act in situations faced by our respondents, we realized where our deepest priorities lay. These revelations had the effect of changing our self-conceptions: although we, at one time, had thought of ourselves as what Rosenbaum (1981) has called "the hippest of nonaddicts" (in this case nondealers), we were suddenly faced with being the straightest members of the crowd. Not only did we not deal, but we also had a stable, long-lasting marriage and family life, and needed the security of a reliable monthly paycheck. Self-insights thus emerged as one of the unexpected outcomes of field research with members of a different cultural group.

The final issue I will discuss involved the various *ethical problems* which arose during this research. Many field-workers have encountered ethical dilemmas or pangs of guilt during the course of their research experiences (Carey, 1972; Douglas, 1976; Humphreys, 1970; Johnson, 1975; Klockars, 1977, 1979; Rochford, 1985). The researchers' role in the field makes this necessary because they can never fully align themselves with their subjects while maintaining their identity and personal commitment to the scientific community. Ethical dilemmas, then, are directly related to the amount of deception researchers use in gathering the data, and the degree to which they have accepted such acts as necessary and therefore neutralized them.

Throughout the research, we suffered from the burden of intimacies and confidences. Guarding secrets which had been told to us during taped interviews was not always easy or pleasant. Dealers occasionally revealed things about themselves or others that we had to pretend not to know when interacting with their close associates. This sometimes meant that we had to lie or build elaborate stories to cover for some people. Their fronts therefore became our fronts, and we had to weave our own web of deception to guard their performances. This became especially disturbing during the writing of the research report, as I was torn by conflicts between using details to enrich the data and glossing over description to guard confidences.[8]

Using the covert research role generated feelings of guilt, despite the fact that our key informants deemed it necessary, and thereby condoned it. Their own covert experiences were far more deeply entrenched than ours, being a part of their daily existence with nondrug world members. Despite the universal presence of covert behavior throughout the setting, we still felt a sense of betrayal every time we ran home to write research notes on observations we had made under the guise of innocent participants.

We also felt guilty about our efforts to manipulate people. While these were neither massive nor grave manipulations, they involved courting people to procure information about them. Our aggressively friendly postures were based on hidden ulterior motives: we did favors for people with the clear expectation that they could only pay us back with research assistance. Manipulation bothered us in two ways: immediately after it was done, and over the long run. At first, we felt awkward, phony, almost ashamed of ourselves, although we believed our rationalization that the end justified the means. Over the long run, though, our feelings were different. When friendship became intermingled with research goals, we feared that people would later look back on our actions and feel we were exploiting their friendship merely for the sake of our research project.

The last problem we encountered involved our feelings of whoring for data. At times, we felt that we were being exploited by others, that we were putting more into the relationship than they, that they were taking us for granted or using us. We felt that some people used a double standard in their relationship with us: they were allowed to lie to us, borrow money and not repay it, and take advantage of us, but we were at all times expected to behave honorably. This was undoubtedly an outgrowth of our initial research strategy where we did favors for people and expected little in return. But at times this led to our feeling bad. It made us feel like we were selling ourselves, our sincerity, and usually our true friendship, and not getting treated right in return.

CONCLUSIONS

The aggressive research strategy I employed was vital to this study. I could not just walk up to strangers and start hanging out with them as Liebow (1967) did, or be sponsored to a member of this group by a social service or reform organization as Whyte (1955) was, and expect to be accepted, let alone welcomed. Perhaps such a strategy might have worked with a group that had nothing to hide, but I doubt it. Our modern, pluralistic society is so filled with diverse subcultures whose interests compete or conflict with each other that each subculture has a set of knowledge which is reserved exclusively for insiders. In order to serve and prosper, they do not ordinarily show this side to just anyone. To obtain the kind of depth insight and information I needed, I had to become like the members in certain ways. They dealt only with people they knew and trusted, so I had to become known and trusted before I could reveal my true self and my research interests. Confronted with secrecy, danger, hidden alliances, misrepresentations, and unpredictable changes of intent, I had to use a delicate combination of overt and covert roles. Throughout, my deliberate cultivation of the norm of reciprocal exchange enabled me to trade my friendship for their knowledge, rather than waiting for the highly unlikely event that information would be delivered into my lap. I thus actively built a web of research contacts, used them to obtain highly sensitive data, and carefully checked them out to ensure validity.

Throughout this endeavor I profited greatly from the efforts of my husband, Peter, who served as an equal partner in this team field research project. It would have been impossible for me to examine this social world as an unattached female and not fall prey to sex role stereotyping which excluded women from business dealings. As a couple, our different genders allowed us to relate in different ways to both men and women (see Warren and Rasmussen, 1977). We also protected each other when we entered the homes of dangerous characters, buoyed each other's initiative and courage, and kept the conversation going when one of us faltered. Conceptually, we helped each other keep a detached and analytical eye on the setting, provided multiperspectival insights, and corroborated, clarified, or (most revealingly) contradicted each other's observations and conclusions. Finally, I feel strongly that to ensure accuracy, research on deviant groups must be conducted in the settings where it naturally occurs. As Polsky (1969: 115–16) has forcefully asserted:

> This means—there is no getting away from it—the study of career criminals *au natural*, in the field, the study of such criminals as they normally go about their work and play, the study of "uncaught" criminals and the study of others who in the past have been caught but are not caught at the time you study them .... Obviously we can no longer afford the convenient fiction that in studying criminals in their natural habitat, we would discover nothing really important that could not be discovered from criminals behind bars.

By studying criminals in their natural habitat I was able to see them in the full variability and complexity of their surrounding subculture, rather than within the artificial environment of a prison. I was thus able to learn about otherwise inaccessible dimensions of their lives, observing and analyzing firsthand the nature of their social organization, social stratification, lifestyle, and motivation.

NOTES

1. Gold (1958) discouraged this methodological strategy, cautioning against overly close friendship or intimacy with informants, lest they lose their ability to act as informants by becoming too much observers. Whyte (1955), in contrast, recommended the use of informants as research aides, not for helping in conceptualizing the data but for their assistance in locating data which supports, contradicts, or fills in the researcher's analysis of the setting.

2. See also Biernacki and Waldorf (1981); Douglas (1976); Henslm (1972); Hoffinan (1980); McCall (1980); and West (1980) for discussions of "snowballing" through key informants.

3. See Kirby and Corzine (1981); Birenbaum (1970); and Henslin (1972)

for more detailed discussion of the nature, problems, and strategies for dealing with courtesy stigmas.

4. We never considered secret tapings because, aside from the ethical problems involved, it always struck us as too dangerous.

5. See Douglas (1976) for a more detailed account of these procedures.

6. A recent court decision, where a federal judge ruled that a sociologist did not have to turn over his field notes to a grand jury investigating a suspicious fire at a restaurant where he worked, indicates that this situation may be changing (Fried, 1984).

7. See Henslin (1972) and Douglas (1972, 1976) for further discussions of this dilemma and various solutions to it.

8. In some cases I resolved this by altering my descriptions of people and their actions as well as their names so that other members of the dealing and smuggling community would not recognize them. In doing this, however, I had to keep a primary concern for maintaining the sociological integrity of my data so that the generic conclusions I drew from them would be accurate. In places, then, where my attempts to conceal people's identities from people who know them have been inadequate, I hope that I caused them no embarrassment. See also Polsky (1969); Rainwater and Pittman (1967); and Humphreys (1970) for discussions of this problem.

REFERENCES

Adler, Patricia A., and Peter Adler. (1987). *Membership Roles in Field Research.* Beverly Hills, CA: Sage.

Becker, Howard. (1963). *Outsiders.* New York: Free Press.

Biernacki, Patrick, and Dan Waldorf. (1981). "Snowball sampling." *Sociological Methods and Research* 10: 141–63.

Birenbaum, Arnold. (1970). "On managing a courtesy stigma." *Journal of Health and Social Behavior* 11: 196–206.

Blumer, Herbert. (1969). *Symbolic Interactionism.* Englewood Cliffs, NJ: Prentice Hall.

Carey, James T. (1972). "Problems of access and risk in observing drug scenes." In Jack D. Douglas, ed., *Research on Deviance*, pp. 71–92. New York: Random House.

Cummins, Marvin, et al. (1972). *Report of the Student Task Force on Heroin Use in Metropolitan Saint Louis.* Saint Louis: Washington University Social Science Institute.

Douglas, Jack D. (1972). "Observing deviance." In Jack D. Douglas, ed., *Research on Deviance*, pp. 3–34. New York: Random House.

———. (1976). *Investigative Social Research.* Beverly Hills, CA: Sage.

Fried, Joseph P. (1984). "Judge protects waiter's notes on fire inquiry." *New York Times*, April 8: 47.

Goffman, Erving. (1963). *Stigma.* Englewood Cliffs, NJ: Prentice Hall.

Gold, Raymond. (1958). "Roles in sociological field observations." *Social Forces* 36: 217–23.

Henslin, James M. (1972). "Studying deviance in four settings: research experiences with cabbies, suicides, drug users and abortionees." In Jack D. Douglas, ed., *Research on Deviance*, pp. 35–70. New York: Random House.

Humphreys, Laud. (1970). *Tearoom Trade.* Chicago: Aldine.

Johnson, John M. (1975). *Doing Field Research.* New York: Free Press.

Kirby, Richard, and Jay Corzine. (1981). "The contagion of stigma." *Qualitative Sociology* 4: 3–20.

Klockars, Carl B. (1977). "Field ethics for the life history." In Robert Weppner, ed., *Street Ethnography*, pp. 201–26. Beverly Hills, CA: Sage.

———. (1979). "Dirty hands and deviant subjects." In Carl B. Klockars and Finnbarr W. O'Connor, eds., *Deviance and Decency*, pp. 261–82. Beverly Hills, CA: Sage.

Liebow, Elliott. (1967). *Tally's Corner.* Boston: Little, Brown.

McCall, Michal. (1980). "Who and where are the artists?" In William B. Shaffir, Robert A. Stebbins, and Allan Turowetz, eds., *Fieldwork Experience*, pp. 145–58. New York: St. Martin's.

Polsky, Ned. (1969). *Hustlers, Beats, and Others*. New York: Doubleday.

Rainwater, Lee R., and David J. Pittman. (1967). "Ethical problems in studying a politically sensitive and deviant community." *Social Problems* 14: 357–66.

Riemer, Jeffrey W. (1977). "Varieties of opportunistic research." *Urban Life* 5: 467–77.

Rochford, E. Burke, Jr. (1985). *Hare Krishna in America*. New Brunswick, NJ: Rutgers University Press.

Rosenbaum, Marsha. (1981). *Women on Heroin*. New Brunswick, NJ: Rutgers University Press.

Warren, Carol A. B., and Paul K. Rasmussen. (1977). "Sex and gender in field research." *Urban Life* 6: 349–69.

Wax, Rosalie. (1952). "Reciprocity as a field technique." *Human Organization* 11: 34–37.

———. (1957). "Twelve years later: An analysis of a field experience." *American Journal of Sociology* 63: 133–42.

West, W. Gordon. (1980). "Access to adolescent deviants and deviance." In William B. Shaffir, Robert A. Stebbins, and Allan Turowetz, eds., *Fieldwork Experience*, pp. 31–44. New York: St. Martin's.

Whyte, William F. (1955). *Street Corner Society*. Chicago: University of Chicago Press.

PART IV

Constructing Deviance

A s we noted in the general introduction, the social constructionist perspective suggests that deviance should be regarded as lodged in a process of
definition, rather than in some objective feature of an object, person, or act.
That perspective therefore recommends that we look at the process by which a
society constructs definitions of deviance and applies them to specific groups of
people associated with deviant objects or acts. The dynamics of these deviance-
defining processes may sometimes eclipse the factual grounding on which they
rest in their significance for the rise of collective moral sentiments.

MORAL ENTREPRENEURS: CAMPAIGNING

The process of constructing and applying definitions of deviance can be understood as a moral enterprise. That is, it involves the constructions of moral meanings and the association of them with specific acts or conditions. The way people
"make" deviance is similar to the way they manufacture anything else, but
because deviance is an abstract concept rather than a tangible product, the process involves individuals drawing on the power and resources of organizations,
institutions, agencies, symbols, ideas, communication, and audiences. Becker
(1963) has suggested that we call the people involved in these activities **moral
entrepreneurs**. The deviance-making enterprise has two facets: rule creation
(without which there would be no deviant behavior) and rule enforcement
(application of the rules that are created to specific groups of people). We thus
have two kinds of moral entrepreneurs: rule creators and rule enforcers. Rule
creators include such people as politicians, crusading public figures, teachers, parents, school administrators, and CEOs of business organizations. When we think

of rule enforcers, we immediately imagine the police, courts, and judges, but the role can also be filled by dormitory resident assistants, members of neighborhood associations, a school's interfraternity council, and parents.

Rule creation can be done by individuals acting either alone or in groups. Prominent individuals who have been influential in campaigning for definitions of deviance include former First Lady Nancy Reagan, for her "Just Say No" and D.A.R.E. antidrug campaigns in the 1980s; actor Charlton Heston, for his presidency of the National Rifle Association (NRA); John Walsh, for founding the National Center for Missing and Exploited Children and the television shows "America's Most Wanted" and "The Hunt"; and filmmaker Michael Moore, for his documentaries about General Motors (*Roger and Me*), the gun lobby (*Bowling for Columbine*), the relation between the oil industry and the Bush administration (*Fahrenheit 9/11*), and the health care industry (*Sicko*). More commonly, however, individuals band together to use their collective energy and resources to change social definitions and to create norms and rules. Groups of moral entrepreneurs represent interest groups that can be galvanized and activated into pressure groups, such as Mothers Against Drunk Driving (MADD), Group Against Smoking Pollution (GASP), National Organization for the Reform of Marijuana Laws (NORML), and Focus on the Family (a right-wing, Christian pro-family group). Rule creators ensure that our society is supplied with a constant stock of deviance by defining the behavior of others as immoral. They do this because they perceive people as threats and feel fearful, distrustful, and suspicious of their behavior. In so doing, they seek to transform private troubles into public issues and their private morality into the normative order.

Moral entrepreneurs manufacture public morality through a multistage process. Their first goal is to generate broad **awareness** of a problem. They do this through a process of what Spector and Kitsuse (1977) called "claimsmaking." Claimsmakers draw our attention to given issues by asserting "danger messages." Not only do they use these messages to create a sense that certain conditions are problematic and pose a present or future danger to society, but they usually also have specific solutions that they recommend. Issues about which we have recently seen danger messages raised include secondhand smoke, drunk driving, hate crimes, college binge drinking, illegal immigrants (and their link to terrorists), outsourcing, guns in schools, junk food, politics in the classroom, and obesity. Because no rules exist to deal with the threatening condition, claimsmakers construct the impression that such rules are necessary.

In so doing, they draw on the testimonials of various "experts" in the field, such as scholars, doctors, eyewitnesses, ex-participants (professional exes), and others with specific knowledge of the situation. Issues are framed in these testimonials and disseminated to society via the media as "typical" in order to promote specific examples, orientations, causes, and solutions. We see the surgeon general warning about the dangers of secondhand smoke, college administrators talking about student drinking and deaths, psychiatrists writing about the dangers of self-injury and eating disorders, sociologists discussing the spread of date and acquaintance rape, national conservative watchdogs such as David Horowitz monitoring liberal bias in the classroom, and ex-FLDS (Fundamentalist Latter Day Saints) members warning about the forced marriage (statutory rape) of young girls, the expulsion of teenage boys, and the isolation and subservience of whole communities to the point of brainwashing.

Several rhetorical techniques are used to package and present these issues in the most compelling way. Statistics may show a rise in the incidence of a given behavior, or its correlation with other social problems. For example, suicide statistics may be contrasted with homicide statistics to draw attention to the importance of suicide, which occurs at a rate nearly 50 percent higher than homicide. Growth in the rates of obesity and its relation to diabetes, heart disease, stroke, and worker absenteeism may be documented. We may hear about the rising tide of illegal immigrants and their financial drain on our public schools and medical services. Free trade agreements may be tied to increasing rates of unemployment. Dramatic case examples can paint a picture of horror in the public's mind, inspiring fear and loathing. Particular cases are usually selected because they have no moral ambiguity and feature the dimension of the problem that is being highlighted.

Thus, for missing children, we want to see stranger abductors rather than snatchings by noncustodial parents, whereas for AIDS, we want to see innocent children who are sick rather than gay men. New syndromes can be advanced, packaging different issues together into a behavioral pattern portrayed as dangerous. Thus, we see "Internet addiction disorder" (caution: People are abandoning their homes and families to spend all their time and money in chat rooms), "centerfold syndrome" (caution: Men who read pornography objectify, commodify, and victimize women, seek unattainable trophy figurines, and are unable to engage in meaningful relationships), "road rage disorder" (caution: Traffic induces people to cut us off and give us the finger), "compulsive hoarding" (caution: Practitioners have excessive clutter; difficulty categorizing, organizing, making decisions about, and throwing away possessions; and fears about needing items that could be thrown away), "chronic procrastination" (caution: You may be an "arousal procrastinator," putting things off for the last-minute rush, or an "avoider," putting off dealing with things because you fear either failure or success), and attention deficit hyperactivity disorder (caution: You could have this medical problem if you daydream, are slow to complete tasks, fidget, have poor concentration, or are distractible, hyperactive, impulsive, and/or reckless). Finally, rhetoric requires that each side seek the (usually competing) "moral high ground" in their assertions and attacks on each other, disavowing special interests and pursuing only the purest public good. For example, people who are opposed to gay marriage are upholding the Bible, decency, and the sanctity of marriage, while people who support it are upholding individual civil liberties and fighting unjust discrimination.

Second, rule creators must bring about a **moral conversion**, convincing others of their views. With the problem outlined, they have to convert neutral parties and previous opponents into supporting partisans. Their successful conversion of others further legitimates their own beliefs. To effect a moral conversion, rule creators must compete for space in the public arena, often a limited resource. Hilgartner and Bosk (1988) have suggested that only so many issues can claim widespread attention, and they do so at the expense of others. As Durkheim noted regarding deviance, only a limited number of public concerns can be supported in society at any given time. Thus, moral entrepreneurs must draw on elements of drama, novelty, politics, and deep mythic themes of the culture to gain the visibility they need. To gain such visibility, moral entrepreneurs must attract the media attention necessary to spread word of their campaign widely. They may orchestrate their campaign to do this through hunger strikes (such as students protesting the manufacturing of university-labeled clothing in Third World "sweatshops"), demonstrations (e.g., antiwar or pro-immigration), civil disobedience (such as environmentalists tying themselves to trees to prevent deforestation or blocking highways to prevent nuclear waste from being stored in their state), marches (e.g., the Gay Pride and Civil Rights movements), and strikes and picketing (e.g., against de-unionization).

Rule creators must also enlist the support of sponsors—opinion leaders who need not have expert knowledge on any particular subject, but who are liked and respected—to provide them with public endorsements. Moral entrepreneurial campaigners often turn to athletes, actors, musicians, religious leaders, and media personalities for such endorsements. Former Democratic presidential candidate Al Gore is notable for his campaign highlighting the growth of global warming (*An Inconvenient Truth*), actor Tom Cruise has actively promoted the dangers of antidepressants, and actor Michael J. Fox campaigned for Democratic candidates to endorse stem cell research.

Finally, rule creators look to different groups in society with which they can form alliances or coalitions to support their campaigns. Alliances are made up of long-term allies, such as the Christian Coalition, Focus on the Family, conservative Republican politicians, and big business. Coalitions, by contrast, represent groups that do not normally lobby together, but that become bonded by their mutual interest in a single issue, such as what we see in the strange union when family groups, conservative Republicans, religious leaders, and, ironically, radical feminists come together to campaign against pornography.

At times, the efforts of moral entrepreneurs are so successful that they create a "**moral panic**." The term was coined by Jock Young, but popularized in the field of deviance by Stanley Cohen's (1972) book on the Mods and the Rockers, making it a widespread and enduring concept. Moral panics arise when a threat to society is depicted, promoting terror and dread with its powerfully persuasive focus on folk devils. Conditions of unsettling social strain make a community ripe for a moral panic to erupt. When it does, expert "claimsmakers" (Spector and Kitsuse, 1977), or issue entrepreneurs, articulate the scope and specific danger of the problem, identifying social conditions that members of some group or other perceived to be offensive and undesirable. To make a claim, it is necessary to engage in a variety of specific activities: naming the problem; distinguishing it from other similar or more encompassing problems; determining the scientific, technical, moral, or legal basis of the claim; and gauging who is responsible for taking ameliorative action. The level of moral anxiety becomes stoked by concerned individuals promoting the severity of the problem, legislators who react and heighten the alarm, and sensationalist news media that whip the public into a "feeding frenzy" through highly emotional claims and fear-based appeals. Folk devils become treated as threats to dominant social interests and values. Moral panics, most recently witnessed in cases of school shootings, priesthood pedophilia, child kidnapping, drinking and rioting on college campuses, Internet predators, obesity, satanic exploitation of children, gangs, drugs, and domestic terrorists, tend to develop a life of their own, irrationally moving in exaggerated propulsion beyond their original impetus. Significant contests may emerge between claimsmakers arguing for the stigmatization of various folk devils and those who attempt to reject the stigmatization or even reverse it onto the original accusers or project it onto some other parties. These stigma contests, central to moral panic theory, tend to play out in the political domain. Their goal is to achieve the dominance of specific moral perceptions and values. Launching a moral panic in a complex and diverse society, in which public attention is being sought by many different claimsmakers and issues, is difficult, and not all would-be moral panics materialize. Moral panics, at their core, are thus power struggles between various groups in society about which of them can dominate and impose their worldviews and values on others as legitimate and true.

Erich Goode and Nachman Ben-Yehuda (1994) articulated a stage-by-stage analysis of moral panics. To be successful, they argued, moral panics usually have to occur during a ripe historical *time*, when a combination of social, ecological, ideological, professional, and/or political forces has contributed to some growing cultural anxiety. They are then *triggered* by a specific incident that precipitates awareness of what is going on and that *targets* the public's attention toward a specific folk devil. Folk devils can take the form of particular individuals, groups, or even behavioral trends. Groups may select their targets specifically for their strategic value, as we see in Chapter 17's discussion of the anti-Disney campaign, to get the most attention out of their efforts. The *content* of allegations that form the moral panic is explored, investigated in detail, repeated unceasingly, and possibly blown out of proportion. Awareness of the danger and of the ripples of the moral panic may be spread through the mass media, grassroots word-of-mouth communication, Internet warnings, public presentations, government hearings, and the testimony of experts (claimsmakers, or what Tuggle and Holmes refer to in Chapter 16 as the "knowledge class" and what Reinarman refers to in Chapter 15 as "professional interest groups"). But all the moral panics fade or *decline* eventually, since they represent inflated fears and cannot be sustained indefinitely. They may founder of their own accord because the problem has been addressed and solved, because the danger has been found to be exaggerated and the claimsmakers grow silent, or because they are replaced by new panics that steal the public's attention. Moral panics usually leave in their wake some residual effect, often taking the form of new institutional arrangements or bureaucratic structures. For example, flying in the United States will never be the same as it was prior to 9/11.

Once the public viewpoint has been swayed and a majority (or a vocal and powerful enough minority) of people have adopted a social definition, it may remain at the level of a norm or become elevated to the status of law through a legislative effort. For example, although obesity has been defined as disgusting and unhealthy, it is not illegal, whereas antismoking campaigns have been successful in legally banning cigarette use in most public places. At the same time, *anti*moral entrepreneurial campaigns arise that seek to sway public opinion in favor of removing an existing moral stigma, as we have seen in the destigmatization and increasing legalization of gay marriage and marijuana.

After norms or rules have been enacted, rule enforcers ensure that they are applied. In our society, this process often tends to be selective. Because of their socioeconomic, racial, religious, gender, political, or other status, various individuals or groups have greater or lesser power to resist the enforcement of rules directed against them. Whole battles may begin anew over individuals' or groups' strength to apply or resist the enforcement of norms and laws, with this arena becoming once again a moral entrepreneurial combat zone. For example,

President George W. Bush's efforts to rebuff the investigations launched by the Democratic Party as a result of charges of manipulated intelligence over his claims about weapons of mass destruction after the invasion of Iraq went sour illustrate core issues of social power.

DIFFERENTIAL SOCIAL POWER: LABELING

Specific behavioral acts are not the only things that can be constructed as deviant; the term can also be applied to a social status, demographic characteristic, or lifestyle. When entire groups of people become relegated to a deviant status through their condition (especially if that status is ascribed by virtue of birth rather than being voluntarily achieved), we see the force of inequality and differential social power in operation. This dynamic has been discussed earlier in reference to both conflict theory and social constructionism: We noted that those who control the resources in society (politics, social status, gender, wealth, religious beliefs, mobilization of the masses) have the ability to dominate the subordinate groups, both materially and ideologically. Thus, certain kinds of laws and enforcement are a product of political action by moral entrepreneurial interest groups that are connected to society's power base. Dominant groups use their strength and position to label and subjugate the weak.

A range of different factors give certain groups greater **social power** in society to construct definitions of deviance and to apply those labels to others. Money is one of the clearest elements, with its potential influence being felt at least two ways. One way is that big businesses can use money to make campaign contributions and sway political candidates, to fund research favorable to their products (studies paid for by the soft-drink industry were recently found to be eight times more likely to find no harmful health influences of drinking soda than studies otherwise funded), to lobby against unfavorable legislation, and to fight restrictive lawsuits. Another way is that money defines individuals' social class, and although rich people do not have as much cash available to do what businesses can afford, it is much harder to define the practices of the middle and upper classes as deviant than those of the lower, working, and under classes.

Second, race and ethnicity influence social power, so the behaviors of the dominant White population are less likely to be labeled, and laws against them enforced, than those of Hispanics and Blacks. Gender is a third element of social power, with men dominating over women politically, economically, historically, religiously, occupationally, culturally, and, hence, interpersonally.

Fourth, people's age affects their relative power in society, with young people (up to the age of 30) and older people (65 and older) holding less respect, influence, attention, and command than their middle-aged counterparts. Fifth, greater numbers and superior organization can empower groups, as positions backed by larger populations often hold sway over smaller ones. Yet, at the same time, well-organized groups, even if they are in the minority, may dominate over bigger, unorganized masses. Sixth, education is acknowledged as a sixth element of social power: As the chapters in this section argue, well-educated professionals have the ability to speak as experts, to organize moral entrepreneurial campaigns, to advocate for their positions, and to argue from a legitimate base of knowledge. Finally, social status (apart from social class) generates power through the prestige, tradition, and respectability associated with various positions in society. For example, religious people have greater social status in contemporary society than atheists have, heterosexual people command greater legitimacy than homosexuals, and married individuals hold greater sway, as a group, than single ones. There are, of course, many more elements of social power that could be articulated as having an influence over labeling individuals, groups, and their characteristics as deviant, but to us, those just presented are the key ones.

In a society characterized by striving for social influence, status, and power, one way to attain those ideals is to pass and enforce rules that define others' behavior as deviant. Thus, taken in this light, the labeling as deviant of attitudes, behaviors, and conditions such as minority ethnic or racial status, female gender, lower social class, youthful age, homosexual orientation, and a criminal record (as some of the readings in this section show), can be seen to reflect the application of differential social power in our society. Individuals in these groups may find themselves discriminated against or blocked from the mainstream of society by virtue of this basic feature of their existence, unrelated to any particular situation or act. This application of the deviant label emphatically illustrates the role of power in the deviance-defining enterprise, as those positioned closer to the center of society, holding the greater social, economic, political, and moral resources, can turn the force of the deviant stigma onto others less fortunately placed. In so doing, they use the definition of deviance to reinforce their own favored position. This politicization of deviance and the power associated with its use serve to remind us that deviance is not a category applied only to those on the marginal outskirts of society: the exotics, erotics, and neurotics. Instead, any group can be pushed into the category by the exercise of another group's greater power.

DIFFERENTIAL SOCIAL POWER:
RESISTING LABELING

On the other side of the coin, powerful groups may be successful in working to resist the application of definitions of deviance to them. Like better looking people, they are granted a "halo effect" that leads others to think highly of them. For example, higher status groups in society are less likely to be perceived as deviant, whether they actively work to fight the label or not.

In some cases, privileged groups undertake proactive collective identity protection. In this regard, many organizations, such as gun owners, pharmaceutical companies, and the manufacturers of cigarettes or alcoholic beverages, work to build and sustain a positive social image. They may hire claimsmakers or lobbyists, contribute to politicians' campaigns, or fund research showing that they are upstanding individuals or organizations. (Recall the aforementioned study showing that research funded by the soft-drink industry was eight times as likely to find no negative correlation between consuming soft drinks and poor health as those otherwise financed.)

But even when they do not become specifically involved in protecting their images, members of more powerful and respected groups in society are less likely to become tainted by deviant labels. Part of the reason is that people hold preconceived biases in favor of those groups and assume that they are responsible and pro-social, whether they are or are not. People also become biased toward them on the basis of their appearances, occupations, behavior, and/or associations, forming instantaneous judgments about them that are positive. Members of such protected groups are often unaware of their privileged status in society and do not realize the discrimination routinely encountered by underprivileged populations.

Differential social power may be applied either directly, as when individuals or groups are judged on their own, or comparatively, as when society judges the behavior of one group against another. Together, these groups of people continue to receive treatment from society as either deviant or nondeviant that reinforces social inequality and the status quo.

15

The Social Construction of Drug Scares

CRAIG REINARMAN

In this overview of U.S. social policies, Reinarman tackles moral and legal attitudes toward illicit drugs. He briefly offers a history of drug scares, the major players engineering them, and the social contexts that have enhanced their development and growth. He then outlines seven factors common to drug scares. Knowledge of these factors enables him to dissect the essential processes in the rule creation and enforcement phases of drug scares, despite the contradictory cultural values of temperance and hedonistic consumption. From this selection, we can see how drugs have been scapegoated to account for a wide array of social problems and used, to keep some groups down by defining their actions as deviant. It is clear that despite our society's views on the negative features associated with all illicit drugs, our moral entrepreneurial and enforcement efforts have been concentrated more stringently against the drugs used by members of the powerless underclass and minority racial groups.

What kind of analysis do you think Reinarman is offering us here of the root causes of drug scares: structural, cultural, or interactionist (or some combination)?

D rug "wars," antidrug crusades, and other periods of marked public concern about drugs are never merely reactions to the various troubles people can have with drugs. These drug scares are recurring cultural and political phenomena *in their own right* and must, therefore, be understood sociologically on their own terms. It is important to understand why people ingest drugs and why some of them develop problems that have something to do with having ingested them. But the premise of this chapter is that it is equally important to understand patterns of acute societal concern about drug use and drug problems. This seems especially so for U.S. society, which has had *recurring* antidrug crusades and a *history* of repressive antidrug laws.

Many well-intentioned drug policy reform efforts in the United States have come face to face with staid and stubborn sentiments against consciousness-altering substances. The repeated failures of such reform efforts cannot be

Reprinted by permission of Craig Reinarman.

explained solely in terms of ill-informed or manipulative leaders. Something deeper is involved, something woven into the very fabric of U.S. culture, something which explains why claims that some drug is the cause of much of what is wrong with the world are *believed* so often by so many. The origins and nature of the *appeal* of antidrug claims must be confronted if we are ever to understand how "drug problems" are constructed in the United States such that more enlightened and effective drug policies have been so difficult to achieve.

In this chapter I take a step in this direction. First, I summarize briefly some of the major periods of antidrug sentiment in the United States. Second, I draw from them the basic ingredients of which drug scares and drug laws are made. Third, I offer a beginning interpretation of these scares and laws based on those broad features of American culture that make *self-control* continuously problematic.

DRUG SCARES AND DRUG LAWS

What I have called drug scares (Reinarman and Levine, 1989a) have been a recurring feature of U.S. society for 200 years. They are relatively autonomous from whatever drug-related problems exist or are said to exist.[1] I call them "scares" because, like Red Scares, they are a form of moral panic ideologically constructed so as to construe one or another chemical bogeyman, à la "communists," as the core cause of a wide array of preexisting public problems.

The first and most significant drug scare was over drink. Temperance movement leaders constructed this scare beginning in the late eighteenth and early nineteenth century. It reached its formal end with the passage of Prohibition in 1919.[2] As Gusfield showed in his classic book *Symbolic Crusade* (1963), there was far more to the battle against booze than long-standing drinking problems. Temperance crusaders tended to be native born, middle-class, non-urban Protestants who felt threatened by the working-class, Catholic immigrants who were filling up U.S. cities during industrialization.[3] The latter were what Gusfield termed "unrepentant deviants" in that they continued their long-standing drinking practices despite middle-class W.A.S.P. norms against them. The battle over booze was the terrain on which was fought a cornucopia of cultural conflicts, particularly over whose morality would be the dominant morality in the United States.

In the course of this century-long struggle, the often wild claims of Temperance leaders appealed to millions of middle-class people seeking explanations for the pressing social and economic problems of industrializing the United States. Many corporate supporters of Prohibition threw their financial and ideological weight behind the Anti-Saloon League and other Temperance and Prohibitionist groups because they felt that traditional working-class drinking practices interfered with the new rhythms of the factory, and thus with productivity and profits (Rumbarger, 1989). To the Temperance crusaders' fear of the bar room as a breeding ground of all sorts of tragic immorality, Prohibitionists added the idea of the saloon as an alien, subversive place where unionists organized and where leftists and anarchists found recruits (Levine, 1984).

This convergence of claims and interests rendered alcohol a scapegoat for most of the nation's poverty, crime, moral degeneracy, "broken" families, illegitimacy, unemployment, and personal and business failure—problems whose sources lay in broader economic and political forces. This scare climaxed in the first two decades of this century, a tumultuous period rife with class, racial, cultural, and political conflict brought on by the wrenching changes of industrialization, immigration, and urbanization (Levine, 1984; Levine and Reinarman, 1991).

The U.S. first real drug law was San Francisco's antiopium den ordinance of 1875. The context of the campaign for this law shared many features with the context of the Temperance movement. Opiates had long been widely and legally available without a prescription in hundreds of medicines (Brecher, 1972; Musto, 1973; Courtwright, 1982; cf. Baumohl, 1992), so neither opiate use nor addiction was really the issue. This campaign focused almost exclusively on what was called the "Mongolian vice" of opium *smoking* by Chinese immigrants (and white "fellow travelers") in dens (Baumohl, 1992). Chinese immigrants came to California as "coolie" labor to build the railroad and dig the gold mines. A small minority of them brought along the practice of smoking opium—a practice originally brought to China by British and American traders in the nineteenth century. When the railroad was completed and the gold dried up, a decade-long depression ensued. In a tight labor market, Chinese immigrants were a target. The white Workingman's Party fomented racial hatred of the low-wage "coolies" with whom they now had to compete for work. The first law against opium smoking was only one of many laws enacted to harass and control Chinese workers (Morgan, 1978).

By calling attention to this broader political–economic context I do not wish to slight the specifics of the local political–economic context. In addition to the Workingman's Party, downtown businessmen formed merchant associations and urban families formed improvement associations, both of which fought for more than two decades to reduce the impact of San Francisco's vice districts on the order and health of the central business district and on family neighborhoods (Baumohl, 1992).

In this sense, the antiopium den ordinance was not the clear and direct result of a sudden drug scare alone. The law was passed against a specific form of drug use engaged in by a disreputable group that had come to be seen as threatening in lean economic times. But it passed easily because this new threat was understood against the broader historical backdrop of long-standing local concerns about various vices as threats to public health, public morals, and public order. Moreover, the focus of attention was dens where it was suspected that whites came into intimate contact with "filthy, idolatrous" Chinese (see Baumohl, 1992). Some local law enforcement leaders, for example, complained that Chinese men were using this vice to seduce white women into sexual slavery (Morgan, 1978). Whatever the hazards of opium smoking, its initial criminalization in San Francisco had to do with both a general context of recession, class conflict, and racism, and with specific local interests in the control of vice and the prevention of miscegenation.

A nationwide scare focusing on opiates and cocaine began in the early twentieth century. These drugs had been widely used for years, but were first criminalized when the addict population began to shift from predominantly white, middle-class, middle-aged women to young, working-class males, African Americans in particular. This scare led to the Harrison Narcotics Act of 1914, the first federal antidrug law (see Duster, 1970).

Many different moral entrepreneurs guided its passage over a six-year campaign: State Department diplomats seeking a drug treaty as a means of expanding trade with China, trade which they felt was crucial for pulling the economy out of recession; the medical and pharmaceutical professions whose interests were threatened by self-medication with unregulated proprietary tonics, many of which contained cocaine or opiates; reformers seeking to control what they saw as the deviance of immigrants and Southern Blacks who were migrating off the farms; and a pliant press which routinely linked drug use with prostitutes, criminals, transient workers (e.g., the Wobblies), and African Americans (Musto, 1973). In order to gain the support of Southern Congressmen for a new federal law that might infringe on "states' rights," State Department officials and other crusaders repeatedly spread unsubstantiated suspicions, repeated in the press, that, for example, cocaine induced African-American men to rape white women (Musto, 1973: 6–10, 67). In short, there was more to this drug scare, too, than mere drug problems.

In the Great Depression, Harry Anslinger of the Federal Narcotics Bureau pushed Congress for a federal law against marijuana. He claimed it was a "killer weed" and he spread stories to the press suggesting that it induced violence— especially among Mexican Americans. Although there was no evidence that marijuana was widely used, much less that it had any untoward effects, his crusade resulted in its criminalization in 1937—and not incidentally a turnaround in his Bureau's fiscal fortunes (Dickson, 1968). In this case, a new drug law was put in place by a militant moral-bureaucratic entrepreneur who played on racial fears and manipulated a press willing to repeat even his most absurd claims in a context of class conflict during the Depression (Becker, 1963). While there was not a marked scare at the time, Anslinger's claims were never contested in Congress because they played upon racial fears and widely held Victorian values against taking drugs solely for pleasure.

In the drug scare of the 1960s, political and moral leaders somehow reconceptualized this same "killer weed" as the "drop out drug" that was leading America's youth to rebellion and ruin (Himmelstein, 1983). Bio-medical scientists also published uncontrolled, retrospective studies of very small numbers of cases suggesting that, in addition to poisoning the minds and morals of youth, LSD produced broken chromosomes and thus genetic damage (Cohen et al., 1967). These studies were soon shown to be seriously misleading if not meaningless (Tijo et al., 1969), but not before the press, politicians, the medical profession, and the National Institute of Mental Health used them to promote a scare (Weil, 1972: 44–46).

I suggest that the reason even supposedly hard-headed scientists were drawn into such propaganda was that dominant groups felt the country was at war—and not merely with Vietnam. In this scare, there was not so much a "dangerous class" or threatening racial group as multi-faceted political and cultural conflict, particularly between generations, which gave rise to the perception that middle-class youth who rejected conventional values were a dangerous threat.[4] This scare resulted in the Comprehensive Drug Abuse Control Act of 1970, which criminalized more forms of drug use and subjected users to harsher penalties.

Most recently we have seen the crack scare, which began in earnest *not* when the prevalence of cocaine use quadrupled in the late 1970s, nor even when thousands of users began to smoke it in the more potent and dangerous form of free-base. Indeed, when this scare was launched, crack was unknown outside of a few neighborhoods in a handful of major cities (Reinarman and Levine, 1989a) and the prevalence of illicit drug use had been dropping for several years (National Institute on Drug Use, 1990). Rather, this most recent scare began in 1986 when freebase cocaine was renamed crack (or "rock") and sold in precooked, inexpensive units on ghetto street corners (Reinarman and Levine, 1989b). Once politicians and the media linked this new form of cocaine use to the inner-city, minority poor, a new drug scare was underway and the solution became more prison cells rather than more treatment slots.

The same sorts of wild claims and Draconian policy proposals of Temperance and Prohibition leaders resurfaced in the crack scare. Politicians have so outdone each other in getting "tough on drugs" that each year since crack came on the scene in 1986 they have passed more repressive laws providing billions more for law enforcement, longer sentences, and more drug offenses punishable by death. One result is that the United States now has more people in prison than any industrialized nation in the world—about half of them for drug offenses, the majority of whom are racial minorities.

In each of these periods more repressive drug laws were passed on the grounds that they would reduce drug use and drug problems. I have found no evidence that any scare actually accomplished those ends, but they did greatly expand the quantity and quality of social control, particularly over subordinate groups perceived as dangerous or threatening. Reading across these historical episodes one can abstract a recipe for drug scares and repressive drug laws that contains the following *seven ingredients:*

1. **A Kernel of Truth** Humans have ingested fermented beverages at least since human civilization moved from hunting and gathering to primitive agriculture thousands of years ago. The pharmacopoeia has expanded exponentially since then. So, in virtually all cultures and historical epochs, there has been sufficient ingestion of consciousness-altering chemicals to provide some basis for some people to claim that it is a problem.

2. **Media Magnification** In each of the episodes I have summarized and many others, the mass media has engaged in what I call the *routinization of caricature*—rhetorically recrafting worst cases into typical cases and the episodic into the epidemic. The media dramatize drug problems, as they do other problems, in the course of their routine news-generating and sales-promoting procedures (see Brecher, 1972: 321–34; Reinarman and Duskin, 1992; and Molotch and Lester, 1974).

3. **Politico-Moral Entrepreneurs I** have added the prefix "politico" to Becker's (1963) seminal concept of moral entrepreneur in order to

emphasize the fact that the most prominent and powerful moral entrepreneurs in drug scares are often political elites. Otherwise, I employ the term just as he intended: to denote the *enterprise,* the work, of those who create (or enforce) a rule against what they see as a social evil.[5]

In the history of drug problems in the United States, these entrepreneurs call attention to drug using behavior and define it as a threat about which "something must be done." They also serve as the media's primary source of sound bites on the dangers of this or that drug. In all the scares I have noted, these entrepreneurs had interests of their own (often financial) which had little to do with drugs. Political elites typically find drugs a functional demon in that (like "outside agitators") drugs allow them to deflect attention from other, more systemic sources of public problems for which they would otherwise have to take some responsibility. Unlike almost every other political issue, however, to be "tough on drugs" in U.S. political culture allows a leader to take a firm stand without risking votes or campaign contributions.

4. **Professional Interest Groups** In each drug scare and during the passage of each drug law, various professional interests contended over what Gusfield (1981: 10–15) calls the "ownership" of drug problems—"the ability to create and influence the public definition of a problem" (1981: 10), and thus to define what should be done about it. These groups have included industrialists, churches, the American Medical Association, the American Pharmaceutical Association, various law enforcement agencies, scientists, and most recently the treatment industry and groups of those former addicts converted to disease ideology.[6] These groups claim for themselves, by virtue of their specialized forms of knowledge, the legitimacy and authority to name what is wrong and to prescribe the solution, usually garnering resources as a result.

5. **Historical Context of Conflict** This trinity of the media, moral entrepreneurs, and professional interests typically interact[s] in such a way as to inflate the extant "kernel of truth" about drug use. But this interaction does not by itself give rise to drug scares or drug laws without underlying conflicts which make drugs into functional villains. Although Temperance crusaders persuaded millions to pledge abstinence, they campaigned for years without achieving alcohol control laws. However, in the tumultuous period leading up to Prohibition, there were revolutions in Russia and Mexico, World War I, massive immigration and impoverishment, and socialist, anarchist, and labor movements, to say nothing of increases in routine problems such as crime. I submit that all this conflict made for a level of cultural anxiety that provided fertile ideological soil for Prohibition. In each of the other scares, similar conflicts— economic, political, cultural, class, racial, or a combination—provided a context in which claims makers could viably construe certain classes of drug users as a threat.

6. **Linking a Form of Drug Use to a "Dangerous Class"** Drug scares are never about drugs *per se,* because drugs are inanimate objects without social consequence until they are ingested by humans. Rather, drug scares are about the use of a drug by particular groups of people who are, typically,

already perceived by powerful groups as some kind of threat (see Duster, 1970; Himmelstein, 1978). It was not so much alcohol problems *per se* that most animated the drive for Prohibition but the behavior and morality of what dominant groups saw as the "dangerous class" of urban, immigrant, Catholic, working-class drinkers (Gusfield, 1963; Rumbarger, 1989). It was *Chinese* opium smoking dens, not the more widespread use of other opiates, that prompted California's first drug law in the 1870s. It was only when smokable cocaine found its way to the African-American and Latino underclass that it made headlines and prompted calls for a drug war. In each case, politico–moral entrepreneurs were able to construct a "drug problem" by linking a substance to a group of users perceived by the powerful as disreputable, dangerous, or otherwise threatening.

7. **Scapegoating a Drug for a Wide Array of Public Problems** The final ingredient is scapegoating, that is, blaming a drug or its alleged effects on a group of its users for a variety of preexisting social ills that are typically only indirectly associated with it. Scapegoating may be the most crucial element because it gives great explanatory power and thus broader resonance to claims about the horrors of drugs (particularly in the conflictual historical contexts in which drug scares tend to occur).

Scapegoating was abundant in each of the cases noted previously. To listen to Temperance crusaders, for example, one might have believed that without alcohol use, the United States would be a land of infinite economic progress with no poverty, crime, mental illness, or even sex outside marriage. To listen to leaders of organized medicine and the government in the 1960s, one might have surmised that without marijuana and LSD there would have been neither conflict between youth and their parents nor opposition to the Vietnam War. And to believe politicians and the media in the past six years is to believe that without the scourge of crack the inner cities and the so-called underclass would, if not disappear, at least be far less scarred by poverty, violence, and crime. There is no historical evidence supporting any of this.

In short, drugs are richly functional scapegoats. They provide elites with fig leaves to place over unsightly social ills that are endemic to the social system over which they preside. And they provide the public with a restricted aperture of attribution in which only a chemical bogeyman or the lone deviants who ingest it are seen as the cause of a cornucopia of complex problems.

TOWARD A CULTURALLY SPECIFIC
THEORY OF DRUG SCARES

Various forms of drug use have been and are widespread in almost all societies comparable to ours. A few of them have experienced limited drug scares, usually around alcohol decades ago. However, drug scares have been *far* less common in other societies, and never as virulent as they have been in the United States

(Brecher, 1972; Levine, 1992; MacAndrew and Edgerton, 1969). There has never been a time or place in human history without drunkenness, for example, but in *most* times and places drunkenness has not been nearly as problematic as it has been in the United States since the late eighteenth century. Moreover, in comparable industrial democracies, drug laws are generally less repressive. Why then do claims about the horrors of this or that consciousness-altering chemical have such unusual power in U.S. culture?

Drug scares and other periods of acute public concern about drug use are not just discrete, unrelated episodes. There is a historical pattern in the United States that cannot be understood in terms of the moral values and perceptions of individual antidrug crusaders alone. I have suggested that these crusaders have benefited in various ways from their crusades. For example, making claims about how a drug is damaging society can help elites increase the social control of groups perceived as threatening (Duster, 1970), establish one class's moral code as dominant (Gusfield, 1963), bolster a bureaucracy's sagging fiscal fortunes (Dickson, 1968), or mobilize voter support (Reinarman and Levine, 1989a, b). However, the recurring character of pharmaco-phobia in U.S. history suggests that there is something about our *culture* which makes citizens more vulnerable to antidrug crusaders' attempts to demonize drugs. Thus, an answer to the question of U.S. unusual vulnerability to drug scares must address why the scapegoating of consciousness-altering substances regularly *resonates* with or appeals to substantial portions of the population.

There are three basic parts to my answer. The first is that claims about the evils of drugs are especially viable in U.S. culture in part because they provide a welcome *vocabulary of attribution* (cf. Mills, 1940). Armed with "DRUGS" as a generic scapegoat, citizens gain the cognitive satisfaction of having a folk devil on which to blame a range of bizarre behaviors or other conditions they find troubling but difficult to explain in other terms. This much may be true of a number of other societies, but I hypothesize that this is particularly so in the United States because in our political culture individualistic explanations for problems are so much more common than social explanations.

Second, claims about the evils of drugs provide an especially serviceable vocabulary of attribution in the United States in part because our society developed from a *temperance culture* (Levine, 1992). The U.S. society was forged in the fires of ascetic Protestantism and industrial capitalism, both of which demand *self-control*. The U.S. society has long been characterized as the land of the individual "self-made man." In such a land, self-control has had extraordinary importance. For the middle-class Protestants who settled, defined, and still dominate the United States, self-control was both central to religious worldviews and a characterological necessity for economic survival and success in the capitalist market (Weber, 1930 [1985]). With Levine (1992), I hypothesize that in a culture in which self-control is inordinately important, drug-induced altered states of consciousness are especially likely to be experienced as "loss of control," and thus to be inordinately feared.[7]

Drunkenness and other forms of drug use have, of course, been present everywhere in the industrialized world. But temperance cultures tend to arise only when industrial capitalism unfolds upon a cultural terrain deeply imbued with the Protestant ethic.[8] This means that only the United States, England, Canada, and parts of Scandinavia have Temperance cultures, the United States being the most extreme case.

It may be objected that the influence of such a Temperance culture was strongest in the nineteenth and early twentieth century and that its grip on the U.S. *Zeitgeist* has been loosened by the forces of modernity and now, many say, post-modernity. The third part of my answer, however, is that on the foundation of a Temperance culture, advanced capitalism has built a *postmodern, mass consumption culture* that exacerbates the problem of self-control in new ways.

Early in the twentieth century, Henry Ford pioneered the idea that by raising wages he could simultaneously quell worker protests and increase market demand for mass-produced goods. This mass consumption strategy became central to modern U.S. society and one of the reasons for our economic success (Marcuse, 1964; Aronowitz, 1973; Ewen, 1976; Bell, 1978). Our economy is now so fundamentally predicated upon mass consumption that theorists as diverse as Daniel Bell and Herbert Marcuse have observed that we live in a mass consumption culture. Bell (1978), for example, notes that while the Protestant work ethic and deferred gratification may still hold sway in the workplace, Madison Avenue, the media, and malls have inculcated a new indulgence ethic in the leisure sphere in which pleasure seeking and immediate gratification reign.

Thus, our economy and society have come to depend upon the constant cultivation of new "needs," the production of new desires. Not only the hardware of social life such as food, clothing, and shelter but also the software of the self—excitement, entertainment, even eroticism—have become mass consumption commodities. This means that our society offers an increasing number of incentives for indulgence—more ways to lose self-control—and a decreasing number of countervailing reasons for retaining it.

In short, drug scares continue to occur in U.S. society in part because people must constantly manage the contradiction between a Temperance culture that insists on self-control and a mass consumption culture which renders self-control continuously problematic. In addition to helping explain the recurrence of drug scares, I think this contradiction helps account for why in the last dozen years millions of Americans have joined 12-Step groups, more than 100 of which have nothing whatsoever to do with ingesting a drug (Reinarman, 1995). "Addiction," or the generalized loss of self-control, has become the meta-metaphor for a staggering array of human troubles. And, of course, we also seem to have a staggering array of politicians and other moral entrepreneurs who take advantage of such cultural contradictions to blame new chemical bogeymen for our society's ills.

NOTES

1. In this regard, for example, Robin Room wisely observes "that we are living at a historic moment when the rate of (alcohol) dependence as a cognitive and existential experience is rising, although the rate of alcohol

consumption and of heavy drinking is falling." He draws from this a more general hypothesis about "long waves" of drinking and societal reactions to them: "[I]n periods of increased questioning of drinking and heavy drinking, the trends in the two forms of dependence, psychological and physical, will tend to run in opposite directions. Conversely, in periods of a "wettening" of sentiments, with the curve of alcohol consumption beginning to rise, we may expect the rate of physical dependence... to rise while the rate of dependence as a cognitive experience falls" (1991: 154).

2. I say "formal end" because Temperance ideology is not merely alive and well in the War on Drugs but is being applied to all manner of human troubles in the burgeoning 12-Step Movement (Reinarman, 1995).

3. From Jim Baumohl I have learned that while the Temperance movement attracted most of its supporters from these groups, it also found supporters among many others (e.g., labor, the Irish, Catholics, former drunkards, women), each of which had its own reading of and folded its own agenda into the movement.

4. This historical sketch of drug scares is obviously not exhaustive. Readers interested in other scares should see, for example, Brecher's encyclopedic work *Licit and Illicit Drugs* (1972), especially the chapter on glue sniffing, which illustrates how the media actually created a new drug problem by writing hysterical stories about it. There was also a PCP scare in the 1970s in which law enforcement officials claimed that the growing use of this horse tranquilizer was a severe threat because it made users so violent and gave them such super-human strength that stun guns were necessary. This, too, turned out to be unfounded

and the "angel dust" scare was short-lived (see Feldman et al., 1979). The best analysis of how new drugs themselves can lead to panic reactions among users is Becker (1967).

5. Becker wisely warns against the "onesided view" that sees such crusaders as merely imposing their morality on others. Moral entrepreneurs, he notes, do operate "with an absolute ethic," are "fervent and righteous," and will use "any means" necessary to "do away with" what they see as "totally evil." However, they also "typically believe that their mission is a holy one," that if people do what they want it "will be good for them." Thus, as in the case of abolitionists, the crusades of moral entrepreneurs often "have strong humanitarian overtones" (1963: 147–8). This is no less true for those whose moral enterprise promotes drug scares. My analysis, however, concerns the character and consequences of their efforts, not their motives.

6. As Gusfield notes, such ownership sometimes shifts over time, for example, with alcohol problems, from religion to criminal law to medical science. With other drug problems, the shift in ownership has been away from medical science toward criminal law. The most insightful treatment of the medicalization of alcohol/drug problems is Peele (1989).

7. See Baumohl's (1990) important and erudite analysis of how the human will was valorized in the therapeutic temperance thought of nineteenth-century inebriate homes.

8. The third central feature of Temperance cultures identified by Levine (1992), which I will not dwell on, is predominance of spirits drinking, that is, more concentrated alcohol than wine or beer and thus greater likelihood of drunkenness.

REFERENCES

Aronowitz, Stanley. (1973). *False Promises: The Shaping of American Working Class Consciousness*. New York: McGraw-Hill.

Baumohl, Jim. (1990). "Inebriate Institutions in North America, 1840–1920." *British Journal of Addidion* 85: 1187–1204.

Baumohl, Jim. (1992). "The 'Dope Fiend's Paradise' Revisited: Notes from Research in Progress on Drug Law Enforcement in San Francisco, 1875–1915." *Drinking and Drug Practices Surveyor* 24: 3–12.

Becker, Howard S. (1963). *Outsiders: Studies in the Sociology of Deviance*. Glencoe, IL: Free Press.

_____. (1967). "History, Culture, and Subjective Experience: An Exploration of the Social Bases of Drug-Induced Experiences." *Journal of Health and Social Behavior* 8: 162–176.

Bell, Daniel. (1978). *The Cultural Contradictions of Capitalism*. New York: Basic Books.

Brecher, Edward M. (1972). *Licit and Illicit Drugs*. Boston: Little Brown.

Cohen, M. M., K. Hirshorn, and W. A. Frosch. (1967). "In Vivo and in Vitro Chromosomal Damage Induced by LSD-25." *New England Journal of Medicine* 227: 1043.

Courtwright, David. (1982). *Dark Paradise: Opiate Addiction in America Before 1940*. Cambridge, MA: Harvard University Press.

Dickson, Donald. (1968). "Bureaucracy and Morality." *Social Problems* 16: 143–156.

Duster, Troy. (1970). *The Legislation of Morality: Law, Drugs, and Moral Judgement*. New York: Free Press.

Ewen, Stuart. (1976). *Captains of Consciousness: Advertising and the Social Roots of Consumer Culture*. New York: McGraw-Hill.

Feldman, Harvey W., Michael H. Agar, and George M. Beschner. (1979). *Angel Dust*. Lexington, MA: Lexington Books.

Gusfield, Joseph R. (1963). *Symbolic Crusade: Status Politics and the American Temperance Movement*. Urbana: University of Illinois Press.

_____. (1981). *The Culture of Public Problems: Drinking–Driving and the Symbolic Order*. Chicago: University of Chicago Press.

Himmelstein, Jerome. (1978). "Drug Politics Theory." *Journal of Drug Issues* 8.

_____. (1983). *The Strange Career of Marihuana*. Westport, CT: Greenwood Press.

Levine, Harry Gene. (1984). "The Alcohol Problem in America: From Temperance to Alcoholism." *British Journal of Addiction* 84: 109–119.

_____. (1992). "Temperance Cultures: Concern About Alcohol Problems in Nordic and English-Speaking Cultures." In G. Edwards et al., eds., *The Nature of Alcohol and Drug Related Problems*. New York: Oxford University Press.

Levine, Harry Gene, and Craig Reinarman. (1991). "From Prohibition to Regulation: Lessons from Alcohol Policy for Drug Policy." *Milbank Quarterly* 69: 461–494.

MacAndrew, Craig, and Robert Edgerton. (1969). *Drunken Comportment*. Chicago: Aldine.

Marcuse, Herbert. (1964). *One-Dimensional Man: Studies in the Ideology of Advanced Industrial Society*. Boston: Beacon Press.

Mills, C. Wright. (1940). "Situated Actions and Vocabularies of Motive." *American Sociological Review* 5: 904–913.

Molotch, Harvey, and Marilyn Lester. (1974). "News as Purposive Behavior: On the Strategic Uses of Routine Events, Accidents, and Scandals." *American Sociological Review* 39: 101–112.

Morgan, Patricia. (1978). "The Legislation of Drug Law: Economic Crisis and Social Control." *Journal of Drug Issues* 8: 53–62.

Musto, David. (1973). *The American Disease: Origins of Narcotic Control.* New Haven, CT: Yale University Press.

National Institute on Drug Abuse. (1990). *National Household Survey on Drug Abuse: Main Findings 1990.* Washington, DC: U.S. Department of Health and Human Services.

Peele, Stanton. (1989). *The Diseasing of America: Addiction Treatment Out of Control.* Lexington, MA: Lexington Books.

Reinarman, Craig. (1995). "The 12-Step Movement and Advanced Capitalist Culture: Notes on the Politics of Self-Control in Postmodernity." In B. Epstein, R. Flacks, and M. Darnovsky, eds., *Contemporary Social Movements and Cultural Politics.* New York: Oxford University Press.

Reinarman, Craig, and Ceres Duskin. (1992). "Dominant Ideology and Drugs in the Media." *International Journal on Drug Policy* 3: 6–15.

Reinarman, Craig, and Harry Gene Levine. (1989a). "Crack in Context: Politics and Media in the Making of a Drug Scare." *Contemporary Drug Problems* 16: 535–577.

———. (1989b). "The Crack Attack: Politics and Media in America's Latest Drug Scare." In Joel Best, ed., *Images of Issues: Typifying Contemporary Social Problems,* pp. 115–137. New York: Aldine de Gruyter.

Room, Robin G. W. (1991). "Cultural Changes in Drinking and Trends in Alcohol Problems Indicators: Recent U.S. Experience." In Walter B. Clark and Michael E. Hilton, eds., *Alcohol in America: Drinking Practices and Problems,* pp. 149–162. Albany: State University of New York Press.

Rumbarger, John J. (1989). *Profits, Power, and Prohibition: Alcohol Reform and the Industrializing of America, 1800–1930.* Albany: State University of New York Press.

Tijo. J. H., W. N. Pahnke, and A. A. Kurland. (1969). "LSD and Chromosomes: A Controlled Experiment." *Journal of the American Medical Association* 210: 849.

Weber, Max. (1985 [1930]). *The Protestant Ethic and the Spirit of Capitalism.* London: Unwin.

Weil, Andrew. (1972). *The Natural Mind.* Boston: Houghton Mifflin.

16

Blowing Smoke: Status Politics and the Smoking Ban

JUSTIN L. TUGGLE AND MALCOLM D. HOLMES

Tuggle and Holmes's chapter on the debate over cigarette smoking in the United States expands Reinarman's consideration to the licit drug realm, examining the struggle and counterstruggle over tobacco between the moral entrepreneurs and the status quo defenders. They note the medical, ethical, and socioeconomic arguments raised to sway public opinion and demonize public consumption of cigarettes. They mention the spate of claims put forward, pitting antismoking groups, which have argued that secondhand smoke is toxic and that smokers should not be allowed to inflict their pollution onto others, against opponents who have argued that the government should not legislate their morality. These issues illustrate the concern that frequently arises over deviance, a concern which manifests itself in the idea that society must balance the right of individual freedoms (the desire to smoke) against the needs of the common good (public health). This selection also shows the relation between making claims and exercising social power, tracing the status of the social groups on each side of the antismoking campaign. Its fundamental message lies in how moral entrepreneurs use their status to attach deviant labels to the behavior of others and, in so doing, to keep those others in a subordinate position. Tuggle and Holmes thus show deviance, as Quinney does, to be a tool by which groups with higher status in society retain and enforce their interests over subordinate groups.

What features of Reinarman's view of how deviance is socially constructed do we see reinforced by Tuggle and Holmes? How does their thesis explain why some groups in society carry out moral crusades against issues that seemingly have little direct impact on them?

O ver the past half century, perceptions of tobacco and its users have changed dramatically. In the 1940s and 1950s, cigarette smoking was socially accepted and commonly presumed to lack deleterious effects (e.g., Ram, 1941). Survey data from the early 1950s showed that a minority believed cigarette smoking caused lung cancer (Viscusi, 1992). By the late 1970s, however,

From "Blowing Smoke: Status Politics and the Smoking Ban," Justin L. Tuggle and Malcolm D. Holmes, Deviant Behavior, 1997, Taylor and Francis. Reprinted with permission by the authors.

estimates from survey data revealed that more than 90 percent of the population thought that this link existed (Roper Organization, 1978). This and other harms associated with tobacco consumption have provided the impetus for an anti-smoking crusade that aims to normatively redefine smoking as deviant behavior (Markle and Troyer, 1979).

There seems to be little question that tobacco is a damaging psychoactive substance characterized by highly adverse chronic health effects (Steinfeld, 1991). In this regard, the social control movement probably makes considerable sense in terms of public policy. At the same time, much as ethnicity and religion played a significant role in the prohibition of alcohol (Gusfield, 1963), social status may well play a part in this latest crusade.

Historically, attempts to control psychoactive substances have linked their use to categories of relatively powerless people. Marijuana use was associated with Mexican Americans (Bonnie and Whitebread, 1970), cocaine with African Americans (Ashley, 1975), opiates with Asians (Ben-Yehuda, 1990), and alcohol with immigrant Catholics (Gusfield, 1963). During the heyday of cigarette smoking, it was thought that

> Tobacco's the one blessing that nature has left for all humans to enjoy.
> It can be consumed by both the "haves" and "have nots" as a common
> leveler, one that brings all humans together from all walks of life
> regardless of class, race, or creed. (Ram, 1941, p. 125)

But in contrast to this earlier view, recent evidence has shown that occupational status (Ferrence, 1989; Marcus et al., 1989; Covey et al., 1992), education (Ferrence, 1989; Viscusi, 1992), and family income (Viscusi, 1992) are related negatively to current smoking. Further, the relationships of occupation and education to cigarette smoking have become stronger in later age cohorts (Ferrence, 1989). Thus we ask, *is the association of tobacco with lower-status persons a factor in the crusade against smoking in public facilities?* Here we examine that question in a case study of a smoking ban implemented in Shasta County, California.

STATUS POLITICS AND THE CREATION
OF DEVIANCE

Deviance is socially constructed. Complex pluralistic societies have multiple, competing symbolic–moral universes that clash and negotiate (Ben-Yehuda, 1990). Deviance is relative, and social morality is continually restructured. Moral, power, and stigma contests are ongoing, with competing symbolic–moral universes striving to legitimize particular lifestyles while making others deviant (Schur, 1980; Ben-Yehuda, 1990).

The ability to define and construct reality is closely connected to the power structure of society (Gusfield, 1963). Inevitably, then, the distribution of deviance is associated with the system of stratification. The higher one's social position, the greater [is] one's moral value (Ben-Yehuda, 1990). Differences in lifestyles and moral beliefs are corollaries of social stratification (Gusfield, 1963; Zürcher and Kirkpatrick, 1976; Luker, 1984). Accordingly, even though grounded in the system of stratification, status conflicts need not be instrumental; they may also be symbolic. Social stigma may, for instance, attach to behavior thought indicative of a weak will (Goffman, 1963). Such moral anomalies occasion status degradation ceremonies, public denunciations expressing indignation not at a behavior per se, but rather against the individual motivational type that produced it (Garfinkel, 1956). The denouncers act as public figures, drawing upon communally shared experience and speaking in the name of ultimate values. In this respect, status degradation involves a reciprocal element: Status conflicts and the resultant condemnation of a behavior characteristic of a particular status category symbolically enhances the status of the abstinent through the degradation of the participatory (Garfinkel, 1956; Gusfield, 1963).

Deviance creation involves political competition in which moral entrepreneurs originate moral crusades aimed at generating reform (Becker, 1963; Schur, 1980; Ben-Yehuda, 1990). The alleged deficiencies of a specific social group are revealed and reviled by those crusading to define their behavior as deviant. As might be expected, successful moral crusades are generally dominated by those in the upper social strata of society (Becker, 1963). Research on the antiabortion (Luker, 1984) and antipornography (Zürcher and Kirkpatrick, 1976) crusades has shown that activists in these movements are of lower socioeconomic status than their opponents, helping explain the limited success of efforts to redefine abortion and pornography as deviance.

Moral entrepreneurs' goals may be either assimilative or coercive reform (Gusfield, 1963). In the former instance, sympathy to the deviants' plight engenders integrative efforts aimed at lifting the repentant to the superior moral plane allegedly held by those of higher social status. The latter strategy emerges when deviants are viewed as intractably denying the moral and status superiority of the reformers' symbolic–moral universe. Although assimilative reform may employ educative strategies, coercive reform turns to law and force for affirmation.

Regardless of aim, the moral entrepreneur cannot succeed alone. Success in establishing a moral crusade is dependent on acquiring broader public support. To that end, the moral entrepreneur must mobilize power, create a perceived threat potential for the moral issue in question, generate public awareness of the issue, propose a clear and acceptable solution to the problem, and overcome resistance to the crusade (Becker, 1963; Ben-Yehuda, 1990).

THE STATUS POLITICS OF CIGARETTE SMOKING

The political dynamics underlying the definition of deviant behaviors may be seen clearly in efforts to end smoking in public facilities. Cigarettes were an insignificant product of the tobacco industry until the end of the nineteenth century, after which they evolved into its staple (U.S. Department of Health and Human Services, 1992). Around the turn of the century, 14 states banned cigarette smoking and all but one other regulated sales to and possession by minors (Nuehring and

Markle, 1974). Yet by its heyday in the 1940s and 1950s, cigarette smoking was almost universally accepted, even considered socially desirable (Nuehring and Markle, 1974; Steinfeld, 1991). Per capita cigarette consumption in the United States peaked at approximately 4,300 cigarettes per year in the early 1960s, after which it declined to about 2,800 per year by the early 1990s (U.S. Department of Health and Human Services, 1992). The beginning of the marked decline in cigarette consumption corresponded to the publication of the report to the surgeon general on the health risks of smoking (U.S. Department of Health, Education, and Welfare, 1964). Two decades later, the hazards of passive smoking were being publicized (e.g., U.S. Department of Health and Human Services, 1986).

Increasingly, the recognition of the apparent relationship of smoking to health risks has socially demarcated the lifestyles of the smoker and nonsmoker, from widespread acceptance of the habit to polarized symbolic–moral universes. Attitudes about smoking are informed partly by medical issues, but perhaps even more critical are normative considerations (Nuehring and Markle, 1974); more people have come to see smoking as socially reprehensible and deviant, and smokers as social misfits (Markle and Troyer, 1979). Psychological assessments have attributed an array of negative evaluative characteristics to smokers (Markle and Troyer, 1979). Their habit is increasingly thought unclean and intrusive.

Abstinence and bodily purity are the cornerstones of the nonsmoker's purported moral superiority (Feinhandler, 1986). At the center of their symbolic–moral universe, then, is the idea that people have the right to breathe clean air in public spaces (Goodin, 1989). Smokers, on the other hand, stake their claim to legitimacy in a precept of Anglo-Saxon political culture—the right to do whatever one wants unless it harms others (Berger, 1986). Those sympathetic to smoking deny that environmental tobacco smoke poses a significant health hazard to the nonsmoker (Aviado, 1986). Yet such arguments have held little sway in the face of counterclaims from authoritative governmental agencies and high status moral entrepreneurs.

The development of the antismoking movement has targeted a lifestyle particularly characteristic of the working classes (Berger, 1986). Not only has there been an overall decline in cigarette smoking, but, as mentioned above, the negative relationships of occupation and education to cigarette smoking have become more pronounced in later age cohorts (Ferrence, 1989). Moreover, moral entrepreneurs crusading against smoking are representatives of a relatively powerful "knowledge class," comprising people employed in areas such as education and the therapeutic and counseling agencies (Berger, 1986).

Early remedial efforts focused on publicizing the perils of cigarette smokers, reflecting a strategy of assimilative reform (Nuehring and Markle, 1974; Markle and Troyer, 1979). Even many smokers expressed opposition to cigarettes and a generally repentant attitude. Early educative efforts were thus successful in decreasing cigarette consumption, despite resistance from the tobacco industry. Then, recognition of the adverse effects of smoking on nonusers helped precipitate a turn to coercive reform measures during the mid-1970s (Markle and Troyer, 1979). Rather than a repentant friend in need of help, a new definition of the smoker as enemy emerged. Legal abolition of smoking in public facilities became one focus of social control efforts, and smoking bans in public spaces have been widely adopted in recent years (Markle and Troyer, 1979; Goodin, 1989).

The success of the antismoking crusade has been grounded in moral entrepreneurs' proficiency at mobilizing power, a mobilization made possible by highly visible governmental campaigns, the widely publicized health risks of smoking, and the proposal of workable and generally acceptable policies to ameliorate the problem. The success of this moral crusade has been further facilitated by the association of deviant characteristics with those in lower social strata, whose stigmatization reinforces existing relations of power and prestige. Despite the formidable resources and staunch opposition of the tobacco industry, the tide of public opinion and policy continues to move toward an antismoking stance.

RESEARCH PROBLEM

The study presented below is an exploratory examination of the link between social status and support for a smoking ban in public facilities. Based on theorizing about status politics, as well as evidence about patterns of cigarette use, it was predicted that supporters of the smoking ban would be of higher status than those who opposed it. Further, it was anticipated that supporters of the ban would be more likely to make negative normative claims denouncing the allegedly deviant qualities of smoking, symbolically enhancing their own status while lowering that of their opponents.

The site of this research was Shasta County, California. The population of Shasta County is 147,036, of whom 66,462 reside in its only city, Redding (U.S. Bureau of the Census, 1990). This county became the setting for the implementation of a hotly contested ban on smoking in public buildings.

In 1988, California voters passed Proposition 99, increasing cigarette taxes by 25 cents per pack. The purpose of the tax was to fund smoking prevention and treatment programs. Toward that end, Shasta County created the Shasta County Tobacco Education Program. The director of the program formed a coalition with officials of the Shasta County chapters of the American Cancer Society and American Lung Association to propose a smoking ban in all public buildings. The three groups formed an organization to promote that cause, Smoke-Free Air For Everyone (SAFE). Unlike other bans then in effect in California, the proposed ban included restaurants and bars, because its proponents considered these to be places in which people encountered significant amounts of secondhand smoke. They procured sufficient signatures on a petition to place the measure on the county's general ballot in November 1992.

The referendum passed with a 56 percent majority in an election that saw an 82 percent turnout. Subsequently, the Shasta County Hospitality and Business Alliance, an antiban coalition, obtained sufficient signatures to force a special election to annul the smoking ban. The special election was held in April 1993. Although the turnout was much lower (48 percent), again a sizable majority (58.4 percent) supported the ban. The ordinance went into effect on July 1, 1993.

ANALYTIC STRATEGY

... [D]ata were analyzed in our effort to ascertain the moral and status conflicts underlying the Shasta County smoking ban ... [based on] interviews with five leading moral entrepreneurs and five prominent status quo defenders.[1] These individuals were selected through a snowball sample, with the original respondents identified through interviews with business owners or political advertisements in the local mass media. The selected respondents repeatedly surfaced as the leading figures in their respective coalitions. Semistructured interviews were conducted to determine the reasons underlying their involvement. These data were critical to understanding how the proposed ban was framed by small groups of influential proponents and opponents; it was expected that their concerns would be reflected in the larger public debate about the ban.

FINDINGS

Moral Entrepreneur/Status Quo Defender Interviews

The moral entrepreneurs and status quo defenders interviewed represented clearly different interests. The former group included three high-level administrators in the county's chapters of the American Cancer Society and American Lung Association. A fourth was an administrator for the Shasta County Tobacco Education Project. The last member of this group was a pulmonary physician affiliated with a local hospital. The latter group included four bar and/or restaurant owners and an attorney who had been hired to represent their interests. Thus, the status quo defenders were small business owners who might see their economic interests affected adversely by the ban. Importantly, they were representatives of a less prestigious social stratum than the moral entrepreneurs.

> The primary concern of the moral entrepreneurs was health. As one stated, I supported the initiative to get the smoking ban on the ballot because of all the health implications that secondhand smoke can create. Smoking and secondhand smoke are the most preventable causes of death in this nation.

Another offered that

> On average, secondhand smoke kills 53,000 Americans each year. And think about those that it kills in other countries! It contains 43 cancer-causing chemical agents that have been verified by the Environmental Protection Agency. It is now listed as a Type A carcinogen, which is the same category as asbestos.

Every one of the moral entrepreneurs expressed concern about health issues during the interviews. This was not the only point they raised, however. Three of the five made negative normative evaluations of smoking, thereby implicitly degrading the status of smokers. They commented that "smoking is no longer an acceptable action," that "smoke stinks," or that "it is just a dirty and annoying habit." Although health was their primary concern, such comments revealed the moral entrepreneurs' negative view of smoking irrespective of any medical issues. Smokers were seen as engaging in unclean and objectionable behavior—stigmatized qualities defining their deviant social status.

The stance of the status quo defenders was also grounded in two arguments. All of them expressed concern about individual rights. As one put it,

> I opposed that smoking ban because I personally smoke and feel that it is an infringement of my rights to tell me where I can and cannot smoke. Smoking is a legal activity, and therefore it is unconstitutional to take that right away from me.

Another argued that

> Many people have died for us to have these rights in foreign wars and those also fought on American soil. Hundreds of thousands of people thought that these rights were worth dying for, and now some small group of people believe that they can just vote away these rights.

Such symbolism implies that smoking is virtually a patriotic calling, a venerable habit for which people have been willing to forfeit their lives in time of war. In the status quo defenders' view, smoking is a constitutionally protected right.

At the same time, each of the status quo defenders was concerned about more practical matters, namely business profits. As one stated, "My income was going to be greatly affected." Another argued,

> If these people owned some of the businesses that they are including in this ban, they would not like it either. By taking away the customers that smoke, they are taking away the mainstay of people from a lot of businesses.

The competing viewpoints of the moral entrepreneurs and status quo defenders revealed the moral issues—health versus individual rights—at the heart of political conflict over the smoking ban. Yet it appears that status issues also fueled the conflict. On the one hand, the moral entrepreneurs denigrated smoking, emphasizing the socially unacceptable qualities of the behavior and symbolically degrading smokers' status. On the other hand, status quo defenders were concerned that their livelihood would be affected by the ban. Interestingly, the occupational status of the two groups differed, with the moral entrepreneurs representing the new knowledge class, the status quo defenders a lower stratum of small business owners. Those in the latter group may not have been accorded the prestige and trust granted those in the former (Berger, 1986). Moreover, the status quo defenders' concern about business was likely seen as self-aggrandizing.

SUMMARY AND DISCUSSION

This research has examined the moral and status politics underlying the implementation of a smoking ban in Shasta County, California. Moral entrepreneurs crusading for the ban argued that secondhand smoke damages health, implicitly grounding their argument in the principle that people have a right to a smoke-free environment. Status quo defenders countered that smokers have a constitutional right to indulge wherever and whenever they see fit. Public discourse echoed these themes, as seen in the letters to the editor of the local newspaper. Thus debate about the smoking ban focused especially on health versus smokers' rights; yet evidence of social status differences between the competing symbolic–moral universes also surfaced. Competing symbolic–moral universes are defined not only by different ethical viewpoints on a behavior, but also by differences in social power—disparities inevitably linked to the system of stratification (Ben-Yehuda, 1990). Those prevailing in moral and stigma contests typically represent the higher socioeconomic echelons of society.

The moral entrepreneurs who engineered the smoking ban campaign were representatives of the prestigious knowledge class, including among their members officials from the local chapters of respected organizations at the forefront of the national antismoking crusade. In contrast, the small business owners who were at the core of the opposing coalition, of status quo defenders, represented the traditional middle class. Clearly, there was an instrumental quality to the restaurant and bar owners' stance, because they saw the ban as potentially damaging to their business interests. But they were unable to shape the public debate, as demonstrated by the letters to the editor.

In many respects, the status conflicts involved in the passage of the Shasta County smoking ban were symbolic. The moral entrepreneurs focused attention on the normatively undesirable qualities of cigarette smoking, and their negative normative evaluations of smoking were reflected in public debate about the ban. Those who wrote in support of the ban more frequently offered negative normative evaluations than antiban writers; their comments degraded smoking and, implicitly, smokers. Since the advent of the antismoking crusade in the United States, smoking has come to be seen as socially reprehensible, smokers as social misfits characterized by negative psychological characteristics (Markle and Troyer, 1979).

Ultimately, a lifestyle associated with the less educated, less affluent, lower occupational strata was stigmatized as a public health hazard and targeted for coercive reform. Its deviant status was codified in the ordinance banning smoking in public facilities, including restaurants and bars. The ban symbolized the deviant status of cigarette smokers, the prohibition visibly demonstrating the community's condemnation of their behavior. Further, the smoking ban symbolically amplified the purported virtues of the abstinent lifestyle. A political victory such as the passage of a law is a prestige-enhancing symbolic triumph that is perhaps even more rewarding than its end result (Gusfield, 1963). The symbolic nature of the ban serendipitously surfaced in another way during one author's unstructured observations in 42 restaurants and 21 bars in the area: Although smoking was not observed in a single restaurant, it occurred without sanction in all but one of the bars. Although not deterring smoking in one of its traditional bastions, the ban called attention to its deviant quality and, instrumentally, effectively halted it in areas more commonly frequented by the abstemious.

Although more systematic research is needed, the findings of this exploratory case study offer a better understanding of the dynamics underlying opposition to smoking and further support to theorizing about the role of status politics in the creation of deviant types. Denunciation of smoking in Shasta County involved not only legitimate allegations about public health, but negative normative evaluations of those engaged in the behavior. In the latter regard, the ban constituted a status degradation ceremony, symbolically differentiating the pure and abstinent from the unclean and intrusive. Not coincidentally, the stigmatized were more likely found among society's lower socioeconomic strata, their denouncers among its higher echelons.

Certainly the class and ethnic antipathies underlying attacks on cocaine and opiate users earlier in the century were more manifest than those revealed in the crusade against cigarette smoking. But neither are there manifest status conflicts in the present crusades against abortion (Luker, 1984) and pornography (Zürcher and Kirkpatrick, 1976); yet the underlying differences of status between opponents in those movements are reflected in their markedly different symbolic–moral universes, as was the case in the present study.

This is not to suggest that smoking should be an approved behavior. The medical evidence seems compelling: Cigarette smoking is harmful to the individual smoker and to those exposed to secondhand smoke. However, the objective harms of the psychoactive substance in question are irrelevant to the validity of our analysis, just as they were to Gusfield's (1963) analysis of the temperance movement's crusade against alcohol use. Moreover, it is not our intention to imply that the proban supporters consciously intended to degrade those of lower social status. No doubt they were motivated primarily by a sincere belief that smoking constitutes a public health hazard. In the end, however, moral indignation and social control flowed down the social hierarchy. Thus, we must ask: Would cigarette smoking be defined as deviant if there were a positive correlation between smoking and socioeconomic status?

NOTE

1. Although the term "moral entrepreneur" is well established in the literature on deviance, there seems to be little attention to or consistency in a corresponding term for the interest group(s) opposing them. Those who have been employed, such as "forces for the status quo" (Markle and Troyer, 1979), tend to be awkward. "Status quo defenders" is used here for lack of a simpler or more common term.

REFERENCES

Ashley, Richard. (1975). *Cocaine: Its History, Uses, and Effects.* New York: St. Martin's Press.

Aviado, Domingo M. (1986). "Health Issues Relating to 'Passive' Smoking." Pp. 137–165 in *Smoking and Society: Toward a More Balanced Assessment,* edited by Robert D. Tollison. Lexington, MA: Lexington Books.

Becker, Howard S. (1963). *Outsiders: Studies in the Sociology of Deviance.* New York: Free Press.

Ben-Yehuda, Nachman. (1990). *The Politics and Morality of Deviance: Moral Panics, Drug Abuse, Deviant Science, and Reversed Stigmatization.* Albany, NY: State University of New York Press.

Berger, Peter L. (1986). "A Sociological View of the Antismoking Phenomenon." Pp. 225–240 in *Smoking and Society: Toward a More Balanced Assessment,* edited by Robert D. Tollison. Lexington, MA: Lexington Books.

Bonnie, Richard J., and Charles H. Whitebread II. (1970). "The Forbidden Fruit and the Tree of Knowledge: An Inquiry into the Legal History of American Marihuana Prohibition." *Virginia Law Review* 56: 971–1203.

Covey, Lirio S., Edith A. Zang, and Ernst L. Wynder. (1992). "Cigarette Smoking and Occupational Status: 1977 to 1990." *American Journal of Public Health* 82: 1230–1234.

Feinhandler, Sherwin J. (1986). *The Social Role of Smoking.* Pp. 167–187 in *Smoking and Society: Toward a More Balanced Assessment,* edited by Robert D. Tollison. Lexington, MA: Lexington Books.

Ferrence, Roberta G. (1989). *Deadly Fashion: The Rise and Fall of Cigarette Smoking in North America.* New York: Garland.

Garfinkel, Harold. (1956). "Conditions of Successful Degradation Ceremonies." *American Journal of Sociology* 61: 402–424.

Goffman, Erving. (1963). *Stigma: Notes on the Management of Spoiled Identity.* Englewood Cliffs, NJ: Prentice Hall.

Goodin, Robert E. (1989). *No Smoking: The Ethical Issues.* Chicago: University of Chicago Press.

Gusfield, Joseph R. (1963). *Symbolic Crusade: Status Politics and the American Temperance Movement.* Urbana, IL: University of Illinois Press.

Luker, Kristin. (1984). *Abortion and the Politics of Motherhood.* Berkeley, CA: University of California.

Marcus, Alfred C., Donald R. Shopland, Lori A. Crane, and William R. Lynn. (1989). "Prevalence of Cigarette Smoking in United States: Estimates from the 1985 Current Population Survey." *Journal of the National Cancer Institute* 81: 409–114.

Markle, Gerald E., and Ronald J. Troyer. (1979). "Smoke Gets in Your Eyes: Cigarette Smoking as Deviant Behavior." *Social Problems* 26: 611–625.

Nuehring, Elaine, and Gerald E. Markle. (1974). "Nicotine and Norms: The Re-Emergence of a Deviant Behavior." *Social Problems* 21: 513–526.

Ram, Sidney P. (1941). *How to Get More Fun Out of Smoking.* Chicago: Cuneo.

Roper Organization. (1978, May). *A Study of Public Attitudes Toward Cigarette Smoking and the Tobacco Industry in 1978, Volume 1.* New York: Roper.

Schur, Edwin M. (1980). *The Politics of Deviance: Stigma Contests and the Uses of Power.* New York: Random House.

Steinfeld, Jesse. (1991). "Combating Smoking in the United States: Progress Through Science and Social Action." *Journal of the National Cancer Institute* 83: 1126–1127.

U.S. Bureau of the Census. (1990). *General Population Characteristics.* Washington, DC: U.S. Government Printing Office.

U.S. Department of Health, Education, and Welfare. (1964). *Smoking and Health: Report of the Advisory Committee to the Surgeon General of the Public Health Service.* Washington, DC: U.S. Government Printing Office.

U.S. Department of Health and Human Services. (1986). *The Health Consequences of Involuntary Smoking. A Report of the Surgeon General.* Washington, DC: U.S. Government Printing Office.

U.S. Department of Health and Human Services. (1992). *Smoking and Health in the Americas. A 1992 Report of the Surgeon General, in Collaboration with the Pan American Health Organization.* Washington, DC: U.S. Government Printing Office.

Viscusi, W. Kip. (1992). *Smoking: Making the Risky Decision.* New York: Oxford University Press.

Zürcher, Louis A. Jr., and R. George Kirkpatrick. (1976). *Citizens for Decency: Antipornography Crusades as Status Defense.* Austin, TX: University of Texas Press.

17

The Disadvantage of a Good Reputation: Disney as a Target for Social Problems Claims

JOEL BEST AND KATHLEEN S. LOWNEY

Definitions of what constitutes a social problem, like definitions of deviance, represent arenas that are morally contested. Moral entrepreneurial campaigns are waged on multiple sides of an issue, driven and supported by ideological, religious, financial, public relations, and turf concerns. Such campaigns are also often driven by claimsmakers who seek to alert the public to their views and attempt to convert others to support them. Some campaigns are successful, while others are not, and some issues are more likely to draw the attention necessary to flourish while others fade into obscurity. In this selection, Best and Lowney break down the moral entrepreneurial campaign against the Disney Company, long a bastion of family orientation and values. Here, they highlight the political nature of the grounds upon which claims against the company are founded, focusing particularly on the role of Disney's good reputation in attracting negative claimsmaking. Their analysis of the rhetoric of three sets of claims, from conservative Christians, political progressives, and social scientists, highlights the role of Disney's social capital in drawing moral entrepreneurial attention and in combating it.

How does this chapter illustrate the dynamics and groups of actors commonly found in moral entrepreneurial campaigns? In what way does it illustrate some of the core elements introduced by the Reinarman and by Tuggle and Holmes in their respective chapters? How does our analysis of social power play into the reception of the aforementioned claims? How does the chapter compare and contrast with Best's chapter (11) on social constructionism?

Sociologists often note the costs of bad reputations. For example, individuals labeled as deviant experience enduring stigma, negative stereotypes are assigned to members of racial and ethnic groups, historical Figures with "difficult

From Joel Best and Kathleen S. Lowney (2009). The Disadvantage of a Good Reputation: Disney as a Target for Social Problems Claims. The Sociological Quarterly, 50(3), 431–449. Copyright © 2009 by John Wiley & Sons, Inc. Reprinted with permission by the publisher.

reputations" are condemned to the roles of villains or fools in collective memory, and so on. In general, bad reputations increase one's vulnerability to having opportunities blocked, to being subjected to tighter social control, to having one's future actions framed in terms of the reputation, or to other disadvantages.

Conversely, good reputations usually are thought to convey advantages; their holders receive the benefit of the doubt. Thus, psychologists speak of a "halo effect" shaping expectations, and Merton (1968) argued that the "Matthew effect" made it easier for esteemed scientists to receive credit for additional work. Good reputations, codified in high grade point averages, impressive résumés, good credit ratings, and other track records of accomplishment, open doors for individuals that remain closed to those of more questionable repute. Similarly, organizations conduct public-relations campaigns to establish good reputations in hopes of warding off critics. In general, a good reputation is a resource that can be drawn upon to improve one's prospects, just as a bad reputation hinders advancement.

Reputations—good and bad—can be used to construct social problems. *Claimsmakers* compete in a social problems marketplace, each seeking to arouse concern about particular issues from the press, the public, and policymakers. In order to attract attention from these audiences, claimsmakers assemble claims that feature a variety of rhetorical elements—statistics, expert opinions, appeals to values, and so on—and, depending on the audiences' reactions, may revise and rework claims to make them more persuasive. This article considers some ways reputations can be incorporated in[to] such claims.

At first glance, it might seem that social problems claimsmaking is particularly likely to target those of bad reputations, and this is often the case. Social problems claims frequently highlight typifying examples that illustrate the problem at its worst; describing particularly troubling incidents serves to direct attention to a problem's most disturbing dimensions so as to make it seem especially serious. These examples establish a problem's bad reputation by becoming rhetorical touchstones over the course of a claimsmaking campaign; references to well-known atrocity tales remind audiences that a problem is serious and demands attention. Moreover, once some condition has been constructed as a social problem, that problem's bad reputation can become the basis for further claims that expand that problem's domain or construct other, analogous problems. Thus, the construction of physical child abuse as a serious problem served as a foundation for claims not just about emotional abuse, sexual abuse, and other forms of child abuse but also about elder abuse, wife abuse, and other problems that could be categorized as types of "abuse."

In short, there is a general presumption that bad reputations increase one's vulnerability to critiques by activists and other claimsmakers, while good reputations can help protect individuals and organizations from such criticism. However, there are scandalous exceptions, instances where individuals or organizations with good reputations stand revealed as involved in problematic behavior. Thus, Nichols (1997) examines how the Bank of Boston—a Firm with a good reputation—became the typifying example for claims about money laundering. Although it was neither the first bank charged with money laundering nor the bank with the greatest number of questionable transactions, attention focused on the Bank of Boston precisely because it was newsworthy to hear that it was implicated in dubious practices. However, such scandals require particular conditions: An individual or organization with a good reputation is discovered to have directly violated standards of propriety implicit in that reputation (e.g., a member of the clergy is exposed as involved in sexual deviance; an accounting Firm allows clients to submit questionable financial statements).

This article examines another way good reputations can be used to construct social problems. It argues that a good reputation can also make an organization or institution generally vulnerable to becoming the target of social problem claims. This is a form of **blowback**. Blowback refers to unanticipated, negative consequences of social action. While achieving a good reputation might be expected to produce benefits such as discouraging criticism, linking those with good reputations to social problems can be a useful rhetorical move in social problems claimsmaking. Demonstrating that even those of good reputation are implicated in some social problem is a way of suggesting that the problem is surprisingly widespread and serious. We illustrate this point with a case study. We argue that the Walt Disney Corporation's close associations with what are widely considered positive moral values serve to make it an attractive target for a broad range of social problems claimsmakers.

By the end of the twentieth century, the Disney Corporation had emerged as one of a handful of giant media conglomerates. In addition to owning Walt Disney Pictures, the Disney theme parks in California, Florida, Paris, Tokyo, and Hong Kong, and various other enterprises that bear the Disney name, Disney controlled the ABC television network, ESPN, and other cable channels, movie production under the Touchstone, Hollywood, and Miramax names, and other subsidiaries. The name Disney has become closely linked in the public mind with decent, family-oriented entertainment. This positive reputation, in turn, makes Disney an attractive target for all sorts of social critiques in a way that its rivals are not.

Our analysis is inevitably selective. We have chosen to examine relatively recent critiques of Disney from three very different sorts of claimsmakers—conservative Christians, political progressives, and social scientists. We chose these cases to demonstrate how a range of critics use Disney's good reputation as a key element in their **claimsmaking rhetoric**. Our goal is not to somehow measure the distribution of claims about Disney but rather to document how some claims incorporate Disney's good reputation and to show that such rhetoric comes from very different claimsmakers.

Because our focus is on the rhetoric of claimsmaking, we are interested only in how claimsmakers use Disney's reputation to construct social problems. Although we argue that Disney has a generally good reputation, as evidenced by the many claims that link Disney to innocent, wholesome family entertainment, we offer no judgment about whether that reputation is deserved. Similarly, when we identify examples of critics using Disney's good reputation as an element in their claimsmaking, the accuracy of those claims is not at issue. The point is not that Disney is good (or bad) or that the critics' claims are correct (or wrong); rather, our point is that Disney's reputation can be and is used to support a variety of social problems claims. Further, we recognize that Disney evokes a range of responses: To say that some conservative Christians or some social scientists attack Disney is not to imply that all—or even most—people in those groups accept those claims. It is that some claimsmakers, belonging to very different sectors of society and making claims about very different issues, find themselves making parallel uses of Disney's reputation in their rhetoric that is of interest.

In the sections that follow, we consider three diverse examples—critiques by conservative Christians, progressives, and social scientists—to illustrate how Disney's good reputation makes it a target for social problems claims. We conclude by reconsidering the relationship between reputation and claimsmaking.

CLAIMS FROM THE CHRISTIAN RIGHT

Religious authorities often offer moral critiques of popular culture. During the late 1980s, the Disney Corporation became a frequent target for claimsmaking by the Christian Right; their campaign peaked in 1996 when the American Family Association (AFA) called upon its members to boycott Disney products:

> For decades Disney was a name American families could trust. Disney meant wholesomeness. Disney meant laughter. Disney meant quality entertainment without the sex, violence and profanity. But more than anything else, Disney meant children. Sadly, "the times they are a changin'." (Wildmon, n.d.:1)

These claims argued that while the Disney Corporation continued to produce some morally suitable entertainment, the audience that consumed that family fare was unknowingly helping the conglomerate to fund other, morally bankrupt activities:

> Disney is making millions of dollars off their family fare and then sinking it into movies, television programs and printed materials that assail the very values of those same families. Disney hopes decent minded Americans never make this connection. AFA hopes and prays that decent minded Americans will make the connection. (Wildmon, n.d.)

The AFA received strong support from the Southern Baptist Convention (SBC), which in 1996 gave Disney one year to alter its pro-gay policies. After the title character on the ABC sitcom Ellen came out as a lesbian, the SBC voted in 1997 to join the boycott. On the "MacNeil/Lehrer News Hour" shortly thereafter, the conservative columnist John Podhoretz said,

> Disney is an interesting target because you have—you have essentially—the accusation is that Disney is a front; that is, that Mickey Mouse and the Little Mermaid and Aladdin and Timon and Pumba are fronts for Miramax which produces Pulp Fiction and "Ellen."... So the idea is that hiding behind the Little Mermaid, using Little Mermaid as a screen, Disney is promoting serious—what we would have considered a generation

ago—counter-cultural values, and that...you've got to go at it because it's a fatter target and a slipperier one. (MacNeil/Lehrer Productions 1997:4)

These critiques highlighted the Disney corporation as a media conglomerate by drawing attention to questionable cultural content produced by firms owned by Disney but not promoted using the Disney name. The AFA, for example, criticized such "objectionable films from Disney subsidiaries" as Priest ("pro-homosexual"), Dogma ("asserts that Christian beliefs are little more than mythology"), Chasing Amy (lesbianism), Pulp Fiction (sex, violence), Color of Night (sex), Clerks (graphic language), Chicks in White Satin (lesbianism), Lie Down with Dogs (homosexuality), [and] The House of Yes (incest). Disney, these critics argued, had tried to have it both ways—maintaining the Disney name as a family-friendly brand while using other brand names to produce morally questionable content.

In particular, the AFA attacked Disney for its "headlong rush to promote homosexuality as normal and to profit enormously from that promotion" (Vitagliano, n.d.:2). In addition to critiques of the content of the corporation's popular culture, Disney was criticized for employing homosexuals (including some in important management positions), for extending benefits coverage to partners of those employees, and for supporting gay-themed events at its parks:

> FACT: Disney helped promote the 6th annual "Gay and Lesbian Day at Walt Disney World." Disney has allowed the homosexual organizers to portray Mickey Mouse and Donald Duck as homosexual lovers; and Minnie Mouse and Daisy Duck as lesbians. (AFA n.d.)

One AFA-affiliated group "even had its spies 'dress in gay garb' and attend gay and lesbian events. 'They found pro-homosexual material being handed out. They found a group discount ticket with a pink Mickey on it. They saw homosexuals in drag, chanting "If you're gay and you know it, clap your hands"'" (Svetkey, 1995:42).

When the AFA called off its boycott in 2006, they claimed that they had achieved enough change to warrant ending their campaign. AFA president Tim Wildmon announced, "We feel after nine years of boycotting Disney we have made our point....Boycotts have always been a last resort for us at AFA, and Disney's attitude, arrogance and embrace of the homosexual lifestyle gave us no choice but to advocate a boycott of the company these last years" (AFA 2006:1). He noted that Disney had made several changes—most notably the upcoming departure of CEO Michael Eisner and the corporate "divorce" between Disney and its Miramax film production company—as evidence of the boycott's effect. A hopeful sign was Disney's decision to coproduce The Lion, the Witch, and the Wardrobe, C. S. Lewis's classic novel with Christian themes. An Orlando Sentinel journalist wrote that "In taking the step of marketing The Lion, the Witch, and the Wardrobe to the Christian community,..., Disney may have disarmed much of the antagonism towards the company that led many evangelicals to boycott the company" (AFA 2006:2). Yet conservative Christians continued to denounce Disney even after the boycott ended (e.g., Garcia, 2007).

In other words, claimsmakers who saw themselves as defenders of traditional morality and family values attacked Disney for betraying the moral principles that the corporation once seemed to represent. This perception of betrayal by critics on the right contrasted with another set of charges coming from the left.

CLAIMS FROM PROGRESSIVES

Just as conservative moralists professed themselves shocked that a company with a reputation for morality might produce popular culture with questionable moral content, progressive critics have challenged Disney's reputation as an exemplar of decency. Their critiques focus on Disney's failure to support various forms of social justice.

Disney has long attracted criticism from the labor movement, beginning with a bitter animators' strike in 1941. Disney's resistance to unionization and the working conditions at its theme parks have attracted considerable attention from labor activists. In addition, a review of Websites critical of Disney notes a substantial cluster of critiques by labor rights organizations focused on low wages in Third World factories producing Disney products. These charges are part of the broader campaign against Third-World working conditions, but Disney's good reputation may make it especially vulnerable to such charges. In 2008, Disney was inducted into the "Sweatshop Hall of Shame"; the announcement juxtaposed Disney's idealistic publicity with the harsh conditions under which its goods were produced:

> This year marks Disney's "Year of a Million Dreams" celebration. But for workers in China who make children's books and toys for the entertainment giant, it's been a year of a zillion labor law violations.... So far, Disney has refused to address these serious allegations of worker abuse and exploitation. (SweatFree Communities, 2008)

Many large corporations' labor records can be questioned, but such challenges may be particularly embarrassing for a company with a good reputation. Even more than labor issues, Disney's progressive critics tend to target what they define as conservative messages promoted by Disney's popular cultural content. In *How to Read Donald Duck* (Dorfman and Mattelart, 1971/1975), two Chilean academics argued that while Disney invites "us all to join the great universal Disney family, which extends beyond all frontiers and ideologies, transcends differences between peoples and nations, and particularities of custom and language" (p. 28), Disney comic books in general—and the character of Scrooge McDuck in particular—presented thinly veiled messages endorsing capitalism and imperialism. The U.S. critics have tended to focus on Disney's willingness to present racial and gender stereotypes and on its presentation of a sanitized version of U.S. history. For instance, Barbara Ehrenreich (2007) charges that the Disney Princess "product line...is saturated with a particularly potent time-release form of the date rape drug." Her argument is that Disney's female characters are portrayed as relatively passive figures and that these products encourage girls to accept traditional gender roles.

Such arguments presume that Disney's products reach large numbers of impressionable children and that the content of those products is harmful. But progressive critics argue that it is precisely this "decent" content that deserves criticism, in that its messages promote a hegemonic, uncritical acceptance of traditional values, so that children exposed to Disney learn to accept capitalism, racism, sexism, and so on. Again, parallel critiques might be leveled at most major media firms, but these claims argue that Disney's good reputation discourages a careful analysis of the values it promotes. If many parents view the Disney label as a guarantee that popular cultural content will be appropriate for children, they may fail to recognize—and therefore unwittingly expose their children to—harmful messages implicit in Disney products. Progressive critics insist that even as Disney's content eschews extreme violence and graphic sexuality, other sorts of objectionable messages permeate its products.

CLAIMS FROM SOCIAL SCIENTISTS

In addition to the moral critiques of conservatives and the social-justice criticisms of progressives, Disney has attracted a good deal of critical attention from social scientists. Often, their focus has been the Disney theme parks. When the parks opened, they struck analysts as something new: Customers entered an all-encompassing, purposefully designed environment within which they would be entertained. While all sorts of businesses (one thinks of restaurants or department stores) seek to create environments that will appeal to customers, the Disney theme parks displayed a more calculated, overarching concern with design. And critics argued that this was a troubling development, that the parks offered an artificial, inauthentic experience. There was the implication that the parks purported to offer fun, but that in fact they constrained and controlled behavior, so that while customers might imagine they were enjoying the experience, they were in fact participating in a degraded, emotionally stunted variety of leisure activity.

In particular, the critics deconstructed the methods used to convey the theme park experience. The parks proved to be the product of a dense system of rules and regulations, beginning with the vocabulary used by Disney to define the principal roles. The people who paid to enter were not customers, let alone the "marks" who visited carnivals; rather, Disney insisted that they be referred to as "guests." Similarly, the people working in the parks were not employees but "cast members." Critics viewed such linguistic conventions as deceptive in that they ignored the parks' underlying economic transactions (customers paying for entertainment, employees working for wages). Even the language used to describe what happened in the parks was artificial and inauthentic.

Consider a trivial example. Employees at the Disney theme parks operate under fairly detailed codes governing dress and behavior; they are, for example, required to smile at the customers, er, guests. The Disney parks sometimes use the slogan "The Happiest Place on Earth." Not surprisingly, low-paid service workers oftentimes do not enjoy their work—including this emotion work—all that much, and they may not always feel like smiling. Thus, there is an opportunity for a Goffmanic backstage revelation: Many of those working—and smiling—at the "happiest place on Earth" are not actually happy! Analysts then go on to detail employees' forms of resistance, such as smiling excessively, so that the customer/ guest realizes that the smile is not sincere. The argument can then be extended to challenge the overall legitimacy of Disney's claims about the parks as a happy experience.

In addition, there are numerous social scientific content analyses of Disney products. Analysts justify their research by noting the broad acceptance of Disney products and the possibility that exposure to these works may prove influential:

> The Disney animated classics are well known to millions of children across the world because of their family-oriented topics, marketing, availability, and success in the home-viewing market....[How] older people are represented in Disney animated films [is of concern because of] the cultivating effect that negative stereotypes of older individuals and the marginalizing of older women and minority groups may have on young children. Disney films may not be the primary source of children's negative perceptions of older people; however, there is evidence that the media do influence children's perceptions. (Robinson et al., 2007:203, 209–10)

In general, social scientists have been critical of Disney as a corporation, as lines of products, and even as social trends. Often, the analysts' rationale for focusing on Disney is the need to debunk its good reputation. Playing off Disney's good reputation allows academics to be shocked—shocked!—when revealing that a major media conglomerate places profit-making above other goals.

DISCUSSION

We have sketched three sets of social problems claims about the Disney Corporation: conservative Christians' complaints that Disney (and particularly its subsidiaries) promote homosexuality and morally questionable values; progressives' complaints about Disney's unfair labor practices as well as sexism and other disturbing positions fostered by the firm's popular culture content; and social scientists' critiques that Disney theme parks promote an artificial, alienating ethos and that its content conveys negative messages. In each case, these claims gain much of their rhetorical power by linking Disney's good name to social problems.

This reveals the importance of Disney's good reputation as a moral exemplar. In part, of course, the Disney name is simply familiar—associated with well-known, well-branded products; one suspects that a very large proportion of the population knows that Mickey Mouse and Disney World come from the Disney corporation. Disney is unusual in the way it combines brand familiarity with corporate awareness. This familiarity makes news about Disney more interesting than news about, say, Time Warner.

Further, the Disney name is associated with childhood, family, innocence, and other positive values. The corporation has carefully protected the name by using it to market only the sorts of general-audience, nonthreatening, noncontroversial material that is widely understood to be Disneyesque. Parents assume that they can take small children to a Disney movie or a Disney park without worry—that the experience would not be too frightening or otherwise disturbing. The result is a sort of idealization: Disney is associated with standing for what is good, decent, and appropriate in a way that few other corporations are.

This imagery creates a recipe for blowback claimsmaking. One simply argues that (1) everyone knows that Disney is associated with what is good; (2) X is good (X being whatever the claimsmakers want to promote: environmentalism, heterosexuality, fair labor practices, sexual purity, gender equality, patriotism, tolerance, the list is endless); but (3) Disney's actions can be seen as somehow inconsistent with X. Therefore, Disney stands accused of hypocrisy, bad-faith, and other problematic shortcomings.

Attacking Disney offers claimsmakers several rhetorical advantages. There is the element of surprise—claiming that a firm with Disney's good reputation is associated with some social problem seems incongruous and therefore interesting. Claims that Disney has fallen short of its own publicly avowed moral principles open the corporation to charges of hypocrisy. Also, because many people are familiar with Disney products (and because purchasing those products [is a] discretionary expense), it is relatively easy to call for action in the form of some sort of boycott. In all of these ways, Disney must seem like an attractive target to claimsmakers.

Why have anti-Disney claims not had more effect? In part, Disney may benefit from its familiarity. Virtually all Americans—and much of the world's population—have been exposed to Disney's products. This means that people can weigh claims that Disney fosters one social problem or another and choose which items they wish to consume.

Disney's good reputation, then, has proven attractive to claimsmakers who imagined that attacking Disney would gain attention. On the one hand, this tactic works: Anti-Disney campaigns often attract media coverage, just as scholarly critiques of Disney seem to have little difficulty finding their way into print. On the other hand, Disney's good reputation seems to have protected it from suffering much damage from these various attacks. Far from insulating Disney from criticism, the firm's good reputation attracts a broad range of social problems claimsmakers. Yet although the company becomes a target for all manner of claims, it seems relatively impervious to this criticism.

This suggests some **limits of blowback claimsmaking**. We take it for granted that some enterprises are deeply morally contaminated. Really bad reputations can infect much of what they touch; not surprisingly, they offer fodder for claimsmakers. Reputations, good and bad, are available to anyone interested in assembling claims. Like typifying examples or statistics, they can be used to make claims more compelling, more competitive in the marketplace for media or public attention.

The best reputations for these rhetorical purposes are those—like Disney's—that are relatively consistent and widely understood. Thus, opposition to Columbus Day celebrations [has] attracted considerable attention precisely because [claims-makers] argue that Columbus, long viewed as an unambiguously heroic figure, should be implicated in slavery, colonialism, and genocide. There may be relatively few such targets. Claims targeting those with ambiguous reputations are likely to attract less attention. Examining how claimsmakers incorporate reputations—good or bad, consistent or ambiguous—offers a way of extending the analysis of social problems rhetoric.

REFERENCES

American Family Association. (2006). "AFA Ends Disney Boycott." Retrieved February 19, 2007 (http://www.afa.net/disney/).

———. (n.d.). "The Case against Disney" (handout). Retrieved March 7, 2008 (http://www.whatyouknowmight-notbeso.com/factpage.html).

Dorfman, Ariel, and Armand Mattelart. (1971/1975). *How to Read Donald Duck: Imperialist Ideology in the Disney Comic*. Translated by David Kunzle. New York: International General.

Ehrenreich, Barbara. (2007). "Bonfire of the Disney Princesses." Retrieved December 20, 2007 (http://www.thenation.com/doc/20071224/ehrenreich).

Garcia, Elena. (2007). "Christian Film-makers Probe Disney's Anti-Family Trend." *Christian Post*, October 25. Retrieved March 6, 2008 (http://www.christianpost.com/pages/print.htm?aid=29828).

MacNeil/Lehrer Productions. (1997). "Not Going to Disneyland." Retrieved February 15, 2007 (http://www.pbs.org/newshour/bb/religion/disney_6-18.html).

Merton, Robert K. (1968). "The Matthew Effect in Science." *Science* 3810:56–63.

Nichols, Lawrence T. (1997). "Social Problems as Landmark Narratives: Bank of Boston, Mass Media and 'Money Laundering.'" *Social Problems* 44:324–41.

Robinson, Tom, Mark Callister, Dawn Magoffin, and Jennifer Moore. (2007). "The Portrayal of Older Characters in Disney Animated Films." *Journal of Aging Studies* 21:203–13.

Svetkey, Benjamin. (1995). "Disney Catches Hell." *Entertainment Weekly*, December 15, 305, pp. 42–43.

SweatFree Communities. (2008). "Sweat-shop Hall of Shame: Inductees for 2008." Retrieved September 13, 2008 (http://www.sweatfree.org).

Vitagliano, Ed. (n.d.). "'Disney Execs in Collusion with Homosexual Activists.' Why American Families Should Boycott Disney" (brochure, American Family Association): 2–3. Retrieved February 28, 2006 (http://www.afa.net/disney/disney.pdf).

Wildmon, Tim. (n.d.). "'How We Are Financing Disney's Depravity.' Why American Families Should Boycott Disney" (brochure, American Family Association): 1. Retrieved February 28, 2006 (http://www.afa.net/disney/disney.pdf).

18

Legitimated Suppression: Inner-city Mexican Americans and the Police

ROBERT J. DURÁN

Beyond moral entrepreneurial campaigns, a range of social factors affects the likelihood of different groups in society being defined as deviant and having definitions of deviance enforced against them. Durán offers us a glimpse into the everyday life realities of a group that holds a low degree of social power in society: poor, young, inner-city Mexican-American men. He describes the way these young people, individually and in groups, are treated by agents of social control. If you've not lived in a poor neighborhood of color, you may not be aware of the routine ways police interact with and harass young people there. Durán's detailed insider account draws us into the lives and perspectives of these people and elaborates the way society, through the police, profiles, tracks, targets, harasses, and violates their basic rights as citizens. Illustrating the concept of dominant and subordinate groups discussed by Quinney's chapter on conflict theory, Durán's narrative shows how deviant status is used by the police to disempower and demean less powerful groups.

How valid do you find the complaints of these young people? What do you think of the proposition of "ecological contamination" or the "minority threat" hypothesis? Does this war on gangs represent a "moral panic" that is really a version of disguised racial prejudice? To what extent do you believe that the way the police treat inner-city young Mexican Americans is based on what they do versus who they are?

During the 1990s and early 2000s, law enforcement agencies nationwide launched an aggressive offensive against gangs. Newspapers and nightly newscasts regularly depicted shootings and murders that were labeled gang related. Police officers reported through the media that gangs were growing in number and increasing in violence. Since the mid-1980s, a large number of cities have created specialized police gang units to support a "war on gangs." As a

Reprinted with permission from the author, Robert J. Durán.

society, we have been bombarded with a "law and order" view of gangs and their communities. Police officers routinely recognize how such a war on gangs is hindered by traditional constitutional protections, so they have developed support to create methods and tactics to sidestep disapproval; in many cases, those methods and tactics have become legitimated.

Yet the effect of all of this antigang law enforcement is more widespread in its reach than to merely gang members. In this article, based on a systematic ethnographic study of Mexican-American gang life in Ogden, Utah, and Denver, Colorado, I describe the way the ethnographic approach serves to systematically disenfranchise and discriminate against a whole segment of U.S. society. The analysis presented benefits from my own experience in gangs and law enforcement to make sense of the data. I report and analyze four important facets of "legitimated suppression": (1) legitimated profiling, (2) interacting with suspected gang members, (3) intelligence gathering, and (4) serious forms of police misconduct. The questionable tactics used by police officers against gangs has created further divisions between law enforcement and the Mexican-American community.

POLICING GANGS

The majority of gang members across the United States have been racially and ethnically labeled by police officers as Latino (47 percent) or African-American (31 percent), and they have been mostly poor (85 percent). Self-reported data indicate that Whites identify as gang members at a higher rate than is captured by police data. Many states legally define gangs as three or more people engaged in criminal activity either collectively or individually. This neutral definition has resulted in an application of the label "gang member" to people who are considered non-White. Acknowledging the racial and ethnic focus is important because policing strategies have created serious social consequences for individuals involved or associated with gangs. Through a survey of 261 police departments, Klein (1995) found that intelligence gathering, crime investigation, and suppression were the most common police actions against gangs and that many states had instituted increased consequences for gang-related crimes. Spergel (1995) agreed that a vigorous "lock-'em-up" approach remained the key action of police departments, particularly in large cities with acknowledged gang problems.

This study began with the research question "How are the lives of people who the police believe are involved in criminal groups affected by law enforcement strategies to suppress gangs?" I chose the aforementioned two cities of Denver, Colorado, and Ogden, Utah because community residents, media, and police departments consider them to have a gang problem. My work focuses specifically on people of Mexican descent who were born in the United States (i.e., the largest percentage of Latinos in the United States). Police departments in both cities consider this ethnic group to have the largest number of gangs and gang members. I begin by discussing the methods I used in gathering my data and the settings in which I did the research. I describe the wide array of factors police gang units use in profiling and initially stopping a myriad of community members. I then provide a portrayal of how the police interact with suspected gang members. Finally, I offer a conclusion that expands on the notions of ecological contamination, moral panic, and minority group threat. This research is unique in its use of lived experience and inside access to a significant number of people who have remained hidden from traditional research on gangs and police.

METHODS AND SETTINGS

The research reported in this article is part of a larger study of Mexican-American gang life in two barrio communities, one in Denver and the other in Ogden. The study was conducted for 5 years (2001–2006) and, informally, throughout 14 years of my life (1992–2006). I used ethnographic research methods, such as direct observation, casual interaction, semistructured interviews, introspection, photography, and videotaping, to collect the data.

A significant part of conducting ethnographic research includes the biography and background of the researcher, because he or she is the research tool. I was a gang member growing up in Ogden, Utah, and I later used my ex–gang member status to keep myself located inside that social world to begin gathering data. I benefited from my work experience and networks in child and family services, law enforcement, and youth corrections. Had I remained an individual gang member, I would have lacked a method that could allow me to gather data and talk with the people directly involved in this lifestyle, and it would have been impossible to understand the nature of gangs.

To gather research data on gangs in Denver, where I was never a gang member, I became involved with a variety of groups focused on gang prevention, high school reform, police observation, and community empowerment. I used my advantage of having special expertise on gangs to begin "opportunistic research" in a new setting. In particular, a significant part of my research in Denver was from a group called Area Support for All People (ASAP). (All names of individuals and groups in this article are pseudonyms.) Current and ex gang members helped start this group in 1991, with the goal of decreasing escalating violence. A group of five to 15 gang members and adult mentors met once a week for 3 hours. The group included a variety of youths from around the city between the ages of 13 and 18. I attended meetings of this group regularly for 2½ years (December 2000 to September 2003).

In the fall of 2001, I began conducting semistructured interviews to capture members' and associates' perspectives on gang culture. I tape-recorded and transcribed 32 interviews with gang members or associates. I paid these respondents $25 each for their time. I maintained 10 "key partners" (six in Denver and four in Ogden) and interviewed them multiple times. Because of the pejorative connotations associated with informants, I found that the term "key partners" better reflected my working collaboration with people who knew gangs from actual experience. I engaged in untaped interviews with 90 additional gang-involved members and associates regarding gangs in both cities. (Overall, I conducted 122 taped and untaped interviews.) The interviews allowed me to maintain my informal insider role, whereas a more formal outsider research posture would have distanced me from the very people I wanted to investigate. All interviews were conducted in English with some minor use of Spanish. Most of my respondents had been in the United States for two to six generations and thus spoke English as their first language. The interviews lasted between 1 and 2 hours. All respondents said that they were associated with Mexican-American or Mexican gangs.

In both cities, there were more associates in the gang scene, with only a small percentage who had actually been "jumped in" and thereby acknowledged as members by the gangs. For this reason, I made little use of snowball sampling and instead followed what I call "judgment sampling": I used my extensive knowledge of people in the communities, a knowledge that was acquired through my participant observation, my inside knowledge of gangs, and the aid of my 10 key partners, to select each person for an interview.

During my research, I interviewed seven gang officers in both states. I regularly requested the two cities' police departments' official statistics relating to gangs and the gang unit. Before pursuing my doctoral work, I had worked a year and a half in two youth correction facilities both prior to graduate studies and during the completion of this research, and one year in law enforcement in the state of Utah. In those capacities, I had been allowed to attend law-enforcement-only sessions that outlined police gang tactics and intelligence gathering. To remain objective about how police interact with possible gang members, I chose to work with an ongoing group of residents who, upon seeing a police stop, walked over and recorded the interactions with camcorders. We worked in teams of two or more people. I call this group People Observing the Police (POP). The group had been meeting regularly in Denver since 2000 and in Ogden during the summer of 2005. After the stop was over, we talked with police officers and the person(s) of interest if possible.

I observed over two hundred police stops, 47 of which included gang units, in all areas of these two cities for 3 years. Most of my time involved patrolling areas with nightlife, such as cruising boulevards and minority communities. The use of police scanners helped me travel to the segregated White communities when an infrequent stop was made. Observation of stops along cruising boulevards allowed me to witness a wide variety of racial and ethnic group encounters with the police. Overwhelmingly, the stops were made in areas adjacent to communities of color, a circumstance that aided my ability to patrol both areas. All of the observed gang stops included Latinos, with lesser numbers of African Americans and Asians. I used the information I obtained to compare and contrast Mexican-American gang members' and associates' claims with those of police officers and the media. I also had official local police documents relating to the purpose and policies of gang enforcement, a police tactic used to suppress gang violence. (See next section.)

According to the U.S. Census, Denver and Ogden are both cities whose Latino population grew from 1940 to 2000 and whose non-Hispanic White population decreased over the same period. Both cities have neighborhoods that, historically, have been primarily Latino (50 percent or more), are located near industrial places of employment, and are segregated from certain other areas of the city. These chiefly Latino areas are known as *barrios*. The numbers of individuals below poverty and in the barrios are similar in the two cities, and so are the median household incomes. The two cities are in geographical areas that were once part of Mexico before westward U.S. expansion, and the majority of Latinos living in those areas were born in the United States.

SYSTEMATIC SUPPRESSION

According to gang unit officers, suppression tactics using intelligence gathering and zero-tolerance policing (hereafter defined as gang enforcement) remained at the heart of both the Denver and Ogden gang units. The premise behind such police tactics is rooted in the broken-windows theory, which focused enforcement efforts on minor offenses in order to prevent social and physical disorder and thus reduce the level of overall crime. Gang officers used a variety of indicators to initiate a stop and develop intelligence through interaction. In Denver and Ogden, barrio community members felt unfairly profiled and poorly treated, on the basis of police perceptions of gang membership, actual or perceived. Mexican-American community claims of harassment were often met with disbelief by middle-class White residents in the city, as well as by authority figures. Police officers continually justified their beliefs by pushing for a higher number of interactions with Mexican Americans in order to substantiate the gang stereotypes that the police held.

Legitimated Profiling

In both Denver and Ogden, police officers were deployed primarily in high-crime districts, which were more often neighborhoods with a higher concentration of Latinos and Blacks (50 to 90 percent) and economic poverty (20 to 70 percent). The police departments' diversity paled in comparison with these neighborhoods. Approximately 20 percent of Denver police officers, and 5 percent of Ogden police officers, were Latino. Because most street crime did not occur in plain sight, police officers had to determine which people were engaging in criminal activity.

Police officers focused on making stops on the basis of the legal justification of reasonable suspicion and probable cause. Probable cause includes a belief based on objective fact which supports the suspicion that a person was committing or about to commit a crime. A lead prosecutor in northern Utah described reasonable suspicion as "facts and circumstances that would lead a reasonable officer to believe that there is a particular problem or indication of criminal activity." Together, reasonable suspicion and probable cause legitimated a wide-ranging assortment of stops.

However, this activity led to a confrontational relationship between police and many residents in Denver and Ogden's barrios because residents believed that police officers were using gang and criminal stereotypes to justify their stops. A little higher than 95 percent (31 out of 32) of the individuals formally interviewed reported that they had been stopped for a variety of reasons that were not based on criminality; in other words, they were profiled. Mexican-American youths reported the following reasons, among others, that the police commonly gave for the stop: "It looked like I was wearing gang clothing" (i.e., sports team or hip-hop clothing); "I was assumed to be out too late"; "people matched my description"; "there were reports of shots fired"; "we had more than three people in the car"; and "we looked suspicious." Other reasons community members were stopped included minor traffic violations that could be detected only with strict scrutiny. If the police found no traffic violations, officers had the option of using vehicle safety ordinances, such as being without a front license plate, violating noise ordinances, having overtinted windows (Utah), hanging rosaries or objects from the rearview mirror (Colorado), standing or driving around in a known gang area, and driving a customized (i.e., lowrider) vehicle. In sum, police officers had a full range of reasons to initiate and later justify a "criminal" stop when speaking with barrio residents.

POP observations supported Mexican-American claims of harassment. During their 5 years of watching the police in Denver, Randolph and Pam, two middle-aged White observers of the police, reported the countless number of times they witnessed gang unit officers searching suspected gang member vehicles for drugs and weapons. Pam reported that officers would stop young men for unclear reasons and take them all out of the vehicle. Randolph said the officers would then ID everyone in the car, check them for outstanding warrants, search their pockets, and then send them on their way. He explained:

> So when that happens over and over again, and it's the same general age group, ethnic group, gender group that it happens to time and time again, and no one is arrested. Like detention and searches are supposed to be based on a reasonable suspicion that a crime has been or is about to be committed, so what is the crime here? It seems that being a Chicano youth for the Denver Gang Unit is reasonable suspicion of criminal activity.

Traffic violations were highly discretionary and also very difficult to prove or disprove. Several researchers have attempted to determine the role and significance of racial profiling in such violations and how the practice is used to further an investigation into the identity of occupants and to search for contraband. Cola, a 27-year-old ex–gang member from Ogden, recalled:

> They stopped me for everything. They even stopped me a couple times to tell me they liked my car. I'm not sure what that had to do with anything. At the time I thought it was nothing, but now that I think back, I realize they would take down all of our names. We were just glad that we weren't in trouble for anything.

The observed and described police discretion produced an elusive standard for establishing reasonable suspicion and probable cause, because it was highly influenced by extralegal factors (i.e., age, class, gender, neighborhood, and race). While researching as a member of POP, I found that gang unit stops were influenced particularly by age, gender, race, and local gang stereotypes in 100 percent (47 out of 47) of the observed police stops. The rationale for many of these stops and for subsequent detention appeared far reaching. Compliance with taking pictures and giving information resulted in the person's release from custody. The rationale of gang officers would leave researchers believing that most people stopped were gang members. However, gang members in both cities were a small percentage of barrio youths. Vigil (2002) found that, in Los Angeles, only 4 to 14 percent of barrio youths joined gangs. Klein (1968) estimated that only 6 percent of Los Angeles youths 10 to 17 years of age were affiliated with gangs. According to my data accumulation over 5 years, there was a greater number of associates than actual members of the gang, probably in the ratio of 80 to 20.

Barrio youths faced greater difficulty than youths from other parts of the city did in entering different parts of the city, because law enforcement often associated such entry with causing problems with rival gangs. Randolph, a middle-aged member of POP from Denver, described a situation in which a car was stopped:

> And the officer will say, I recognized the people in the backseat of that car as being from east Denver, and I wanted to know why they were in west Denver. Now, that's not a reasonable suspicion of a crime. People in the United States are supposed to have freedom of movement. Obviously, the reason they were there was because they were cruising during Cinco de Mayo. It's a famous event for a lot of youth, and so they will go cruise Federal [Boulevard] because it's a big thing. So, it's a ridiculous reason to say someone is from another jurisdiction, and that's why I stopped them.

A third demographic factor community members believed that they were stopped for was their gender. Men or teenage boys were perceived as more highly targeted than women or teenage girls. Thus, the seven women interviewed described fewer negative interactions with the police than the men described, but still believed that the police would try to use them to gather information. The young women who attended ASAP thought that the men were stopped and harassed more by the police. Randolph, the middle-aged member of POP from Denver, commented on this pattern:

> If it's a car full of girls, they are far less likely, we saw, to be stopped. So, like, for every car full of girls they stopped, they stopped 10 cars full of guys, and we know that multiple passengers were much more likely to be targeted.

When police were unable to profile individuals as gang members, they relied on other criteria, which many Mexican-American youths satisfied in the clothes they wore (including clothing with numbers on them), the cut of their hair, or the tattoos they had gotten. Most youths would dress in clothes that were fashionable with their peers. Such dress, however, created great confusion for the police and even for gang members when the majority of youths dressed in baggy clothes from urban hip-hop brands (e.g., Ben Davis, Dickies, Johnny Blaze, Karl Kani, Phat Farm, Roca Wear, and Sean John) and clothing with numbers (e.g., Fubu 05, Joker 77, and Sports Jerseys). Lucita, a 25-year-old gang associate from Ogden, recalled seeing when the police would approach her Mexican-American friends:

> They get harassed, they get questioned, they get pulled over for any reason because they got their sunglasses on, or their windows are too tinted, or because they are wearing their pants a certain way, which is funny because you catch these White kids trying to do the same thing but they never get asked those questions. They never get asked "Why you dress like that?" or "Where are you going?"

With police officers making a high percentage of their stops on the basis of extralegal factors relating to age, gender, neighborhood, race, and gang stereotypes, community members were skeptical about the officers' stated primary objective as that of targeting crime. Instead, a large number of residents believed that police officers were there to enforce social control over the neighborhood and the people who lived there.

Interacting with Suspected Gang Members

The interactions between police and the Mexican-American community members were very tense because of the vagueness of the encounters and their unknown outcomes. Several people who observed police officers interacting with suspected gang members noted that the police used techniques of domination and suppression in dealing with them. Because police often lacked for a reason to interact with, search, or arrest the urban youths they stopped, they commonly attempted to incite provocations to justify a search or arrest.

A number of researchers have explored officers' negative preconceptions toward minorities and gang members, as well as police perceptions that stricter enforcement is required. Bridges and Steen (1998) reported that probation officers' divergent beliefs about White and Black criminality shaped both their assessment of dangerousness and their recommendations regarding sentencing. Bayley and Mendelsohn (1968) reported that officers' beliefs are similar to those held in the wider society. Through experience, individuals who associated with or were in gangs came to feel that they were treated the worst.

Intelligence Gathering

Intelligence gathering was a key component of police suppression tactics. Donner (1980) reported that police officers justified surveillance conducted on people and groups on the grounds that it prevented violence. However, police intelligence gathering has allowed the labeling of entire racial and ethnic groups, especially men, as gang members. Once people land on such lists, it becomes more likely that their future acts will be discovered, prosecuted, and dealt with punitively. Denver and Ogden gang lists do not require criminal activity for inclusion, and they remain in the file for at least 5 years.

To create lists of gang members, police officers repeatedly asked Mexican-American youths to what gang they belonged. According to the police department gang protocol, people who admitted gang membership satisfied the first and primary requirement for being placed on a gang list. Most people asked, however, denied membership. The police used different tactics, ranging from talking nonchalantly to coercion, to discover gang involvement. Most respondents interviewed who were not involved with gangs and who therefore denied membership, believed that officers suspected them of lying. Officers would search for clues to possible gang membership by asking individuals to pull up their shirts so that the officers could look for tattoos. Or they asked what high school the youths attended, in the hope of associating them with a high school that many gang members attended. Anne, the 24-year-old gang associate from Ogden, said:

> I was pregnant, and me and my friend were cruising, and we were just sitting there parked, and the cops came over, and a couple other people were parked there, and they were in a gang, but they said we couldn't be loitering around there. And right away, they were yelling at us what gang were we in. I said, "I'm not in a gang," and he said, "Don't lie," like yelling at me, "don't lie." I was like, "I'm not in a gang." And then he asked, "Why you around all of these gang members if you're not in a gang?" I was pregnant and a girl hanging out with another girl, and so it made me pretty mad. And frustrated, too, because I kept telling him, but he wouldn't let it go. He kept saying, "You're in a gang! Tell me!"

Although Anne was associating with gang members, she was not a member. A large number of people in the barrio know someone in a gang, but that acquaintanceship does not make them a member. Individual gang membership created a stereotype of this or that particular racial or ethnic background, and the stereotype spread to everyone living in the barrio. Police officers could then use the gang label to legitimate their interactions with the Mexican-American community. Gang labels were then maintained by the presumption of clear and precise policies and guidelines that countered all forms of legal challenges and complaints, yet no one outside the gang unit had access to the police files to verify their accuracy.

Others whose family members were involved in gangs were often treated as members of the gang. Monique, a 22-year-old from Ogden who had two brothers involved in gangs, mentioned how the police automatically assumed that she was a member and accordingly treated her poorly.

Although acting civil and cordial may not be a requirement for policing, stops during which the police were anything but civil and cordial produced feelings of anger, distrust, and hopelessness, particularly when police could do whatever they wished and get away with it. Everyone interviewed cited examples of being treated like dirt and then simply told to go on their way once officers found no reason to take the stop further (the majority of the time). Interestingly, police officers' "fishing expeditions" would not always pay off. Anne, the 24-year-old gang associate from Ogden, said:

They're dicks; they don't care, and they don't care if you're a girl. I had one of the gang cops search me, and I know that is against the law. Not search me but pat me down, like really pat me down! I know they are not supposed to do that, and I told him, "You can't pat me down." I'm like, "You're supposed to have a female officer." He was all, "You don't tell me what to do." You know, just their little attitude, they'll put you down to your face. You're nothing, you're a piece of shit, They totally don't have any respect for anybody who is a gang member or who they think is a gang member. I don't know how they choose the gang task force, but they don't seem to understand anything about gangs. All they focus on is getting them off the street and into jail. It's awful.

Gang-labeled Mexican Americans—those who were on the gang list—were approached differently because they were seen and treated as criminals even when following the law. Police perceptions shaped gang membership as a "master status" (Hughes, 1945) that combines ascribed and achieved statuses with the belief that the member would have lifelong gang involvement. Changing this image was very difficult for gang members, particularly those attempting to leave the gang lifestyle. Many police stops of gang members would begin with the police ordering these Mexican Americans out of their vehicle and telling them to put their hands up in the air or lie face down on the ground. The officers drew their guns more frequently than they did when stopping other people and attempted to investigate assumed gang involvement and planned activities.

On the one hand, Klein (2004) found that gang officer perceptions of gangs often did not match research findings. He found that (1) most gang crime is minor, (2) most gang activity is noncriminal, (3) street gangs were social groups, (4) street life becomes a part of gang culture, and (5) the community context in which gangs arise was often ignored. The police, on the other hand, portrayed gangs as violent criminal organizations, fundamentally different from other social groups and divorced from local community problems. Raúl, an 18-year-old ex–gang member from Denver, said:

They [the police] mess with you all of the time. Like, if you are a gang member, they be stopping you all of the time. Checking to see if you have any weapons, some of these police officers are racist, they think we are all violent and do bad crimes, but I think we are different.

Police officers' repeated profiling of Mexican Americans, which was based on stereotypical perceptions, and their coercive intelligence-gathering tactics justified increased funding, increased gang-related legislation, and a movement to relocate greater numbers of this population to the penitentiary.

Serious Forms of Police Misconduct

Respondents interviewed claimed that both Denver's and Ogden's police departments often used excessive physical force. Although researchers for the Bureau of

Justice Statistics reported that the use of force occurred in less than 1 percent of all encounters with citizens during the year of their survey (Greenfeld, Langan, and Smith, 1999), at least 34 percent (11 out of 32) of my respondents had experienced physical abuse one or more times. Eighty percent of this misconduct occurred during an arrest and in an isolated area during evening hours. People were more likely to be victimized by police in impoverished Black and Latino neighborhoods. The level of abuse and misconduct in the two cities was different from that noted in the infamous Los Angeles Rampart Division case, but currently remains underresearched and veiled in secrecy. Denver had a high rate of police shootings from 1980 to 2007, with 222 people shot, and 103 people killed, by the Denver Police Department. The captain of the Denver Gang Unit assured me that many policies and protections were instituted to prevent misconduct from happening within his city. Nevertheless, two Denver gang officers were charged for not logging at least 80 pieces of drug evidence into the police department property bureau after making numerous arrests and giving tickets for possession of marijuana and related paraphernalia (Vaughan, 2000). One of these officers was accused of harassing and brutalizing gang members within the Denver area, and that was the likely reason he was shot by a suspected gang member during a questionable traffic stop. The alleged gang member was also shot to death during the incident. Nevertheless, the gang unit claimed that they had few complaints. Rodney, an African-American Latino resident who was a gang associate from Denver, told me:

> I wish every gang member would actually report the abuse that they
> would go through by the Denver Police Department. Then we would
> have a better picture of what the role is that unit plays. But the gang
> members don't feel like they have a right to report when they have
> been beat up. If they actually took the time to document this stuff, we
> would actually see the Denver Police is putting in more work than
> anybody. They function as a gang.

Human Rights Watch (1998, 2) argued that "race continues to play a central role in police brutality in the United States. Indeed, despite gains in many areas since the civil rights movement of the 1950s and 1960s, one area that has been stubbornly resistant to change has been the treatment afforded racial minorities by the police." Cyclone, a 25-year-old ex–gang member and ex–prison inmate, said:

> I've been beaten by cops before. I was running from the police and I
> was drunk. I wrecked a car and I got out and started to run and I
> noticed there were five different counties of cops. There were cops
> from every district surrounding me. I laid down on the ground, and the
> cop that jumped on me started punching me in back of the head. I went
> into County [Jail] and let them know that I was having serious
> migraines, and I showed them the bumps on my head. They took a
> report, and that's all that was ever said. They didn't do anything to the
> officer that whupped my ass. I got charged with resisting arrest and was

tied to the bumper of a car. [Question: How many times do you think he hit you in the back of the head?] Probably about four of five. [What were you doing?] I was in handcuffs on my stomach while his knee was in my back and the other cops were watching. They know something happened. If I wasn't in cuffs when he was hitting me, I would have defended myself. They have the reports on the bumps on my head, severe handcuff marks on my arms; I couldn't feel my left hand for nearly an hour after they took the cuffs off.

Several of the interviewees also believed that illegal immigrants were treated worse than others by police. Mirandé (1987) suggested that undocumented immigrants were especially vulnerable because they lacked resources and familiarity with the justice system. Immigrants also reported fewer instances of abuse, because they feared deportation. Problematic urban conditions and a minority presence together have resulted in police violence being used proactively rather than the police simply reacting to a criminal threat. The Mexican-American community recognized that a simple stop or interaction with the police had a variety of outcomes that were seen as legally permissible but that law enforcement officers were not going to treat Whites living in their racially segregated neighborhoods with the same type of aggression.

CONCLUSION

In many ways, the gang enforcement experiences of Mexican-American youths living in the barrios can be difficult to believe compared with traditional notions of policing and studies on gangs. The research reported here runs contrary to all of the legitimated claims given by law enforcement and routinely heard in the media. In these segregated barrios, police gang units systematically targeted people of Mexican descent by disguising the racial and ethnic implications of the units' actions in a new, nonracial rhetoric discussed by Omi and Winant (1994) as "code-words." The higher number of stops and the coercive questioning dramatically increased the number of people labeled as gang members and associates. Donner (1980) reported that the language used by law enforcement officers shielded their tactics from attack. Suppression and intelligence gathering against the Mexican-American community became legitimated when it was justified with the term "gang." The barrio residents were not antipolice; rather, they were against the profiling and the demeaning treatment. The end result of aggressive differential policing was a greater divide between the barrio and law enforcement.

The 1990s brought a new policing effort involving gang units. The endeavor spread to cities around the nation, targeting perceived criminality in poor and minority communities. The problem became, not that of law, but rather the *enforcement* of laws. Police gang units legitimated the social control of people beyond involvement in crime to include merely *perceived* criminality. Heymann (2000) found the effectiveness of this new policing problematic.

He reported that the cumulative effect of stopping more Blacks and Latinos is to diminish their equal protection status in this country. An increasing number of research studies related to gangs and police are questioning the suppression tactics used by law enforcement to target groups of people living in racially and economically segregated social environments. Jackson and Rudman (1993) suggested that these suppression strategies were bound to fail because they had a negative impact on impoverished communities and then spread gang influence by creating a link between the community and prison.

The research presented in this paper supports Werthman and Piliavin's (1967) theoretical proposition of "ecological contamination": people living in neighborhoods high in alleged gang membership are often suspected of being gang members because gang membership is "contagious" in these high-density areas. The results highlight how people living in those neighborhoods became viewed as poor, young Mexican-American males and not just anyone living within the city. This focus supports the "minority group threat" hypothesis that the rise in numbers of people of color in U.S. cities has created a disjuncture between the perceived threat and the actual threat and has generated an unfounded "moral panic" about the danger of such populations. In both cities examined here, Latinos were increasing in number, and their growth produced heightened perceptions of gang involvement.

In sum, gang enforcement by gang units and patrol officers involved several theoretical patterns. First, police activities pushed more youths into joining gangs. Second, structurally vulnerable areas became targeted with concentrated aggressive gang enforcement that supported unsubstantiated assumptions about gangs and that fueled a moral panic by labeling nongang members as gang members (the ecological contamination). Third, aggressive policing of marginalized and oppressed communities did not eliminate crime, but led instead to greater divisions between barrio residents and law enforcement.

REFERENCES

Bayley, D. H., and Mendelsohn, H. (1968). *Minorities and the police: Confrontation in America.* New York: Free Press.

Bridges, G. S., and S. Steen, S. (1998). Racial disparities in official assessments of juvenile offenders: Attributional stereotypes as mediating mechanisms. *American Sociological Review, 63,* 554–570.

Donner, F. J. (1980). *The age of surveillance: The aims and methods of America's political intelligence system.* New York: Knopf.

Greenfeld, L. A., Langan, P. A., and Smith, S. K. (1999). Police use of force: Collection of national data. NCJ 165040. Washington, DC: U.S. Department of Justice, Bureau of Justice Statistics and National Institute of Justice.

Heymann, P. B. (2000). The new policing. *Fordham Urban Law Journal, 28,* 405–456.

Hughes, E. C. (1945). Dilemmas and contradictions of status. *American Journal of Sociology, 50,* 353–359.

Human Rights Watch. (1998). *Shielded from justice: Police brutality and accountability in the United States.* New York: Human Rights Watch.

Jackson, P., and Rudman, C. (1993). Moral panics and the response to gangs in California. In S. Cummings and D. J. Monti (Eds.), *Gangs: The origins and impact of contemporary youth gangs in the United States.* Albany, NY: State University of New York Press.

Klein, M. W. (1968). *From association to guilt: The group guidance project in juvenile gang intervention.* Los Angeles, CA: Youth Studies Center, University of Southern California and the Los Angeles County Probation Department.

Klein, M. W. (1995). *The American street gang: Its nature, prevalence, and control.* New York: Oxford University Press.

Klein, M. W. (2004). *Gang cop: The words and ways of Officer Paco Domingo.* Walnut Creek, CA: AltaMira.

Mirandé, A. (1987). *Gringo justice.* Notre Dame, IN: University of Notre Dame.

Omi, M., and Winant, H. (1994). *Racial formation in the United States: From the 1960s to the 1990s.* New York: Routledge.

Spergel, I. A. (1995). *The youth gang problem: A community approach.* New York: Oxford University Press.

Vaughan, K. (2000). Charges filed against two cops. Veteran gang officers accused of destroying evidence in "at least" 80 criminal cases. *Rocky Mountain News*, local section, July 20:4A.

Vigil, J. D. (2002). *A rainbow of gangs: Street cultures in the mega-city.* Austin, TX: University of Texas Press.

Werthman, C., and Piliavin, I. (1967). "Gang members and the police." In D. Bordua (Ed.), *The police: Six sociological essays.* New York: Wiley.

19

Homophobia and Women's Sport

ELAINE M. BLINDE AND DIANE E. TAUB

*Blinde and Taub explore the role of attributed sexual orientation in
disempowering women who violate gender norms: varsity female collegiate
athletes. By challenging the gender order and opposing male domination, these
women intrude into a traditional male sanctum and threaten the male domain of
physicality and strength. By casting the lesbian label on women athletes, society
stigmatizes them as masculine and as sexual perverts. While the homosexual
label is routinely used to degrade male athletes who fail to live up to the
hypermasculine ideal, the lesbian label is used to divide and silence female
athletes. As a result, they may adopt the perspective of their oppressors and
demean their teammates as lesbians, thus destroying team solidarity.
Alternatively, they may shun the label but be forced to acknowledge its
demeaning power as they attempt to escape it. The forceful effect of the lesbian
label applied to women athletes shows the dominance not only of heterosexuals
over homosexuals, but also of men over women.*

*Why do you think the women athletes in this article passively accept the
stigma cast onto them? What factors lead some stigmatized groups to fight their
social labels while others accept such labels and hide their deviance?*

Central to the preservation of a patriarchal and heterosexist society is a well-established gender order with clearly defined norms and sanctions governing the
behavior of men and women. This normative gender system is relayed to and installed
in members of society through a pervasive socialization network that is evident in
both everyday social interaction and social institutions (Schur, 1984). Conformity
to established gender norms contributes to the reproduction of male dominance and
heterosexual privilege (Lenskyj, 1991; Stockard and Johnson, 1980).

Despite gender role socialization, not all individuals engage in behavior consistent with gender expectations. Recognizing the potential threat of such aberrations, various mechanisms exist that encourage compliance with the normative
gender order. Significant in such processes are the stigmatization and devaluation
of those whose behavior deviates from the norm (Schur, 1984).

From Elaine M. Blinde and Diane E. Taube, "Homophobia and Women's Sport: The
Disempowerment of Athletes." *Journal of the North Central Sociological Association*, Vol. 25,
No. 2. Copyright © 1992. Reproduced by permission of Taylor & Francis, LLC,
http://www.taylorandfrancis.com.

Women's violation of traditional gender role norms represents a particularly serious threat to the patriarchal and heterosexist society because this deviant behavior resists women's subordinate status (Schur, 1984). When women engage in behavior that challenges the established gender order, and thus opposes male domination, attempts are often made by those most threatened to devalue these women and ultimately control their actions. One means of discrediting women who violate gender norms and thereby questioning their "womanhood" is to label them lesbian (Griffin, 1987).

The accusation of lesbianism is a powerful controlling mechanism given the homophobia that exists within U.S. society. Homophobia, representing a fear of or negative reaction to homosexuality (Pharr, 1988), results in stigmatization directed at those assumed to violate sexuality norms. Lesbianism, in particular, is viewed as threatening to the established patriarchal order and heterosexual family structure since lesbians reject their "natural" gender role, as well as resist economic, emotional, and sexual dependence on men (Gartrell, 1984; Lenskyj, 1991).

As a means for both discouraging homosexuality and maintaining a patriarchal and heterosexist gender order (Pharr, 1988), homophobia controls behavior through contempt for purported norm violators (Koedt, Levine, and Rapone, 1973). One method of control is the frequent application of the lesbian label to women who move into traditional male-dominated fields such as politics, business, or the military (Lenskyj, 1991). This "lesbian baiting" (Pharr, 1988:19) suggests that women's advancement into these arenas is inappropriate. Such messages are particularly potent since they are lodged in a society that condemns, devalues, oppresses, and victimizes individuals labeled as homosexuals (Lenskyj, 1990).

Another male arena in which women have made significant strides, and thus risk damaging accusation and innuendo, is that of sport (Blinde and Taub, 1992; Lenskyj, 1990). Sport is a particularly susceptible arena for lesbian labeling due to the historical linkage of masculinity with athleticism (Birrell, 1988). When women enter the domain of sport they are viewed as violating the docile female gender role, and therefore extending culturally constructed boundaries of femininity (Cobban, 1982; Lenskyj, 1986; Watson, 1987). The attribution of masculine qualities to women who participate in sport leads to a questioning of their sexuality and subsequently makes athletes targets of homophobic accusations (Lenskyj, 1986)....

Therefore, the present study explores the stereotyping of women athletes as lesbians and the accompanying homophobia fostering this label. General themes and processes which inform us of how these individuals handle the lesbian issue are identified. These dynamics are grounded in the contextual experiences of women athletes and relayed through their voices.

Athletic directors at seven large Division I universities were contacted by telephone and asked to participate in a study examining various aspects of the sport experience of female college athletes. These administrators were requested to provide a list of the names and addresses of all varsity women athletes for the purpose of contacting them for telephone interviews. [Three of the seven universities—two in the Midwest and one in the South—responded with lists.]... Interested athletes were encouraged to return an informed consent form indicating their willingness to participate in a tape-recorded telephone interview. Based on this initial contact, a total of 16 athletes agreed to be in the study.

In order to increase the sample size to the desired 20 to 30 respondents, the names of 30 additional athletes were randomly and proportionately selected from the three lists. Eight of these athletes agreed to be interviewed, resulting in a final sample size of 24. Athletes in the sample were currently participating in a variety of women's intercollegiate varsity sports—basketball (n = 5), track and field (n = 4), volleyball (n = 3), swimming (n = 3), softball (n = 3), tennis (n = 2), diving (n = 2), and gymnastics (n = 2). With an average age of 20.2 years and overwhelmingly Caucasian (92 percent), the sample contained 2 freshmen, 9 sophomores, 5 juniors, and 8 seniors. A majority of the athletes (n = 22) were recipients of an athletic scholarship....

Semistructured telephone interviews were conducted by two trained female interviewers. All interviews were tape-recorded and lasted from 50 to 90 minutes. Questions were open-ended in nature so that athletes would not feel constrained in discussing those issues most relevant to their experiences. Follow-up questions were utilized to probe how societal perceptions of women athletes impact their behavior and experiences.

RESULTS

Examination of the responses of athletes revealed two prevailing themes related to the presence of the lesbian stereotype in women's sport—(a) a silence surrounding the issue of lesbianism in women's sport, and (b) athletes' internalization of societal stereotypes concerning lesbians and women athletes. It is suggested that these two processes disempower women athletes and thus are counterproductive to the self-actualizing capability of sport participation (Theberge, 1987).

Silence Surrounding Lesbianism in Women's Sport

One of the most pervasive themes throughout the interviews related to the general silence associated with the lesbian stereotype in women's sport. Although a topic of which athletes are cognizant, reluctance to discuss and address lesbianism in women's sport was evident. Based on the responses of athletes, this silence was manifested in several ways: (a) athletes' difficulty in discussing lesbian topic, (b) viewing lesbianism as a personal and irrelevant issue, (c) disguising athletic identity to avoid lesbian label, (d) team difficulty in addressing lesbian issue, and (e) administrative difficulty in addressing lesbian issue.

Athletes' Difficulty in Discussing Lesbian Topic

Initial indication of silencing was illustrated by the difficulty and uneasiness many athletes experienced in discussing the lesbian stereotype. Some respondents were initially reluctant to mention the topic of lesbianism; discussion of the issue was frequently preceded by awkward or long pauses suggesting feelings of uneasiness or discomfort. Athletes were most likely to introduce this topic when questions were asked about societal perceptions of women's sport and female athletes, as well as inquiries about the existence of stereotypes associated with women athletes. Moreover, the lesbian issue was sometimes discussed without specifically using the term lesbian. For example, some athletes evaded the issue by making indirect references to lesbianism (e.g., using the word "it" rather than a more descriptive term)....

Respondents' approach to the topic of lesbianism indicates the degree to which women athletes have been socialized into a cycle of silence. Such silence highlights the suppressing effects of homophobia. Moreover, athletes' reluctance to discuss topics openly related to lesbianism may be to avoid what Goffman (1963) has termed "courtesy stigma," a stigma conferred despite the absence of usual qualifying behavior.

Viewing Lesbianism as a Personal and Irrelevant Issue

A second indicator of the silence surrounding the lesbian stereotype was reflected in athletes' general comments about lesbianism. Many respondents indicated that sexual orientation was a very personal issue and thus represented a private and extraneous aspect of an individual's life. These athletes felt it was inappropriate for others to be concerned about the sexual orientation of women athletes.

Although such a manifestation of silence might reflect the path of least resistance by relieving athletes of the need to discuss or disclose their sexual orientation (Lenskyj, 1991), it does not eliminate the stigma and stress experienced by women athletes. Also, making lesbianism a private issue does not confront or challenge the underlying homophobia that allows the label to carry such significance. The strategy of making sexual orientation a personal issue depoliticizes lesbianism and ignores broader societal issues.

Disguising Athletic Identity to Avoid Lesbian Label

A third form of silence surrounding the lesbian stereotype was the tendency for athletes to hide their athletic identities. Nearly all the respondents indicated that despite feeling pride in being an athlete, there were situations where they preferred that others not know their athletic identity. Although not all athletes indicated that this concealment was to prevent being labeled a lesbian, it was obvious that there was a perceived stigma associated with athletics that many women wanted to avoid (e.g., masculine women, women trying to be men, jock image). In most cases, respondents indicated that disguising their athletic identity was either directly or indirectly related to the lesbian stereotype....

Athletes also stated that they (or other athletes they knew) accentuated certain behaviors in order to reduce the possibility of being labeled a lesbian. Being seen with men, having a boyfriend, or even being sexually promiscuous with men were commonly identified strategies to reaffirm an athlete's heterosexuality.

As one athlete commented: "If you are a female athlete and do not have a boy-friend, you are labeled [lesbian]."

As reflected in the responses of athletes, the role of sport participant was often intentionally de-emphasized in order to reduce the risk of being labeled lesbian. Modification of athletes' behavior, even to the point of denying critical aspects of self, was deemed necessary for protection from the negativism attached to the lesbian label. This disguising of athletic identity exemplifies what Kitzinger (1987:92) termed "role inversion." In such a situation individuals attempt to demonstrate that their group stereotype is inaccurate by accentuating traits that are in opposition to those commonly associated with the group (in the case of women athletes, stressing femininity and heterosexuality).

Team Difficulty in Addressing Lesbian Issue

Not only did the silence surrounding lesbianism impact certain aspects of the lives of individual athletes, but it also affected interpersonal relationships among team members. This silence was often counterproductive to the development of positive group dynamics (e.g., team cohesion, open lines of communication).

As was often true at the individual level, women's sport teams were unable collectively to discuss, confront, or challenge the labeling of women athletes as lesbians. One factor complicating the ability of women athletes to confront the lesbian stereotype was the divisive nature of the label itself (Gentile, 1982); the lesbian issue sometimes split teams into factions or served as the basis for clique formation.

Heterosexual and lesbian athletes often had limited interaction with each other outside the sport arena. Moreover, athletes established distance between themselves and those athletes most likely to be labeled lesbian (i.e., those possessing "masculine" physical or personality characteristics)....

From the interviews, there was little evidence that lesbian and nonlesbian athletes collectively pooled their efforts to confront or challenge the lesbian stereotype so prevalent in women's sport. The silence surrounding lesbianism creates divisions among women athletes; this dissension has the effect of preventing female bonding and camaraderie (Lenskyj, 1986). Rather than recognizing their shared interests, women athletes focus on their differences and thus deny the formation of "alliance" (Pheterson, 1986:149). This difficulty in attaining team cohesion is unfortunate since women's sport is an activity where women as a group can strive for common goals (Lenskyj, 1990). The lesbian stereotype not only limits female solidarity, but also minimizes women's ability to challenge collectively the patriarchal and heterosexist system in which they reside (Bennett et al., 1987).

Administrative Difficulty in Addressing Lesbian Issue

Another manifestation of silence relayed in the responses of athletes was the apparent unwillingness of coaches and athletic directors to confront openly the lesbian stereotype. As was found with individual athletes and teams, those in leadership positions in women's sport refused to address or challenge this stereotype. Reluctance to confront the lesbian issue at the administrative level undoubtedly influenced the manner in which athletes handled the stereotype....

Because the women's intercollegiate sport system is homophobic and predominately male-controlled (i.e., over half of coaches and four-fifths of administrators are men) (Acosta and Carpenter, 1992), it is assumed that survival in women's sport requires collusion in a collective strategy of silence about and denial of lesbianism (Griffin, 1987). Coaches and administrators fear that openly addressing the lesbian issue may result in women's sport losing the recent gains made in such areas as fan support, budgets, sponsorship, and credibility (Griffin, 1987). Therefore, leaders yield to this fear as they strive to achieve acceptability for women's sport. Such accommodation to the patriarchal, heterosexist sport structure not only contributes to isolation as coaches and administrators are afraid to discuss lesbianism, but also limits their identification with feminist and women's issues (Duquin, 1981; Hargreaves, 1990; Pharr, 1988; Zipter, 1988).

Conclusions

Based on athletes' responses, it was evident that the silence surrounding the lesbian issue in women's sport was deeply ingrained at all levels of the women's intercollegiate sport structure. Such widespread silencing reflects the negativism and fear associated with lesbianism that are so prevalent in a homophobic society. This strategy of silence or avoidance, however, is counterproductive to efforts to dispel or minimize the impact of the lesbian stereotype. Not only does silence disallow a direct confrontation with those who label athletes lesbian, but it also perpetuates the power of the label by leaving unchallenged rumors and insinuations. Moreover, the fear, ignorance, and negative images that are frequently associated with women athletes are reinforced by this silence (Zipter, 1988).

Numerous aspects of women's experience in sport are ignored due to the silence surrounding the subject of lesbianism. For example, refusing to address this issue has limited understanding of the dimensionality and complexity of women's sport participation. Moreover, since the stigma associated with the lesbian label inhibits athletes from discussing this topic with each other, these women frequently do not realize that they possess shared experiences that would provide the foundation for female bonding. Without an "alliance" among athletes, little progress is made in improving their plight (Pheterson, 1986). Finally, as a result of this preoccupation with silence, women athletes often engage in self-denial as they hide their athletic identity.

Athletes' Internalization of Societal Stereotypes

A second major theme reflected in the responses of athletes was a general internalization of stereotypic representations of lesbians and women athletes. As argued by Kitzinger (1987) and Pheterson (1986), members of oppressed and socially marginalized groups often find themselves accepting the stereotypes and prejudices held by the dominant society. Representing "internalized oppression"

(Pheterson, 1986:148), the responses of athletes revealed an identification with the aggressor, self-concealment, and dependence on others for self-definition (Kitzinger, 1987; Pheterson, 1986). Acceptance of these societal representations by a disadvantaged group (in this case women athletes) grants legitimacy to the position of those who oppress and contributes to the continued subordination of the oppressed (Wolf, 1986). Based on our interviews, athletes' internalization of stereotypes and prejudices were reflected by three categories of responses: (1) acceptance of lesbian stereotypes, (2) acceptance of women's sport team stereotypes, and (3) acceptance of negative images of lesbianism.

Acceptance of Lesbian Stereotypes

In response to various open-ended questions, it was apparent that athletes were able to identify a variety of factors that they felt led others to label women athletes as lesbians (e.g., physical appearance, dress, personality characteristics, nature of sport activity). Given that the attribution of homosexuality is most likely to be associated with traits and behaviors judged to be more appropriate for members of the opposite sex (Dunbar, Brown, and Amoroso, 1973; Dunkle and Francis, 1990), it was not surprising that athletes' rationale for the lesbian label included such attributes as muscularity, short hair, masculine clothing, etc.

When athletes were asked about the validity of the lesbian label in women's sport, affirmative replies were frequently based on conjecture. For example, to provide support for why they felt there was a basis for labeling women athletes as lesbians, respondents made such comments as "there are masculine girls on some teams," "it is really obvious," or "you can just tell that some athletes are lesbians."

These explanations tend to reflect an acceptance of societal definitions of lesbianism—beliefs that are largely male-centered and supportive of a patriarchal, heterosexist system (e.g., "girls who look like guys"). Indeed, previous research has shown that people associate physical appearance with homosexuality (Levitt and Klassen, 1974; McArthur, 1982; Unger, Hilderbrand, and Madar, 1982). For example, attractiveness is equated with heterosexuality and a larger, muscular body build is identified with lesbianism.

Moreover, the remarks of athletes demonstrate that the very group that is oppressed (in this case women athletes) accepts societal stereotypes about lesbians and has incorporated these images into their managing of the situation. As suggested by Gartrell (1984) and certainly evident in this sample of women athletes, cultural myths about lesbianism perpetuated in a homophobic society are often firmly ingrained in the thinking of affected individuals.

Acceptance of Women's Sport Team Stereotypes

Relative to providing a rationale for why the lesbian label was more likely to be associated with athletes in certain sports, respondents again demonstrated an understanding and internalization of societal stereotypes. The sports most commonly identified with the lesbian label were softball, field hockey, and basketball.

In attempting to explain why these team sports were singled out, athletes mentioned such factors as the nature of bodily contact or amount of aggression in the sport, as well as the body build, muscularity, or athleticism needed to play the sport.

Respondents often relied on the "masculine" and "feminine" stereotypes to differentiate sports in which participating women were more or less likely to be subjected to the lesbian label. Although participants in team sports were more likely than individual sports (e.g., gymnastics, swimming, tennis, golf) to be associated with the lesbian label, it was interesting to note that volleyball was often exempt from the connotations of lesbianism.

The higher incidence of lesbian labeling found in team sports (as opposed to individual sports) may also be related to the potential that team sports provide for interpersonal interactions. As mentioned earlier, emphasizing teamwork and togetherness, team sports allow women rare opportunities to bond collectively in pursuit of a group goal (Lenskyj, 1990). Recognition of this power of female bonding is often reflected by male opposition to women-only activities (Lenskyj, 1990).

Acceptance of Negative Images of Lesbianism

During the course of the interviews, a large majority of athletes made comments about lesbians which reflected an internalization of the negativism associated with lesbianism. Respondents also demonstrated a similar acceptance when they relayed conversations they had had with both teammates and outsiders.

One form of negativism was reflected by statements that specifically "put down" lesbians. Athletes' negative comments about lesbians were included in conversations with outsiders so others would not associate the lesbian label with them. Representing a form of projection (Gross, 1978), some athletes attempted to disassociate from traits that they saw in themselves (e.g., strength, muscularity, aggressiveness)....

It is ironic that athletes rarely directed their anger or condemnations at the homophobic society that restricts the actions of women athletes, including the nonlesbian athlete. Rather, by focusing on athletes as lesbians, a blame-the-victim approach diverts attention from the cause of the oppression (Pharr, 1988). As is often true of oppressed groups, a blame-the-victim philosophy results in an acceptance of the belief system of the oppressor (in this case a patriarchal, heterosexist society) (Pharr, 1988). Like other marginalized groups, women athletes accept the normative definitions of their deviance (Kitzinger, 1987); in effect, such responses represent a form of collusion with the oppressive forces (Pheterson, 1986). Interestingly, no mention was made by respondents about attempts to engage the assistance or support of units on campus sympathetic to gay and lesbian issues (e.g., feminist groups, gay and lesbian organizations, affirmative action offices).

Conclusion

From the interview responses, it was evidence that athletes had internalized societal stereotypes related to lesbians and women athletes, as well as the negativism directed toward lesbianism. This acceptance was so ingrained in these athletes that they were generally unaware of the political ramifications of both lesbianism and the accompanying lesbian stereotype as applied to women athletes. Despite their gender norm violation as athletes, these women often had a superficial understanding of gender issues. Such a lack of awareness may be due in part to the absence of a feminist consciousness in athletes (Boutilier and SanGiovanni, 1983; Kaplan, 1979) and their open disavowal of being a "feminist," "activist," or "preacher of women's liberation." Accepting societal definitions of their deviance, as well as the inability to see their personal experiences as political in nature, attests to this limited consciousness (Boutilier and SanGiovanni, 1983). Athletes' responses are indicative of the degree to which they exhibit internal homophobia so common in U.S. society.

Only a few athletes possessed deeper insight into factors that may underlie the labeling of women athletes as lesbians. For example, one respondent felt women athletes were a "threat to men since they can stand on their own feet." Or, in another situation, an athlete viewed lesbian labeling as a means to devalue women athletes or successful women in general. Still another respondent suggested the label stemmed from jealousy and thus was used as a means to "get back" at women athletes. These rare remarks by respondents transcend the blame-the-victim view held by the majority of athletes. Such commitments indicate a deeper understanding of how homophobia and patriarchal ideology limit or control women's activities and their bodies.

DISEMPOWERMENT

Given the silence surrounding the lesbian issue and the degree to which athletes have internalized societal images of lesbians and women athletes, the presence of the lesbian stereotype has negative ramifications for women athletes. Although sport participation possesses the potential for creativity and physical excellence (Theberge, 1987), women modify their behavior so they will not be viewed as "stepping out of line." Women athletes become disempowered (Pharr, 1988) through processes that detract from or reduce the self-actualizing potential of the sport experience.

Attaching the label of lesbian to women who engage in sport diminishes the sporting accomplishments of athletes. Women athletes are seen as something less than "real women" because they do not exemplify traditional female qualities (e.g., dependency, weakness, passivity); thus their accomplishments are not viewed as threatening to men (Birrell, 1988). Interestingly, the athletes interviewed believed that the specific group most likely to engage in lesbian labeling was male athletes.

Discrediting women with the label of lesbian works further to control the number of females in sport, particularly in a homophobic society where prejudice against lesbians is intense (Birrell, 1988; Zipter, 1988). Keeping women out of sport, in turn, prevents females from discovering the power and joy of their own physicality (Birrell, 1988) and experiencing the potential of their body. Moreover, discouraging women from participating in sport disempowers them by removing an arena where women can bond together (Birrell, 1988; Cobban, 1982)....

Another form of disempowerment occurs for those athletes who are lesbians. Intense homophobia often forces lesbians to deny their very essence, thus making the lesbian athlete invisible. Concealment, although protecting these lesbian athletes' identity, imposes psychological strain and can undermine positive self-conceptions (Schur, 1984). Misrepresenting their sexuality, lesbian athletes are not in a position to confront the homophobia so prevalent in women's sport. Consequently, this ideology not only remains intact, but also is strengthened (Ettore, 1980).

REFERENCES

Acosta, R. Vivian, and Linda Jean Carpenter. (1992). "Women in Intercollegiate Sport: A Longitudinal Study—Fifteen Year Update 1977–1992." Unpublished manuscript, Brooklyn College, Department of Physical Education, Brooklyn.

Bennett, Roberta S., K. Gail Whitaker, Nina Jo Woolley Smith, and Anne Sablove. (1987). "Changing the Rules of the Game: Reflections Toward a Feminist Analysis of Sport." *Women's Studies International Forum* 10: 369–386.

Birrell, Susan. (1988). "Discourses on The Gender/Sport Relationship: From Women in Sport to Gender Relations." Pp. 459–502 in *Exercise and Sport Science Reviews*, vol. 16, edited by K. B. Pandolf. New York: MacMillan.

Blinde, Elaine M., and Diane E. Taub. (1992). "Women Athletes as Falsely Accused Deviants: Managing the Lesbian Stigma." *The Sociological Quarterly* 33: 521–533.

Boutilier, Mary A., and Lucinda SanGiovanni. (1983). *The Sporting Woman*. Champaign, IL: Human Kinetics.

Cobban, Linn Ni. (1982). "Lesbians in Physical Education and Sport."

Pp. 179–186 in *Lesbian Studies: Present and Future*, edited by M. Cruikshank. New York: Feminist Press.

Dunbar, John, Marvin Brown, and Donald M. Amoroso. (1973). "Some Correlates of Attitudes Toward Homosexuality." *Journal of Social Psychology* 89: 271–279.

Dunkle, John H., and Patricia L. Francis. (1990). "The Role of Facial Masculinity/Feminity in The Attribution of Homosexuality." *Sex Roles* 23: 157–167.

Duquin, Mary E. (1981). "Feminism and Patriarchy in Physical Education." Paper presented at the annual meetings of the North American Society for the Sociology of Sport, Fort Worth, TX.

Ettore, E. M. (1980). *Lesbians, Women and Society*. London: Routledge and Kegan Paul.

Gartrell, Nanette. (1984). "Combating Homophobia in the Psychotherapy of Lesbians." *Women and Therapy* 3: 13–29.

Gentile, S. (1982). "Out of The Kitchen." *City Sports Monthly* 8: 27.

Goffman, Erving. (1963). *Stigma: Notes on the Management of Spoiled Identity*. Englewood Cliffs, NJ: Prentice Hall.

Griffin, Patricia S. (1987). "Homophobia, Lesbians, and Women's Sports: An Exploratory Analysis." Paper presented at the annual meetings of the American Psychological Association, New York.

Gross, Martin L. (1978). *The Psychological Society*. New York: Simon & Schuster.

Hargreaves, Jennifer A. (1990). "Gender on the Sports Agenda." *International Review for Sociology of Sport* 25: 287–308.

Kaplan, Janice. (1979). *Women and Sports*. New York: Viking.

Kitzinger, Celia. (1987). *The Social Construction of Lesbianism*. London: Sage.

Koedt, Anne, Ellen Levine, and Anita Rapone. (1973). *Radical Feminism*. New York: Quadrangle.

Lenskyj, Helen. (1986). *Out of Bounds: Women, Sport and Sexuality*. Toronto: Women's Press.

———. (1990). "Power and Play: Gender and Sexuality Issues in Sport and Physical Activity." *International Review for Sociology of Sport* 25: 235–245.

———. (1991). "Combatting Homophobia in Sport and Physical Education." *Sociology of Sport Journal* 8: 61–69.

Levitt, Eugene E., and Albert D. Klassen, Jr. (1974). "Public Attitudes toward Homosexuality: Part of the 1970 National Survey by the Institute for Sex Research." *Journal of Homosexuality* 1: 29–43.

McArthur, Leslie Z. (1982). "Judging a Book by Its Cover: A Cognitive Analysis of the Relationship between Physical Appearance and Stereotyping." Pp. 149–211 in *Cognitive Social Psychology*, edited by Albert H. Hastorf

and Alice M. Isen. New York: Elsevier/North-Holland.

Pharr, Suzanne. (1988). *Homophobia: A Weapon of Sexism*. Inverness, CA: Clurdon.

Pheterson, Gail. (1986). "Alliances between Women: Overcoming Internalized Oppression and Internalized Domination." *Signs: Journal of Women in Culture and Society* 12: 146–160.

Schur, Edwin M. (1984). *Labeling Women Deviant: Gender, Stigma, and Social Control*. New York: McGraw-Hill.

Stockard, Jean, and Miriam M. Johnson. (1980). *Sex Roles: Sex Inequality and Sex Role Development*. Englewood Cliffs, NJ: Prentice Hall.

Theberge, Nancy. (1987). "Sport and Women's Empowerment." *Women's Studies International Forum* 10: 387–393.

Unger, Rhoda K., Marcia Hilderbrand, and Theresa Madar. (1982). "Physical Attractiveness and Assumptions about Social Deviance: Some Sex-by-Sex Comparisons." *Personality and Social Psychology Bulletin* 8: 293–301.

Watson, Tracey. (1987). "Women Athletes and Athletic Women: The Dilemmas and Contradictions of Managing Incongruent Identities." *Sociological Inquiry* 57: 431–446.

Wolf, Charlotte. (1986). "Legitimation of Oppression: Response and Reflexivity." *Symbolic Interaction* 9: 217–234.

Zipter, Yvonne. (1988). *Diamonds Are a Dyke's Best Friend: Reflections, Reminiscences, and Reports from the Field on the Lesbian National Pastime*. Ithaca, NY: Firebrand Books.

20

The Mark of a Criminal Record

DEVAH PAGER

To measure concretely the interactive effects of race and a criminal record, Pager devised an experimental design in which she constructed a fabricated pair of job applicants who were matched on all features except their criminal history. She then sent out these matched pairs, of Whites and Blacks, to apply for real jobs in Milwaukee, and noted how far the candidates got in the interview process. Along the way, she recorded the employer's likelihood of dismissing the applicants right away, checking their references, calling them back for further interviews, and offering them the job. In a real demonstration of the effects of race and a criminal record on employment opportunities, Pager found that, while Whites were offered more jobs than Blacks and applicants with no criminal history were offered more jobs than those who had served time, even Whites with criminal pasts were more likely to be hired than Blacks who had led law-abiding lives. She also found that employers were more likely to hold stereotypes suspecting Blacks, especially young Black men, of being prone to crime and of being unreliable employees.

While stratification researchers typically focus on schools, labor markets, and the family as primary institutions affecting inequality, a new institution has emerged as central to the sorting and stratifying of young and disadvantaged men: the criminal justice system. With over 2 million individuals currently incarcerated, and over half a million prisoners released each year, the large and growing numbers of men being processed through the criminal justice system raises important questions about the consequences of this massive institutional intervention.

This article focuses on the consequences of incarceration for the employment outcomes of black and white men. While previous survey research has demonstrated a strong *association* between incarceration and employment, there remains little understanding of the mechanisms by which these outcomes are produced. In the present study, I adopt an experimental audit approach to test formally the degree to which a criminal record affects subsequent employment opportunities. By using matched pairs of individuals to apply for real entry-level jobs, it becomes possible to directly measure the extent to which a criminal

From "The Mark of a Criminal Record," by Devah Pager in *American Journal of Sociology* 108(5). Copyright © 2003 by the University of Chicago Press. Reprinted by permission of the publisher and author.

record—in the absence of other disqualifying characteristics—serves as a barrier to employment among equally qualified applicants. Further, by varying the race of the tester pairs, we can assess the ways in which the effects of race and criminal record interact to produce new forms of labor market inequalities.

TRENDS IN INCARCERATION

Over the past three decades, the number of prison inmates in the United States has increased by more than 600 percent, leaving it the country with the highest incarceration rate in the world (Bureau of Justice Statistics, 2002; Barclay, Tavares, and Siddique, 2001). During this time, incarceration has changed from a punishment reserved primarily for the most heinous offenders to one extended to a much greater range of crimes and a much larger segment of the population. Recent trends in crime policy have led to the imposition of harsher sentences for a wider range of offenses, thus casting an ever-widening net of penal intervention.[1]

While the recent "tough on crime" policies may be effective in getting criminals off the streets, little provision has been made for when they get back out. Of the nearly 2 million individuals currently incarcerated, roughly 95 percent will be released, with more than half a million being released each year (Slevin, 2000). According to one estimate, there are currently over 12 million ex-felons in the United States, representing roughly 8 percent of the working-age population (Uggen, Thompson, and Manza, 2000). Of those recently released, nearly two-thirds will be charged with new crimes and over 40 percent will return to prison within three years (Bureau of Justice Statistics, 2000). Certainly, some of these outcomes are the result of desolate opportunities or deeply ingrained dispositions, grown out of broken families, poor neighborhoods, and little social control (Sampson and Laub, 1993; Wilson, 1997). Besides these contributing factors, there is evidence that experience with the criminal justice system in itself has adverse consequences for subsequent opportunities. In particular, incarceration is associated with limited future employment opportunities and earnings potential (Freeman, 1987; Western, 2002), which themselves are among the strongest predictors of recidivism (Shover, 1996; Sampson and Laub, 1993; Uggen, 2000).

The expansion of the prison population has been particularly consequential for blacks. The incarceration rate for young black men in the year 2000 was nearly 10 percent, compared to just over 1 percent for white men in the same age group (Bureau of Justice Statistics, 2001). Young black men today have a 28 percent likelihood of incarceration during their lifetime (Bureau of Justice Statistics, 1997), a figure that rises above 50 percent among young black high school dropouts (Pettit and Western, 2001). These vast numbers of inmates translate into a large and increasing population of black ex-offenders returning to communities and searching for work. The barriers these men face in reaching economic self-sufficiency are compounded by the stigma of minority status and criminal record. The consequences of such trends for widening racial disparities are potentially profound (see Western and Pettit, 1999; Freeman and Hölzer, 1986).

The objective of this study is to assess whether the effect of a criminal record differs for black and white applicants. Most research investigating the differential impact of incarceration on blacks has focused on the differential *rates* of incarceration and how those rates translate into widening racial disparities. In addition to disparities in the rate of incarceration, however, it is also important to consider possible racial differences in the *effects* of incarceration. Almost none of the existing literature to date has explored this issue, and the theoretical arguments remain divided as to what we might expect.

On one hand, there is reason to believe that the signal of a criminal record should be less consequential for blacks. Research on racial stereotypes tells us that Americans hold strong and persistent negative stereotypes about blacks, with one of the most readily invoked contemporary stereotypes relating to perceptions of violent and criminal dispositions (Smith, 1991; Sniderman and Piazza, 1993; Devine and Elliot, 1995). If it is the case that employers view all blacks as potential criminals, they are likely to differentiate less among those with official criminal records and those without. Actual confirmation of criminal involvement then will provide only redundant information, while evidence against it will be discounted. In this case, the outcomes for all blacks should be worse, with less differentiation between those with criminal records and those without.

On the other hand, the effect of a criminal record may be worse for blacks if employers, already wary of black applicants, are more hesitant when it comes to taking risks on blacks with proven criminal tendencies. The literature on racial stereotypes also tells us that stereotypes are most likely to be activated and reinforced when a target matches on more than one dimension of the stereotype (Quillian and Pager, 2002; Darley and Gross, 1983; Fiske and Neuberg, 1990). While employers may have learned to keep their racial attributions in check through years of heightened sensitivity around employment discrimination, when combined with knowledge of a criminal history, negative attributions are likely to intensify.

A third possibility, of course, is that a criminal record affects black and white applicants equally. The results of this audit study will help to adjudicate between these competing predictions.

STUDY DESIGN

The basic design of this study involves the use of four male auditors (also called testers), two blacks and two whites. The testers were 23-year-old college students from Milwaukee who were matched on the basis of physical appearance and general style of self-presentation. Objective characteristics that were not already identical between pairs—such as educational attainment and work experience—were made similar for the purpose of the applications. Within each team, one auditor was randomly assigned a "criminal record" for the first week; the pair then rotated which member presented himself as the ex-offender for each successive week of employment searches, such that each tester served in the criminal record condition for an equal number of cases. By varying which member of the pair presented himself as having a criminal record, unobserved differences within the pairs of applicants were effectively controlled. No significant differences were found for the outcomes of individual testers or by month of testing.

Job openings for entry-level positions (defined as jobs requiring no previous experience and no education greater than high school) were identified from the Sunday classified advertisement section of the *Milwaukee Journal Sentinel*.[2] In addition, a supplemental sample was drawn *from Jobnet,* a state-sponsored Web site for employment listings, which was developed in connection with the W-2 Welfare-to-Work initiatives.[3]

The audit pairs were randomly assigned 15 job openings each week. The white pair and the black pair were assigned separate sets of jobs, with the same-race testers applying to the same jobs. One member of the pair applied first, with the second applying one day later (randomly varying whether the ex-offender was first or second). A total of 350 employers were audited during the course of this study: 150 by the white pair and 200 by the black pair. Additional tests were performed by the black pair because black testers received fewer callbacks on average, and there were thus fewer data points with which to draw comparisons. A larger sample size enabled me to calculate more precise estimates of the effects under investigation.

Immediately following the completion of each job application, testers filled out a six-page response form that coded relevant information from the test. Important variables included type of occupation, metropolitan status, wage, size of establishment, and race and sex of employer.[4] Additionally, testers wrote narratives describing the overall interaction and any comments made by employers (or included on applications) specifically related to race or criminal records.

Tester Profiles

In developing the tester profiles, emphasis was placed on adopting characteristics that were both numerically representative and substantively important. In the present study, the criminal record consisted of a felony drug conviction (possession, with intent to distribute, cocaine) and 18 months of (served) prison time. A drug crime (as opposed to a violent or property crime) was chosen because of its prevalence, its policy salience, and its connection to racial disparities in incarceration.[5] It is important to acknowledge that the effects reported here may differ depending on the type of offense.

THE EFFECT OF A CRIMINAL RECORD FOR WHITES

I begin with an analysis of the effect of a criminal record among whites. White noncriminals can serve as our baseline in the following comparisons, representing the presumptively nonstigmatized group relative to blacks and those with criminal records. Given that all testers presented roughly identical credentials, the differences experienced among groups of testers can be attributed fully to the effects of race or criminal status.

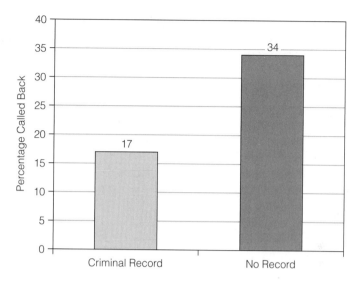

F I G U R E 20.1 The effect of a criminal record on employment opportunities for whites. The effect of a criminal record is statistically significant (P < .01).

Figure 20.1 shows the percentage of applications submitted by white testers that elicited callbacks from employers, by criminal status. As illustrated below, there is a large and significant effect of a criminal record, with 34 percent of whites without criminal records receiving callbacks, relative to only 17 percent of whites with criminal records. A criminal record thereby reduces the likelihood of a callback by 50 percent.

There were some fairly obvious examples documented by testers that illustrate the strong reaction among employers to the signal of a criminal record. In one case, a white tester in the criminal record condition went to a trucking service to apply for a job as a dispatcher. The tester was given a long application, including a complex math test, which took nearly 45 minutes to fill out. During the course of this process, there were several details about the application and the job that needed clarification, some of which involved checking with the supervisor about how to proceed. No concerns were raised about his candidacy at this stage. When the tester turned the application in, the secretary brought it into a back office for the supervisor to look over, so that an interview could perhaps be conducted. When the secretary came back out, presumably after the supervisor had a chance to look over the application more thoroughly, he was told the position had already been filled. While, of course, isolated incidents like this are not conclusive, this was not an infrequent occurrence. Often testers reported seeing employers' levels of responsiveness change dramatically once they had glanced down at the criminal record question.

Clearly, the results here demonstrate that criminal records close doors in employment situations. Many employers seem to use the information as a screening mechanism, without attempting to probe deeper into the possible context or complexities of the situation. As we can see here, in 50 percent of cases, employers were unwilling to consider equally qualified applicants on the basis of their criminal record.

Of course, this trend is not true among all employers, in all situations. There were, in fact, some employers who seemed to prefer workers who had been recently released from prison. One owner told a white tester in the criminal record condition that he "like[d] hiring people who ha[d] just come out of prison because they tend to be more motivated, and are more likely to be hard workers [not wanting to return to prison]." Another employer for a cleaning company attempted to dissuade the white noncriminal tester from applying because the job involved "a great deal of dirty work." The tester with the criminal record, on the other hand, was offered the job on the spot. A criminal record is thus not an obstacle in all cases, but on average, as we see above, it reduces employment opportunities substantially.

THE EFFECT OF RACE

A second major focus of this study concerns the effect of race. African Americans continue to suffer from lower rates of employment relative to whites, but there is tremendous disagreement over the source of these disparities. The idea that race itself—apart from other correlated characteristics—continues to play a major role in shaping employment opportunities has come under question in recent years (e.g., D'Souza, 1995; Steele, 1991). The audit methodology is uniquely suited to address this question. While the present study design does not provide the kind of cross-race matched-pair tests that earlier audit studies of racial discrimination have used, the between-group comparisons (white pair vs. black pair) can nevertheless offer an unbiased estimate of the effect of race on employment opportunities.

Figure 20.2 presents the percentage of callbacks received for both categories of black testers relative to those for whites. The effect of race in these findings is strikingly large. Among blacks without criminal records, only 14 percent received callbacks, relative to 34 percent of white noncriminals (P < .01). In fact, even whites *with* criminal records received more favorable treatment (17 percent) than blacks *without* criminal records (14 percent). The rank ordering of groups in this graph is painfully revealing of employer preferences: race continues to play a dominant role in shaping employment opportunities, equal to or greater than the impact of a criminal record.

The magnitude of the race effect found here corresponds closely to those found in previous audit studies directly measuring racial discrimination. Bendick et al. (1994), for example, found that blacks were 24 percentage points less likely to receive a job offer relative to their white counterparts, a finding very close to the 20 percentage point difference (between white and black nonoffenders) found here. Thus in the eight years since the last major employment audit of race was conducted, very little has changed in the reaction of employers to minority applicants. Despite the many rhetorical arguments used to suggest that

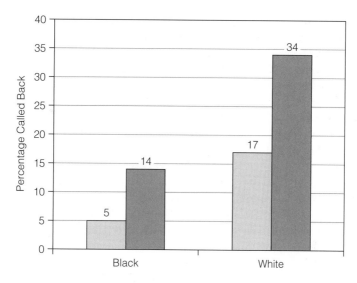

F I G U R E   20.2 The effect of a criminal record for black and white job applicants. The main effects of race and criminal record are statistically significant (P < .01). The interaction between the two is not significant in the full sample. Light bars represent criminal record; dark bars represent no criminal record.

direct racial discrimination is no longer a major barrier to opportunity (e.g., D'Souza, 1995; Steele, 1991), as we can see here, employers, at least in Milwaukee, continue to use race as a major factor in hiring decisions.

RACIAL DIFFERENCES IN THE EFFECTS OF A CRIMINAL RECORD

The final question this study sought to answer was the degree to which the effect of a criminal record differs depending on the race of the applicant. Based on the results presented in Figure 20.2, the effect of a criminal record appears more pronounced for blacks than it is for whites. While this interaction term is not statistically significant, the magnitude of the difference is nontrivial. While the ratio of callbacks for nonoffenders relative to ex-offenders for whites is 2:1, this same ratio for blacks is nearly 3:1. The effect of a criminal record is thus 40 percent larger for blacks than for whites.

This evidence is suggestive of the way in which associations between race and crime affect interpersonal evaluations. Employers, already reluctant to hire blacks, appear even more wary of blacks with proven criminal involvement. Despite the fact that these testers were bright articulate college students with effective styles of self-presentation, the cursory review of entry-level applicants leaves little room for these qualities to be noticed. Instead, the employment barriers of minority status and criminal record are compounded, intensifying the stigma toward this group.

The salience of employers' sensitivity toward criminal involvement among blacks was highlighted in several interactions documented by testers. On three separate occasions, for example, black testers were asked in person (before submitting their applications) whether they had a prior criminal history. None of the white testers were asked about their criminal histories up front.

DISCUSSION

There is serious disagreement among academics, policy makers, and practitioners over the extent to which contact with the criminal justice system—in itself— leads to harmful consequences for employment. The present study takes a strong stand in this debate by offering direct evidence of the causal relationship between a criminal record and employment outcomes. While survey research has produced noisy and indirect estimates of this effect, the current research design offers a direct measure of a criminal record as a mechanism producing employment disparities. Using matched pairs and an experimentally assigned criminal record, this estimate is unaffected by the problems of selection, which plague observational data. While certainly there are additional ways in which incarceration may affect employment outcomes, this finding provides conclusive evidence that mere contact with the criminal justice system, in the absence of any transformative or selective effects, severely limits subsequent employment opportunities. And while the audit study investigates employment barriers to ex-offenders from a microperspective, the implications are far-reaching. The finding that ex-offenders are only one-half to one-third as likely as nonoffenders to be considered by employers suggests that a criminal record indeed presents a major barrier to employment. With over 2 million people currently behind bars and over 12 million people with prior felony convictions, the consequences for labor market inequalities are potentially profound.

Second, the persistent effect of race on employment opportunities is painfully clear in these results. Blacks are less than half as likely to receive consideration by employers, relative to their white counterparts, and black nonoffenders fall behind even whites with prior felony convictions. The powerful effects of race thus continue to direct employment decisions in ways that contribute to persisting racial inequality. In light of these findings, current public opinion seems largely misinformed. According to a recent survey of residents in Los Angeles, Boston, Detroit, and Atlanta, researchers found that just over a quarter of whites believe there to be "a lot" of discrimination against blacks, compared to nearly two-thirds of black respondents (Kluegel and Bobo, 2001). Over the past decade, affirmative action has come under attack across the country based on the argument that direct racial discrimination is no longer a major barrier to opportunity. According to this study, however, employers, at least in Milwaukee, continue to use race as a major factor in their hiring decisions.

When we combine the effects of race and criminal record, the problem grows more intense. Not only are blacks much more likely to be incarcerated than whites; based on the findings presented here, but they may also be more strongly affected by the impact of a criminal record. Previous estimates of the aggregate consequences of incarceration may therefore underestimate the impact on racial disparities.

Finally, in terms of policy implications, this research has troubling conclusions. In our frenzy of locking people up, our "crime control" policies may in fact exacerbate the very conditions that lead to crime in the first place. Research consistently shows that finding quality steady employment is one of the strongest predictors of desistance from crime (Shover, 1996; Sampson and Laub, 1993; Uggen, 2000). The fact that a criminal record severely limits employment opportunities—particularly among blacks—suggests that these individuals are left with few viable alternatives.[6]

As more and more young men enter the labor force from prison, it becomes increasingly important to consider the impact of incarceration on the job prospects of those coming out. No longer a peripheral institution, the criminal justice system has become a dominant presence in the lives of young disadvantaged men, playing a key role in the sorting and stratifying of labor market opportunities. This article represents an initial attempt to specify one of the important mechanisms by which incarceration leads to poor employment outcomes. Future research is needed to expand this emphasis to other mechanisms (e.g., the transformative effects of prison on human and social capital), as well as to include other social domains affected by incarceration (e.g., housing, family formation, political participation, etc.); in this way, we can move toward a more complete understanding of the collateral consequences of incarceration for social inequality.

At this point in history, it is impossible to tell whether the massive presence of incarceration in today's stratification system represents a unique anomaly of the late twentieth century, or part of a larger movement toward a system of stratification based on the official certification of individual character and competence. Whether this process of negative credentialing will continue to form the basis of emerging social cleavages remains to be seen.

NOTES

1. For example, the recent adoption of mandatory sentencing laws, most often used for drug offenses, removes discretion from the sentencing judge to consider the range of factors pertaining to the individual and the offense that would normally be taken into account. As a result, the chances of receiving a state prison term after being arrested for a drug offense rose by 547 percent between 1980 and 1992 (Bureau of Justice Statistics, 1995).

2. Occupations with legal restrictions on ex-offenders were excluded from the sample. These include jobs in the health care industry, work with children and the elderly, jobs requiring the handling of firearms (i.e., security

guards), and jobs in the public sector. An estimate of the collateral consequences of incarceration would also need to take account of the wide range of employment formally off-limits to individuals with prior felony convictions.

3. Employment services like *Jobnet* have become a much more common method of finding employment in recent years, particularly for difficult-to-employ populations such as welfare recipients and ex-offenders. Likewise, a recent survey by Hölzer and Stoll (2001) found that nearly half of Milwaukee employers (46 percent) use *Jobnet* to advertise vacancies in their companies.

4. See Pager (2002) for a discussion of the variation across each of these dimensions.

5. Over the past two decades, drug crimes were the fastest growing class of offenses. In 1980, roughly one out of every sixteen state inmates was incarcerated for a drug crime; by 1999, this figure had jumped to one out of every five (Bureau of Justice Statistics, 2000). In federal prisons, nearly three out of every five inmates are incarcerated for a drug crime (Bureau of Justice Statistics, 2001). A significant portion of this increase can be attributed to changing policies concerning drug enforcement. By 2000, every state in the country had adopted some form of truth-in-sentencing laws, which impose mandatory sentencing minimums for a range of offenses. These laws have been applied most frequently to drug crimes, leading to more than a fivefold rise in the number of drug arrests that result in incarceration and a doubling of the average length of sentences for drug convictions (Mauer, 1999; Blumstein and Beck, 1999). While the steep rise in drug enforcement has been felt across the population, this "war on drugs"

has had a disproportionate impact on African Americans. Between 1990 and 1997, the number of black inmates serving time for drug offenses increased by 60 percent, compared to a 46 percent increase in the number of whites (Bureau of Justice Statistics, 1995). In 1999, 26 percent of all black state inmates were incarcerated for drug offenses, relative to less than half that proportion of whites (Bureau of Justice Statistics, 2001).

6. There are two primary policy recommendations implied by these results. First and foremost, the widespread use of incarceration, particularly for non-violent drug crimes, has serious, long-term consequences for the employment problems of young men. The substitution of alternatives to incarceration, therefore, such as drug treatment programs or community supervision, may serve to better promote the well-being of individual offenders as well as to improve public safety more generally through the potential reduction of recidivism. Second, additional thought should be given to the widespread availability of criminal background information. As criminal record databases become increasingly easy to access, this information may be more often used as the basis for rejecting otherwise qualified applicants. If instead criminal history information were suppressed—except in cases that were clearly relevant to a particular kind of job assignment— ex-offenders with appropriate credentials might be better able to secure legitimate employment. While there is some indication that the absence of official criminal background information may lead to a greater incidence of statistical discrimination against blacks (see Bushway, 1997; Hölzer et al. 2001), the net benefits of this policy change may in fact outweigh the potential drawbacks.

REFERENCES

Barclay, Gordon, Cynthia Tavares, and Arsalaan Siddique. (2001). "International Comparisons of Criminal Justice Statistics, 1999." London: U.K. Home Office for Statistical Research.

Bendick, Marc, Jr., Charles Jackson, and Victor Reinoso. (1994). "Measuring Employment Discrimination through Controlled Experiments." *Review of Black Political Economy* 23:25–48.

Blumstein, Alfred, and Allen J. Beck. (1999). "Population Growth in U.S. Prisons, 1980–1996." pp. 17–62 in *Prisons: Crime and Justice: A Review of Research*, vol. 26. Edited by Michael Tonry and J. Petersilia. Chicago: University of Chicago Press.

Bureau of Justice Statistics. (1995). *Prisoners in 1994*, by Allen J. Beck and Darrell K. Gilliard. Special report. Washington, DC: Government Printing Office.

———. (1997). *Lifetime Likelihood of Going to State or Federal Prison*, by Thomas P. Bonczar and Allen J. Beck. Special report, March.

———. (2000). Bulletin. *Key Facts at a Glance: Number of Persons in Custody of State Correctional Authorities by Most Serious Offense 1980–99.* Washington, DC: Government Printing Office.

———. (2001). *Prisoners in 2000*, by Allen J. Beck and Paige M. Harrison. August. Bulletin. Washington, DC: NCJ 188207.

———. (2002). *Sourcebook of Criminal Justice Statistics.* (Last accessed March 1, 2003.) Available www.albany.edu/sourcebook.

Bushway, Shawn D. (1997). "Labor Market Effects of Permitting Employer Access to Criminal History Records." Working paper. University of Maryland, Department of Criminology.

Darley, J. M., and P. H. Gross. (1983). "A Hypothesis-Confirming Bias in Labeling Effects." *Journal of Personality and Social Psychology* 44:20–33.

Devine, P. G., and A. J. Elliot. (1995). "Are Racial Stereotypes Really Fading? The Princeton Trilogy Revisited." *Personality and Social Psychology Bulletin* 21 (11):1139–50.

D'Souza, Dinesh. (1995). *The End of Racism: Principles for a Multiracial Society.* New York: Free Press.

Fiske, Susan, and Steven Neuberg. (1990). "A Continuum of Impression Formation, from Category-Based to Individuating Processes." pp. 1–63 in *Advances in Experimental Social Psychology*, vol. 23. Edited by Mark Zanna. New York: Academic Press.

Freeman, Richard B. (1987). "The Relation of Criminal Activity to Black Youth Employment." *Review of Black Political Economy* 16 (1–2):99–107.

Freeman, Richard B., and Harry J. Hölzer, eds. (1986). *The Black Youth Employment Crisis.* Chicago: University of Chicago Press for National Bureau of Economic Research.

Hölzer, Harry, and Michael Stoll. (2001). *Employers and Welfare Recipients: The Effects of Welfare Reform in the Workplace.* San Francisco: Public Policy Institute of California.

Kluegel, James, and Lawrence Bobo. (2001). "Perceived Group Discrimination and Policy Attitudes: The Sources and Consequences of the Race and Gender Gaps." pp. 163–216 in *Urban Inequality: Evidence from Four Cities*, edited by Alice O'Connor, Chris Tilly, and Lawrence D. Bobo. New York: Russell Sage Foundation.

Mauer, Marc. (1999). *Race to Incarcerate.* New York: New Press.

Pager, Devah. (2002). "The Mark of a Criminal Record." *Doctoral dissertation.* Department of Sociology, University of Wisconsin–Madison.

Pettit, Becky, and Bruce Western. (2001). "Inequality in Lifetime Risks of Imprisonment." Paper presented at the annual meetings of the American Sociological Association. Anaheim, August.

Quillian, Lincoln, and Devah Pager. (2002). "Black Neighbors, Higher Crime? The Role of Racial Stereotypes in Evaluations of Neighborhood Crime." *American Journal of Sociology* 107 (3):717–67.

Sampson, Robert J., and John H. Laub. (1993). *Crime in the Making: Pathways and Turning Points through Life.* Cambridge, Mass: Harvard University Press.

Shover, Neil. (1996). *Great Pretenders: Pursuits and Careers of Persistent Thieves.* Boulder, Colo: Westview.

Slevin, Peter. (2000). "Life after Prison: Lack of Services Has High Price." *Washington Post,* April 24.

Smith, Tom W. (1991). *What Americans Say about Jews.* New York: American Jewish Committee.

Sniderman, Paul M., and Thomas Piazza. (1993). *The Scar of Race.* Cambridge, Mass: Harvard University Press.

Steele, Shelby. (1991). *The Content of Our Character: A New Vision of Race in America.* New York: Harper Perennial.

Uggen, Christopher. (2000). "Work as a Turning Point in the Life Course of Criminals: A Duration Model of Age, Employment, and Recidivism." *American Sociological Review* 65 (4): 529–46.

Uggen, Christopher, Melissa Thompson, and Jeff Manza. (2000). "Crime, Class, and Reintegration: The Socioeconomic, Familial, and Civic Lives of Offenders." Paper presented at the American Society of Criminology meetings, San Francisco, November 18.

Western, Bruce. (2002). "The Impact of Incarceration on Wage Mobility and Inequality." *American Sociological Review* 67 (4):526–46.

Western, Bruce, and Becky Pettit. (1999). "Black–White Earnings Inequality, Employment Rates, and Incarceration." Working Paper no. 150. New York: Russell Sage Foundation.

Wilson, William Julius. (1997). *When Work Disappears: The World of the New Urban Poor.* New York: Vintage Books.

21

The Saints and the Roughnecks

WILLIAM J. CHAMBLISS

Chambliss's description of the Saints and the Roughnecks shows how the power of social class can operate to facilitate groups' resistance to deviant labels. In this classic selection from the sociological literature, Chambliss describes how the Saints engage in as many or more delinquent acts than the Roughnecks, yet are perceived as "good boys" merely engaging in typical adolescent high jinks. On the one hand, the greater social power contained in their higher class background enables the definition of their behavior as socially normative, allowing the police, teachers, community members, and parents to look the other way. On the other hand, the Roughnecks, who come from the "wrong side of the tracks," are perceived to be troublemakers, rabble-rousers, and delinquents. We see conflict and labeling theories in effect here because social class is the determinant of society's reactions. Behavior done by teenagers from upstanding, middle-class families is tolerated, while similar behavior engaged in by lower class youths is reinforced as deviant. Once again, labels are applied on the basis of status, not patterns of behavior.

How does Chambliss's concept of reinforcement compare with Erikson's use of the self-fulfilling prophecy? How does it affect the Saints and the Roughnecks differently?

Eight promising young men—children of good, stable, white, upper-middle-class families, active in school affairs, good precollege students—were some of the most delinquent boys at Hanibal High School. While community residents and parents knew that these boys occasionally sowed a few wild oats, they were totally unaware that sowing wild oats completely occupied the daily routine of these young men. The Saints were constantly occupied with truancy, drinking, wild driving, petty theft, and vandalism. Yet not one was officially arrested for any misdeed during the two years I observed them.

From Society, V. 11. No. 1, 1973, "The Saints and the Roughnecks," by William J. Chambliss. Copyright © 1973, with kind permission of Springer Science and Business Media.

This record was particularly surprising in light of my observations during the same two years of another gang of Hanibal High School students, six lower-class white boys known as the Roughnecks. The Roughnecks were constantly in trouble with police and community even though their rate of delinquency was about equal with that of the Saints. What was the cause of this disparity? The result? The following consideration of the activities, social class, and community perceptions of both gangs may provide some answers.

THE SAINTS

The Saints From Monday to Friday

The Saints' principal daily concern was with getting out of school as early as possible. The boys managed to get out of school with minimum danger that they would be accused of playing hookey through an elaborate procedure for obtaining "legitimate" release from class. The most common procedure was for one boy to obtain the release of another by fabricating a meeting of some committee, program, or recognized club. Charles might raise his hand in his 9:00 chemistry class and ask to be excused—a euphemism for going to the bathroom. Charles would go to Ed's math class and inform the teacher that Ed was needed for a 9:30 rehearsal of the drama club play. The math teacher would recognize Ed and Charles as "good students" involved in numerous school activities and would permit Ed to leave at 9:30. Charles would return to his class, and Ed would go to Tom's English class to obtain his release. Tom would engineer Charles's escape. The strategy would continue until as many of the Saints as possible were freed. After a stealthy trip to the car (which had been parked in a strategic spot), the boys were off for a day of fun.

Over the two years I observed the Saints, this pattern was repeated nearly every day. There were variations on the theme, but in one form or another, the boys used this procedure for getting out of class and then off the school grounds. Rarely did all eight of the Saints manage to leave school at the same time. The average number avoiding school on the days I observed them was five.

Having escaped from the concrete corridors the boys usually went either to a pool hall on the other (lower-class) side of town or to a cafe in the suburbs. Both places were out of the way of people the boys were likely to know (family or school officials), and both provided a source of entertainment. The pool hall entertainment was the generally rough atmosphere, the occasional hustler, the sometimes drunk proprietor, and, of course, the game of pool. The cafe's entertainment was provided by the owner. The boys would "accidentally" knock a glass on the floor or spill cola on the counter—not all the time, but enough to be sporting. They would also bend spoons, put salt in sugar bowls, and generally tease whoever was working in the cafe. The owner had opened the cafe recently and was dependent on the boys' business which was, in fact, substantial since between the horsing around and the teasing they bought food and drinks.

The Saints on Weekends

On weekends the automobile was even more critical than during the week, for on weekends the Saints went to Big Town—a large city with a population of over a million 25 miles from Hanibal. Every Friday and Saturday night most of the Saints would meet between 8:00 and 8:30 and would go into Big Town. Big Town activities included drinking heavily in taverns or nightclubs, driving drunkenly through the streets, and committing acts of vandalism and playing pranks.

By midnight on Fridays and Saturdays the Saints were usually thoroughly high, and one or two of them were often so drunk they had to be carried to the cars. Then the boys drove around town, calling obscenities to women and girls; occasionally trying (unsuccessfully so far as I could tell) to pick girls up; and driving recklessly through red lights and at high speeds with their lights out. Occasionally they played "chicken." One boy would climb out the back window of the car and across the roof to the driver's side of the car while the car was moving at high speed (between 40 and 50 miles an hour); then the driver would move over and the boy who had just crawled across the car roof would take the driver's seat.

Searching for "fair game" for a prank was the boys' principal activity after they left the tavern. The boys would drive alongside a foot patrolman and ask directions to some street. If the policeman leaned on the car in the course of answering the question, the driver would speed away, causing him to lose his balance. The Saints were careful to play this prank only in an area where they were not going to spend much time and where they could quickly disappear around a corner to avoid having their license plate number taken.

Construction sites and road repair areas were the special province of the Saints' mischief. A soon-to-be-repaired hole in the road inevitably invited the Saints to remove lanterns and wooden barricades and put them in the car, leaving the hole unprotected. The boys would find a safe vantage point and wait for an unsuspecting motorist to drive into the hole. Often, though not always, the boys would go up to the motorist and commiserate with him about the dreadful way the city protected its citizenry.

Leaving the scene of the open hole and the motorist, the boys would then go searching for an appropriate place to erect the stolen barricade. An "appropriate place" was often a spot on a highway near a curve in the road where the barricade would not be seen by an oncoming motorist. The boys would wait to watch an unsuspecting motorist attempt to stop and (usually) crash into the wooden barricade. With saintly bearing the boys might offer help and understanding.

A stolen lantern might well find its way onto the back of a police car or hang from a street lamp. Once a lantern served as a prop for a reenactment of the "midnight ride of Paul Revere" until the "play," which was taking place at 2:00 A.M. in the center of a main street of Big Town, was interrupted by a police car several blocks away. The boys ran, leaving the lanterns on the street, and managed to avoid being apprehended.

Abandoned houses, especially if they were located in out-of-the-way places, were fair game for destruction and spontaneous vandalism. The boys would break windows, remove furniture to the yard and tear it apart, urinate on the walls, and scrawl obscenities inside.

Through all the pranks, drinking, and reckless driving the boys managed miraculously to avoid being stopped by police. Only twice in two years was I aware that they had been stopped by a Big City policeman. Once was for speeding (which they did every time they drove whether they were drunk or sober), and the driver managed to convince the policeman that it was simply an error. The second time they were stopped they had just left a nightclub and were walking through an alley. Aaron stopped to urinate and the boys began making obscene remarks. A foot patrolman came into the alley, lectured the boys, and sent them home. Before the boys got to the car one began talking in a loud voice again. The policeman, who had followed them down the alley, arrested this boy for disturbing the peace and took him to the police station where the other Saints gathered. After paying a $5.00 fine, and with the assurance that there would be no permanent record of the arrest, the boy was released.

The boys had a spirit of frivolity and fun about their escapades. They did not view what they were engaged in as "delinquency," though it surely was by any reasonable definition of that word. They simply viewed themselves as having a little fun and who, they would ask, was really hurt by it? The answer had to be no one, although this fact remains one of the most difficult things to explain about the gang's behavior. Unlikely though it seems, in two years of drinking, driving, carousing, and vandalism no one was seriously injured as a result of the Saints' activities.

The Saints in School

The Saints were highly successful in school. The average grade for the group was "B," with two of the boys having close to a straight "A" average. Almost all of the boys were popular and many of them held offices in the school. One of the boys was vice-president of the student body one year. Six of the boys played on athletic teams.

At the end of their senior year, the student body selected ten seniors for special recognition as the "school wheels"; four of the ten were Saints. Teachers and school officials saw no problem with any of these boys and anticipated that they would all "make something of themselves."

How the boys managed to maintain this impression is surprising in view of their actual behavior while in school. Their technique for covering truancy was so successful that teachers did not even realize that the boys were absent from school much of the time. Occasionally, of course, the system would backfire and then the boy was on his own. A boy who was caught would be most contrite, would plead guilty and ask for mercy. He inevitably got the mercy he sought.

Cheating on examinations was rampant, even to the point of orally communicating answers to exams as well as looking at one another's papers. Since none of the group studied, and since they were primarily dependent on one another for help, it is surprising that grades were so high. Teachers contributed to the deception in their admitted inclination to give these boys (and presumably others like them) the benefit of the doubt. When asked how the boys did in school, and when pressed on specific examinations, teachers might admit that they were disappointed in John's performance, but would quickly add that they "knew that he was capable of doing better," so John was given a higher grade than he had actually earned. How often this happened is impossible to know. During the time that I observed the group, I never saw any of the boys take homework home. Teachers may have been "understanding" very regularly.

One exception to the gang's generally good performance was Jerry, who had a "C" average in his junior year, experienced disaster the next year, and failed to graduate. Jerry had always been a little more nonchalant than the others about the liberties he took in school. Rather than wait for someone to come get him from class, he would offer his own excuse and leave. Although he probably did not miss any more classes than most of the others in the group, he did not take the requisite pains to cover his absences. Jerry was the only Saint whom I ever heard talk back to a teacher. Although teachers often called him a "cut up" or a "smart kid," they never referred to him as a troublemaker or as a kid headed for trouble. It seems likely, then, that Jerry's failure his senior year and his mediocre performance his junior year were consequences of his not playing the game the proper way (possibly because he was disturbed by his parents' divorce). His teachers regarded him as "immature" and not quite ready to get out of high school.

The Police and the Saints

The local police saw the Saints as good boys who were among the leaders of the youth in the community. Rarely, the boys might be stopped in town for speeding or for running a stop sign. When this happened the boys were always polite, contrite, and pled for mercy. As in school, they received the mercy they asked for. None ever received a ticket or was taken into the precinct by the local police.

The situation in Big City, where the boys engaged in most of their delinquency, was only slightly different. The police there did not know the boys at all, although occasionally the boys were stopped by a patrolman. Once they were caught taking a lantern from a construction site. Another time they were stopped for running a stop sign, and on several occasions they were stopped for speeding. Their behavior was as before: contrite, polite, and penitent. The urban police, like the local police, accepted their demeanor as sincere. More important, the urban police were convinced that these were good boys just out for a lark.

THE ROUGHNECKS

Hanibal townspeople never perceived the Saints' high level of delinquency. The Saints were good boys who just went in for an occasional prank. After all, they were well dressed, well mannered, and had nice cars. The Roughnecks were a different story. Although the two gangs of boys were the same age, and both groups engaged in an equal amount of wild-oat sowing, everyone agreed that the not-so-well-dressed, not-so-well-mannered, not-so-rich boys were heading for trouble. Townspeople would say, "You can see the gang members at the drug-store, night after night, leaning against the storefront (sometimes drunk) or slouch-ing around inside buying cokes, reading magazines, and probably stealing old Mr. Wall blind. When they are outside and girls walk by, even respectable girls, these boys make suggestive remarks. Sometimes their remarks are downright lewd."

From the community's viewpoint, the real indication that these kids were in for trouble was that they were constantly involved with the police. Some of them had been picked up for stealing, mostly small stuff, of course, "but still it's stealing small stuff that leads to big time crimes." "Too bad," people said. "Too bad that these boys couldn't behave like the other kids in town: stay out of trouble, be polite to adults, and look to their future."

The community's impression of the degree to which this group of six boys (ranging in age from 16 to 19) engaged in delinquency was somewhat distorted. In some ways the gang was more delinquent than the community thought; in other ways they were less.

The fighting activities of the group were fairly readily and accurately per-ceived by almost everyone. At least once a month, the boys would get into some sort of fight, although most fights were scraps between members of the group or involved only one member of the group and some peripheral hanger-on. Only three times in the period of observation did the group fight together: once against a gang from across town, once against two blacks, and once against a group of boys from another school. For the first two fights the group went out "looking for trouble"—and they found it both times. The third fight followed a football game and began spontaneously with an argument on the football field between one of the Roughnecks and a member of the opposition's football team.

Jack had a particular propensity for fighting and was involved in most of the brawls. He was a prime mover of the escalation of arguments into fights.

More serious than fighting, had the community been aware of it, was theft. Although almost everyone was aware that the boys occasionally stole things, they did not realize the extent of the activity. Petty stealing was a frequent event for the Roughnecks. Sometimes they stole as a group and coordinated their efforts; other times they stole in pairs. Rarely did they steal alone.

The thefts ranged from very small things like paperback books, comics, and ballpoint pens to expensive items like watches. The nature of the thefts varied from time to time. The gang would go through a period of systematically sho-plifting items from automobiles or school lockers. Types of thievery varied with the whim of the gang. Some forms of thievery were more profitable than others, but all thefts were for profit, not just thrills.

Roughnecks siphoned gasoline from cars as often as they had access to an automobile, which was not very often. Unlike the Saints, who owned their own cars, the Roughnecks would have to borrow their parents' cars, an event which occurred only eight or nine times a year. The boys claimed to have stolen cars for joy rides from time to time.

Ron committed the most serious of the group's offenses. With an unidentified associate the boy attempted to burglarize a gasoline station. Although this station had been robbed twice previously in the same month, Ron denied any involvement in either of the other thefts. When Ron and his accomplice approached the station, the owner was hiding in the bushes beside the station. He fired both barrels of a double-barreled shotgun at the boys. Ron was severely injured; the other boy ran away and was never caught. Though he remained in critical condition for several months, Ron finally recovered and served six months of the following year in reform school. Upon release from reform school, Ron was put back a grade in school, and began running around with a different gang of boys. The Roughnecks considered the new gang less delinquent than themselves, and during the following year Ron had no more trouble with the police.

The Roughnecks, then, engaged mainly in three types of delinquency: theft, drinking, and fighting. Although community members perceived that this gang of kids was delinquent, they mistakenly believed that their illegal activities were primarily drinking, fighting, and being a nuisance to passersby. Drinking was limited among the gang members, although it did occur, and theft was much more prevalent than anyone realized.

Drinking would doubtless have been more prevalent had the boys had ready access to liquor. Since they rarely had automobiles at their disposal, they could not travel very far, and the bars in town would not serve them. Most of the boys had little money, and this, too, inhibited their purchase of alcohol. Their major source of liquor was a local drunk who would buy them a fifth if they would give him enough extra to buy himself a pint of whiskey or a bottle of wine.

The community's perception of drinking as prevalent stemmed from the fact that it was the most obvious delinquency the boys engaged in. When one of the boys had been drinking, even a casual observer seeing him on the corner would suspect that he was high.

There was a high level of mutual distrust and dislike between the Roughnecks and the police. The boys felt very strongly that the police were unfair and corrupt. Some evidence existed that the boys were correct in their perception.

The main source of the boys' dislike for the police undoubtedly stemmed from the fact that the police would sporadically harass the group. From the standpoint of the boys, these acts of occasional enforcement of the law were whimsical and uncalled-for. It made no sense to them, for example, that the police would come to the corner occasionally and threaten them with arrest for loitering when the night before the boys had been out siphoning gasoline from cars and the police had been nowhere in sight. To the boys, the police were stupid on the one hand, for not being where they should have been and catching the boys in a serious offense, and unfair on the other hand, for trumping up "loitering" charges against them.

From the viewpoint of the police, the situation was quite different. They knew, with all the confidence necessary to be a policeman, that these boys were engaged in criminal activities. They knew this partly from occasionally catching them, mostly from circumstantial evidence ("the boys were around when those tires were slashed"), and partly because the police shared the view of the community in general that this was a bad bunch of boys. The best the police could hope to do was to be sensitive to the fact that these boys were engaged in illegal acts and arrest them whenever there was some evidence that they had been involved. Whether or not the boys had in fact committed a particular act in a particular way was not especially important. The police had a broader view: their job was to stamp out these kids' crimes; the tactics were not as important as the end result.

Over the period that the group was under observation, each member was arrested at least once. Several of the boys were arrested a number of times and spent at least one night in jail. While most were never taken to court, two of the boys were sentenced to six months' incarceration in boys' schools.

The Roughnecks in School

The Roughnecks' behavior in school was not particularly disruptive. During school hours they did not all hang around together, but tended instead to spend most of their time with one or two other members of the gang who were their special buddies. Although every member of the gang attempted to avoid school as much as possible, they were not particularly successful and most of them attended school with surprising regularity. They considered school a burden—something to be gotten through with a minimum of conflict. If they were "bugged" by a particular teacher, it could lead to trouble. One of the boys, Al, once threatened to beat up a teacher and, according to the other boys, the teacher hid under a desk to escape him.

Teachers saw the boys the way the general community did, as heading for trouble, as being uninterested in making something of themselves. Some were also seen as being incapable of meeting the academic standards of the school. Most of the teachers expressed concern for this group of boys and were willing to pass them despite poor performance, in the belief that failing them would only aggravate the problem.

The group of boys had a grade point average just slightly above "C." No one in the group failed a grade, and no one had better than a "C" average. They were very consistent in their achievement or, at least, the teachers were consistent in their perception of the boys' achievement.

Two of the boys were good football players. Herb was acknowledged to be the best player in the school and Jack was almost as good. Both boys were criticized for their failure to abide by training rules, for refusing to come to practice as often as they should, and for not playing their best during practice. What they lacked in sportsmanship they made up for in skill, apparently, and played every game no matter how poorly they had performed in practice or how many practice sessions they had missed.

TWO QUESTIONS

Why did the community, the school, and the police react to the Saints as though they were good, upstanding, nondeliquent youths with bright futures but to the Roughnecks as though they were tough, young criminals who were headed for trouble? Why did the Roughnecks and the Saints in fact have quite different careers after high school—careers which, by and large, lived up to the expectations of the community?

The most obvious explanation for the differences in the community's and law enforcement agencies' reactions to the two gangs is that one group of boys was "more delinquent" than the other. Which group was more delinquent? The answer to this question will determine in part how we explain the differential responses to these groups by the members of the community and, particularly, by law enforcement and school officials.

In sheer number of illegal acts, the Saints were the more delinquent. They were truant from school for at least part of the day almost every day of the week. In addition, their drinking and vandalism occurred with surprising regularity. The Roughnecks, in contrast, engaged sporadically in delinquent episodes. While these episodes were frequent, they certainly did not occur on a daily or even a weekly basis.

The difference in frequency of offenses was probably caused by the Roughnecks' inability to obtain liquor and to manipulate legitimate excuses from school. Since the Roughnecks had less money than the Saints, and teachers carefully supervised their school activities, the Roughnecks' hearts may have been as black as the Saints', but their misdeeds were not nearly as frequent.

There are really no clear-cut criteria by which to measure qualitative differences in antisocial behavior. The most important dimension of the difference is generally referred to as the "seriousness" of the offenses.

If seriousness encompasses the relative economic costs of delinquent acts, then some assessment can be made. The Roughnecks probably stole an average of about $5.00 worth of goods a week. Some weeks the figure was considerably higher, but these times must be balanced against long periods when almost nothing was stolen.

The Saints were more continuously engaged in delinquency but their acts were not for the most part costly to property. Only their vandalism and occasional theft of gasoline would so qualify. Perhaps once or twice a month they would siphon a tankful of gas. The other costly items were street signs, construction lanterns, and the like. All of these acts combined probably did not quite average $5.00 a week, partly because much of the stolen equipment was abandoned and presumably could be recovered. The difference in cost of stolen property between the two groups was trivial, but the Roughnecks probably had a slightly more expensive set of activities than did the Saints.

Another meaning of seriousness is the potential threat of physical harm to members of the community and to the boys themselves. The Roughnecks were more prone to physical violence; they not only welcomed an opportunity to fight; they went seeking it. In addition, they fought among themselves frequently. Although the fighting never included deadly weapons, it was still a menace, however minor, to the physical safety of those involved.

The Saints never fought. They avoided physical conflict both inside and outside the group. At the same time, though, the Saints frequently endangered their own and other people's lives. They did so almost every time they drove a car, especially if they had been drinking. Sober, their driving was risky; under the influence of alcohol it was horrendous. In addition, the Saints endangered the lives of others with their pranks. Street excavations left unmarked were a very serious hazard.

Evaluating the relative seriousness of the two gangs' activities is difficult. The community reacted as though the behavior of the Roughnecks was a problem, and they reacted as though the behavior of the Saints was not. But the members of the community were ignorant of the array of delinquent acts that character- ized the Saints' behavior. Although concerned citizens were unaware of much of the Roughnecks' behavior as well, they were much better informed about the Roughnecks' involvement in delinquency than they were about the Saints'.

Visibility

Differential treatment of the two gangs resulted in part because one gang was infinitely more visible than the other. This differential visibility was a direct func- tion of the economic standing of the families. The Saints had access to automo- biles and were able to remove themselves from the sight of the community. In as routine a decision as to where to go to have a milkshake after school, the Saints stayed away from the mainstream of community life. Lacking transportation, the Roughnecks could not make it to the edge of town. The center of town was the only practical place for them to meet since their homes were scattered through- out the town and any noncentral meeting place put an undue hardship on some members. Through necessity the Roughnecks congregated in a crowded area where everyone in the community passed frequently, including teachers and law enforcement officers. They could easily see the Roughnecks hanging around the drugstore.

The Roughnecks, of course, made themselves even more visible by making remarks to passersby and by occasionally getting into fights on the corner. Mean- while, just as regularly, the Saints were either at the cafe on one edge of town or in the pool hall at the other edge of town. Without any particular realization that they were making themselves inconspicuous, the Saints were able to hide their time-wasting. Not only were they removed from the mainstream of traffic, but they were also almost always inside a building.

On their escapades the Saints were also relatively invisible, since they left Hanibal and travelled to Big City. Here, too, they were mobile, roaming the city, rarely going to the same area twice.

Demeanor

To the notion of visibility must be added the difference in the responses of group members to outside intervention with their activities. If one of the Saints was confronted with an accusing policeman, even if he felt he was truly innocent of a wrongdoing, his demeanor was apologetic and penitent. A Roughneck's attitude was almost the polar opposite. When confronted with a threatening adult authority, even one who tried to be pleasant, the Roughneck's hostility and disdain were clearly observable. Sometimes he might attempt to put up a veneer of respect, but it was thin and was not accepted as sincere by the authority.

School was no different from the community at large. The Saints could manipulate the system by feigning compliance with the school norms. The availability of cars at school meant that once free from the immediate sight of the teacher, the boys could disappear rapidly. And this escape was well enough planned that no administrator or teacher was nearby when the boys left. A Roughneck who wished to escape for a few hours was in a bind. If it were possible to get free from class, downtown was still a mile away, and even if he arrived there, he was still very visible. Truancy for the Roughnecks meant almost certain detection, while the Saints enjoyed almost complete immunity from sanctions.

Bias

Community members were not aware of the transgressions of the Saints. Even if the Saints had been less discreet, their favorite delinquencies would have been perceived as less serious than those of the Roughnecks.

In the eyes of the police and school officials, a boy who drinks in an alley and stands intoxicated on the street corner is committing a more serious offense than is a boy who drinks to inebriation in a nightclub or a tavern and drives around afterwards in a car. Similarly, a boy who steals a wallet from a store will be viewed as having committed a more serious offense than a boy who steals a lantern from a construction site.

Perceptual bias also operates with respect to the demeanor of the boys in the two groups when they are confronted by adults. It is not simply that adults dislike the posture affected by boys of the Roughneck ilk; more important is the conviction that the posture adopted by the Roughnecks is an indication of their devotion and commitment to deviance as a way of life. The posture becomes a cue, just as the type of the offense is a cue, to the degree to which the known transgressions are indicators of the youths' potential for other problems.

Visibility, demeanor, and bias are surface variables that explain the day-to-day operations of the police. Why do these surface variables operate as they do? Why did the police choose to disregard the Saints' delinquencies while breathing down the backs of the Roughnecks?

The answer lies in the class structure of U.S. society and the control of legal institutions by those at the top of the class structure. Obviously, no representative of the upper class drew up the operational chart for the police which led them to look in the ghettoes and on street corners—which led them to see the demeanor of lower-class youth as troublesome and that of upper-middle-class youth as tolerable. Rather, the procedure simply developed from experience—experience with irate and influential upper-middle-class parents insisting that their son's vandalism was simply a prank and his drunkenness only a momentary "sowing of wild oats"—experience with cooperative or indifferent, powerless, lower-class parents who acquiesced to the law's definition of their son's behavior.

ADULT CAREERS OF THE SAINTS
AND THE ROUGHNECKS

The community's confidence in the potential of the Saints and the Roughnecks apparently was justified. If anything the community members underestimated the degree to which these youngsters would turn out "good" or "bad."

Seven of the eight members of the Saints went on to college immediately after high school. Five of the boys graduated from college in four years. The sixth one finished college after two years in the army, and the seventh spent four years in the air force before returning to college and receiving a B.A. degree. Of these seven college graduates, three went on for advanced degrees. One finished law school and is now active in state politics, one finished medical school and is practicing near Hanibal, and one boy is now working for a Ph.D. The other four college graduates entered submanagerial, managerial, or executive training positions with larger firms.

The only Saint who did not complete college was Jerry. Jerry had failed to graduate from high school with the other Saints. During his second senior year, after the other Saints had gone on to college, Jerry began to hang around with what several teachers described as a "rough crowd"—the gang that was heir apparent to the Roughnecks. At the end of his second senior year, when he did graduate from high school, Jerry took a job as a used car salesman, got married, and quickly had a child. Although he made several abortive attempts to go to college by attending night school, when I last saw him (ten years after high school) Jerry was unemployed and had been living on unemployment for almost a year. His wife worked as a waitress.

Some of the Roughnecks have lived up to community expectations. A number of them were headed for trouble. A few were not.

Jack and Herb were the athletes among the Roughnecks and their athletic prowess paid off handsomely. Both boys received unsolicited athletic scholarships to college. After Herb received his scholarship (near the end of his senior year), he apparently did an about-face. His demeanor became very similar to that of the Saints. Although he remained a member in good standing of the Roughnecks, he stopped participating in most activities and did not hang on the corner as often.

Jack did not change. If anything, he became more prone to fighting. He even made excuses for accepting the scholarship. He told the other gang members that the school had guaranteed him a "C" average if he would come to play football—an idea that seems far-fetched, even in this day of highly competitive recruiting.

During the summer after graduation from high school, Jack attempted suicide by jumping from a tall building. The jump would certainly have killed most people trying it, but Jack survived. He entered college in the fall and played four years of football. He and Herb graduated in four years, and both are teaching and coaching in high schools. They are married and have stable families. If anything, Jack appears to have a more prestigious position in the community than does Herb, though both are well respected and secure in their positions.

Two of the boys never finished high school. Tommy left at the end of his junior year and went to another state. That summer he was arrested and placed on probation on a manslaughter charge. Three years later he was arrested for murder; he pleaded guilty to second degree murder and is serving a 30-year sentence in the state penitentiary.

Al, the other boy who did not finish high school, also left the state in his senior year. He is serving a life sentence in a state penitentiary for first-degree murder.

Wes is a small-time gambler. He finished high school and "bummed around." After several years he made contact with a bookmaker who employed him as a runner. Later he acquired his own area and has been working it ever since. His position among the bookmakers is almost identical to the position he had in the gang; he is always around but no one is really aware of him. He makes no trouble and he does not get into any. Steady, reliable, capable of keeping his mouth closed, he plays the game by the rules, even though the game is an illegal one.

That leaves only Ron. Some of his former friends reported that they had heard he was "driving a truck up north," but no one could provide any concrete information.

REINFORCEMENT

The community responded to the Roughnecks as boys in trouble, and the boys agreed with that perception. Their pattern of deviancy was reinforced, and breaking away from it became increasingly unlikely. Once the boys acquired an image of themselves as deviants, they selected new friends who affirmed that self-image. As that self-conception became more firmly entrenched, they also became willing to try new and more extreme déviances. With their growing alienation came freer expression of disrespect and hostility for representatives of the legitimate society. This disrespect increased the community's negativism, perpetuating the entire process of commitment to deviance. Lack of a commitment to deviance works the same way. In either case, the process will perpetuate itself unless some event (like a scholarship to college or a sudden failure) external to the established relationship intervenes. For two of the Roughnecks (Herb and Jack), receiving college athletic scholarships created new relations and culminated in a break with the established pattern of deviance. In the case of one of the Saints (Jerry), his parents' divorce and his failing to graduate from high school changed some of his other relations. Being held back in school for a year and losing his place among the Saints had sufficient impact on Jerry to alter his self-image and virtually to assure that he would not go on to college as his peers did. Although the experiments of life can rarely be reversed, it seems likely in view of the behavior of the other boys who did not enjoy this special treatment by the school that Jerry, too, would have "become something" had he graduated as anticipated. For Herb and Jack outside intervention worked to their advantage; for Jerry it was his undoing.

Selective perception and labeling—finding, processing, and punishing some kinds of criminality and not others—means that visible, poor, nonmobile, outspoken, undiplomatic "tough" kids will be noticed, whether their actions are seriously delinquent or not. Other kids, who have established a reputation for being bright (even though underachieving), disciplined and involved in respectable activities, who are mobile and monied, will be invisible when they deviate from sanctioned activities. They'll sow their wild oats—perhaps even wider and thicker than their lower-class cohorts—but they won't be noticed. When it's time to leave adolescence most will follow the expected path, settling into the ways of the middle class, remembering fondly the delinquent but unnoticed fling of their youth. The Roughnecks and others like them may turn around, too. It is more likely that their noticeable deviance will have been so reinforced by police and community that their lives will be effectively channeled into careers consistent with their adolescent background.

22

Doctors and the Context of Medical Crime and Deviance

JOHN LIEDERBACH

Doctors are another group with the social status and power to rise above the deviant acts they commit and maintain a prestigious reputation. Malfeasance and misconduct are rampant in the medical professions, as we constantly read in the newspaper, yet we are less likely to suspect doctors of wrongdoing, Liederbach suggests, because of their positive reputation and role as altruistic healers. They are, nonetheless, as subject to the temptations to commit fraud and abuse as are any other individuals. Liederbach describes such common deviant and criminal practices as billing schemes, prescription violations, unnecessary treatments, kickbacks, and fraud. Liederbach notes that occupational crimes such as these, in which people engage in deviance for their own benefit, not that of their organizations, will occur wherever opportunities can be found. This chapter is especially relevant to today's political debates about universal health care, the precarious financial footing of Medicaid and Medicare, the corporatization of health care, and the rise of for-profit hospitals, which have led both the government and patients to mistrust medical providers.

What sources of social power are discussed in this chapter that give doctors the ability to refute their potential labeling as deviant?

Health care is big business in the United States. The delivery of medicine involves not only physicians, but also health maintenance organizations (HMOs), large-scale insurance conglomerates, for-profit hospital chains, and government-sponsored medical benefit programs. Health care spending on a national basis was recently estimated to top $2.8 trillion, or roughly 18 percent of the nation's gross domestic product (GDP). The exponential growth of the health care industry and spending on medical care has undoubtedly expanded opportunities for fraud and abuse in the system. Doctors, with their position of status and trust, have a special place in it.

Recent estimates of the cost of health care fraud ranged from $42 billion to $80 billion annually. The annual cost of fraud and abuse associated with the government-sponsored Medicare program alone has been estimated to be

Reprinted with permission from the author, John Liederbach.

$60 billion to $90 billion, including $48 billion in improper payments to doctors. The enormous cost of medical crime and deviance has been defined as a form of *"white-collar wilding"* perpetrated by medical professionals, a reference usually reserved for the most heinous street criminals (Watkin, Friedman, and Doran, 1992). The consequences are not limited to financial fraud and abuse: Unnecessary medical procedures, negligent care, prescription violations, and the sexual abuse of patients exact an enormous physical toll as well. An estimated 400,000 patients are said to be victims of negligent mistakes or misdiagnoses each year. Almost 180,000 patients die every year, at least in part because of negligent care. Up to two million patients are needlessly subjected to physical risks through unnecessary operations each year. The resulting cost of unnecessary medical procedures approaches $4 billion. These startling statistics provide an overview of the vast costs—including both financial and physical costs—that result from medical crime and deviance.

THE "OPPORTUNITY CONTEXT" OF MEDICAL CRIME: TRUST, STATUS, AND SELF-REGULATION

The term "opportunity context" has routinely been used in the research literature to describe how crime and deviance occur within and among different groups of offenders, depending on their social, cultural, or occupational circumstances. Among those to whom the term has been applied are juvenile delinquents and traditional street criminals, various white-collar or high-status offenders, and even those who work within the criminal justice system, such as police officers. The idea is that crime does not result solely from the motivations of people who are unscrupulous, maladjusted, or otherwise defined as "defective"; crimes occur in situations or contexts that allow and even encourage these individuals to perpetrate them. In the case of doctors, the context within which they practice medicine provides both opportunities and incentives for rogue doctors to engage in deviance and crime, as well as receive some degree of protection in the event that they are identified and become the subject of potential punishment.

The first factor in the opportunity context for medical crime is doctors' *altruistic* or trustworthy public image. The altruistic or trustworthy image of doctors originates with the Hippocratic Oath, or the explicit profession of a doctor's commitment to the best interest of the patient. The oath contains statements on the avoidance of "overtreatment" and nihilistic or useless treatment, as well as a description of medicine as an "art" that requires warmth, sympathy, and understanding toward the patient. The oath defines doctors as selfless professionals whose priorities lie with the well-being of the patient rather than financial gain or profit. The oath and related altruistic image fosters opportunities for crime and deviance in at least two ways. First, the image promotes a certain degree of trust on the part of patients. Patients who believe that their wellness is the most important or even the *only* priority may not hold doctors accountable in cases where the patients are clearly the victims of negligent care or financial fraud. Second, the trustworthy image of doctors potentially hinders the identification and prosecution of medical crimes because these kinds of accusations are difficult to make against individuals like doctors, who are highly trusted and respected. Taken together, these two considerations seem to create what Bucy (1989) described as a "pattern of deference" or reverence toward doctors that both affords them deviant and criminal opportunities and protects them from scrutiny and potential punishment.

The second factor in the opportunity context for medical crime is the high *social status* of doctors. Doctors have consistently been recognized as high-status professionals primarily because of their lucrative salaries and degree of occupational prestige. The median salary of family physicians was recently estimated to be $175,000 per year. Doctors who specialize can earn considerably more. For example, the average salary range for oncologists is $315,000–$457,000, and the average salary for neurosurgeons is $767,627. Doctors are consistently ranked among the most prestigious occupations, and they often outrank other "very prestigious" professionals, including members of the clergy, business executives, and college professors.

Criminologists who specialize in the study of white-collar crime have always emphasized the general reluctance to criminally prosecute high-status offenders, as well as the criminal opportunities that are afforded to those with economic power and prestige. High-status professionals possess the financial and political clout to influence how criminal statutes are written and enforced, and they are more likely to "escape arrest and conviction...than those who lack such power" (Sutherland, 1949). The relative absence of criminal prosecutions of Wall Street and other financial executives associated with the recent mortgage default crisis provides a clear example of how culpable high-status professionals often escape criminal penalties. Previous scholarship that focused on the specific crimes of physicians demonstrates the lenient criminal penalties—or the complete absence of them—in sanctions imposed on doctors who pillage the Medicaid program, provide negligent patient care, and/or physically abuse their patients.

The third factor in the opportunity context for medical crime is *self-regulation*. Professional groups, including doctors, lawyers, accountants, and professors, maintain member organizations designed to identify and punish misconduct within the group. These mechanisms of "self-regulation" are populated by members of the professional group who are responsible for investigating misconduct and imposing penalties against members of their own profession. The justification for self-regulation relies on two main assumptions. First, members of the professional group are presumed to be in the best position to identify misconduct within the group; and second, members of the professional group are uniquely qualified to judge the scope of misconduct and adjudicate appropriate penalties. In the medical profession, state medical review boards populated primarily by doctors are supposed to provide a "first line of defense" against medical deviance and crime. These boards can revoke medical licenses or otherwise discipline doctors who fail to meet professional or legal standards. The medical community regards the imposition of civil or criminal penalties as both unwarranted and unnecessary, especially in cases that involve errors in clinical judgment.

The profession's reliance on self-regulation, however, may facilitate criminal opportunities by shielding its members from more effective punishments. State medical boards have continually failed to identify doctors who are chronically incompetent, and they often punish them with "slaps on the wrist." Liederbach, Cullen, Sundt, and Geis (2001) describe how self-regulation and other traditional systems of control, including state medical boards, failed to adequately discipline doctors who were obviously incompetent and, in some cases, had maimed and or killed patients during one or more medical procedures. The potential effectiveness of self-regulation in medicine is also limited by the well-documented "code of silence" among medical professionals, who may sometimes be hesitant to report cases of deviance and crime because they fear retaliation.

TYPES OF MEDICAL DEVIANCE AND CRIME

Taken together, the factors that make up the context of medical crime provide rogue doctors ample opportunities to perpetrate deviance and crime. Medical deviance and crime includes a range of acts committed by medical professionals within the context of their occupational role. More generally, medical crimes have been included in the category of "professional occupational crimes" because they derive from opportunities provided through trust given only to occupational elites (Green, 1997). This section provides an overview of selected medical offenses, including fraudulent billing schemes, prescription violations, unnecessary treatments and procedures, medical "kickbacks," and Medicaid fraud.

Billing Schemes

Fraudulent billing schemes comprise a wide variety of acts, including, but not limited to, (a) billing in cases where no medical service was provided, (b) billing for services that were not rendered as described in the claim for payment, (c) billing for services that have been previously billed and paid, and (d) duplicate claims. The Federal Bureau of Investigation (FBI) has become the primary agency for exposing cases of health care fraud, including billing schemes, and the agency maintains jurisdiction over federal and private insurance programs. Recent FBI initiatives designed in collaboration with the health care industry utilize sophisticated technology and data-mining techniques to identify patterns of fraud in medical billing and other electronic databases. Through 2011, the FBI had investigated 2,690 cases of health care fraud that resulted in 736 convictions. These investigations yielded $1 billion in fines and over $1 billion in civil settlements against fraudulent health care providers.

Prescription Violations

Medical doctors are the only persons entrusted to prescribe dangerous drugs and addictive drugs, including narcotics, amphetamines, tranquilizers, and opioid pain relievers (OPRs). Doctors are required by law to limit access to prescription drugs on the basis of medical need. An alarming number of physicians violate this trust. The Public Citizen Health Research Group (PCHRG) identified 1,521 doctors who were disciplined for misprescribing or overprescribing drugs between 1988 and 1996. At least 69 percent of those doctors were not even temporarily suspended from practicing medicine. The PCHRG described various types of scenarios with prescription violations. Among these scenarios were doctors caught selling blank prescriptions to known drug addicts, one physician who dispensed expired drugs from an old, unlabeled spice jar, and another doctor who prescribed dangerous weight loss pills to a patient for 4 years without any medical exam. The patient eventually had a stroke.

More recently, significant public attention has focused on problems related to opioid analgesic prescription offenses (offenses involving OPRs). OPRs are a class of drugs that include oxycodone, methadone, and hydrocodone. The Centers for Disease Control (CDC) have documented significant increases in overdose deaths involving OPR's; deaths involving OPRs now exceed deaths involving heroin and cocaine combined, and it is believed that prescription drugs account for most of the increase in those death rates since 1999. A recent study by Goldenbaum et al. (2008) identified and described 725 doctors who were charged with criminal and/or administrative offenses related to prescribing OPRs between 1998 and 2006. The study provides composite descriptions representing the typical offense and charges involved in these cases, including those which involved (a) medically unnecessary and clinically inappropriate prescriptions, (b) the prescription of high-dosage units of oxycodone together with muscle relaxants, (c) the prescription of OPRs without prior medical exams, and (d) doctors who purposively supplemented their income by selling presigned blank prescription pads and dispensing OPR samples without examinations or prescriptions.

Unnecessary Treatments, Tests, and Surgery

Doctors who subject patients to noninvasive medically unnecessary treatments and tests commit fraud in that they receive compensation based on deception: They get paid for services that were not needed. Recent studies demonstrate the scope of the problem and the staggering financial and physical costs associated with unnecessary medical treatments and tests. Unnecessary and inappropriate tests and treatment, including unnecessary blood tests, urinalyses, electrocardiograms, and bone density scans, costs roughly $6.8 billion per year. The aforementioned studies focused only on costs derived from unnecessary treatments and tests associated with primary care practices and did not consider costs derived from more costly specialty care practices. The recent publicized case of one New York area cardiologist provides an egregious case of unnecessary and costly specialty care. The doctor admitted to ordering diagnostic tests regardless of patient symptoms. The tests cost over $19 million over a 9-year period. The doctor pleaded guilty to health care fraud, and his practice was characterized by prosecutors as a "medical mill," or a medical practice intentionally designed to prescribe unnecessary treatments or engage in fraudulent billing schemes.

Doctors who perform unnecessary invasive medical procedures, including surgery, can be considered to perpetrate a form of assault because they expose the patient to needless physical risks. Unnecessary surgery, including cardiac procedures and spinal surgeries, is estimated to cost at least $150 billion per year and account for 10 to 20 percent of all operations in some specialties. Spinal fusion surgery has recently come under increased scrutiny in this regard. The number of spinal fusion surgeries performed in the United States doubled from 2002 to 2008, to 413,000, generating $34 billion in bills. One prominent spine surgeon, who chairs the Department of Orthopedics at Dartmouth Medical School, publicly commented, "It's amazing how much evidence there is that fusions don't work, yet surgeons do them anyway...The only one who isn't benefitting from the equation is the patient" (quoted in Waldman and Armstrong, 2010). The implication is that the fee-for-service nature of the U.S. health care system rewards doctors who perform more surgeries, whether or not they are necessary or in the best interest of the patient.

Medical "Kickbacks"

"Kickbacks" are defined as payments from one party to another in exchange for referred business or other income-producing deals. Two primary types of medical "kickbacks" are recognized in the research literature: fee splitting and self-referrals. *Fee splitting* occurs when one doctor receives payment from another doctor in exchange for referring patients. These arrangements typically occur with a primary care physician receiving a referral fee from some type of medical specialist. Fee splitting artificially inflates medical costs and can also endanger the quality of patient care. The American Medical Association (AMA) recognizes fee splitting as an unethical practice. Fee splitting was recognized as a common form of deviance among physicians as early as the 1940s, and the problem was the focus of congressional investigations during the 1970s. *Self-referrals* involve sending patients to specialized medical facilities in which the physician has a financial interest. Studies have shown that self-referring doctors who own a financial stake in clinical laboratories, rehabilitation facilities, or diagnostic imaging centers tend to refer patients more often to these facilities. The practice is presumed to increase both costs and insurance premiums. In 1992, Congress enacted the "Stark Law," which prohibits physicians from referring Medicare patients to treatments that involve the doctor's own medical scanning equipment; however, exceptions to the law have limited its effectiveness.

Medicaid Fraud

The Medicaid program originated during the 1960s with the goal of extending health coverage to Americans who could not otherwise afford it. Medicaid covers over 62 million people (roughly 20 percent of all Americans), including low-income individuals, children and families, pregnant women, and those with disabilities. The program fills obvious and enormous gaps in the national health care coverage landscape, but fraud and abuse have been endemic to the program since its inception. Jesilow, Pontell, and Geis (1993) provide a detailed overview of the origins and structure of the Medicaid program, and they identify some factors that explain how the Medicaid program greatly expanded opportunities for medical deviance and crime.

Many within the medical profession opposed the creation of the Medicaid program because it introduced an unwelcome influence—the government—into the practice of medicine. Many physicians were dissatisfied with the intrusion of government into medical practice, and the situation led to both flaws in the initial design of the program and opposition on the part of some doctors against the rules and regulations of the program. For example, the original opposition of doctors helped to influence the creation of a program that initially did not include *any* provisions for punishing doctors who violated the rules. Doctors who had opposed the system were unlikely to feel guilty about violating the rules or even to view acts of deviance and crime perpetrated within the system as unethical or wrong.

In 2012, improper payments through the program were estimated to total $19 billion in federal Medicaid funds and $11 billion in state funds. For example, a dentist operating out of a Brooklyn storefront was indicted on charges that he had fraudulently billed Medicaid for more than $1 million dollars after he claimed to have performed as many as 991 dental procedures per day. Also, one Buffalo area school district referred 4,434 students to Medicaid-provided speech therapy sessions in a single day without talking to them or even reviewing their medical records, and a former New York State fraud prosecutor characterized the state's Medicaid program as a "honeypot" for unscrupulous medical providers.

The recent movement toward in-home health care has expanded opportunities for crime in the system. For example, investigations in New Orleans and in Washington, DC, uncovered fraud perpetrated by operators of home-care agencies and the so-called personal care assistants. In some cases, medical providers recruited Medicaid recipients, told them to claim that they needed expensive Medicaid-provided in-home health care, and instructed them on how to exaggerate their claims and convince doctors to approve treatment plans that were never actually necessary or implemented. Other cases involved personal care attendants who billed for services that were never provided to disabled seniors and for Medicaid-provided transportation services to medical facilities that never occurred. The number of people covered by Medicaid has recently expanded, in part because of unemployment and economic hardship caused by the Great Recession, and implementation of the Affordable Care Act in 2014 will likely quicken the pace of this expansion, increasing the potential for fraud and abuse.

The Medicaid program posed one of the first significant challenges to the professional autonomy of physicians. The program not only successfully expanded health coverage to many of the nation's poorest citizens, but it also changed the traditional opportunity context of medical crime and created ways for rogue doctors and other medical care providers to plunder and loot the system.

CORPORATIZED MEDICINE AND THE RESPONSE
TO MEDICAL NEGLIGENCE

The health care landscape has changed dramatically since the 1980s, with the emergence of what has been referred to as the medical–industrial complex, or the enormous *for-profit* industry in medical care that now involves investor-owned for-profit hospital chains, HMOs, the health care insurance industry, and the pharmaceutical industry. The emergence of these and other for-profit entities has driven a revolution in the provision of health care (McKinlay and Stoeckle, 1988) and affected the opportunity context of medical deviance and crime in at least two important ways. First, the nature and character of corporatized medicine and the profit-oriented strategies associated with the model seem to create incentives to reduce the quality of patient care and thereby the cost of doing business. Second, corporatized medicine may change the way in which society responds to cases of medical negligence and may increase the vulnerability of doctors to legal attacks and the imposition of criminal penalties against doctors who harm patients through negligent or reckless medical care.

The most obvious impact of these trends can be demonstrated by a review of the for-profit strategies pursued by some hospital chains and HMOs. For-profit hospitals are owned by private investors or shareholders and may distribute profits derived from the medical care they provide to owners and/or shareholders. There is evidence to demonstrate that for-profit hospitals are both more costly to patients and provide worse quality of care. One of the largest for-profit hospital chains is the Hospital Corporation of America (HCA), a conglomerate of 165 hospitals and 115 surgery centers located in 20 U.S. states and England. In 2001, the company was fined $840 million for Medicaid fraud and eventually also agreed to pay an additional $631 million in civil penalties and damages to the U.S. government for submitting false medical claims. The case was then the largest health care fraud in U.S. history.

For their part, HMOs utilize a number of procedures that are designed to increase the economic productivity of physicians. For example, HMOs reward doctors for *not* using diagnostic tests and for *avoiding* patient referrals. A twist on the traditional problems associated with medical kickbacks and fee splitting is the *"premium split,"* whereby the HMO *compensates doctors who refuse to make referrals.* There is also evidence to suggest that profits are increasing at the expense of quality patient care. The use of financial incentives by some HMOs has been found to alter physician treatment decisions significantly. The situation suggests an inherent conflict of interest between the profit-driven strategies of HMOs and the best interests of the patient. Conflicted doctors may not perform effectively or may exploit their position and harm patients.

The emergence of corporatized medicine and incentives that could reduce the quality of patient care may also change the way society responds to cases of medical negligence, increasing the vulnerability of doctors to legal attacks and to the imposition of criminal penalties against doctors who harm patients through negligent or reckless medical care. Corporatization can thus significantly alter the opportunity context of medical crime in terms of the degree of trust that patients afford to doctors. As Liederbach et al. (2001) explain, "Doctors and patients have been replaced by 'providers' and 'enrollees' in the new managed care environment. Physicians must now divide their loyalties between the patient and the organization, and the patient is fast assuming a secondary role." This development has the potential to shift the public image of doctors from that of the selfless, altruistic professional committed to the best interest of the patient to one that is more similar to the way the public views corporate business executives. Corporatization ties the medical enterprise more clearly to money, and when the profit motive competes with doctors' commitment to quality patient care, victims of medical negligence may be more likely to demand the imposition of criminal penalties against doctors.

REFERENCES

Bucy, P. H. (1989). "Fraud By Fright: White Collar Crime by Health Care Providers." *North Carolina Law Review* 67: 855–937.

Goldenbaum, D. M., M. Christopher, R. M. Gallager, S. Fishman, R. Payne, D. Joranson, D. Edmonson, J. McKee, and A. Thexton. (2008). "Physicians Charged with Opioid Analgesic-Prescribing Offenses." *Pain Medicine* 9(6): 737–747.

Green, G. S. (1997). *Occupational Crime*, 2nd ed. Chicago: Nelson Hall.

Jesilow, P., Pontell, H. N., and G. Geis. (1993). *Prescription For Profit: How Doctors Defraud Medicaid*. Berkeley: University of California Press.

Liederbach, J., F. T. Cullen, J. Sundt, and G. Geis. (2001). "The Criminalization of Physician Violence: Social Control in Transformation?" *Justice Quarterly* 18(1): 141–167.

McKinlay, J. B., and J. D. Stoeckle. (1988). "Corporatization and the Social Transformation of Medicine." *International Journal of Health Services* 18: 191–200.

Sutherland, E. H. (1949). *White-Collar Crime*. New Haven: Yale University Press.

Waldman, P., and D. Armstrong. (2010). "Highest Paid US Doctors Get Rich with Fusion Surgery Debunked by Studies." *Bloomberg*, December 30.

Watkin, G., D. Friedman, and G. Doran. (1992). "Health Care Fraud." *US News and World Report*, February 24, pp. 34–43.

PART V

Deviant Identity

We have just looked at how some categories of people and behavior become defined as deviant. Yet social constructionism suggests that a deviant classification floating around abstractly in society is not meaningful unless it gets attached to people. Groups in society work not only to create definitions of deviance, but also to create situations in which deviance occurs and is labeled. In this section, we will examine the way deviance is evoked and shaped. Having a definition of deviance and an environment in which it can occur is not enough to apply the label; also required is that that people accept the identity and make it their own. Identities are the way people think of themselves. The study of deviant identities has focused on how people develop and manage nonnormative self-conceptions. In Part V, we will examine how the concept of deviance becomes applied to individuals and how it affects their self-conception.

IDENTITY DEVELOPMENT

We mentioned earlier that, although many people engage in deviance, the label is applied to only a small percentage of them. Such labeling is tied to their formerly "secret deviance" (Becker, 1963) becoming exposed or to an abstract status coming to bear on their personal experience. Thus, Jews may not feel stigmatized unless they experience anti-Semitism, and embezzlers may not think of themselves as thieves until they are caught. When those things happen, the group involved enters the pathway to the deviant identity, a pathway that follows a certain trajectory. The process of acquiring a deviant identity unfolds as a "deviant" (Becker, 1963) or "moral" (Goffman, 1961) career, with people passing through stages that move them out of their innocent identities and toward one labeled as "different" by society. In our own work (Adler and Adler, 2006), we proposed a model of the seven stages of the **deviant identity career**.

Stage 1 begins once people are *caught and publicly identified* as deviant; their lives change in several ways. Others start to think of them differently. For example, suppose there has been a rash of thefts in a college dormitory, and Jessica, a first-year student, is finally caught and identified as the culprit. She may or may not be reported to authorities and charged with theft, but regardless, she will experience an informal labeling process. Once she is caught, the news about her is likely to spread. In stage 2, people will probably change their attitudes toward her, as they find themselves talking about her behind her back. They may look at her behavior and engage in *"retrospective interpretation"* (Schur, 1971) as they think about her differently, reflecting about her past to see if her current and earlier behavior can be recast differently in light of their new information. Where did she say she was when the last theft occurred? Where did she say she got the money to buy that new sweater?

In stage 3, as the news about Jessica spreads, either informally or through official agencies of social control, she may develop what Goffman (1963) has called a *"spoiled identity"*: an identity with a tarnished reputation. Erikson (Chapter 1) noted that news about deviance is of high interest in a community, commanding intense focus from a wide audience. Deviant labeling is hard to reverse, he suggested, and once people's identities are spoiled, they are hard to rehabilitate socially. Erikson discussed "commitment ceremonies," such as trials or psychiatric hearings, in which individuals are officially labeled as deviant. Few corresponding ceremonies exist, he remarked, to mark the *cleansing* of people's identities and welcome them back into the normative fold. Individuals may thus find it hard to recover from the lasting effect of such identity labeling, and despite their best efforts, they often find that society expects them to commit further acts of deviance. Merton (1938) referred to this expectation as a "self-fulfilling prophesy," whereby people tend to enact the behavior associated with the labels placed upon them despite possible intentions otherwise.

Jessica's dorm mates and former friends may then engage in stage 4 behavior: what Lemert (1951) has called *"the dynamics of exclusion."* In this stage, Jessica's friends deride her and ostracize her from their social group. When she enters the room, she may notice that a sudden hush falls over the conversation. People may not feel comfortable leaving her alone in their room. They may exclude her from their meal plans and study groups. She may become progressively shut out from nondeviant activities and circles, such as honors societies, professional associations, relationships, or jobs. At the same time, in stage 5, others may welcome or *include* Jessica in their deviant circles or activities. She may find that she has developed a reputation that, though repelling to some groups, is attractive to others, who may welcome her as "cool" and invite her into their circles. Thus, individuals may find that, as they move down the pathway of their deviant careers, they shift friendship circles, being pushed away from the company of some while being simultaneously welcomed into the company of others.

Sixth, others usually begin to *treat differently* those defined as deviant, indicating through their actions that their feelings and attitudes toward the newly deviant have shifted, often in a negative sense. They may not accord Jessica the same level of credibility they previously accorded her, and they may tighten the margin of social allowance they allot her. Seventh, and finally, people react to this treatment by using what Charles Horton Cooley referred to as their "looking-glass selves." In the culminating stage of the identity career, they *internalize the deviant label* and come to think of themselves differently. This new perspective is likely to affect their future behavior. Although not all people who get caught in deviance progress completely through the full set of seven stages, Becker (1963) described the process as the effects of labeling.

Once people are labeled as deviant and accept that label into their self-conceptions, a variety of outcomes may ensue. We all juggle a range of identities and social selves through which we relate to people, including those of sibling, child, friend, student, neighbor, and customer. Identities also derive from some of our demographic or occupational features, such as race, gender, age, religion, and social class. Hughes (1945) suggested that some statuses are highly dominant, overpowering others and coloring the way people are viewed. Having a known deviant identity may become one of these "**master statuses**," rising to the top of the hierarchy, infusing people's self-concept and others' reactions, and taking precedence over all other statuses. Many social statuses fade in and out of relevance as people move through various situations, but a master status accompanies them into all their contexts, forming the key identity through which others see them. Deviant attributes such as a minority race, heroin addiction, and homosexuality are prime examples of this kind of master status. Others then may think of an individual, for instance, as a Hispanic person, as a heroin addict, or as gay, with the individual's occupation or hobbies being seen as secondary to his or her deviant attribute. Hughes noted that master statuses are linked in society to **auxiliary traits**, the common social preconceptions that people associate with these statuses. Self-injurers, for example, may be widely assumed to be either adolescent White women from middle-class backgrounds or disadvantaged youths whose lives are unhappy. They may be thought of as lacking impulse control, seeking attention, survivors of abuse, or mentally unstable, or as people who seek to cry for help or inject control into their lives. Heroin addicts may be suspected of being prostitutes or thieves, and homosexuals may be suspected of being sexually promiscuous or infected with AIDS. This type of identification spreads the image of deviance to cover the person as a whole and not just one part of him or her.

The relationship between master statuses and their auxiliary traits in society is reciprocal. When people learn that others have a certain deviant master status, they may impute the associated auxiliary traits onto them. Inversely, when people begin to recognize a few traits that they can put together to form the pattern of auxiliary traits associated with a particular deviant master status, they are likely to attribute that master status to others. For example, if parents notice that their children are staying out late with their friends, wearing "alternative" clothing styles, growing dreadlocks in their hair, dropping out of after-school activities, and hanging out with the "wrong" crowd of friends, they may suspect them of using drugs or committing crimes.

Lemert (1967) asserted another processual depiction of the deviant identity career with his concepts of **primary** and **secondary deviance**. Primary deviance consists of a stage when people commit deviant acts but their deviance goes unrecognized. As a result, others do not cast the deviant label onto them, and they neither assume it nor perform a deviant role. Their self-conceptions are free of this image. Some people remain at the primary deviance stage throughout the time they are committing deviance, never advancing further. Yet, a percentage of them do progress to secondary deviance. The seven stages of the identity career move people from primary to secondary deviance; their infractions become discovered, others identify them as deviant, and the labeling process ensues, with all of its identity consequences. Others come to regard these individuals as deviant, and they do as well. As they move into secondary deviance, individuals initially deny the label but eventually come to accept it reluctantly as it becomes increasingly pressed upon them. They recognize their own deviance as they are forced to interact with others through the stigma with which they have been labeled. Sometimes this internalization comes as a justification of, or social defense against, the problems associated with their deviant label, as individuals use the label to take the offensive. At any rate, the label becomes an identity that significantly affects their role performance. Some people may compartmentalize their deviant identity, but others exhibit "role engulfment" (Schur, 1971), becoming totally caught up in this master status.

Most individuals who progress to secondary deviance advance no further, but a subset of them moves on to what Kitsuse (1980) called **tertiary deviance**. In contrast to primary deviants, who engage in deviance denial, and secondary deviants, who accept their deviant identities, tertiary deviants, as seen by Kitsuse, are those who embrace their deviance. These are people who decide that their deviance is not a bad thing. They may adopt a relativist perspective and decide that their deviant label is socially constructed by society, not intrinsic to their behavior. For example, individuals with learning differences consider themselves more creative than "typical" people. Or they may hold to an absolutist perspective and embrace their deviant category as intrinsically real, as when gays who "discover" their underlying homosexuality accept it as natural. They therefore strongly identify with their deviance and fight, usually with the organized help of like others, to combat the deviant label that is applied to them. They may engage in "identity politics" and speak publicly, protest, rally, pursue civil disobedience, educate, raise funds, lobby, and practice various other forms of political advocacy to change society's view of their deviance. Examples include people who fight to destigmatize labels applied to obesity, prostitution, and race or ethnicity.

All of these identity career concepts encompass a progression through several stages. They begin with the commission of the deviant act and lead to individuals' apprehension and public identification. They move through the changing expectations of others toward them, marked by shifting social acceptance or rejection by their friends and acquaintances. The breadth, seriousness, and longevity of the deviant identity label are significantly more profound when individuals undergo official labeling processes than when they are merely informally labeled. With their internalization of the deviant label, adoption of the associated self-identity, and public interaction through that self-identity, they ultimately move into groups of different deviant associates and commit further acts of deviance.

ACCOUNTS

When people say or do things that appear odd to others, they risk being labeled as deviant. We all engage in instances of deviant behavior, but at the same time we desire to maintain a positive self-image in both our own eyes and the eyes of others. In order to avoid the negative consequences of being labeled as deviant and to preserve their untarnished identities, individuals may engage in a variety of interactional strategies designed to normalize their behavior. Mills (1940) suggested that people use "vocabularies of motive" in conversation, presenting legitimate reasons to others around them that explain the meaning of their actions. This motive talk restores a sense of normalcy to interactions that are disrupted by questionable events.

Sykes and Matza (1957, 666) suggested that people commonly make "justifications for deviance that are seen as valid by the delinquent but not by the legal system or society at large." Individuals using these justifications are attempting to resolve the contradictions between what people say and what they do. Sykes and Matza offered five **"techniques of neutralization"** through which people rationalize their behavior, either prospectively or retrospectively. Through *denials of responsibility*, individuals suggest that their deviance was due to acts beyond their control ("I couldn't help myself," "It was not my fault"). In *denying injury*, they mitigate their offense by alluding to the absence of consequences, arguing that no one was hurt ("No harm, no foul"). When they make a *denial of the victim*, they legitimate their behavior by suggesting either that no specific victim can be identified ("It's a huge corporation; nobody will notice it") or that the persons who are hurt do not deserve victim status ("Gays deserve to be beaten up"). Some people *appeal to higher loyalties* by rationalizing their behavior as serving a greater good (loyalty to a friend, to higher principles, to God). Finally, in *condemning the condemners*, people turn the table on the accusers, throwing attention away from themselves by focusing on things their accusers have done wrong ("Oh, you think you're so easy to live with?" "Police are nothing but pigs").

Scott and Lyman (1968) further refined our conception of accounts by suggesting that all accounts can be seen as either **excuses** or **justifications**. In offering excuses, individuals admit the wrongfulness of their actions but distance themselves from the blame. Their excuses are often fairly standard phrases or ideas designed to soften the deviance and relieve individuals of their accountability. Excuses may include *appeals to accidents* ("My computer malfunctioned and lost my file"), *appeals to defeasibility* or misinformation ("I thought my roommate turned my paper in"), *appeals to biological drives* ("Men will be men") and *scapegoating* ("She borrowed my notes, and I couldn't get them back in time to study for the test").

With justifications, individuals accept responsibility for their actions but seek to have specific instances excused. In so doing, they try to legitimate the acts or its consequences. In drawing on justifications, individuals may invoke *sad tales* ("I am a prostitute so that I can afford to put food on the table to feed my kids" or "I turn tricks because I was sexually abused as a child") or the need for *self-fulfillment* ("Taking hallucinogenic drugs expands my consciousness and makes me a more caring person").

Hewitt and Stokes (1975) added to our understanding of accounts by presenting a set of verbal explanations designed specifically to precede the deviant acts that people saw as imminent in their future. They suggested that Lyman and Scott's accounts were primarily retrospective in nature, whereas their own "**disclaimers**" were fundamentally prospective. People *hedge*, they suggested, by prefacing their remarks to indicate a measure of uncertainty about what they are going to do ("I'm not sure this is going to work, but…"). They use *credentializing* when they know their act will be discredited but they are attempting to give a purpose or legitimacy to it ("I'm not prejudiced; in fact some of my best friends are XXX, but …"). Sometimes people invoke *sin licenses* when they know that their behavior will be poorly received but they want to suggest that this is a time when the ordinary rules might be suspended ("I realize you might think this is wrong, but…"). *Cognitive disclaimers* try to make sense out of something that looks like it might not be well understood ("This may seem strange to you…"). Finally, *appeals for the suspension of judgment* aim to deflect the negative consequences of acts or remarks that may be offensive or angering ("Hear me out before you explode…"). Disclaimers, then, are conversational tactics that people invoke specifically before they launch ahead into something commonly judged as inappropriate.

STIGMA MANAGEMENT

When people are labeled as deviant, it marks them with a stigma in the eyes of society. As we have seen, this label may lead to devaluation and exclusion. Consequently, people with deviant features learn how to "manage" their stigma so that they are not shamed or ostracized. This effort requires considerable social skills.

Goffman (1963) has suggested that people with potential deviant stigma fall into two categories: "**the discreditable**" and "**the discredited**." The former are those with easily concealable deviant traits (ex-convicts, secret homosexuals) who may manage themselves so as to avoid the deviant stigma. The latter are either members of the former category who have revealed their deviance or those who cannot hide their deviance (the obese, racial minorities, the physically disabled). The lives of discreditables are characterized by a constant focus on secrecy and information control. Goffman observed that most discreditables engage in "**passing**" as "normals" in their everyday lives, concealing their deviance and fitting in with regular people. They may do this by avoiding contacts with "stigma symbols," those objects or behaviors that would tip people off to their deviant condition. Thus, an anorectic avoids family meals, and mental patients take their medications surreptitiously. Another technique for passing includes using "disidentifiers" such as props, actions, or verbal expressions to distract people and fool them into thinking that one does not have the deviant stigma. Thus, homosexuals brag about heterosexual conquests or take a date to the company picnic, and members of ethnic minorities laugh at ethnic slurs about their group. Finally, discreditables may "lead a double life," maintaining two different lifestyles with two distinct groups of people, one that knows about their deviance and one that does not.

In their endeavor to conceal their deviance, people may employ the aid of others to help "**cover**" for them. In these team performances, friends and family members may assist the deviants by concealing their identities, their whereabouts, their deficiencies, or their pasts. They may even coach the deviants on how to construct stories designed to hide their deviance.

Another form of stigma management, sometimes adopted when concealment fails, involves disclosing the deviance. People may do this for cathartic reasons (alleviating their burden of secrecy), therapeutic reasons (casting the deviance in a positive light), or preventive reasons (so that others don't find out in negative ways later). Although many people's disclosures lead to rejection, others, such as people with sexually transmitted diseases, may find that some people even sympathize with them about their condition.

Disclosures of deviance can follow two courses. In observing the interactions between discredited deviants and nondeviants, Davis (1961) noted that nondeviants often interact with people who possess recognizable deviant traits in patterned ways. He proposed a normalization process that begins with "**deviance disavowal**": The nondeviants ignore the others' deviance and act as if it does not exist. After this conspicuous and stilted ignoring of the individual's deviance, if they spend more time in the company of the deviants, nondeviants progress to a stage of limited engagement in which more relaxed interaction begins and is directed at features of the people other than their deviant stigma. The relationship can achieve full normalization—the point at which the deviant stigma is overlooked and almost forgotten—only when the deviance is broached by someone (usually the deviant him- or herself) and discussed enough so that all the questions are answered and set to rest. Davis illustrated this form of stigma management by discussing the case of physically disabled people who, at first, are shunned by their coworkers or fellow students, but who gradually fit into the crowd when others realize that there is more to them than their disability. (They root for the same teams, listen to the same music, share the same major, etc.).

In contrast, deviant people can strive to normalize their relationships with nondeviants through "**deviance avowal**" (Turner 1972), in which the deviants openly acknowledge their stigma but try to present themselves in a positive light. This avowal often takes the form of humor, with the deviants "breaking the ice" by joking about their deviant attribute. In that way, they show others that they can take the perspective of nondeviants and see themselves as deviant too, thus forming a bridge to others. This action demonstrates further that deviants have nondeviant aspects and that they can see the world as others do. An example is when members of ethnic minority groups make self-deprecating comments about themselves based on stereotypes.

Thus far, we have considered individual modes of adaptation to the stigma of deviance. Yet these stigma can also be managed through a group or collective effort. Many voluntary associations of stigmatized individuals exist, from the early organizations of prostitutes (COYOTE: Call Off Your Old Tired Ethics) to more recent ones such as the Gay Liberation Front, the Little People of America, the National Stuttering Project, and the Gray Panthers. Most well known are the 12-step programs modeled after the tremendous success of Alcoholics Anonymous, including such groups as Overeaters Anonymous, Narcotics Anonymous, and Gamblers Anonymous.

These groups vary in character. Some are organized along what Lyman (1970) calls an **expressive** dimension, whose primary function is to provide support for their members. This support can take the form of organizing social and recreational activities, dispersing legal or medical information, or offering services such as shopping, meals, or transportation. Expressive groups tend to be apolitical, helping their members adapt to their social stigma rather than evade it. They also serve their members by permitting them to come together in the company of other deviants, avoid the censure of nonstigmatized nondeviants, and seek collective solutions to their common problems. It is within such groups that they can make disclosures to others without fear of rejection. For instance, the Little People of America have an annual conference that not only provides a social gathering for people of similar height, but also offers support and advice regarding practical problems (e.g., setting up one's house) and social concerns (such as dating).

Lyman has also described groups with an **instrumental** dimension, whereby members gather together not only to accomplish the expressive functions, but also to organize for political activism. This dimension embodies Kitsuse's tertiary deviation, in which individuals reject the societal conception and treatment of their stigma and organize to change social definitions. They fight to get others to modify their views of the status or behavior in question so that society, like they, will no longer regard it as deviant. Examples of such groups include ACT UP, an AIDS organization whose members have tried to change social attitudes toward AIDS patients; National Organization for Women (NOW); and Disabled in Action.

On another continuum, Lyman noted that groups may vary between **conformity** and **alienation**. *Conformative* groups fundamentally adhere to the norms and values of society. They accept most conventional views, with the occasional exception of their own deviance. They generally use their backstage arenas to counsel members on how to fit in with others who may neither accept nor understand them. Thus, support groups for bipolar people may offer advice to members on how to find the right doctor and on how to find information on various benefits and side effects of different drugs and on the risks and benefits of going drug free. But they do not generally glorify either mania or depression. When these groups do break with society in the way they regard their own deviance and when their members instrumentally try to fight for the legitimation of their deviance, they use conventional means to attain their goals.

Groups may be *alienative* for one of two reasons: Either they are willing to step outside of conventional means to fight for changed definitions of their single form of deviance, or they have multiple values that conflict with society's values. There are many examples of such groups. Activists such as the Black Panthers, a single-issue group, were willing to break the law to fight for improved social opportunities and status for African Americans. Radical feminists might not resort to violating laws, but their dissatisfaction with the social structures that disempower women is grounded in multiple dimensions of society. Modern-day descendants of the Ku Klux Klan, such as the skinheads, Aryan Nation, and various militia groups, may incorporate both elements, rejecting social attitudes of acceptance of Blacks, Jews, immigrants, gays, and others, and, at the same time, resorting to violence to attain their ends, such as blowing up the Murrah Federal Building in Oklahoma City to avenge the Waco siege in which members of the Branch Davidian cult and their leader, David Koresh, perished during an Federal Bureau of Investigation (FBI) assault. Members of other groups that exhibit alienation, such as the Amish, nudists, and hippie communes, simply want to take their radically different values and form communities removed from conventional society.

23

The Adoption and Management of a "Fat" Identity

DOUGLAS DEGHER AND GERALD HUGHES

Degher and Hughes's selection on the way people come to think of themselves as fat is a study in identity transformation. The authors posit a model in which individuals align their self-conception with cues that they derive from their external environment. Although the subjects in this study originally do not hold a view of themselves as obese, they receive active status cues (people say things) and passive status cues (their clothes no longer fit) that jar them away from their former self-conceptions. They follow a process of recognizing that they can no longer be considered to have a normal build, after which they reconceptualize themselves as fat, a category that they judge fits them more appropriately. The fat status has a new, negative identity that they adopt, devaluing them and locating them within the deviant realm.

Do you accept Degher and Hughes's suggestion that people change their identities by going through this sequential process, or do you think it is something that happens in other ways: more suddenly or through another process? How important do you consider the role of others versus the individuals themselves in shifting identities?

The interactionist perspective has come to play an important part in contemporary criminological and deviance theory. Within this approach, deviance is viewed as a subjectively problematic identity rather than an objective condition of behavior. At the core is the emphasis on "process" rather than on viewing deviance as a static entity. To paraphrase vintage Howard Becker, "... social groups create deviance by making the rules whose infraction constitutes deviance. Consequently, deviance is not a quality of the act ... but rather a consequence of the application by others of rules and sanctions to an offender"

(Becker, 1963, p. 9). Attention is focused upon the *interaction* between those being labeled deviant and those promoting the deviant label. In the interactionist literature, emphases are in two major areas: (a) the conditions under which the label "deviant" comes to be applied to an individual and the consequences for the individual of having adopted that label (Tannenbaum, 1939; Lemert, 1951; Kitsuse, 1962, p. 247; Goffman, 1963; Baum, 1987, p. 96; Greenberg, 1989, p. 79), and/or (b) the role of social control agents[1] in contributing to the application of deviant labels (Becker, 1963; Piliavin & Briar, 1964, p. 206; Cicourel, 1968; Schur, 1971; Conrad, 1975, p. 12).

Much of this literature frequently assumes that once an individual has been labeled, the promoted label and attendant identity [are] either internalized or rejected. As Lemert proposes, the shift from primary to secondary deviance is a categorical one, and is primarily a response to problems created by the societal reaction (Lemert, 1951, p. 40).

What is most often neglected is an examination of the mechanistic features of this identity shift. Our focus is on this "identity change process," which is what we have chosen to call this identity shift. Of interest is how individuals come to make some personal sense out of proffered labels and their attendant identities.

METHODOLOGY

The primary methodological tool employed to construct our identity change process model comes from "grounded" analysis.... The model presented in this paper emerged from comments and codes appearing in interviews with obese members of a weight reduction organization that had weekly meetings. The frequency of attendance allowed us to consider the members typical, and allowed us to suggest that major issues of obesity are transsituational and temporally durable. If obesity disappeared tomorrow, we would still be able to apply the generic concepts generated from our data to make statements about the process of "identity change." As suggested by Hadden, Degher, and Fernandez, our focus is on process rather than on unit characteristics of social phenomena (Hadden, Degher, & Fernandez, 1989, p. 9). This provides us with insights that have import for major issues in sociological theory.

SITE SELECTION

Because obese individuals suffer both internally (negative self-concepts) and externally (discrimination), they possess what Goffman refers to as a "spoiled identity" (Goffman, 1963). This seems to be the case particularly in contemporary United States with what may be described as an almost pathological emphasis on fitness. The boom in health clubs, sales of videotapes on fitness, diet books, and so forth promote a definition of the "healthy" physical presence. As Kelly (1990) sees it, the boom in physical fitness in the mid-1980s is an attempt by many people to create a specific image of an ideal body. Thus, body build becomes a crucial element in self-appraisal. Consequently, fat people are an ideal strategic group within which to study the "identity change process."

Obese people are not only the subject of negative stereotypes; but they are also actively discriminated against in college admissions (Canning & Mayers, 1966, p. 1172), pay more for goods and services (Petit, 1974), receive prejudicial medical treatment (Maddox, Back, & Liederman, 1968, p. 287; Maddox & Liederman, 1969, p. 214), are treated less promptly by salespersons (Pauley, 1989, p. 713), have higher rates of unemployment (Laslett & Warren, 1975, p. 69), and receive lower wages (Register, 1990, p. 130). The obese label is one that seems to clearly fit Becker's description of a "master status," that is,

> Some statuses in our society, as in others, override all other statuses and have a certain priority... the status deviant (depending on the kind of deviance) is this kind of master status ... one will be identified as a deviant first, before other identifications are made. (Becker, 1963, p. 33)

Obese people are "fat" first, and only secondarily are seen as possessing ancillary characteristics.

The site for the field observations had to meet two requirements: (a) it had to contain a high proportion of obese, or formerly obese individuals; and (b) these individuals had to be identifiable by the observer. The existence of a large number of national weight control organizations (a) whose membership is composed of individuals who have internalized an obese identity, and (b) who emphasize a radical program of identity change, make these organizations an excellent choice as strategic sites for study and analysis. The local franchise chapter of one of these national weight loss organizations satisfied both of our requirements, and was selected as the site for our study.

Attendance at the weekly meetings of this national weight control group is restricted to individuals who are current members of the organization. Since one requirement for membership is that the individual be at least 10 pounds over the maximum weight for his or her sex and height (according to New York Life tables), all of the people attending the meetings are, or were, overweight, and a high proportion of them are, or were, sufficiently overweight to be classified as obese.[2]

During the period of the initial field observations, the weekly membership of the group varied from 30 to 100 members, with an average attendance of around 60 members. Although there was a considerable turnover in membership, the greatest part of this turnover consisted of "rejoins" (individuals who had been members previously, and were joining again).

Although we have no quantitative data from which to generalize, the group membership appeared to represent a cross section of the larger community. The group included both male and female members, although females did constitute about three-fourths of the membership. Although the membership was predominantly white, a range of ethnicities, notably Hispanic and Native American, existed within the group. The majority of the members appeared to fall within the 30 to 50 age range, although there was a member as young as 11, and one over 70.

DATA COLLECTION

Two types of data were gathered for this study: field observations and in-depth interviews. The field observations were performed while [we attended] meetings of a local weight control organization. The insights gained from these observations were used primarily to develop interview guides. There were two major sources of observation during this period: premeeting conversations; and exchanges during the meeting itself.[3] The observations were recorded in note form and served to provide an orientation for the subsequent interviews. The goal during this period of observation was to gain insight into the basic processes of obesity and the obese career.

The in-depth interviews were carried out with 29 members from the local group. The interviews were solicited on a voluntary basis, and each individual was assured anonymity. The interviewees were representative of the group membership. Although most were middle-aged, middle-income white females, various age groups, ethnicities, marital statuses, genders, and social classes were represented.

These interviews lasted in length from ½ to 2½ hours, with the average interview being about 1 hour and 15 minutes in duration. The interviews produced almost 40 hours of taped discussion, which yielded more than 600 pages of typed transcript for coding.

THE IDENTITY CHANGE PROCESS

In conceptualizing the "identity changes" process, the concept of "career" was employed. As Goffman notes, "career" refers "... to any social strand of any person's course through life" (Goffman, 1961, p. 127). In the present paper, our concern is the change process that takes place as individuals come to see definitions of self in light of specific transmitted information.

An important aspect of this career model is what Becker referred to as "career contingencies," or "... those factors on which mobility from one position to another depends. Career contingencies include both the objective facts of social structure, and changes in the perspectives, motivations, and desires of the individual" (Becker, 1963, p. 24).

Thus, the "identity change" process must be viewed on two levels: a public (external) and a private (internal) level. As Goffman has stated, "One value of the concept of career is its two-sidedness. One side is linked to internal matters held dearly and closely, such as image of self and felt identity; the other [to a] publicly acceptable institutional complex" (Goffman, 1961, p. 127; Adler & Adler interpolation).

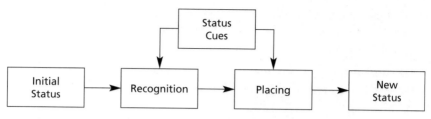

FIGURE 23.1   Visualization of the Identity Change Process (ICP)

On the public level, social status exists as part of the public domain; social status is socially defined and promoted. The social environment not only contains definitions and attendant stereotypes for each status, [but] also contains information, in the form of *status cues*, about the applicability of that status for the individual.

On the internal level, two distinct cognitive processes must take place for the identity change process to occur: first, the individual must come to recognize that the current status is inappropriate; and second, the individual must locate a new, more appropriate status. Thus, in response to the external status cues, the individual comes to recognize internally that the initial status is inappropriate; and then he or she uses the cues to locate a new, more appropriate status. The identity change occurs in response to, and is mediated through, the status cues that exist in the social environment (see Figure 23.1).

STATUS CUES: THE EXTERNAL COMPONENT

Status cues make up the public or external component of the identity change process. A status cue is some feature of the social environment that contains information about a particular status or status dimension. Because this paper is about obesity, the cues of interest are about "fatness." Such status cues provide information about whether or not the individual is "fat," and if so, how "fat."

"Recognizing" and "placing" comprise the internal component of the identity change process and occur in response to, and are mediated through, the status cues that exist in the social environment. In order to understand completely the identity change process, it is necessary to explain the interaction between outer and inner processes (Scheff, 1988, p. 396), or in our case, external and internal components of the process.

Status cues are transmitted in two ways: actively and passively. Active cues are communicated through interaction. For example, people are informed by peers, friends, spouses, etc., that they are overweight. The following are some typical comments that occurred repeatedly in the interviews in response to the question, "How did you know that you were fat?"

> I was starting to be called chubby, and being teased in school.
> When my mother would take me shopping, she'd get angry because the clothes that were supposed to be in my age group wouldn't fit me. She would yell at me.

> Well, people would say, "When did you put on all your weight, Bob?" You know, something like that. You know, you kind of get the message, that, you know, I did put on weight.

A second category of cues might accurately be described as passive in form. The information in these cues exists within the environment, but the individual must be sensitized in some way to that information. For example, standing on a scale will provide an individual with information about weight. It is up to the individual to get on the scale, look at the numbers, and then make some sense out of them. Other passive cues might involve seeing one's reflection in a mirror, standing next to others, fitting in chairs, or, as frequently mentioned by respondents, the sizing of clothes. The comments below, all made in response to the question, "How did you know that you were fat?" are representative of passive cue statements.

> I think that it was not being able to wear the clothes that the other kids wore.
> How did I know? Because when we went to get weighed, I weighed more than my, uh, a girl my height should have weighed, according to the chart, according to all the charts that I used to read. That's when I first noticed that I was overweight.
> I would see all these ladies come in and they could wear size 11 and 12, and I thought, Why can't I do that? I should be able to do that.

Both active and passive cues serve as mechanisms for communicating information about a specific status. As can be seen from the data, events occur that force the individual to evaluate his or her conception of self.

RECOGNIZING

The term "recognizing" refers to the cognitive process by which an individual becomes aware that a particular status is no longer appropriate. As shown in the figure, the process assumes the individual's acceptance of some initial status. For obese individuals, the initial status is that of "normal body build."[4] "This assumption is based on the observation that none of our interviewees assumed that they were "always fat." Even those who were fat as children could identify when they became aware that they were "fat." Through the perception of discrepant status cues, the individual comes to recognize that the initial status is inappropriate. It is possible that the person will perceive the discrepant cues and will either ignore or reject them, in which case the initial status is retained. The factors regulating such a failure to recognize are important, but are not dealt with in this paper. Further research on this point is called for.

Status cues are the external mechanisms through which the recognizing takes place, but it is paying attention to the information contained in these cues that triggers the internal cognitive process of recognizing.

An important point is that the acceptance (or rejection) of a particular status does not occur simply because the individual possesses a set of objective characteristics. For example, two people may have similar body builds, but one may have a self-definition of "fat" whereas the other may not. There appears to be a rather tenuous connection between objective condition and subjective definition. The following comments are supportive of this disjunction.

> I was really, as far as pounds go, very thin, but I had a feeling about myself that I was huge.
>
> Well, I don't remember ever thinking about it until I was about in eighth grade. But I was looking back at pictures when I was little. I was always chunky, chubby.

This lack of necessary connection between objective condition and subjective definition points up an important and frequently overlooked feature of social statuses: the extent to which they are *self-evident*. Self-evidentiality refers to the degree to which a person who possesses certain objective status characteristics is *aware* that a particular status label applies to them.[5]

Some statuses possess a high degree of evidentiality: gender identification is one of these.[6] On the other hand, being beautiful or intelligent is somewhat nonself-evident. This is not to imply that individuals are either ignorant of these statuses or of the characteristics upon which they are assigned. People may know that other people are intelligent, but they may be unaware that the label is equally applicable to them.

One idea that emerged quite early from the interviews was that being "fat" is a relatively nonself-evident status. Individuals do not recognize that "fat" is a description that applies to them.[7] The objective condition of being overweight is not sufficient, in itself, to promote the adoption of a "fat" identity. This nonself-evidentiality is demonstrated in the following excerpts.

> I think that I just thought that it was a little bit here and there. I didn't think of it, and I didn't think of myself as looking bad. But you know, I must have.
>
> I have pictures of me right after the baby was born. I had no idea that I was that fat.

The self-evidentiality of a status is important in the discussion of the identity change process. The less self-evident a status, the more difficult the recognizing process becomes. Moreover, because recognition occurs in response to status cues, the self-evidentiality of a status will influence the type of cues that play the most prominent role in identity change.

A somewhat speculative observation should be made about status cues in the recognizing process. For our subjects, recognizing occurred primarily through active cues. When passive cues were involved, they typically were highly visible and unambiguous. In general, active cues appear to be more potent in forcing the individual's attention to the information that the current status is inappropriate. The predominance of these active cues is possibly a consequence of the relatively nonself-evident character of the "fat" status. It is probable that the less self-evident a status, the more likely that the recognizing process will occur through active rather than passive cues.

Once the individual comes to recognize the inappropriateness of the initial status, it becomes necessary to locate a new, more appropriate status. This search for a more appropriate status is referred to as the "placing" process.[8]

PLACING

Placing refers to a cognitive process whereby an individual comes to identify an appropriate status from among those available. The number of status categories along a status dimension influences the placing process. A status dimension may contain any number of status categories. If there are only two status dimensional categories, such as in the case of gender, the placement process is more or less automatic. When individuals recognize that they do not belong in one category, the remaining category becomes the obvious alternative. The greater the number of status categories, the more difficult the placing process becomes.

The body build dimension contains an extremely large number of categories. When an individual recognizes that he or she does not possess a "normal" body build, there are innumerable alternatives open. The knowledge that one's status lies toward the "fat" rather than the "thin" end of the continuum still presents a wide range of choices. In everyday conversation, we hear terms that describe these alternatives: chubby, porky, plump, hefty, full-figured, beer belly, etc. All are informal descriptions reflecting the myriad categories along the body build dimension.

> I wasn't real fat in my eyes. I don't think. I was just chunky.
>
> Not fat. I didn't exactly classify it as fat. I just thought, I'm, you know, I am a pudgy lady.
>
> I don't think that I have ever called myself fat. I have called myself heavy.

Even when individuals adopt a "fat" identity, they attempt to make distinctions about how fat they are. Because being fat is a devalued status, individuals attempt to escape the full weight of its negative attributes while still acknowledging the nonnormal status. The following responses exemplify this attempt to neutralize the pejorative connotations of having a "fat" status. The practice of differentiating one's status from others becomes vital in managing a fat identity.

Q. How did you know that you weren't that fat?

A. Well, comparing myself to others at the time, I didn't really feel that I was that fat. But I knew maybe because they didn't treat me the same way they treated people who were heavier than me. You know, I got teased lightly, but I was still liked by a lot of people, and the people that were heavy weren't.

As is apparent from this excerpt, the individual neutralized self-image by linking "fatness" with the level of teasing done by peers.

NEW STATUS

The final phase of the identity change process involves the acceptance of a new status. For our informants, it was the acceptance of a "fat" status, along with its previously mentioned pejorative characterizations.[9]

> I hate to look in mirrors. I hate that. It makes me feel so self-conscious. If I walk into a store, and I see my reflection in the glass, I just look away.
>
> We'd go somewhere and I would think, "I never look as good as everybody else." You know everybody always looks better. I'd cry before we'd go bowling because I'd think, "Oh, I just look awful."

As is clear, the final phase of the identity change process involves the internalization of a negative (deviant) definition of self. For many fat people accepting a new status means starting on the merry-go-round of weight reduction programs.[10] Many of these programs or organizations attempt to get members to accept a devalued status fully, and then work to change it. Consequently, individuals are forced to "admit" that they are fat and to "witness" in front of others.[11] The new identity becomes that of a "fat" person, which the weight reduction programs then attempt to transform. A further analysis of the impact of informal organizations on the identity change process will be attempted in another paper.

CONCLUSION

In this paper, we have attempted to fill a void within the interactionist literature by presenting an inductively generated model of the identity change process. The proposed model treats the change process from a career focus, and thus addresses both the external (public) and the internal (cognitive) features of the identity change.

We have suggested that the adoption of a new status takes place through two sequential cognitive processes, "recognizing" and "placing." First, the individual must come to recognize that a current status is no longer appropriate. Second, the individual must locate a new, more appropriate status from among those available. We have further suggested that these internal or cognitive processes are triggered by and mediated through status cues, which exist in the external environment. These cues can be either active or passive. Active cues are transmitted through interaction, whereas passive cues must be sought out by the individual.

We also found a relationship between the evidentiality of the status, that is, how obvious that status is to the individual, and the role of the different types of cues in the identity change process. Finally, we have suggested that the adoption of a new status is a trigger for further career changes.

Although the model presented in this paper was generated inductively from field data on obese individuals, we are confident that it may be fruitfully applied to the study of other deviant careers. It seems particularly appropriate where the identity involved has a low degree of self-evidentiality.

In addition, we feel that the focus upon the different types of status cues and their differing roles in the recognizing and placing processes can lead to a better understanding of how institutionally promoted identity change occurs.

NOTES

1. Included here is research on both rule creators and rule enforcers. We have not made an attempt to analytically separate the two types of investigation.

2. Some of the members had successfully lost their excess weight. When these people were present at a meeting, a leader was careful to introduce them to the other members of the class and to tell how much weight they had lost. This was done to uphold their claim to acceptance by the other group members.

3. Access to this information was gained from an "insider" perspective because one of the researchers was well known among the membership, being an "off and on" member of the organization for three years. Thus, he was not confronted with the problem of gaining entry into a semiclosed social setting. Similarly, because the observer had "been an ongoing participant of the group," he did not have to desensitize the other members of the group to his presence.

4. It is important to note that this process can operate generically. That is, it is not only applicable to the "identity change" from a "normal" to a "deviant" identity, but can also encompass the reverse process as well. In a forthcoming project, we will use the process to analyze how various rehabilitation programs attempt to get individuals back to the initial status.

5. This concept is different in an important way from what Goffman calls "visibility." He uses the term to refer to "... how well or how badly the stigma is adapted to provide means of communicating that the individual possesses it" (Goffman, 1963, p. 48). The focus of the concept is on how readily the social environment can identify that the individual possesses a stigmatized trait. The concept of self-evidentiality deals with how readily the individual can internalize possession of the stigmatized trait. The focus is upon the actor's perceptions, not on the audience.

6. We are referring here to the physiological description of being male or female. We realize that sex roles are much less self-evident.

7. Conversely, a number of individuals thought of themselves as "fat" or "obese," and were objectively "normal." In this case, the existence of objective indicators was insufficient to prevent the individual from adopting a "fat" identity.

8. In some instances, recognizing and placing occur simultaneously. This is especially true when the cue involved is an active one, and contains information about both the initial and new statuses. For example, if peers call a child "fatty," this interaction informs the child that the "normal" status is inappropriate. At the same time, it informs the child that being "fat" is the appropriate status. Even here however, the individual must recognize before it is possible to place.

9. This phase corresponds closely to that presented in much of the "subcultural" research. (See Schur, 1971; Becker, 1963; Sykes & Matza, 1957, p. 664.)

10. Weight Watchers, TOPS, Overeaters Anonymous, Diet Center, and

OptiFast are typical examples of this type of program.

11. By witnessing, we are referring to the process whereby individuals come to renounce, in front of others, a former self and former behaviors associated with that self. Some religious groups, Synanon, Alcoholics Anonymous, etc., seem to encourage this type of degradation of self.

REFERENCES

Baum, L. (1987, August 3). Extra pounds can weigh down your career. *Business Week*, p. 96.

Becker, H. S. (1963). *Outsiders: Studies in the sociology of deviance.* New York: Free Press.

Canning, H., and J. Mayers. (1966). Obesity: Its possible effects on college acceptance. *New England Journal of Medicine*, 275(24) 1172–1174.

Cicourel, A. (1968). *The social organization of juvenile justice.* New York: Wiley.

Conrad, P. (1975). The discovery of hyper-kinesis: Notes on the medicalization of deviant behavior. *Social Problems*, 23(1), 12–21.

Goffman, E. (1961). *Asylums.* Garden City, NY: Anchor.

————. (1963). *Stigma: Notes on the management of spoiled identity.* Englewood Cliffs, NJ: Prentice Hall.

Greenberg, D. (1989). The antifat conspiracy. *New Scientist*, 22 (April 22): 79.

Hadden, S. O., D. Degher, and R. Fernandez. (1989). Sports as a strategic ethnographic arena. *Arena Review* 13(1): 9–19.

Kelly, J. R. (1990). *Leisure*, 2nd ed. Englewood Cliffs, NJ: Prentice Hall.

Kitsuse, J. (1962). Societal reactions to deviant behavior: Problems of theory and method. *Social Problems*, 9 (Winter): 247–256.

Laslett, B., and C. A. B. Warren. (1975). Losing weight: The organizational promotion of behavior change. *Social Problems*, 23(1): 69–80.

Lemert, E. (1951). *Social Pathology.* New York: McGraw-Hill.

Maddox, G. L., K. W. Back, and V. Liederman. (1968). Overweight as social deviance and disability. *Journal of Health and Social Behavior*, 9(4): 287–298.

Maddox, G. L., and V. Liederman. (1969). Overweight as a social disability with medical implications. *Journal of Medical Education*, 9(4): 287–298.

Pauley, L. L. (1989). Customer weights as a variable in salespersons' response time. *Journal of Social Psychology*, 129: 713–714.

Petit, D. W. (1974). The ills of the obese. In G. A. Gray and J. E. Bethune, *Treatment and management of obesity.* New York: Harper & Row.

Piliavin, I., and Briar, S. (1964). Police encounters with juveniles. *American Journal of Sociology* (September): 206–214.

Register, C. A. (1990). Wage effects of obesity among young workers. *Social Science Quarterly*, 71 (March): 130–141.

Scheff, T. (1988). Shame and conformity: The deference emotion system. *American Journal of Sociology*, 53 (June): 395–406.

Schur, E. M. (1971). *Labeling deviant behavior: Its sociological implications.* New York: Harper & Row.

Sykes, G., and D. Matza. (1957). Techniques of neutralization: A theory of delinquency. *American Sociological Review* (December): 664–670.

Tannenbaum, F. (1939). *Crime and the community.* New York: Columbia University Press.

24

The Paradox of the Bisexual Identity

MARTIN S. WEINBERG, COLIN J. WILLIAMS, AND DOUGLAS W. PRYOR

Weinberg, Williams, and Pryor's study of the identity career followed by individuals who become bisexual illustrates a much more complex identity trajectory than Degher and Hughes's portrayal of their fat subjects. As the selection sets forth, the attraction of individuals to members of the same gender leads them to question and reject their affiliation with the heterosexual identity. But their continuing attraction to members of the other gender leads them to question the appropriateness, for them, of the homosexual designation. Aware that they do not fit comfortably into either of these common sexual identity labels, they can become confused about their identity label. After struggling with their ambivalence about their dual sexual orientations, they discover the bisexual option and begin the process of reconceptualizing themselves. Although finding this label and discovering bisexual others aids their process of self-acceptance, they remain doubly stigmatized because of their rejection by both the heterosexual and gay and lesbian communities, and they may not feel completely comfortable with this liminal identity. Over time, as they age, they experience a decrease in the salience of the bisexual identity, yet their dual attractions do not wane; thus, they experience an increase in the certainty and stability of their bisexual identity. The social processes underlying such an identity trajectory are described in ""this piece as well.

How important do you consider the role of individuals, the people they encounter, the individuals' subcultures, and/or society in affecting deviant identities? How likely do you think it is for people to form and re-form their identity over time?

Given that taking on a bisexual identity involves the rejection of not one but two recognized categories of sexual identity, heterosexual and homosexual, how is it that some people come to identify themselves as bisexuals? To our knowledge, no previous model of *bisexual* identity formation exists. We present such a model based on the following questions: (1) How far is the label "bisexual" clearly recognized, understood, and available to people as an identity? and (2) For the study participants, what are the problems in finding the "bisexual" label, understanding what the label means, dealing with social disapproval from

Reprinted with permission from the author, Martin S. Weinberg.

straight and gay/lesbian people, and continuing to use the label once it is adopted? From the fieldwork and interviews, we found that four stages captured the participants' most common experiences when dealing with questions of identity: initial confusion, finding and applying the label, settling into the identity, and continued uncertainty.

METHODS

At the outset of the research, to be included as a study participant, a person had to self-identify as bisexual, have more than incidental sexual feelings toward and/ or sexual activity with both sexes, and have attended the Bisexual Center, a support center for bisexuals established in the mid-1970s. The first wave of interviews was completed in 1983 with 100 people. This research was followed up with a second-wave study 5 years later (in 1988), when 61 of the original group were reinterviewed. A third wave, conducted 8 years later (in 1996), contained 37 of the initial 100 individuals. Furthermore, for the second wave, besides the 61 repeat cases, we interviewed 39 people, none of whom were in the first wave of the study. In the third wave, we reinterviewed 19 of this latter subgroup a second time. Thus, we analyze what happened in the lives of 56 study participants over a period of 8 years, as well as what happened for about two-thirds of the participants over a period of 13 years. The timing of the periods is also significant, occurring before, during, and after the emergence of the AIDS crisis and during the height, as well as after the demise, of the Bisexual Center.

THE STAGES

Initial Confusion

Many people interviewed said that they had experienced a period of considerable confusion, doubt, and struggle regarding their sexual identity before defining themselves as bisexual. This perplexity was ordinarily the first step in the process of becoming bisexual.

They described a number of major sources of early confusion about their sexual identity. For some, it was the experience of having **strong sexual feelings for both sexes** that was unsettling, disorienting, and sometimes frightening. Often, they said that they did not know how to easily handle or resolve these sexual feelings:

> In the past, I couldn't reconcile different desires I had. I didn't understand them. I didn't know what I was. And I ended up feeling really mixed up, unsure, and kind of frightened. (Female, hereafter F)

> I thought I was gay, and yet I was having these intense fantasies and feelings about fucking women. I went through a long period of confusion. (Male, hereafter, M)

Others were confused because they thought strong sexual feelings for, or sexual behavior with, the same sex meant an *end to their long-standing heterosexuality*:

> I was afraid of my sexual feelings for men and ... that if I acted on them that would negate my sexual feelings for women. I knew absolutely no one else who had... sexual feelings for both men and women, and didn't realize that was an option. (M)

A third source of confusion in this initial stage stemmed from attempts by participants to categorize their feelings for, and/or behaviors with, both sexes, yet not being able to do so. Unaware of the term "bisexual," some tried to organize their sexuality by using readily available labels of "heterosexual" or "homosexual"—but these did not seem to fit. No sense of sexual identity jelled; an aspect of themselves remained unclassifiable:

> I thought I had to be either gay or straight. That was the big lie. It was confusing.... That all began to change in the late sixties. It was a long and slow process. (F)

Finally, others suggested that they experienced a great deal of confusion because of their "*homophobia*"—their difficulty in facing up to the same-sex component of their sexuality. The consequence was often long-term denial:

> I thought I might be able to get rid of my homosexual tendencies through religious means—prayer, belief, counseling—before I came to accept it as part of me. (M)

This experience was more common among the men than the women, but not exclusively so.

The intensity of the confusion and the extent to which it existed in the lives of the people we met at the Bisexual Center, whatever its particular source, was summed up by Bill, who said he thinks this sort of thing happens a lot at the Bi Center. People come in "very confused" and experience some really painful stress.

Finding and Applying the Label

Following the initial period of confusion, which often spanned years, was the experience of finding and applying the label. We asked the people we interviewed for specific factors or events in their lives that led them to define themselves as bisexual.

For many who were unfamiliar with the term "bisexual," the *discovery* that the category in fact existed was a turning point. This happened simply by their hearing the word, reading about it somewhere, or learning of a place called the Bisexual Center. The discovery provided a means of making sense of long-standing feelings for both sexes:

> Early on I thought I was just gay, because I was not aware there was another category, bisexual. I always knew I was interested in men and

women. But I did not realize there was a name for these feelings and behaviors until I took Psychology 101 and read about it, heard about it there. That was in college. (F)

The first time I heard the word, which was not until I was 26, I realized that was what fit for me. What it fit was that I had sexual feelings for both men and women. Up until that point, the only way that I could define my sexual feelings was that I was either a latent homosexual or a confused heterosexual. (M)

In the case of others, the turning point was their first *same-sex or other-sex experience*, coupled with the recognition that sex was pleasurable with both sexes. These were people who already seemed to have knowledge of the label "bisexual" yet, without experiences with both men and women, could not label themselves accordingly:

The first time I had actual intercourse, an orgasm with a woman, it led me to realize I was bisexual, because I enjoyed it as much as I did with a man, although the former occurred much later on in my sexual experiences.... I didn't have an orgasm with a woman until twenty-two, while with males, that had been going on since the age of thirteen. (M)

After my first involved sexual affair with a woman, I also had feelings for a man, and I knew I did not fit the category dyke. I was also dating gay-identified males. So I began looking at gay/lesbian and heterosexual labels as not fitting my situation. (F)

Still others reported not so much a specific experience as a turning point, but emphasized the recognition that their sexual feelings for both sexes were simply *too strong to deny.* They eventually came to the conclusion that it was unnecessary to choose between them:

I found myself with men but couldn't completely ignore my feelings for women. When involved with a man, I always had a close female relationship. When one or the other didn't exist at any given time, I felt I was really lacking something. I seem to like both. (F)

The last factor that was instrumental in leading people to initially adopt the label "bisexual" was the *encouragement and support of others*. Encouragement sometimes came from a partner who already defined himself or herself as bisexual:

Encouragement from a man I was in a relationship with. We had been together 2 or 3 years at the time—he began to define as bisexual.... [He] encouraged me to do so as well. He engineered a couple of threesomes with another woman. Seeing one other person who had bisexuality as an identity that fit them seemed to be a real encouragement. (F)

Encouragement also came from sex-positive organizations, primarily the Bisexual Center, but also places like San Francisco Sex Information (SFSI), the Pacific Center, and the Institute for Advanced Study of Human Sexuality.

> At the gay pride parade, I had seen the brochures for the Bisexual Center. Two years later, I went to a Tuesday night meeting. I immediately felt that I belonged and that, if I had to define myself, that this was what I would use: (M)

> Through SFSI and the Bi Center, I found a community of people... [who] were more comfortable for me than were the exclusive gay or heterosexual communities.... [It was] beneficial for myself to be... in a sex-positive community. I got more strokes and came to understand myself better.... I felt that it was necessary to express my feelings for males and females without having to censor them, which is what the gay and straight communities pressured me to do. (F)

Thus, the participants became familiar with, and came to the point of adopting, the label "bisexual" in a variety of ways: through reading about it on their own, being in therapy, talking to friends, having experiences with sex partners, learning about the Bi Center, visiting SFSI or the Pacific Center, and coming to accept their sexual feelings.

Settling into the Identity

Usually, it took years from the time of their first sexual attractions to, or behaviors with, both sexes before people came to think of themselves as bisexual. The next stage then was one of settling into the identity, a stage that was characterized by a more complete transition in self-labeling.

Most reported that this settling-in stage was the consequence of **becoming more self-accepting**. They became less concerned with the negative attitudes of others about their sexual preferences:

> I realized that the problem of bisexuality isn't mine. It's society's. They are having problems dealing with my bisexuality. So I was then thinking, if they had a problem dealing with it, so should I. But I don't. (F)

> I learned to accept the fact that there are a lot of people out there who aren't accepting. They can be intolerant, selfish, shortsighted, and so on. Finally, in growing up, I learned to say "So what, I don't care what others think." (M)

The increase in self-acceptance was often attributed to the **continuing support** from friends, counselors, and the Bi Center, to reading, and to just being in San Francisco:

> I think going to the Bi Center really helped a lot. I think going to the gay baths and realizing there were a lot of men who sought the same

outlet I did really helped. Talking about it with friends has been helpful, and being validated by female lovers that approve of my bisexuality. Also, the reaction of people whom I've told, many of whom weren't even surprised. (M)

The majority of the people we came to know through the interviews seemed settled in their sexual identity. Ninety percent said that they did not think they were currently in transition from being homosexual to being heterosexual or from being heterosexual to being homosexual. However, when we probed further by asking this group "Is it possible, though, that someday you could define yourself as either lesbian/gay or heterosexual?" about 40 percent answered yes. About two-thirds of these indicated that the change could be in either direction, although almost 70 percent said that such a change was not probable.

We asked those who thought that a change was possible what it might take to bring it about. The most common responses referred to becoming involved in a meaningful relationship that was monogamous or very intense. Often, the sex of the hypothetical partner was not specified, underscoring the fact that the overall quality of the relationship was what really mattered:

If I should meet a woman and want to get married, and if she was not open to my relating to men, I might become heterosexual again. (M)

Getting involved in a longer term relationship like marriage where I wouldn't need a sexual involvement with anyone else. The sex of the... partner wouldn't matter. It would have to be someone who[m] I could commit my whole life to exclusively, a lifelong relationship. (F)

Thus, "settling into the identity" must be seen in relative terms. We were struck by the absence of closure that characterized the study's participants—even those who appeared most committed to their identity. This absence of closure led us to posit a next stage in the formation of sexual identity, one that seems unique to bisexuals.

Continued Uncertainty

The belief that bisexuals are confused about their sexual identity is quite common. The conception has been promoted especially by those lesbians and gays who see bisexuality as being an inauthentic identity in and of itself. One evening, a facilitator at a Bisexual Center rap group put this belief in a slightly different form:

One of the myths about bisexuality is that you can't be bisexual without somehow being "schizoid." The lesbian and gay communities do not see being bisexual as a crystallized or complete sexual identity. The gay and lesbian community believes there is no such thing as bisexuality. They think that bisexuals are people who are in transition [to becoming gay/lesbian] or that they are people afraid of being stigmatized [as gay/lesbian] by the heterosexual majority.

We addressed the issue directly in the interviews with two questions: "Do you *presently* feel confused about your bisexuality?" and "Have you ever felt confused about your sexuality?" For the men, a quarter and 84 percent answered "yes." For the women, it was about a quarter and 56 percent.

When asked to provide the details about this uncertainty, the participants' primary response was that, *even after having discovered and applied the label "bisexual" to themselves and having come to the point of apparent self-acceptance, they still experienced continued intermittent periods of doubt and uncertainty regarding the use of this label to make sense of their sexuality.* One reason was the **lack of social validation** and support that came with being a self-identified bisexual. The social reaction people received made it difficult to sustain the identity over the long haul.

Although the heterosexual world was said to be completely intolerant of any degree of homosexuality, the reaction of the gay and lesbian world mattered more. Many bisexuals referred to the persistent pressures they experienced to relabel themselves "gay" or "lesbian" and to engage in sexual activity exclusively with the same sex. It was asserted that no one was *really* bisexual and that calling oneself "bisexual" was a politically incorrect and inauthentic identity:

> Sometimes, [the pressure comes from] the repeated denial the gay community directs at us. Their negation of the concept and of the term "bisexual" has sometimes made me wonder whether I was just imagining the whole thing. (M)

> [The pressure comes from] my involvement with the gay community. There was extreme political pressure. The lesbians said that bisexuals didn't exist. To them, I had to make up my mind and identify as lesbian.... I was really questioning my identity—that is, about defining myself as bisexual. (F)

Lack of support also came from the **absence of bisexual role models, no real bisexual community** aside from the Bisexual Center, and **nothing in the way of public recognition** of bisexuality, all of which bred uncertainty and confusion:

> I went through a period of dissociation, of being very alone and isolated. That was due to my bisexuality. People would ask, well, what was I? I wasn't gay and I wasn't straight. So I didn't fit. (F)

> I don't feel like I belong in a lot of situations because society is so polarized as heterosexual or homosexual. There are not enough bi organizations or public places to go to, like bars, restaurants, clubs.... (F)

For some, continuing uncertainty about their sexual identity was related to their inability to translate their sexual feelings into sexual behaviors (some of the women had *never* engaged in sex with a woman):

> Should I try to have a sexual relationship with a woman? ... Should I just back off and keep my distance, just try to maintain a friendship? I question whether I am really bisexual because I don't know if I will ever act on my physical attractions for females. (F)

> I know I have strong sexual feelings towards men, but then I don't
> know how to get close to or be sexual with a man. I guess that what
> happens is I start wondering how genuine my feelings are.... (M)

For the men, confusion stemmed more from the practical concerns of
implementing and managing multiple partners or from questions about how to
find an involved same-sex relationship and what that might mean on a social and
personal level:

> I felt very confused about how I was going to manage my life in terms
> of developing relationships with both men and women. I still see it as a
> difficult lifestyle to create for myself because it involves a lot of hard
> work and understanding on my part and that of the men and women
> I'm involved with. (M)

Many men and women felt doubts about their bisexual identity because of
being in an exclusive sexual relationship. After being exclusively involved with
another-sex partner for a time, some of the participants questioned the homosexual
side of their sexuality. Conversely, after being exclusively involved with a partner of
the same sex, other participants called into question the heterosexual component of
their sexuality:

> In the last relationship I had with a woman, my heterosexual feelings
> were very diminished. Being involved in a lesbian lifestyle put stress on
> my self-identification as a bisexual. It seems confusing to me because I
> am monogamous for the most part; monogamy determines my lifestyle
> to the extremes of being heterosexual or homosexual. (F)

Others made reference to a lack of sexual activity together with weaker sexual feelings and affections for one sex. Such awareness did not fit in with the
perception that bisexuals should have balanced desires and behaviors. The consequence was doubt about "really" being bisexual:

> On the level of sexual arousal and deep romantic feelings, I feel them
> much more strongly for women than for men. I've gone so far as
> questioning myself when this is involved. (M)

> I definitely am attracted to, and it is much easier to deal with, males.
> Also, guilt for my attraction to females has led me to wonder if I am just
> really toying with the idea. Is the sexual attraction I have for females
> something I constructed to pass time, or what? (F)

Just as "settling into the identity" is a relative phenomenon, so, too, is "continued uncertainty," which can involve a lack of closure as part and parcel of
what it means to be bisexual.

Reaching Midlife: Diminished Salience, Yet Increased Certainty

When we interviewed the participants in the final wave, we found that their
identity trajectories had taken a surprising turn. Four areas stood out as arenas

of change as they grew older. The first was a *change in sexual involvement*: Their sexual activity and interest had declined. This change was often the first step in the weakened relevance of their bisexual identity. The participants attributed it both to the aging process itself and to the effect of increasing responsibilities associated with age. One woman captured the totality of the decline most vividly: "Less activity, less partners, less interest, less time, less energy." Women were also more likely to attribute their loss of sexual interest to either the onset of menopause or a perceived decrease in sexual attractiveness connected with aging.

For men, outside factors played a larger part. Steve (age 49), who wanted us to know that his bisexual feelings had not changed, said it had become harder to act on them: "The drive just isn't there as much anymore. The libido is less and I view people more paternally." Charlie (age 44) was currently married and had had sexual relations with only one other woman (an old friend) in the previous 8 years:

> It [sex] seems less important most of the time. It's a combination of
> being older and having other stuff in our lives: work, all the stuff people
> do, a house to take care of. Our energy just goes in other directions.

The passage of time often meant that these distractions—notably, increasing social commitments for people as they entered into new occupational and familial roles—loomed larger.

In the past, a majority of the participants had described themselves as having had more energy and having been less encumbered with occupational and familial commitments. This meant that most had previously experienced more freedom and felt a greater desire to explore their sexuality and to make a greater commitment to a bisexual lifestyle. Since then, new commitments and identities competed with the salience of a bisexual identity in organizing their lives.

A second factor, which also affected the salience of the bisexual identity, lay in participants' *changes in sexual direction*. Nearly half the group was now sexually active with only one sex.

Almost a third of the participants had become exclusively heterosexual in their sexual behavior. A move in the heterosexual direction was often linked to meeting a new other-sex partner. Heidi (age 41) said, "[I've] become more heterosexual because I've been involved with a new man since 1988. So I'm more heterosexual by default." A move toward monogamy often hindered continued participation in same-sex activity. Sometimes, this had to do with a desire for a simpler life. For a few, it was linked to a partner's sexual preference or demands. Ingrid (age 40) used to be very involved in the swing community but reported, "In the last seven years, I've been involved with the same man.... I don't think it's okay for him to be with other women. He's straight. So I don't think I should be with others either."

Just as often, a reason offered for sexual drift in the heterosexual direction was a decrease in opportunities for same-sex partners. This change frequently was because of some mundane life-course factor, such as a move, a new job, or a lack of time due to the changing demands and circumstances of everyday life. Frank, age 58, noted, "It's occurred through lack of opportunity since moving from the Bay. I'm not around homosexual people here, which has led me to be more heterosexual. That part of my life may be over."

Previous waves in the research showed that a major factor leading participants to call their bisexuality into question was whether they were having sex with both sexes. Fewer were doing so as they aged, because of a change in commitments and a decrease in the salience of a bisexual identity. Nonetheless, dual attractions did not necessarily disappear.

A third factor that characterized the lives of participants was a *change in their community ties*. In 1984, the Bisexual Center closed, cutting off many participants from easy access to other bisexuals. Even though new groups for bisexuals emerged, participants found them to be more youth oriented and radical. For example, the queer and transgender movements challenged older notions of what it meant to be bisexual. One of these groups, Queer Nation, was a movement of young activists against conventional gay and lesbian politics. Two-fifths of the study participants said that they had absolutely no knowledge of the current use of the term *queer*. The remainder said that they had heard of the term and the movement, which they saw as being positive, especially in its emphasis on inclusiveness, which they felt affirmed bisexuality, yet, at the same time, they were not involved in queer politics.

Aging, lifestyle, and historical changes, then, had detached the study participants from the bisexual community as much as the bisexual community had become detached from them. This outcome could mean that a particular person's bisexual identity would lose the support necessary to maintain it.

Finally, we directly examined whether the study participants experienced a *change in self-identity*. The earlier waves of the research found participants who had, often with some uncertainty, adopted a self-identity as bisexual. By the third wave, many of the study's participants, now in midlife, had become *more certain* of their bisexual identity and had obtained closure. In addition, four-fifths of those who identified as bisexual in the second wave in 1988 continued to self-define as bisexual in the final wave in 1996. This finding shows that, contrary to popular belief, the bisexual identity can be stable and that people who self-define as bisexual are not necessarily "in transition" toward another sexual preference identity. Even though life-course events and generational changes may have decreased bisexual commitments, the identity nonetheless survived intact, although the role it played in the participants' lives was reduced. The processes of *decreasing salience, yet increasing certainty and stability of identity,* may thus coexist.

As the participants aged, there were often confirmatory experiences that made the certainty of their bisexuality stand out in relief, even for those who were not having sex with both sexes. This certainty made their bisexuality difficult to deny. Penny (age 41) reported, "Age has made me stop wondering [about my identity]. I accepted the way I am [bi], even though others said I would change. I'm more sure my identity won't change now." Harry (age 46) said, "I've become more positive and more accepting of the fact that that's who I am." Paul (age 53) said that he continued to identify as bisexual "[because of] growing older. The longer your history, the more sure you are. The longer you live with something, the more you're sure of it. I've a stronger sense of my bisexuality as part of my core being." And Colleen (age 54), summed it up in the statement, "I would like—'Still bisexual, damn you; it wasn't a transition'—written on my tombstone." Thus, as the experience of dual attractions continued across the life course, people could look back and see a continuing pattern in their sex lives.

CONCLUSION

We do not wish to claim too much for the model we present of bisexual identity formation. There are limits to its general application. The people we interviewed were unique in that not only did *all* the participants define themselves as bisexual (a consequence of the selection criteria), but they were also all members of a bisexual social organization in a city that, perhaps more than any other in the United States, could be said to provide a bisexual subculture of some sort. Bisexuals in places other than San Francisco surely must move through the early phases of the identity process with a great deal more difficulty. Many probably never reach the later stages.

Finally, the phases of the model we present are very broad and somewhat simplified. Although the particular problems we detail within different phases may be restricted to the type of bisexuals interviewed in this study, the broader phases can form the basis for the development of more sophisticated models of bisexual identity formation.

Still, not all bisexuals will follow these patterns. Indeed, given the relative weakness of the bisexual subculture, there may be more varied ways of acquiring a bisexual identity. Also, the involvement of bisexuals in the heterosexual world means that various changes in heterosexual lifestyles (e.g., a decrease in open marriages or swinging) will be a continuing, and as yet unexplored, influence on bisexual identity. Wider societal changes, notably the existence of AIDS and the rise of the queer movement, may also make for changes in the overall identity process. Being used to choice and being open to both sexes can give bisexuals a range of adaptations in their sexual life that is not available to others.

Finally, although *doing* may be necessary to *being* for a bisexual person, a continuing sexual attraction to both women and men may suffice when the salience of the sexual preference identity is lower and the time, energy, and opportunities to act on the attraction decrease.

25

Anorexia Nervosa and Bulimia

PENELOPE A. McLORG AND DIANE E. TAUB

McLorg and Taub's study of eating disorders describes and analyzes women's progression along an identity career from their initial stage of hyperconformity through Lemert's stages of primary and secondary deviance. They illustrate how the intense societal preoccupation about weight leads women to the kind of deviant behavior that initially maintains their positive external status while they are deteriorating internally. Along the way, these women move through a progression of more common fixations about dieting to frustration with dieting and movement toward more radical solutions, such as bingeing, purging, compulsive exercising, and stopping eating. These behaviors stand apart from the individuals' identities, McLorg and Taub argue, and enable them to avoid the deviant label and self-conception. As a result, those affected with anorexia nervosa or bulimia can remain in the primary deviance stage until they get caught and labeled as having an eating disorder. Once that happens, however, they move to secondary deviance, reconceptualized by others as anorectic or bulimic. With this label cast on them, they are forced to interact with others through the vehicle of their deviance, thus reinforcing their eating disorders.

How might you relate this article to the typology of deviance proposed by Heckert and Heckert?

Current appearance norms stipulate thinness for women and muscularity for men; these expectations, like any norms, entail rewards for compliance and negative sanctions for violations. Fear of being overweight—of being visually deviant—has led to a striving for thinness, especially among women. In the extreme, this avoidance of overweight engenders eating disorders, which themselves constitute deviance. Anorexia nervosa, or purposeful starvation, embodies visual as well as behavioral deviation; bulimia, binge-eating followed by vomiting, and/or laxative abuse, is primarily behaviorally deviant.

Besides [having] a fear of fatness, anorexics and bulimics exhibit distorted body images. In anorexia nervosa, a 20 to 25 percent loss of initial body weight occurs, resulting from self-starvation alone or in combination with excessive exercising, occasional binge-eating, vomiting and/or laxative abuse. Bulimia denotes cyclical

From Penelope A. McLorg and Diane E. Taub, "Anorexia Nervosa and Bulimia: The Development of Deviant Identities." *Deviant Behavior* 8(2). Copyright © 1987. Reproduced by permission of Taylor & Francis, LLC, http://www.taylorandfrancis.com.

(daily, weekly, for example) binge-eating followed by vomiting or laxative abuse; weight is normal or close to normal (Humphries et al., 1982). Common physical manifestations of these eating disorders include menstrual cessation or irregularities and electrolyte imbalances; among behavioral traits are depression, obsessions/ compulsions, and anxiety (Russell, 1979; Thompson and Schwartz, 1982).

Increasingly prevalent in the past two decades, anorexia nervosa and bulimia have emerged as major health and social problems. Termed an epidemic on college campuses (Brody, as quoted in Schur, 1984: 76), bulimia affects 13 percent of college students (Halmi et al., 1981). Less prevalent, anorexia nervosa was diagnosed in 0.6 percent of students utilizing a university health center (Stangler and Printz, 1980). However, the overall mortality rate of anorexia nervosa is 6 percent (Schwartz and Thompson, 1981) to 20 percent (Humphries et al., 1982); bulimia appears to be less life-threatening (Russell, 1979).

Particularly affecting certain demographic groups, eating disorders are most prevalent among young, white, affluent (upper-middle to upper class) women in modern, industrialized countries (Crisp, 1977; Willi and Grossmann, 1983). Combining all of these risk factors (female sex, youth, high socioeconomic status, and residence in an industrialized country), [the] prevalence of anorexia nervosa in upper class English girls' schools is reported at 1 in 100 (Crisp et al., 1976). The age of onset for anorexia nervosa is bimodal at 14.5 and 18 years (Humphries et al., 1982); the most frequent age of onset for bulimia is 18 (Russell, 1979).

Eating disorders have primarily been studied from psychological and medical perspectives.[1] Theories of etiology have generally fallen into three categories: the ego psychological (involving an impaired child–maternal environment); the family systems (implicating enmeshed, rigid families); and the endocrinological (involving a precipitating hormonal defect). Although relatively ignored in previous studies, the sociocultural components of anorexia nervosa and bulimia (the slimness norm and its agents of reinforcement, such as role models) have been postulated as accounting for the recent, dramatic increases in these disorders (Schwartz et al., 1982; Boskind-White, 1985).[2]

Medical and psychological approaches to anorexia nervosa and bulimia obscure the social facets of the disorders and neglect the individuals' own definitions of their situations. Among the social processes involved in the development of an eating disorder is the sequence of conforming behavior, primary deviance, and secondary deviance. Societal reaction is the critical mediator affecting the movement through the deviant career (Becker, 1963). Within a framework of labeling theory, this study focuses on the emergence of anorexic and bulimic identities, as well as on the consequences of being career deviants.

METHODOLOGY

Sampling and Procedures

Most research on eating disorders has utilized clinical subjects or nonclinical respondents completing questionnaires. Such studies can be criticized for simply counting and describing behaviors and/or neglecting the social construction of the disorders. Moreover, the work of clinicians is often limited by therapeutic orientation. Previous research may also have included individuals who were not in therapy on their own volition and who resisted admitting that they had an eating disorder.

Past studies thus disregard the intersubjective meanings respondents attach to their behavior and emphasize researchers' criteria for definition as anorexic or bulimic. In order to supplement these sampling and procedural designs, the present study utilizes participant observation of a group of self-defined anorexics and bulimics.[3] As the individuals had acknowledged their eating disorders, frank discussion and disclosure were facilitated.

Data are derived from a self-help group, BANISH, Bulimics/Anorexics In Self-Help, which met at a university in an urban center of the mid-South. Founded by one of the researchers (D. E. T.), BANISH was advertised in local newspapers as offering a group experience for individuals who were anorexic or bulimic. Despite the local advertisements, the campus location of the meetings may have selectively encouraged university students to attend. Nonetheless, in view of the modal age of onset and socioeconomic status of individuals with eating disorders, college students have been considered target populations (Crisp et al., 1976; Halmi et al., 1981).

The group's weekly two-hour meetings were observed for two years. During the course of this study, 30 individuals attended at least one of the meetings. Attendance at meetings was varied: 10 individuals came nearly every Sunday; 5 attended approximately twice a month; and the remaining 15 participated once a month or less frequently, often when their eating problems were "more severe" or "bizarre." The modal number of members at meetings was 12. The diversity in attendance was to be expected in self-help groups of anorexics and bulimics.

> [Most] people's involvement will not be forever or even [for] a long time. Most people get the support they need and drop out. Some take the time to help others after they themselves have been helped but even they may withdraw after a time. It is a natural and in many cases *necessary* process. (Emphasis in original.) (American Anorexia/Bulimia Association, 1983)

Modeled after Alcoholics Anonymous, BANISH allowed participants to discuss their backgrounds and experiences with others who empathized. For many members, the group constituted their only source of help; these respondents were reluctant to contact health professionals because of shame, embarrassment, or financial difficulties.

In addition to field notes from group meetings, records of other encounters with all members were maintained. Participants visited the office of one of the researchers (D. E. T.), called both researchers by phone, and invited them to their homes or out for a cup of coffee. Such interaction facilitated genuine communication and mutual trust. Even among the 15 individuals who did not attend the meetings regularly, contact was maintained with 10 members on a monthly basis.

Supplementing field notes were informal interviews with 15 group members, lasting from two to four hours. Because they appeared to represent more extensive experience with eating disorders, these interviewees were chosen to amplify their comments about the labeling process, made during group meetings. Conducted near the end of the two-year observation period, the interviews focused on what the respondents thought antedated and maintained their eating disorders. In addition, participants described others' reactions to their behaviors as well as their own interpretations of these reactions. To protect the confidentiality of individuals quoted in the study, pseudonyms are employed.

Description of Members

The demographic composite of the sample typifies what has been found in other studies (Fox and James, 1976; Crisp, 1977; Herzog, 1982; Schlesier-Stropp, 1984). Group members' ages ranged from 19 to 36, with the modal age being 21. The respondents were white, and all but one were female. The sole male and three of the females were anorexic; the remaining females were bulimic.[4]

Primarily composed of college students, the group included four nonstudents, three of whom had college degrees. Nearly all the members derived from upper-middle or lower-upper class households. Eighteen students and two nonstudents were never-marrieds and uninvolved in serious relationships; two nonstudents were married (one with two children); two students were divorced (one with two children); and six students were involved in serious relationships. The duration of eating disorders ranged from 3 to 15 years.

CONFORMING BEHAVIOR

In the backgrounds of most anorexics and bulimics, dieting figures prominently, beginning in the teen years (Crisp, 1977; Johnson et al., 1982; Lacey et al., 1986). As dieters, these individuals are conformist in their adherence to the cultural norms emphasizing thinness (Garner et al., 1980; Schwartz et al., 1982). In our society, slim bodies are regarded as the most worthy and attractive; overweight is viewed as physically and morally unhealthy—"obscene," "lazy," "slothful," and "gluttonous" (Dejong, 1980; Ritenbaugh, 1982; Schwartz et al., 1982).

Among the agents of socialization promoting the slimness norm is advertising. Female models in newspaper, magazine, and television advertisements are uniformly slender. In addition, product names and slogans exploit the thin orientation; examples include "Ultra Slim Lipstick," "Miller Lite," and "Virginia Slims." While retaining pressures toward thinness, an Ayds commercial attempts a compromise for those wanting to savor food: "Ayds… so you can taste, chew, and enjoy, while you lose weight." Appealing particularly to women, a nationwide fast-food restaurant chain offers low-calorie selections, so individuals can have a "license to eat." In the latter two examples, the notion of enjoying food is combined with the message to be slim. Food and restaurant advertisements overall convey the pleasures of eating, whereas advertisements for other products, such as fashions and diet aids, reinforce the idea that fatness is undesirable.

Emphasis on being slim affects everyone in our culture, but it influences women especially because of society's traditional emphasis on women's appearance. The slimness norm and its concomitant narrow beauty standards exacerbate the objectification of women (Schur, 1984). Women view themselves as visual entities and recognize that conforming to appearance expectations and "becoming [an] attractive object [is a] role obligation" (Laws, as quoted in Schur, 1984: 66). Demonstrating the beauty motivation behind dieting, a recent Nielsen survey indicated that of the 56 percent of all women aged 24 to 54 who dieted during the previous year, 76 percent did so for cosmetic, rather than health, reasons (Schwartz et al., 1982). For most female group members, dieting was viewed as a means of gaining attractiveness and appeal to the opposite sex. The male respondent, as well, indicated that "when I was fat, girls didn't look at me, but when I got thinner, I was suddenly popular."

In addition to responding to the specter of obesity, individuals who develop anorexia nervosa and bulimia are conformist in their strong commitment to other conventional norms and goals. They consistently excel at school and work (Russell, 1979; Bruch, 1981; Humphries et al., 1982), maintaining high aspirations in both areas (Theander, 1970; Lacey et al., 1986). Group members generally completed college-preparatory courses in high school, aware from an early age that they would strive for a college degree. Also, in college as well as high school, respondents joined honor societies and academic clubs.

Moreover, preanorexics and prebulimics display notable conventionality as "model children" (Humphries et al., 1982: 199), "the pride and joy" of their parents (Bruch, 1981: 215), accommodating themselves to the wishes of others. Parents of these individuals emphasize conformity and value achievement (Bruch, 1981). Respondents felt that perfect or near-perfect grades were expected of them; however, good grades were not rewarded by parents, because "A's" were common for these children. In addition, their parents suppressed conflicts, to preserve the image of the "all-American family" (Humphries et al., 1982). Group members reported that they seldom, if ever, heard their parents argue or raise their voices.

Also conformist in their affective ties, individuals who develop anorexia nervosa and bulimia are strongly, even excessively, attached to their parents. Respondents' families appeared close-knit, demonstrating palpable emotional ties. Several group members, for example, reported habitually calling home at prescribed times, whether or not they had any news. Such families have been termed "enmeshed" and "overprotective," displaying intense interaction and concern for members' welfare (Minuchin et al., 1978; Selvini-Palazzoli, 1978). These qualities could be viewed as marked conformity to the norm of familial closeness.[5]

Another element of notable conformity in the family milieu of preanorexics and prebulimics concerns eating, body weight/shape, and exercising (Kalucy et al., 1977; Humphries et al., 1982). Respondents reported their fathers' preoccupation with exercising and their mothers' engrossment in food preparation. When group members dieted and lost weight, they received an extraordinary amount of approval. Among the family, body size became a matter of "friendly rivalry." One bulimic informant recalled that she, her mother, and her coed sister all strived to wear a size five, regardless of their heights and body frames. Subsequent to this study, the researchers learned that both the mother and sister had become bulimic.

As preanorexics and prebulimics group members thus exhibited marked conformity to cultural norms of thinness, achievement, compliance, and parental attachment. Their families reinforced their conformity by adherence to norms of family closeness and weight/body shape consciousness.

PRIMARY DEVIANCE

Even with familial encouragement, respondents, like nearly all dieters (Chernin, 1981), failed to maintain their lowered weights. Many cited their lack of willpower to eat only restricted foods. For the emerging anorexics and bulimics, extremes such as purposeful starvation or bingeing accompanied by vomiting and/or laxative abuse appeared as "obvious solutions" to the problem of retaining weight loss. Associated with these behaviors was a regained feeling of control in lives that had been disrupted by a major crisis. Group members' extreme weight-loss efforts operated as coping mechanisms for entering college, leaving home, or feeling rejected by the opposite sex.

The primary inducement for both eating adaptations was the drive for slimness: with slimness came more self-respect and a feeling of superiority over "unsuccessful dieters." Brian, for example, experienced a "power trip" upon consistent weight loss through starvation. Binges allowed the purging respondents to cope with stress through eating while maintaining a slim appearance. As former strict dieters, Teresa and Jennifer used bingeing/purging as an alternative to the constant self-denial of starvation. Acknowledging their parents' desires for them to be slim, most respondents still felt it was a conscious choice on their part to continue extreme weight-loss efforts. Being thin became the "most important thing" in their lives—their "greatest ambition."

In explaining the development of an anorexic or bulimic identity, Lemert's (1951; 1967) concept of primary deviance is salient. Primary deviance refers to a transitory period of norm violations which do not affect an individual's self-concept or performance of social roles. Although respondents were exhibiting anorexic or bulimic behavior, they did not consider themselves to be anorexic or bulimic.

At first, anorexics' significant others complimented their weight loss, expounding on their new "sleekness" and "good looks." Branch and Eurman (1980: 631) also found anorexics' families and friends describing them as "well-groomed," "neat," "fashionable," and "victorious." Not until the respondents approached emaciation did some parents or friends become concerned and withdraw their praise. Significant others also became increasingly aware of the anorexics' compulsive exercising, preoccupation with food preparation (but not consumption), and ritualistic eating patterns (such as cutting food into minute pieces and eating only certain foods at prescribed times).

For bulimics, friends or family members began to question how the respondents could eat such large amounts of food (often in excess of 10,000 calories a day) and stay slim. Significant others also noticed calluses across the bulimics' hands, which were caused by repeated inducement of vomiting. Several bulimics were "caught in the act," bent over commodes. Generally, friends and family required substantial evidence before believing that the respondents' bingeing or purging was no longer sporadic.

SECONDARY DEVIANCE

Heightened awareness of group members' eating behavior ultimately led others to label the respondents "anorexic" or "bulimic." Respondents differed in their histories of being labeled and accepting the labels. Generally, first termed anorexic by friends, family, or medical personnel, the anorexics initially vigorously denied the label. They felt they were not "anorexic enough," not skinny enough; Robin did not regard herself as having the "skeletal" appearance she associated with anorexia nervosa. These group members found it difficult to differentiate between socially approved modes of weight loss—eating less and exercising more—and the extremes of those behaviors. In fact, many of their activities— cheerleading, modeling, gymnastics, aerobics—reinforced their pursuit of thinness. Like other anorexics, Chris felt she was being "ultra-healthy" with "total control" over her body.

For several respondents, admitting they were anorexic followed the realization that their lives were disrupted by their eating disorder. Anorexics' inflexible eating patterns unsettled family meals and holiday gatherings. Their regimented lifestyle of compulsively scheduled activities—exercising, school, and meals—precluded any spontaneous social interactions. Realization of their adverse behaviors preceded the anorexics' acknowledgment of their subnormal body weight and size.

Contrasting with anorexics, the binge/purgers, when confronted, more readily admitted that they were bulimic and that their means of weight loss was "abnormal." Teresa, for example, knew "very well" that her bulimic behavior was "wrong and unhealthy," although "worth the physical risks." While the bulimics initially maintained that their purging was only a temporary weight-loss method, they eventually realized that their disorder represented a "loss of control." Although these respondents regretted the self-indulgence, "shame," and "wasted time," they acknowledged their growing dependence on bingeing/purging for weight management and stress regulation.

The application of anorexic or bulimic labels precipitated secondary deviance, wherein group members internalized these identities. Secondary deviance refers to norm violations which are a response to society's labeling: "secondary deviation ... becomes a means of social defense, attack or adaptation to the overt and covert problems created by the societal reaction to primary deviance" (Lemert, 1967: 17). In contrast to primary deviance, secondary deviance is generally prolonged, alters the individual's self-concept, and affects the performance of his/her social roles.

As secondary deviants, respondents felt that their disorders "gave a purpose" to their lives. Nicole resisted attaining a normal weight because it was not "her"—she accepted her anorexic weight as her "true" weight. For Teresa, bulimia became a "companion"; and Julie felt [that] "every aspect of her life," including time management and social activities, was affected by her bulimia. Group members' eating disorders became the salient element of their self-concepts, so that they related to familiar people and new acquaintances as anorexics or bulimics. For example, respondents regularly compared their body shapes and sizes with those of others. They also became sensitized to comments about their appearance, whether or not the remarks were made by someone aware of their eating disorder.

With their behavior increasingly attuned to their eating disorders, group members exhibited role engulfment (Schur, 1971). Through accepting anorexic or bulimic identities, individuals centered activities around their deviant role, downgrading other social roles. Their obligations as students, family members, and friends became subordinate to their eating and exercising rituals. Socializing, for example, was gradually curtailed because it interfered with compulsive exercising, bingeing, or purging.

Labeled anorexic or bulimic, respondents were ascribed a new status with a different set of role expectations. Regardless of other positions the individuals occupied, their deviant status, or master status (Hughes, 1958; Becker, 1963), was identified before all others. Among group members, Nicole, who was known as the "school's brain," became known as the "school's anorexic." No longer viewed as conforming model individuals, some respondents were termed "starving waifs" or "pigs."

Because of their identities as deviants, anorexics' and bulimics' interactions with others were altered. Group members' eating habits were scrutinized by friends and family and used as a "catchall" for everything negative that happened to them. Respondents felt self-conscious around individuals who knew of their disorders; for example, Robin imagined people "watching and whispering" behind her. In addition, group members believed others expected them to "act" anorexic or bulimic. Friends of some anorexic group members never offered them food or drink, assuming continued disinterest on the respondents' part. While being hospitalized, Denise felt she had to prove to others she was not still vomiting, by keeping her bathroom door open. Other bulimics, who lived in dormitories, were hesitant to use the restroom for normal purposes lest several friends be huddling at the door, listening for vomiting. In general, individuals interacted with the respondents largely on the basis of their eating disorder; in doing so, they reinforced anorexic and bulimic behaviors.

Bulimic respondents, whose weight-loss behavior was not generally detectable from their appearance, tried earnestly to hide their bulimia by bingeing and purging in secret. Their main purpose in concealment was to avoid the negative consequences of being known as a bulimic. For these individuals, bulimia connoted a "cop-out": like "weak anorexics," bulimics pursued thinness but yielded to urges to eat. Respondents felt other people regarded bulimia as "gross" and had little sympathy for the sufferer. To avoid these stigmas or "spoiled identities," the bulimics shrouded their behaviors.

Distinguishing types of stigma, Goffman (1963) describes discredited (visible) stigmas and discreditable (invisible) stigmas. Bulimics, whose weight was approximately normal or even slightly elevated, harbored discreditable stigmas. Anorexics, on the other hand, suffered both discreditable and discredited stigmas—the latter due to their emaciated appearance. Certain anorexics were more reconciled than the bulimics to their stigmas: for Brian, the "stigma of anorexia was better than the stigma of being fat." Common to the stigmatized individuals was an inability to interact spontaneously with others. Respondents were constantly on guard against [the] topics of eating and body size.

Both anorexics and bulimics were held responsible by others for their behavior and presumed able to "get out of it if they tried." Many anorexics reported being told to "just eat more," while bulimics were enjoined to simply "stop eating so much." Such appeals were made without regard for the complexities of the problem. Ostracized by certain friends and family members, anorexics and bulimics felt increasingly isolated. For respondents, the self-help group presented a nonthreatening forum for discussing their disorders. Here, they found mutual understanding, empathy, and support. Many participants viewed BANISH as a haven from stigmatization by "others."

Group members, as secondary deviants, thus endured negative consequences, such as stigmatization, from being labeled. As they internalized the labels anorexic or bulimic, individuals' self-concepts were significantly influenced. When others interacted with the respondents on the basis of their eating disorders, anorexic or bulimic identities were encouraged. Moreover, group members' efforts to counteract the deviant labels were thwarted by their master statuses.

DISCUSSION

Previous research on eating disorders has dwelt almost exclusively on medical and psychological facets. Although necessary for a comprehensive understanding of anorexia nervosa and bulimia, these approaches neglect the social processes involved. The phenomena of eating disorders transcend concrete disease entities and clinical diagnoses. Multifaceted and complex, anorexia nervosa and bulimia require a holistic research design, in which sociological insights must be included.

A limitation of medical/psychiatric studies, in particular, is researchers' use of a priori criteria in establishing salient variables. Rather than utilizing predetermined standards of inclusion, the present study allows respondents to construct their own reality. Concomitant to this innovative approach to eating disorders is the selection of a sample of self-admitted anorexics and bulimics. Individuals' perceptions of what it means to become anorexic or bulimic are explored. Although based on a small sample, findings can be used to guide researchers in other settings.

With only 5 to 10 percent of reported cases appearing in males (Crisp, 1977; Stangler and Printz, 1980), eating disorders are primarily a women's aberrance. The deviance of anorexia nervosa and bulimia is rooted in the visual objectification of women and [its] attendant slimness norm. Indeed, purposeful starvation and bingeing/purging reinforce the notion that "a society gets the deviance it deserves" (Schur, 1979: 71). As recently noted (Schur, 1984), the sociology of deviance has generally bypassed systematic studies of women's norm violations. Like male deviants, females endure label applications, internalizations, and fulfillments.

The social processes involved in developing anorexic or bulimic identities comprise the sequence of conforming behavior, primary deviance, and secondary deviance. With a background of exceptional adherence to conventional norms, especially the striving for thinness, respondents subsequently exhibit the primary deviance of starving or bingeing/purging. Societal reaction to these behaviors leads to secondary deviance, wherein respondents' self-concepts and master statuses become anorexic or bulimic. Within this framework of labeling theory, the persistence of eating disorders, as well as the effects of stigmatization, is elucidated.

Although during the course of this research some respondents alleviated their symptoms through psychiatric help or hospital treatment programs, no one was labeled "cured." An anorexic is considered recovered when weight is normal for two years; a bulimic is termed recovered after being symptom-free for one and one-half years (American Anorexia/Bulimia Association Newsletter, 1985). Thus, deviance disavowal (Schur, 1971), or efforts after normalization to counteract the deviant labels, remains a topic for future exploration.

NOTES

1. Although instructive, an integration of the medical, psychological, and sociocultural perspectives on eating disorders is beyond the scope of this paper.

2. Exceptions to the neglect of sociocultural factors are discussions of sex-role socialization in the development of eating disorders. Anorexics' girlish appearance has been interpreted as a rejection of femininity and womanhood (Orbach, 1979; Bruch, 1981; Orbach, 1985). In contrast, bulimics have been characterized as overconforming to traditional female sex roles (Boskind-Lodahl, 1976).

3. Although a group experience for self-defined bulimics has been reported (Boskind-Lodahl, 1976), the researcher, from the outset, focused on Gestalt and behaviorist techniques within a feminist orientation.

4. One explanation for fewer anorexics than bulimics in the sample is that, in the general population, anorexics are outnumbered by bulimics at 8 or 10 to 1

(Lawson, as reprinted in American Anorexia/Bulimia Association Newsletter, 1985: 1). The proportion of bulimics to anorexics in the sample is 6.5 to 1. In addition, compared to bulimics, anorexics may be less likely to attend a self-help group as they have a greater tendency to deny the existence of an eating problem (Humphries et al., 1982). However, the four anorexics in the present study were among the members who attended the meetings most often.

5. Interactions in the families of anorexics and bulimics might seem deviant in being inordinately close. However, in the larger societal context, the family members epitomize the norms of family cohesiveness. Perhaps unusual in their occurrence, these families are still within the realm of conformity. Humphries and colleagues (1982: 202) refer to the "highly enmeshed and protective" family as part of the "idealized family myth."

REFERENCES

American Anorexia/Bulimia Association. (1983). Correspondence. April.

American Anorexia/Bulimia Association Newsletter. (1985). 8(3).

Becker, Howard S. (1963). *Outsiders*. New York: Free Press.

Boskind-Lodahl, Marlene. (1976). Cinderella s stepsisters: A feminist perspective on anorexia nervosa and bulimia." *Signs, Journal of Women in Culture and Society* 2: 342–56.

Boskind-White, Marlene. (1985). "Bulimarexia: A socio-cultural perspective." In S. W. Emmett (ed.), *Theory and Treatment of Anorexia Nervosa and Bulimia: Biomédical, Sociocultural, and Psychological Perspective,* pp. 113–26. New York: Brunner/Mazel.

Branch, C., H. Hardin, and Linda J. Eurman. (1980). "Social attitudes toward patients with anorexia nervosa." *American Journal of Psychiatry* 137: 631–32.

Bruch, Hilde. (1981). "Developmental considerations of anorexia nervosa and obesity." *Canadian Journal of Psychiatry* 26: 212–16.

Chernin, Kim. (1981). *The Obsession: Reflections on the Tyranny of Slenderness.* New York: Harper & Row.

Crisp, A. H. (1977). "The prevalence of anorexia nervosa and some of its associations in the general population." *Advances in Psychosomatic Medicine* 9: 38–47.

Crisp, A. H., R. L. Palmer, and R. S. Kalucy. (1976). "How common is anorexia nervosa? A prevalence study." *British Journal of Psychiatry* 128: 549–54.

Dejong, William. (1980). "The stigma of obesity: The consequences of naive assumptions concerning the causes of physical deviance." *Journal of Health and Social Behavior* 21: 75–87.

Fox, K. C., and N. Mel. James. (1976). "Anorexia nervosa: A study of 44 strictly defined cases." *New Zealand Medical Journal* 84: 309–12.

Garner, David M., Paul E. Garfinkel, Donald Schwartz, and Michael Thompson. (1980). "Cultural expectations of thinness in women." *Psychological Reports* 47: 483–91.

Goffman, Erving. (1963). *Stigma.* Englewood Cliffs, NJ: Prentice Hall.

Halmi, Katherine A., James R. Falk, and Estelle Schwartz. (1981). "Binge-eating and vomiting: A survey of a college population." *Psychological Medicine* 11: 697–706.

Herzog, David B. (1982). "Bulimia: The secretive syndrome." *Psychosomatics* 23: 481–83.

Hughes, Everett C. (1958). *Men and Their Work.* New York: Free Press.

Humphries, Laurie L., Sylvia Wrobel, and H. Thomas Wiegert. (1982). "Anorexia nervosa." *American Family Physician* 26: 199–204.

Johnson, Craig L., Marilyn K. Stuckey, Linda D. Lewis, and Donald M. Schwartz. (1982). "Bulimia: A descriptive survey of 316 cases." *International Journal of Eating Disorders* 2(1): 3–16.

Kalucy, R. S., A. H. Crisp, and Britta Harding. (1977). "A study of 56 families with anorexia nervosa." *British Journal of Medical Psychology* 50: 381–95.

Lacey, Hubert J., Sian Coker, and S. A. Birtchnell. (1986). "Bulimia: Factors associated with its etiology and maintenance." *International Journal of Eating Disorders* 5: 475–87.

Lemert, Edwin M. (1951). *Social Pathology.* New York: McGraw-Hill.

_____. (1967). *Human Deviance, Social Problems and Social Control.* Englewood Cliffs, NJ: Prentice Hall.

Minuchin, Salvador, Bernice L. Rosman, and Lester Baker. (1978). *Psychosomatic Families: Anorexia Nervosa in Context.* Cambridge, MA: Harvard University Press.

Orbach, Susie. (1979). *Fat Is a Feminist Issue.* New York: Berkeley.

_____. (1985). "Visibility/invisibility: Social considerations in anorexia nervosa—a feminist perspective." In S. W. Emmett (ed.), *Theory and Treatment of Anorexia Nervosa and Bulimia: Biomédical, Sociocultural, and Psychological Perspective*, pp. 127–38. New York: Brunner/Mazel.

Ritenbaugh, Cheryl. (1982). "Obesity as a culture-bound syndrome." *Culture, Medicine and Psychiatry* 6: 347–61.

Russell, Gerald. (1979). "Bulimia nervosa: An ominous variant of anorexia nervosa." *Psychological Medicine* 9: 429–48.

Schlesier-Stropp, Barbara. (1984). "Bulimia: A review of the literature." *Psychological Bulletin* 95: 247–57.

Schur, Edwin M. (1971). *Labeling Deviant Behavior.* New York: Harper & Row.

_____. (1979). *Interpreting Deviance: A Sociological Introduction.* New York: Harper & Row.

_____. (1984). *Labeling Women Deviant: Gender, Stigma, and Social Control.* New York: Random House.

Schwartz, Donald M., and Michael G. Thompson. (1981). "Do anorectics get well? Current research and future needs." *American Journal of Psychiatry* 138: 319–23.

Schwartz, Donald M., Michael G. Thompson, and Craig L. Johnson. (1982). "Anorexia nervosa and bulimia: The socio-cultural context." *International Journal of Eating Disorders* 1(3): 20–36.

Selvini-Palazzoli, Mara. (1978). *Self-Starvation: From Individual to Family Therapy in the Treatment of Anorexia Nervosa.* New York: Jason Aronson.

Stangler, Ronnie S., and Adolph M. Printz. (1980). "DSM-III: Psychiatric diagnosis in a university population." *American Journal of Psychiatry* 137: 937–10.

Theander, Sten. (1970). "Anorexia nervosa." *Acta Psychiatrica Scandinavica Supplement* 214: 24–31.

Thompson, Michael G., and Donald M. Schwartz. (1982). "Life adjustment of women with anorexia nervosa and anorexic-like behavior." *International Journal of Eating Disorders* 1(2): 47–60.

Willi, Jurg, and Samuel Grossmann. (1983). "Epidemiology of anorexia nervosa in a defined region of Switzerland." *American Journal of Psychiatry* 140: 564–67.

26

Challenging a Marginalized Identity: The Female Parolee

TARA D. OPSAL

In this chapter on the deviant identity, Opsal looks at how women who have been released from prison navigate the complex waters of parole and the stigma of its deviant identity. As we saw in Pager's chapter on the stigma of a criminal record, incarceration leaves a mark that makes life extremely difficult. The women in Opsal's carefully researched study discuss the way they use the rhetoric of their narrative accounts to frame their identity (to themselves and others) in a positive way. They align themselves with more socially acceptable behaviors and positions, and they distance themselves from those whom they morally condemn. At the same time, they are forced to constantly navigate the complex waters of parole, meeting its demands and managing its restrictions.

How would you compare the deviant status and social power of the women described in this chapter with the Mexican-American youths, the women athletes, the Black criminals, the Saints and Roughnecks, and the doctors discussed in other chapters? How does the women's outcasting, or "othering," affect the way they can be treated in society? How effective would you rate their claims to more prosocial identities? From reading this chapter, on what basis do you think that people can remake their identities?

I was cautiously optimistic about Liz's circumstances after our first interview, even knowing the challenges women face postincarceration. After being released from prison as a parolee to a community unfamiliar to her, Liz had found subsidized and stable housing through an organization that offers support to individuals postincarceration. And she had found full-time work that not only allowed her to meet her basic needs but also seemed to sustain her in more meaningful ways. Liz explained:

> I don't get paid all that much.... Building from scratch is a little diffi-
> cult, to try to get out on your own like that. But, it'll happen. Like,
> I—the harder I work, the more I feel like I'm working towards that in a
> positive way, you know?

Reprinted with permission from the author Tara D. Opsal.

There was, however, an unusual aspect to Liz's work that made it tenuous. A day laborer agency had placed her at an apartment complex as a temporary employee doing ground work and building maintenance. Impressed by her work ethic, the boss there requested Liz on a long-term basis. She explained to me that, even though she outworked all the other employees (who were men), the corporation that owned the complex was unlikely to hire her on a more permanent basis because they had a policy against hiring felons. "Hiring" Liz via the day laborer agency largely dissolved the company of any legal liability that could arise from hiring her directly. Liz was aware of how being a felon limited her work opportunities and set her back. Despite these experiences, she explained adamantly that she did not let people's preconceived notions limit her: "I've learned not to let people's opinions affect me."

Three months later, at our second interview, Liz's employment circumstances had changed dramatically. Because of her reliability and continued hard work, her boss "decided that they were gonna go to the corporate office and ask them to overlook my felony" in order to hire her on a permanent basis. She explained what happened next:

> Soon as corporate found out I had a felony, there was a person on the
> property who said, "Get her off the property right this minute."
> They didn't even know me. They didn't even know how much I had
> done for them. My boss, I thought he was gonna cry ... they have
> my back. They wanted me. But, corporate said, "A felony."

Although Liz understood herself in a very different way, her employer ultimately defined her as a felon. Situations like this provide indisputable evidence that felons are stigmatized. The U.S. culture clearly demonstrates the social meaning it attaches to this stigmatized identity. Our cultural stories position felons at or near the bottom rung of our social order; Americans view and discuss them primarily in ways that point to their deficits and the problems those deficits cause. These cultural stories also frame offenders as irredeemable criminals. Hence, felons, as social deviants, are regarded as fundamentally different from the rest of "us," and whether they have served their time on the inside or not, they remain both culpable and suspect.

Despite a diverse and developed body of research that illustrates the strategies and identity work of individuals with stigmatized identities, and despite growing interest in understanding the effects of living with a criminal record on formerly incarcerated individuals, as well as the experiences of women postincarceration, we still know little about the identity work of the formerly incarcerated. The focus of this article is on understanding how individuals negotiate a negative social identity that is premised on their culpability.

STIGMA AND IDENTITY TALK

In his classic work on stigma, Erving Goffman (1963, 2–3) explains that, when people consider an attribute of an individual "bad, or dangerous, or weak," they reduce that individual "from a whole and usual person to a tainted, discounted one." Goffman also posits that individuals who inhabit a stigmatized identity adhere to the same beliefs about their own identities and characteristics; they see and understand themselves as holding a stigmatized identity or attribute. However, other scholars argue that individuals do not passively accept the stigmatized identities others bestow upon them and, instead, actively resist them, using a variety of different strategies.

Although these scholars challenge Goffman's framing of stigma management, they do recognize that, because stigma is a social construct, attributes of individuals become stigmatized only in comparison to what collectives define as normal and good within the current structural and cultural boundaries of a particular society. Hence, stigma is not a fixed or immutable trait. Rather, we create it via interaction and, therefore, can also contest or resist it in the same manner. One way that individuals do so is through "narrative identity work," a strategy that provides them the opportunity to create a vivid version of how they see themselves or even to construct new identities. In this article, I examine the narrative strategies that women returning to their communities as parolees utilize to contest and resist their connection to a stigmatized identity. Specifically, I describe how they draw on conventional scripts and story lines to repair their identity and create a socially valued prosocial self. First, however, I briefly describe the social meaning connected to the felon identity, as well as its consequences.

The Stigma of a Criminal Record

Researchers have established that individuals reintegrating into their communities after spending time in prison experience a host of difficulties. After walking out of the prison gates, they struggle to meet their basic needs, such as finding housing and employment and reuniting with their families. They must restart their lives, often in similar or worse structural circumstances than prior to their incarceration. Furthermore, they do so—as illustrated at the outset of this article—while contending with the stigma of a criminal record.

Criminal behavior is widespread among the general population, suggesting that the line between "offenders" and "nonoffenders" is blurry at best. Yet, most social groups in the United States consider a criminal record deeply discrediting. "Criminal" is not just a reflection of one's past misdeeds but is also a prediction of future behavior, and because "the idea that people can become essentially good seems to contradict a fundamental belief of contemporary society" (Maruna, 2001, 5), individuals branded "criminal" by the state struggle to escape the label. As others have pointed out, this dichotomy creates two distinct categories of people: us (the nonoffenders) and them (the offenders). The women in this study were well aware that many viewed them as part of an underclass because they were felons; they explained that people believed that "felons are trash" or are a "menace to society."

It is important to understand that there are social, economic, and political costs to being labeled a felon. For example, employers are significantly less likely to hire ex-offenders, and it is exponentially more difficult for felons to find safe and affordable housing. Moreover, the government revokes a number of rights and privileges; for example, many convicted felons are deemed ineligible for education loans, public assistance, driving privileges, and public housing.

The Parolee Identity

People and social policy, then, stigmatize felons. The participants in this study acknowledged being cast as "other" and experiencing tangible consequences as a result. These women, however, were also forced to contend with an additional stigmatized identity connected to their past misdeeds: the identity of parolee. Some states release individuals from prison unconditionally; however, many individuals leave prison and reenter their communities under the continued supervision of the state. On top of having to meet their basic needs, these individuals are also required to meet a variety of technical conditions deemed necessary by the state parole board. Typical conditions include submitting to regular drug or alcohol testing, meeting regularly with their parole officer, attending mandated counseling sessions, holding down employment, not associating with other individuals with felony records, and abstaining from committing new crimes. Utilizing methods of surveillance, the parole officer is responsible for making sure that the parolee consistently meets these conditions; if the parolee fails to meet a condition, the parole officer is able to use his or her discretion to revoke the individual's parole, with the ultimate consequence that the parolee is returned to prison. Participants in this study were persistently aware that they lived under a system of surveillance that bounded their behavior. Zaria explained:

> These people have control over my life right now. They know that. Little do they know—no they know it, and I know it, that they have complete control over my life right now, they do. I can't go anywhere, I can't talk to anybody, I can't do nothin'. I have to do everything they tell me to do.

Their connection to the institution of parole reminded the women not only they held a stigmatized identity, but also that they were a part of the criminal processing system and, hence, not actually free. Any potential misstep could be the one that sent them back to prison.

METHOD

This article derives from data gathered through a series of interviews with 43 women who were newly released from prison onto parole in the Denver metropolitan area. I conducted up to three semistructured interviews over a period of 1 year with each woman and focused on understanding the challenges they faced as they were released from prison and returned to their communities as parolees. The interviews lasted, on average, 90 minutes and took place in public locations of the participant's choosing or in their homes. I digitally recorded each interview and then transcribed all the interviews verbatim.

Initially, I recruited participants by working directly with prison officials from a local women's prison. Officials welcomed me into the prison each week and spoke about the study directly with women who were about to be released to the Denver metropolitan area. As the study progressed, I used additional recruitment methods, including advertising at community organizations that offered resources to the recently released, as well as snowball sampling that occurred both inside and outside the prison.

I interviewed each participant up to three times over a period of 1 year. The first interview occurred as soon as possible after the person's release from prison. I attempted to conduct the second interview as close as possible to 3 months after the initial interview. The manifest purpose of this interview was to track change among the participants in the study. Retaining contact with newly released prisoners proved to be a formidable task. Women struggled to secure stable housing, so their contact information often changed once or several times relatively quickly after our first interview. I conducted second interviews with 30 women.

The final interview occurred 1 year after the initial contact or 1 year after the woman was released on parole. This last interview occurred with a much smaller subset ($n = 9$) of the original sample. I utilized the last interview primarily to develop and fine-tune themes that emerged as central during data analysis.

The demographic characteristics of the sample were racially and ethnically diverse. However, in comparing the racial composition of the sample with the racial makeup of all women released onto parole in Colorado during the study's recruitment period, it turned out that Blacks are overrepresented in the sample. A racial composition similar to this one should be anticipated, because the sample for the current study was limited to the Denver metropolitan area and offenders who are Black are more likely to be released to this area than are members of other racial or ethnic categories. The age of the participants ranged from 23 to 54 years, with a mean and median of 37 years of age.

Consistent with other research on female offenders, a significant number of the participants were mothers ($n = 31$). In addition, most were in prison because of drug-related offenses; even those women who officially went to prison for offenses unrelated to drugs often reported that the offense occurred because of their drug use. Almost all of the women ($n = 39$) reported a history of drug or alcohol dependency. The role of drug use was important because of the increasing use of drug testing as a form of supervising parolees. Parole officers required each woman in the sample—including women who were not incarcerated for a drug offense—to take regular random drug tests.

RECRAFTING PERSONAL IDENTITY
BY RESISTING STIGMA

Goffman (1963) noted that individuals with stigmatized identities use information management strategies to "*pass*" or "*cover*." The women in this study reported responding in both of these ways, particularly in their search for employment.

These kinds of strategies enabled the women to mitigate or at least manage potential harm that might have come to them by virtue of the stigma attached to their identity. The stories they told about inhabiting an identity that was stigmatized centered on separating the suspect meanings attached to being a felon from how they viewed their own selves. Through their narratives, the women worked to detach the meanings associated with being a felon—"untrustworthy," "trash," "dysfunctional," "a menace," "negative,"—from their self-conception.

The women I interviewed actively resisted the stigma associated with being incarcerated by refusing to internalize its meaning; Riessman (2000) calls this type of resistance strategy *"resistant thinking."* The major way these women resisted this label was by complicating the premise that the stigma of a criminal record relies on: that bad people who do bad things end up in prison. Women pointed out that the situations that end with somebody in prison are not, as Ronda states, "black and white." Drawing from her own experiences with being incarcerated, she explained:

> They [society] don't see that a person who's doing good, something can happen and their life can change. It happens. It can happen to anybody. They can go through a really hard time in their life and decide to do something really stupid, and that one step and now you're in front of a judge and you're gonna get in trouble for it.

Similar to this passage, many participants told stories that illuminated how good people (i.e., not criminals) often end up in prison after experiencing something that could happen to anybody at any time. These stories allowed the women in the study to alleviate part of the social distance that existed between them and those who did not have to deal with the stigma of a criminal record.

Another type of resistant thinking that some women employed paralleled the strategy just explained but uses different reasoning. Specifically, some women resisted stigma by refusing to acknowledge that there were any significant differences between themselves and individuals who had never been a part of the criminal justice system. Vie described this approach when she stated that it does not "bother" her when people know that she has a criminal record. She explained, "I look at it like this: If you don't have a record, I guarantee you're doin' somethin' you ain't got no business doin' and you haven't got caught for it. So you just be a little smarter than others." Similarly, Tamara stated,

> Most people have done a lot of dirt, they just didn't get caught, you know? You just got away with it. You can thank God that you did that. But when the shoe is on the other foot, they can look down at the other person, and they do. . . . Some of them are very judgmental, and if not for the mercy of God they could have been sitting where I'm sitting. Some people don't remember that.

This latter resistance strategy parallels the technique of neutralization that Sykes and Matza (1957) described as *"condemnation of the condemners."* This neutralization technique, like the others Sykes and Matza identified, allows individuals to rationalize their own criminal behavior so that they can align themselves with conventional society and moderate potential damage to their self-image. The technique enables women to craft self-stories that resist the stigma associated with being a felon, rather than simply abdicating responsibility for their actions.

By resisting the stigma associated with a criminal record, the women in this study simultaneously refashion their identity in a way that aligns them with conventional society and actors; using identity work, they are able to craft themselves as an "us" rather than a "them" or "other."

Building a Postdrug-using Self

Parolees are legally mandated to inhabit a stigmatized social identity that exists solely to make sure that they do not pose continued risk to the larger community of "noncriminals." Although parole is also supposed to assist individuals postincarceration, the women in the current study clearly understood that the purpose of the institution was to "keep an eye on them." The most common way that parole did this was through regular drug testing.

All of the women who had previously used drugs or alcohol explained that, after their time away from the streets, from drugs, and from old acquaintances, they realized that they deeply desired to change their relationship with illegal drugs by "being done with all that illegal stuff." Not only did they state their desire to change, but they also posited themselves as individuals who were already changed and fundamentally different from their past selves. By identifying not just as clean, but as different, the women were able to frame their present selves as individuals who no longer made the kinds of choices they did when they were addicted. Nisha's story illustrates this process well.

Nisha had a criminal justice record related to her off-and-on drug use that began for her as a teenager. She spent over 2 years in prison, was released, and was quickly revoked because she started "meeting the wrong people, getting in with the dope dealers." Nisha, however, looks at her second stint in prison that resulted from this revocation as the time she changed herself. She explained, "Going back the second time, I decided I wanted to do this sober. I wanted a better life. I wanted to be free from misery."

Nisha used this new and improved refashioned identity self-centered on difference and change to support her belief that she would be able to complete parole successfully this time. To be able to finish parole without going back to prison, she explained, you really have to be a different person:

> If you want change, you'll do change. But if you don't want change, you can't change. You can only temporarily change. Before, I thought I wanted change, and then I chose to go back on the path I used to lead, because it was easier for me, but I knew, and I chose not to change, so therefore my parole officer sent me back. So I got another chance, and I'm gonna do it this time, because I want change. I want to go home and be with my kids, I want to live a drug-free life. I want to be able to be an abiding citizen and do what I need to do and not always be in trouble and be bad-ass. That's not me.

In this final passage, Nisha configures a *"replacement self,"* clearly separating her "new" self from her old drug-using and criminal self. As the women constructed these replacement selves, they also demonized their past drug-using behavior by identifying why those former selves needed to be replaced. They explained that, while they used drugs or alcohol, they sacrificed relationships with family, experienced violence, or were consumed with getting their next hit. Most of the women who talked about their drug or alcohol addictions told extremely painful stories about how they believed that these behaviors were detrimental to their lives in some way. For example, Tamara explained that being on drugs and out on the streets using drugs was dangerous:

> Man, when I think about how many chances I took with my life and
> didn't care, didn't care, really, really didn't care. That's pretty much
> how at the bottom I was. It was kind of like I didn't care whether I
> lived or died today, pretty much.

By distancing themselves from their drug-using behavior and their past drug-addicted selves, the women decenter the necessity of being labeled a parolee, because it was their past drug-addicted selves who acquired that label. It was their former selves whom the criminal justice system deemed necessary to be monitored, and thus it is only these former selves who justify the appellation of the parolee identity. Moreover, their identity work described in this section that fashions themselves as changed also connects them to a valued prosocial identity, one that is drug and crime free.

Negotiating "Slipups"

Time played an important role in the women's identity work because occasionally some who identified strongly with the narrative of acquiring a new, nonparolee identity at an initial interview "slipped up" and in subsequent interviews reported using drugs or alcohol. These slipups were almost always characterized as big mistakes and weighed quite heavily on the minds of the women, who expressed a great deal of guilt and remorse. After being drug free for several months, LouLou attended a family party where there was coke. She stated, "I don't know what got into me or what, but I did it. I just tasted it." About this lapse in her postdrug self, LouLou explained:

> I really felt little, like, I just—I—[long pause] don't know. Like, I just
> let everybody down, that's how I felt. Like stupid, you know? And I
> know it was stupid, but you know, I just, I'm trying to live with it.

When they relapsed, many women expressed significant concern that their parole officer would find out and they would be returned to prison for violating a parole condition. However, the few women that used and remained on the outside typically emphasized how, by using drugs, they sacrificed relationships with other people who were important to them. LouLou, for example, had restarted a relationship with her teenage daughter before using and was scared that her daughter and her other children would find out about her lapse: "I just don't want to lose my kids, and I'm scared of that." Although some women, like LouLou, "slipped" after presenting themselves as drug free and changed, most continued to identify as changed. As Zaria explained about her slip,

> That was the first I ever, ever used and came home, ever in my whole entire life. Back in the day, I'd use and use and use and use and days and days … and I didn't do that. So I know that it really wasn't what I wanted to do. It wasn't because I knew I had to be home. I just went home. And that's somethin' I would have never done back then.

Connecting to a Culturally Coveted Social Identity

Because motherhood is a readily available and widely accepted identity, being a mom allows many women access to a culturally valued identity. The final way that the women in this study narratively repaired their damaged identity was by identifying strongly as mothers.

Contemporary (socially constructed) U.S. middle-class norms indicate that mothers must be wholly committed to rearing their children and dedicated to meeting those children's needs, always at the sacrifice of their own. Most "real-life" mothering practices do not coincide with this model of mothering. With certainty, the women in the study fall outside of these constructed boundaries. For example, the state often rescinded the mothers' custody of their children because of drug addictions, family members or foster parents cared for these women's children because the women were in prison, or the mothers never really had a role in their children's lives because of their addiction histories or having gotten caught up in the system.

Some of the women expressed feeling remorse over the way they believed that their drug history had been detrimental to their children. Some explained that they felt regret being behind bars while their children were on the outside. They were aware, for example, that their actions before becoming incarcerated had exposed their children to the "wrong kinds" of people and left some children temporarily homeless. Some mothers also spoke of feeling guilty for having to leave their children behind when they spent time in jail or prison knowing that few custody arrangements provided their children healthy environments where they received the kinds of caretaking they needed. However, although many of the mothers spoke briefly about their mothering practices they identified as harmful and how those practices resulted in feelings of guilt and remorse, when I attempted to probe these stories further, they changed. Women entirely resisted continuing to talk about stories that might qualify them as bad mothers and challenge their connection to their "new" identity as a mother. Instead, they either discontinued their narrative about their identity as a mother or, more commonly, they reconstructed their experiences so that they could be understood as "acceptable mothering practices." By engaging in this kind of narrative, they were able to legitimate their connection to the motherhood identity.

The following remarks from Dee, who used illegal substances off and on while she raised two children into adulthood, illustrate the process. She explained, "I took care of my kids all the time. I was a good mother." She pointed to the fact that her daughter was now a registered nurse and her son was a "good father" as evidence of her own quality parenting practices. More so, she expressed that she taught them "you gotta get into this world, you gotta take care, you gotta grow up and do your own thing. You've gotta be responsible for yourself, regardless of me, because I'm your mother." Later, she stated that, although, growing up, her children "didn't like" the fact that she was using drugs, "they learned some things about bein' strong because of it. I told them, 'I gotta do me. I love you guys, but you guys gotta do you too.' They grew up that way. So I pushed them to do good."

This narrative and others like it pointed to how their children developed strength, autonomy, and responsibility as a result of these women's mothering practices—clearly traits that were valued in the mothers' eyes. Reconstructing the boundaries of good mothering allowed the women a more valid connection to the socially coveted role of mother because it gave them an opportunity to recast their past and present their mother self on their own terms.

Motherhood as a Source of Motivation

Although the women explained that their mothering practices benefited their children, they also explained how being mothers benefited them. The women in this study often reported that the prospect of reuniting with children and having a presence in their lives served as a motivating factor to do well on parole or stay away from drugs. Nisha, a mother of two children, illustrates this attitude when she explains,

> I know that the sooner I get this [parole] done with, the sooner I can go home to my kids, and that's my main and most important focus right now, getting my life straight so I can go be a mother to my kids.

Several months after being released from prison, Freesia found out that she was pregnant. She saw the future possibility of being a mother as a reason to stay off of the streets. She explained that being pregnant "makes me want to be more responsible." Further, she stated that she was done with drugs and drinking because she was not going to "jeopardize this little kid." Freesia's outlook is consistent with other research which suggests that motherhood may provide a strong incentive for desistance from crime and from drug use.

Challenges to the Women's Identity Work

Although it may benefit formerly incarcerated women to connect with their children, regaining custody postprison can be an arduous process. Many individuals who return to their communities from prison struggle to meet their own basic needs; bringing a child into this financially volatile equation is simply not an option for many caregivers. Indeed, although many women in the study narratively worked to connect with their mother identity and deeply desired to "be" mothers on the outside and directly reconnect with their children in a permanent way, none of the women realized that desire during the study period.

A few women had full custody of their children prior to going to prison and, when sentenced, handed custody over to various family members. Hence, upon their release, they knew where their children were, they could have some form of contact with them, and they knew that regaining custody of their children was largely a matter of negotiating with their parole officer. Negotiating, however, was not easy or straightforward. Ronda's children, for example, were in the custody of her parents, who lived several states away. Ronda had hoped that she would be paroled to California, where her parents resided; if that happened, she explained, it would make the custody transition smoother and would also provide her with built-in emotional and financial support. She noted, however, that, upon her release, her parole officer was not interested in transferring her parole to California: "She's like, 'No, you have to get established here,'" said Ronda, "'you're not going back to California.'" Similarly, Linda took custody of her son after her mother decided that she could not watch him any longer because he was acting aggressively toward her. Linda explained that, when her parole officer found out, he said, "You can either send your son back to your mom, he can go into a foster home, or you're goin' back to prison."

Unlike Ronda and Linda, most women in this study did not have custody of their children prior to incarceration. Often, custody had been rescinded by the state because of drug addiction or time spent in prison, or else at some point the state had determined the women to be negligent parents. As these women worked to understand what their relationship could look like with their children, they worked through greater levels of ambiguity surrounding that relationship. Indeed, although some of the women looked forward to the day when they would connect with their children, most simply hoped that they would be able to see, speak with, or live with their children again.

Lola, a mother of three, strongly identified as a mother despite not having had contact with her two youngest children for over 4 years or with her 13-year-old daughter for even longer. Notwithstanding this passage of time, and not knowing where her second ex-husband and two youngest children lived, she emphatically explained at our first interview together, "I want to start a relationship with my kids as soon as possible. I've wasted too much time." Lola said that her first husband took her oldest daughter away from her "because I was drinking pretty heavily" and that she lost her two youngest kids to social services. Expanding on this statement, she explained that, after she and her second husband started using crack, she was incarcerated for drug-related charges at the same time that she was supposed to be in court for her custody hearing. Because of the missed court date, she lost custody of her children. After spending over 2 years in prison, she was trying to figure out how to negotiate the court system and paperwork in order to restore her custody rights. About this process, she stated,

> I don't know, it's confusing, and it's hard. It's complicated and it hurts. But I'm not gonna let that alter what I'm doing. . . . It's like everyone is offering you help as far as the [criminal justice] system, but they don't take it further than that for a mother, a mom.

This story points to the idea that the resources available to parolees are likely based on a male model and do not consider the role of gender.

As their time on the outside passed, some women continued to narrate a strong connection to their mother identities at follow-up interviews; these women were more likely to have some sort of contact with their children and were also significantly more likely to remain optimistic about their desisting selves, as well as their reentry process more generally. In contrast, women who became less likely to narrate themselves as mothers were not only less optimistic about reconnecting with their children, but also more likely to develop a sense of generalized hopelessness about being on the outside. In fact, several of these women—including Lola—became reengaged with illegal activity—in particular, the use of illegal drugs.

CONCLUSION

In this article, I have focused on a group of formerly incarcerated women as they returned to their communities as parolees living under the continued supervision and surveillance of the state. I described how they challenged the stigma of a criminal record through narrative identity talk that relied on prosocial cultural values, stories, and characters. For this group of women, negotiating their stigmatized identity did not just mean refashioning an identity in the present; it also meant engaging with their past selves. For example, as they created "new" postdrug-using selves by narrating stories of self-transformation, the women drew on their past experiences as drug users as evidence of the need for their new selves. In addition, some women reconstructed the boundaries of good mothering to redefine their past mother selves. These claims about their past selves provided them an opportunity to narrate a more credible connection to prosocial conventional story lines and characters in the present and future. By drawing on their experiences with stigma postincarceration, this article ultimately demonstrates how these women used their own experiences to reject the cultural story that casts them as irredeemable offenders of our moral code.

REFERENCES

Goffman, Erving. (1963). *Stigma: Notes on the management of spoiled identity.* Englewood Cliffs, NJ: Prentice Hall.

Maruna, Shadd. (2001). *Making good: How ex-convicts reform and rebuild their lives.* Washington, DC: American Psychological Association.

Riessman, Catherine Kohler. (2000). Stigma and everyday resistance practices: Childless women in South India. *Gender & Society* 14: 111–35.

Sykes, Gresham, and David Matza. (1957). "Techniques of neutralization: A theory of delinquency." *American Sociological Review* 22: 664–70.

27

Convicted Rapists'
Vocabulary of Motive

DIANA SCULLY AND JOSEPH MAROLLA

Scully and Marolla's study of the way rapists rationalize their behavior offers a fascinating glimpse into the accounts offered by criminals. The authors interview the most hard-core segment of the rapist population, those sentenced to prison time. In analyzing these men's rationalizations, Scully and Marolla draw on Scott and Lyman's (1968) classic typology of accounts: excuses and justifications. In using excuses, men acknowledge the wrongfulness of the act but deny full responsibility. The authors find that excuses are used primarily by those who admit to their deviant acts. Men who deny having committed rape (over 80 percent of the population) are more prone to use justifications, accepting responsibility for their act but providing reasons that legitimate their behavior as not wrong. Scully and Marolla examine the various disavowal techniques, shed light on the repertoire of culturally available neutralizing accounts, and analyze the connection between types of accounts used and the way offenders locate blame.

Which types of accounts do you find more compelling, excuses or justifications? How does this article make you feel about the effectiveness of accounts in neutralizing people's views of themselves as deviant? How effective do you think they are in neutralizing others' views of them?

Psychiatry has dominated the literature on rapists since "irresistible impulse" (Glueck, 1925: 243) and "disease of the mind" (Glueck, 1925: 243) were introduced as the causes of rape. Research has been based on small samples of men, frequently the clinicians' own patient population. Not surprisingly, the medical model has predominated: rape is viewed as an individualistic, idiosyncratic symptom of a disordered personality. That is, rape is assumed to be a

From Diana Scully and Joseph Marolla, "Convicted Rapists' Vocabulary of Motive: Excuses and Justifications." *Social Problems*, Vol. 31, No. 5, 1984. Copyright © 1984 Society for the Study of Social Problems. All rights reserved. Reprinted by permission of University of California Press Journals and Diana Scully.

psycho-pathologic problem and individual rapists are assumed to be "sick." However, advocates of this model have been unable to isolate a typical or even predictable pattern of symptoms that are causally linked to rape. Additionally, research has demonstrated that fewer than 5 percent of rapists were psychotic at the time of their rape (Abel et al., 1980).

We view rape as behavior learned socially through interaction with others; convicted rapists have learned the attitudes and actions consistent with sexual aggression against women. Learning also includes the acquisition of culturally derived vocabularies of motive, which can be used to diminish responsibility and to negotiate a nondeviant identity.

Sociologists have long noted that people can, and do, commit acts they define as wrong and, having done so, engage various techniques to disavow deviance and present themselves as normal. Through the concept of "vocabulary of motive," Mills (1940: 904) was among the first to shed light on this seemingly perplexing contradiction. Wrongdoers attempt to reinterpret their actions through the use of a linguistic device by which norm-breaking conduct is socially interpreted. That is, anticipating the negative consequences of their behavior, wrongdoers attempt to present the act in terms that are both culturally appropriate and acceptable.

Following Mills, a number of sociologists have focused on the types of techniques employed by actors in problematic situations (Hall and Hewitt, 1970; Hewitt and Hall, 1973; Hewitt and Stokes, 1975; Sykes and Matza, 1957). Scott and Lyman (1968) describe excuses and justifications, linguistic "accounts" that explain and remove culpability for an untoward act after it has been committed. *Excuses* admit [that] the act was bad or inappropriate but deny full responsibility, often through appeals to accident, or biological drive, or through scapegoating. In contrast, *justifications* accept responsibility for the act but deny that it was wrong—that is, they show [that] in this situation the act was appropriate. *Accounts* are socially approved vocabularies that neutralize an act or its consequences and are always a manifestation of an underlying negotiation of identity.

Stokes and Hewitt (1976: 837) use the term "aligning actions" to refer to those tactics and techniques used by actors when some feature of a situation is problematic. Stated simply, the concept refers to an actor's attempt, through various means, to bring his or her conduct into alignment with culture. Culture in this sense is conceptualized as a "set of cognitive constraints—objects—to which people must relate as they form lines of conduct" (1976: 837), and includes physical constraints, expectations and definitions of others, and personal biography. Carrying out aligning actions implies both awareness of those elements of normative culture that are applicable to the deviant act and, in addition, an actual effort to bring the act into line with this awareness. The result is that deviant behavior is legitimized.

This paper presents an analysis of interviews we conducted with a sample of 114 convicted, incarcerated rapists. We use the concept of accounts (Scott and Lyman, 1968) as a tool to organize and analyze the vocabularies of motive which this group of rapists used to explain themselves and their actions. An analysis of their accounts demonstrates how it was possible for 83 percent (n = 114)[1] of these convicted rapists to view themselves as nonrapists.

When rapists' accounts are examined, a typology emerges that consists of admitters and deniers. Admitters (n = 47) acknowledged that they had forced sexual acts on their victims and defined the behavior as rape. In contrast, deniers[2] either eschewed sexual contact or all association with the victim (n = 35),[3] or admitted to sexual acts but did not define their behavior as rape (n = 32).... By and large, the deniers used justifications while the admitters used excuses. In some cases, both groups relied on the same themes, stereotypes, and images: some admitters, like most deniers, claimed that women enjoyed being raped. Some deniers excused their behavior by referring to alcohol or drug use, although they did so quite differently than admitters. Through these narrative accounts, we explore convicted rapists' own perceptions of their crimes....

JUSTIFYING RAPE

Deniers attempted to justify their behavior by presenting the victim in a light that made her appear culpable, regardless of their own actions. Five themes run through rapists' attempts to justify their rapes: (1) women as seductresses; (2) women mean "yes" when they say "no"; (3) most women eventually relax and enjoy it; (4) nice girls don't get raped; and (5) guilty of a minor wrongdoing.

(1) Women as Seductresses

Men who rape need not search far for cultural language which supports the premise that women provoke or are responsible for rape. In addition to common cultural stereotypes, the fields of psychiatry and criminology (particularly the sub-field of victimology) have traditionally provided justifications for rape, often by portraying raped women as the victims of their own seduction (Albin, 1977; Marolla and Scully, 1979). For example, Hollander (1924: 130) argues:

> Considering the amount of illicit intercourse, rape of women is very rare indeed. Flirtation and provocative conduct, i.e. tacit (if not actual) consent is generally the prelude to intercourse.

Since women are supposed to be coy about their sexual availability, refusal to comply with a man's sexual demands lacks meaning and rape appears normal. The fact that violence and, often, a weapon are used to accomplish the rape is not considered. As an example, Abrahamsen (1960: 61) writes:

> The conscious or unconscious biological or psychological attraction between man and woman does not exist only on the part of the offender toward the woman but, also, on her part toward him, which in many instances may, to some extent, be the impetus for his sexual attack. Often a women [sic] unconsciously wishes to be taken by force—consider the theft of the bride in Peer Gynt.

Like Peer Gynt, the deniers we interviewed tried to demonstrate that their victims were willing and, in some cases, enthusiastic participants. In these accounts, the rape became more dependent upon the victim's behavior than upon their own actions.

Thirty-one percent (n = 10) of the deniers presented an extreme view of the victim. Not only willing, but she was also the aggressor, a seductress who lured them, unsuspecting, into sexual action. Typical was a denier convicted of his first rape and accompanying crimes of burglary, sodomy, and abduction. According to the presentence reports, he had broken into the victim's house and raped her at knife point. While he admitted to the breaking and entry, which he claimed was for altruistic purposes ("to pay for the prenatal care of a friend's girlfriend"), he also argued that when the victim discovered him, he had tried to leave but she had asked him to stay. Telling him that she cheated on her husband, she had voluntarily removed her clothes and seduced him. She was, according to him, an exemplary sex partner who "enjoyed it very much and asked for oral sex.[4] Can I have it now?" he reported her as saying. He claimed they had spent hours in bed, after which the victim had told him he was good looking and asked to see him again. "Who would believe I'd meet a fellow like this?" he reported her as saying.

In addition to this extreme group, 25 percent (n = 8) of the deniers said the victim was willing and had made some sexual advances. An additional 9 percent (n = 3) said the victim was willing to have sex for money or drugs. In two of these three cases, the victim had been either an acquaintance or picked up, which the rapists said led them to expect sex.

(2) Women Mean "Yes" When They Say "No"

Thirty-four percent (n = 11) of the deniers described their victim as unwilling, at least initially, indicating either that she had resisted or that she had said no. Despite this, and even though (according to presentence reports) a weapon had been present in 64 percent (n = 7) of these 11 cases, the rapists justified their behavior by arguing that either the victim had not resisted enough or that her "no" had really meant "yes." For example, one denier who was serving time for a previous rape was subsequently convicted of attempting to rape a prison hospital nurse. He insisted he had actually completed the second rape, and said of his victim: "She semistruggled but deep down inside I think she felt it was a fantasy come true." The nurse, according to him, had asked a question about his conviction for rape, which he interpreted as teasing. "It was like she was saying, 'rape me.'" Further, he stated that she had helped him along with oral sex and "from her actions, she was enjoying it." In another case, a 34-year-old man convicted of abducting and raping a 15-year-old teenager at knife point as she walked on the beach, claimed it was a pickup. This rapist said women like to be overpowered before sex, but to dominate after it begins.

> A man's body is like a Coke bottle, shake it up, put your thumb over
> the opening and feel the tension. When you take a woman out, woo
> her, then she says "no, I'm a nice girl," you have to use force. All men

do this. She said "no" but it was a societal no, she wanted to be coaxed. All women say "no" when they mean "yes" but it's a societal no, so they won't have to feel responsible later.

Claims that the victim didn't resist or, if she did, didn't resist enough, were also used by 24 percent (n = 11) of admitters to explain why, during the incident, they believed the victim was willing and that they were not raping. These rapists didn't redefine their acts until sometime after the crime. For example, an admitter who used a bayonet to threaten his victim, an employee of the store he had been robbing, stated:

> At the time I didn't think it was rape. I just asked her nicely and she didn't resist. I never considered prison. I just felt like I had met a friend. It took about five years of reading and going to school to change my mind about whether it was rape. I became familiar with the subtlety of violence. But at the time, I believed that as long as I didn't hurt anyone it wasn't wrong. At the time, I didn't think I would go to prison, I thought I would beat it.

Another typical case involved a gang rape in which the victim was abducted at knife point as she walked home about midnight. According to two of the rapists, both of whom were interviewed, at the time they had thought the victim had willingly accepted a ride from the third rapist (who was not interviewed). They claimed the victim didn't resist and one reported her as saying she would do anything if they would take her home. In this rapist's view, "She acted like she enjoyed it, but maybe she was just acting. She wasn't crying, she was engaging in it." He reported that she had been friendly to the rapist who abducted her and, claiming not to have a home phone, she gave him her office number—a tactic eventually used to catch the three. In retrospect, this young man had decided, "She was scared and just relaxed and enjoyed it to avoid getting hurt." Note, however, that while he had redefined the act as rape, he continued to believe she enjoyed it.

Men who claimed to have been unaware that they were raping viewed sexual aggression as a man's prerogative at the time of the rape. Thus, they regarded their act as little more than a minor wrongdoing even though most possessed or used a weapon. As long as the victim survived without major physical injury, from their perspective, a rape had not taken place. Indeed, even U.S. courts have often taken the position that physical injury is a necessary ingredient for a rape conviction.

(3) Most Women Eventually Relax and Enjoy It

Many of the rapists expected us to accept the image, drawn from cultural stereotype, that once the rape began, the victim relaxed and enjoyed it.[5] Indeed, 69 percent (n = 22) of deniers justified their behavior by claiming not only that the victim was willing, but also that she enjoyed herself, in some cases to an immense degree. Several men suggested that they had fulfilled their victims' dreams.

Additionally, while most admitters used adjectives such as "dirty," "humiliated," and "disgusted" to describe how they thought rape made women feel, 20 percent (n = 9) believed that their victim enjoyed herself. For example, one denier had posed as a salesman to gain entry to his victim's house. But he claimed he had had a previous sexual relationship with the victim, that she agreed to have sex for drugs, and that the opportunity to have sex with him produced "a glow, because she was really into oral stuff and fascinated by the idea of sex with a black man. She felt satisfied, fulfilled, wanted me to stay, but I didn't want her." In another case, a denier who had broken into his victim's house but who insisted the victim was his lover and let him in voluntarily, declared "She felt good, kept kissing me and wanted me to stay the night. She felt proud after sex with me." And another denier, who had hid in his victim's closet and later attacked her while she slept, argued that while she was scared at first, "once we got into it, she was OK." He continued to believe he hadn't committed rape because "she enjoyed it and it was like she consented."

(4) Nice Girls Don't Get Raped

The belief that "nice girls don't get raped" affects perception of fault. The victim's reputation, as well as characteristics or behavior which violate normative sex role expectations, are perceived as contributing to the commission of the crime. For example, Nelson and Amir (1975) defined hitchhike rape as a victim-precipitated offense.

In our study, 69 percent (n = 22) of deniers and 22 percent (n = 10) of admitters referred to their victims' sexual reputation, thereby evoking the stereotype that "nice girls don't get raped." They claimed that the victim was known to have been a prostitute, or a "loose" woman, or to have had a lot of affairs, or to have given birth to a child out of wedlock. For example, a denier who claimed he had picked up his victim while she was hitchhiking stated, "To be honest, we [his family] knew she was a damn whore and whether she screwed one or 50 guys didn't matter." According to presentence reports this victim didn't know her attacker and he abducted her at knife point from the street. In another case, a denier who claimed to have known his victim by reputation stated:

> If you wanted drugs or a quick piece of ass, she would do it. In court she said she was a virgin, but I could tell during sex [rape] that she was very experienced.

When other types of discrediting biographical information were added to these sexual slurs, a total of 78 percent (n = 25) of the deniers used the victim's reputation to substantiate their accounts. Most frequently, they referred to the victim's emotional state or drug use. For example, one denier claimed his victim had been known to be loose and, additionally, had turned state's evidence against her husband to put him in prison and save herself from a burglary conviction. Further, he asserted that she had met her current boyfriend, who was himself in and out of prison, in a drug rehabilitation center where they were both clients.

Evoking the stereotype that women provoke rape by the way they dress, a description of the victim as seductively attired appeared in the accounts of

22 percent (n = 7) of deniers and 17 percent (n = 8) of admitters. Typically, these descriptions were used to substantiate their claims about the victim's reputation. Some men went to extremes to paint a tarnished picture of the victim, describing her as dressed in tight black clothes and without a bra; in one case, the victim was portrayed as sexually provocative in dress and carriage. Not only did she wear short skirts, but she was observed to "spread her legs while getting out of cars." Not all of the men attempted to assassinate their victim's reputation with equal vengeance. Numerous times they made subtle and offhand remarks like, "She was a waitress and you know how they are."

The intent of these discrediting statements is clear. Deniers argued that the woman was a "legitimate" victim who got what she deserved. For example, one denier stated that all of his victims had been prostitutes; presentence reports indicated they were not. Several times during his interview, he referred to them as "dirty sluts," and argued "anything I did to them was justified." Deniers also claimed their victim had wrongly accused them and was the type of woman who would perjure herself in court.

(5) Only a Minor Wrongdoing

The majority of deniers did not claim to be completely innocent and they also accepted some accountability for their actions. Only 16 percent (n = 5) of deniers argued that they were totally free of blame. Instead, the majority of deniers pleaded guilty to a lesser charge. That is, they obfuscated the rape by pleading guilty to a less serious, more acceptable charge. They accepted being oversexed, accused of poor judgment or trickery, even some violence, or guilty of adultery or contributing to the delinquency of a minor, charges that are hardly the equivalent of rape.

Typical of this reasoning is a denier who met his victim in a bar when the bartender asked him if he would try to repair her stalled car. After attempting unsuccessfully, he claimed the victim drank with him and later accepted a ride. Out riding, he pulled into a deserted area "to see how my luck would go." When the victim resisted his advances, he beat her and he stated:

> I did something stupid. I pulled a knife on her and I hit her as hard as I would hit a man. But I shouldn't be in prison for what I did. I shouldn't have all this time [sentence] for going to bed with a broad.

This rapist continued to believe that while the knife was wrong, his sexual behavior was justified.

In another case, the denier claimed he picked up his under-age victim at a party and that she voluntarily went with him to a motel. According to presentence reports, the victim had been abducted at knife point from a party. He explained:

> After I paid for a motel, she would have to have sex but I wouldn't use a weapon. I would have explained. I spent money and, if she still said no, I would have forced her. If it had happened that way, it would have been rape to some people but not to my way of thinking. I've done that kind of thing before. I'm guilty of sex and contributing to the delinquency of a minor, but not rape.

In sum, deniers argued that, while their behavior may not have been completely proper, it should not have been considered rape. To accomplish this, they attempted to discredit and blame the victim while presenting their own actions as justified in the context. Not surprisingly, none of the deniers thought of himself as a rapist. A minority of the admitters attempted to lessen the impact of their crime by claiming the victim enjoyed being raped. But despite this similarity, the nature and tone of admitters' and deniers' accounts were essentially different.

EXCUSING RAPE

In stark contrast to deniers, admitters regarded their behavior as morally wrong and beyond justification. They blamed themselves rather than the victim, although some continued to cling to the belief that the victim had contributed to the crime somewhat, for example, by not resisting enough.

Several of the admitters expressed the view that rape was an act of such moral outrage that it was unforgivable. Several admitters broke into tears at intervals during their interviews. A typical sentiment was,

> I equate rape with someone throwing you up against a wall and tearing your liver and guts out of you.... Rape is worse than murder ... and I'm disgusting.

Another young admitter frequently referred to himself as repulsive and confided:

> I'm in here for rape and in my own mind, it's the most disgusting crime, sickening. When people see me and know, I get sick.

Admitters tried to explain their crime in a way that allowed them to retain a semblance of moral integrity. Thus, in contrast to deniers' justifications, admitters used excuses to explain how they were compelled to rape. These excuses appealed to the existence of forces outside of the rapists' control. Through the use of excuses, they attempted to demonstrate that either intent was absent or responsibility was diminished. This allowed them to admit rape while reducing the threat to their identity as a moral person. Excuses also permitted them to view their behavior as idiosyncratic rather than typical and, thus, to believe they were not "really" rapists. Three themes run through these accounts: (1) the use of alcohol and drugs; (2) emotional problems; and (3) nice guy image.

(1) The Use of Alcohol and Drugs

A number of studies have noted a high incidence of alcohol and drug consumption by convicted rapists prior to their crime (Groth, 1979; Queen's Bench Foundation, 1976). However, more recent research has tentatively concluded that the connection between substance use and crime is not as direct as previously thought (Ladouceur, 1983). Another facet of alcohol and drug use mentioned in the literature is its utility in disavowing deviance. McCaghy (1968) found that child molesters used alcohol as a technique for neutralizing their deviant identity. Marolla and Scully (1979), in a review of psychiatric literature, demonstrated how alcohol consumption is applied differently as a vocabulary of motive. Rapists can use alcohol both as an excuse for their behavior and to discredit the victim and make her more responsible. We found the former common among admitters and the latter common among deniers.

Alcohol and/or drugs were mentioned in the accounts of 77 percent (n = 30) of admitters and 84 percent (n = 21) of deniers and both groups were equally likely to have acknowledged consuming a substance—admitters, 77 percent (n = 30); deniers, 72 percent (n = 18). However, admitters said they had been affected by the substance; if not the cause of their behavior, it was at least a contributing factor. For example, an admitter who estimated his consumption to have been eight beers and four "hits of acid" reported:

Rapists' Accounts of Own and Victims' Alcohol and/or Drug (A/D) Use and Effect

	Admitters n = 39%	Deniers n = 25%
Neither self nor victim used A/D	23	16
Self used A/D	77	72
Of self used, no victim use	51	12
Self affected by A/D	69	40
Of self affected, no victim use or effect	54	24
Self A/D users who were affected	90	56
Victim used A/D	26	72
Of victim used, no self use	0	0
Victim affected by A/D	15	56
Of victim affected, no self use or effect	0	40
Victim A/D users who were affected	60	78
Both self and victim used and affected by A/D	15	16

Straight, I don't have the guts to rape. I could fight a man but not that. To say, "I'm going to do it to a woman," knowing it will scare and hurt her, takes guts or you have to be sick.

Another admitter believed that his alcohol and drug use,

... brought out what was already there but in such intensity it was uncontrollable. Feelings of being dominant, powerful, using someone for my own gratification, all rose to the surface.

In contrast, deniers' justifications required that they not be substantially impaired. To say that they had been drunk or high would cast doubt on their ability to control themselves or to remember events as they actually happened.

Consistent with this, when we asked if the alcohol and/or drugs had had an effect on their behavior, 69 percent (n = 27) of admitters, but only 40 percent (n = 10) of deniers, said they had been affected.

Even more interesting were references to the victim's alcohol and/or drug use. Since admitters had already relieved themselves of responsibility through claims of being drunk or high, they had nothing to gain from the assertion that the victim had used or been affected by alcohol and/or drugs. On the other hand, it was very much in the interest of deniers to declare that their victim had been intoxicated or high: that fact lessened her credibility and made her more responsible for the act. Reflecting these observations, 72 percent (n = 18) of deniers and 26 percent (n = 10) of admitters maintained that alcohol or drugs had been consumed by the victim. In addition, while 56 percent (n = 14) of deniers declared she had been affected by this use, only 15 percent (n = 6) of admitters made a similar claim. Typically deniers argued that the alcohol and drugs had sexually aroused their victim or rendered her out of control. For example, one denier insisted that his victim had become hysterical from drugs, not from being raped, and it was because of the drugs that she had reported him to the police. In addition, 40 percent (n = 10) of deniers argued that while the victim had been drunk or high, they themselves either hadn't ingested or weren't affected by alcohol and/or drugs. None of the admitters made this claim. In fact, in all of the 15 percent (n = 6) of cases where an admitter said the victim was drunk or high, he also admitted to being similarly affected.

These data strongly suggest that whatever role alcohol and drugs play in sexual and other types of violent crime, rapists have learned the advantage to be gained from using alcohol and drugs as an account. Our sample were aware that their victim would be discredited and their own behavior excused or justified by referring to alcohol and/or drugs.

(2) Emotional Problems

Admitters frequently attributed their acts to emotional problems. Forty percent (n = 19) of admitters said they believed an emotional problem had been at the root of their rape behavior, and 33 percent (n = 15) specifically related the problem to an unhappy, unstable childhood or a marital–domestic situation. Still others claimed to have been in a general state of unease. For example, one admitter said that at the time of the rape he had been depressed, feeling he couldn't do anything right, and that something had been missing from his life. But he also added, "being a rapist is not part of my personality." Even admitters who could locate no source for an emotional problem evoked the popular image of rapists as the product of disordered personalities to argue they also must have problems:

> The fact that I'm a rapist makes me different. Rapists aren't all there.
> They have problems. It was wrong so there must be a reason why
> I did it. I must have a problem.

Our data do indicate that a precipitating event, involving an upsetting problem of everyday living, appeared in the accounts of 80 percent (n = 38) of admitters and 25 percent (n = 8) of deniers. Of those experiencing a precipitating event, including deniers, 76 percent (n = 35) involved a wife or girlfriend. Over and over, these men described themselves as having been in a rage because of an incident involving a woman with whom they believed they were in love.

Frequently, the upsetting event was related to a rigid and unrealistic double standard for sexual conduct and virtue which they applied to "their" woman but which they didn't expect from men, didn't apply to themselves, and, obviously, didn't honor in other women. To discover that the "pedestal" didn't apply to their wife or girlfriend sent them into a fury. One especially articulate and typical admitter described his feeling as follows. After serving a short prison term for auto theft, he married his "childhood sweetheart" and secured a well-paying job. Between his job and the volunteer work he was doing with an ex-offender group, he was spending long hours away from home, a situation that had bothered his wife. In response to her request, he gave up his volunteer work, though it was clearly meaningful to him. Then, one day, he discovered his wife with her former boyfriend "and my life fell apart." During the next several days, he said his anger had made him withdraw into himself and, after three days of drinking in a motel room, he abducted and raped a stranger. He stated:

> My parents have been married for many years and I had high expectations about marriage. I put my wife on a pedestal. When I walked in on her, I felt like my life had been destroyed, it was such a shock. I was bitter and angry about the fact that I hadn't done anything to my wife for cheating. I didn't want to hurt her [victim], only to scare and degrade her.

It is clear that many admitters, and a minority of deniers, were under stress at the time of their rapes. However, their problems were ordinary—the types of upsetting events that everyone experiences at some point in life. The overwhelming majority of the men were not clinically defined as mentally ill in court-ordered psychiatric examinations prior to their trials. Indeed, our sample is consistent with Abel et al. (1980) who found fewer than 5 percent of rapists were psychotic at the time of their offense.

As with alcohol and drug intoxication, a claim of emotional problems works differently depending upon whether the behavior in question is being justified or excused. It would have been counter-productive for deniers to have claimed to have had emotional problems at the time of the rape. Admitters used psychological explanations to portray themselves as having been temporarily "sick" at the time of the rape. Sick people are usually blamed for neither the cause of their illness nor for acts committed while in that state of diminished capacity. Thus, adopting the sick role removed responsibility by excusing the behavior as having been beyond the ability of the individual to control. Since the rapists were not "themselves," the rape was idiosyncratic rather than typical behavior. Admitters asserted a nondeviant identity despite their self-proclaimed disgust with what they had done. Although admitters were willing to assume the sick role, they did not view their problem as a chronic condition, nor did they believe themselves to be insane or permanently impaired. Said one admitter, who believed that he needed psychological counseling: "I have a mental disorder, but I'm not crazy." Instead, admitters viewed their "problem" as mild, transient, and curable. Indeed, part of the appeal of this excuse was that not only did it relieve responsibility, but, as with alcohol and drug addiction, it allowed the rapist to "recover." Thus, at the time of their interviews, only 31 percent (n = 14) of admitters indicated that "being a rapist" was part of their self-concept. Twenty-eight percent (n = 13) of admitters stated they had never thought of themselves as rapists, 8 percent (n = 4) said they were unsure, and 33 percent (n = 16) asserted they had been a rapist at one time but now were recovered. A multiple "ex-rapist," who believed his "problem" was due to "something buried in my subconscious" that was triggered when his girlfriend broke up with him, expressed a typical opinion:

> I was a rapist, but not now. I've grown up, had to live with it. I've hit the bottom of the well and it can't get worse. I feel born again to deal with my problems.

(3) Nice Guy Image

Admitters attempted to further neutralize their crime and negotiate a nonrapist identity by painting an image of themselves as a "nice guy." Admitters projected the image of someone who had made a serious mistake but, in every other respect, was a decent person. Fifty-seven percent (n = 27) expressed regret and sorrow for their victim indicating that they wished there were a way to apologize for or amend their behavior. For example, a participant in a rape–murder, who insisted his partner did the murder, confided, "I wish there was something I could do besides saying 'I'm sorry, I'm sorry.' I live with it 24 hours a day and, sometimes, I wake up crying in the middle of the night because of it."

Schlenker and Darby (1981) explain the significance of apologies beyond the obvious expression of regret. An apology allows a person to admit guilt while at the same time seeking a pardon by signaling that the event should not be considered a fair representation of what the person is really like. An apology separates the bad self from the good self, and promises more acceptable behavior in the future. When apologizing, an individual is attempting to say "I have repented and should be forgiven," thus making it appear that no further rehabilitation is required.

The "nice guy" statements of the admitters reflected an attempt to communicate a message consistent with Schlenker's and Darby's analysis of apologies. It was an attempt to convey that rape was not a representation of their "true" self. For example,

> It's different from anything else I've ever done. I feel more guilt about this. It's not consistent with me. When I talk about it, it's like being assaulted myself. I don't know why I did it, but once I started, I got into it. Armed robbery was a way of life for me, but not rape. I feel like I wasn't being myself.

Admitters also used "nice guy" statements to register their moral opposition to violence and harming women, even though, in some cases, they had seriously injured their victims. Such was the case of an admitter convicted of gang rape:

> I'm against hurting women. She should have resisted. None of us were the type of person that would use force on a woman. I never positioned myself on a woman unless she showed an interest in me. They would play to me, not me to them. My weakness is to follow. I never would have stopped, let alone pick her up without the others. I never would have let anyone beat her. I never bothered women who didn't want sex; never had a problem with sex or getting it. I loved her—like all women.

Finally, a number of admitters attempted to improve their self-image by demonstrating that, while they had raped, it could have been worse if they had not been a "nice guy." For example, one admitter professed to being especially gentle with his victim after she told him she had just had a baby. Others claimed to have given the victim money to get home or make a phone call, or to have made sure the victim's children were not in the room. A multiple rapist, whose pattern was to break in and attack sleeping victims in their homes, stated:

> I never beat any of my victims and I told them I wouldn't hurt them if they cooperated. I'm a professional thief. But I never robbed the women I raped because I felt so bad about what I had already done to them.

Even a young man, who raped his five victims at gun point and then stabbed them to death, attempted to improve his image by stating:

> Physically they enjoyed the sex [rape]. Once they got involved, it would be difficult to resist. I was always gentle and kind until I started to kill them. And the killing was always sudden, so they wouldn't know it was coming.

SUMMARY AND CONCLUSIONS

Convicted rapists' accounts of their crimes include both excuses and justifications. Those who deny what they did was rape justify their actions; those who admit it was rape attempt to excuse it or themselves. This study does not address why some men admit while others deny, but future research might address this question. This paper does provide insight on how men who are sexually aggressive or violent construct reality, describing the different strategies of admitters and deniers.

Admitters expressed the belief that rape was morally reprehensible. But they explained themselves and their acts by appealing to forces beyond their control, forces which reduced their capacity to act rationally and thus compelled them to rape. Two types of excuses predominated: alcohol/drug intoxication and emotional problems. Admitters used these excuses to negotiate a moral identity for themselves by viewing rape as idiosyncratic rather than typical behavior. This allowed them to reconceptualize themselves as recovered or "ex-rapists," [people] who had made a serious mistake which did not represent their "true" [selves].

In contrast, deniers' accounts indicate that these men raped because their value system provided no compelling reason not to do so. When sex is viewed as a male entitlement, rape is no longer seen as criminal. However, the deniers had been convicted of rape, and like the admitters, they attempted to negotiate an identity. Through justifications, they constructed a "controversial" rape and attempted to demonstrate how their behavior, even if not quite right, was appropriate in the situation. Their denials, drawn from common cultural rape stereotypes, took two forms, both of which ultimately denied the existence of a victim.

The first form of denial was buttressed by the cultural view of men as sexually masterful and women as coy but seductive. Injury was denied by portraying the victim as willing, even enthusiastic, or as politely resistant at first but eventually yielding to "relax and enjoy it." In these accounts, force appeared merely as a seductive technique. Rape was disclaimed: rather than harm the woman, the rapist had fulfilled her dreams. In the second form of denial, the victim was portrayed as the type of woman who "got what she deserved." Through attacks on the victim's sexual reputation and, to a lesser degree, her emotional state, deniers attempted to demonstrate that since the victim wasn't a "nice girl," they were not rapists. Consistent with both forms of denial was the self-interested use of alcohol and drugs as a justification. Thus, in contrast to admitters, who accentuated their own use as an excuse, deniers emphasized the victim's consumption in an effort to both discredit her and make her appear more responsible for the rape. It is important to remember that deniers did not invent these justifications. Rather, they reflect a belief system that has historically victimized women by promulgating the myth that women both enjoy and are responsible for their own rape.

While admitters and deniers present an essentially contrasting view of men who rape, there were some shared characteristics. Justifications particularly, but also excuses, are buttressed by the cultural view of women as sexual commodities, dehumanized and devoid of autonomy and dignity. In this sense, the sexual objectification of women must be understood as an important factor contributing to an environment that trivializes, neutralizes, and, perhaps, facilitates rape.

Finally, we must comment on the consequences of allowing one perspective to dominate thought on a social problem. Rape, like any complex continuum of behavior, has multiple causes and is influenced by a number of social factors. Yet, dominated by psychiatry and the medical model, the underlying assumption that rapists are "sick" has pervaded research. Although methodologically unsound, conclusions have been based almost exclusively on small clinical populations of rapists—that extreme group of rapists who seek counseling in prison and are the most likely to exhibit psychopathology. From this small, atypical group of men, psychiatric findings have been generalized to all men who rape. Our research, however, based on volunteers from the entire prison population, indicates that some rapists, like deniers, viewed and understood their behavior from a popular cultural perspective. This strongly suggests that cultural perspectives, and not an idiosyncratic illness, motivated their behavior. Indeed, we can argue that the psychiatric perspective has contributed to the vocabulary of motive that rapists use to excuse and justify their behavior (Scully and Marolla, 1984).

Efforts to arrive at a general explanation for rape have been retarded by the narrow focus of the medical model and the preoccupation with clinical populations. The continued reduction of such complex behavior to a singular cause hinders, rather than enhances, our understanding of rape.

NOTES

1. These numbers include pretest interviews. When the analysis involves either questions that were not asked in the pretest or that were changed, they are excluded and thus the number changes.

2. There is, of course, the possibility that some of these men really were innocent of rape. However, while the U.S. criminal justice system is not without flaw, we assume that it is highly unlikely that this many men could have been unjustly convicted of rape, especially since rape is a crime with traditionally low conviction rates. Instead, for purposes of this research, we assume that these men were guilty as charged and that their attempt to maintain an image of nonrapist springs from some psychologically or sociologically interprétable mechanism.

3. Because of their outright denial, interviews with this group of rapists did not contain the data being analyzed here and, consequently, they are not included in this paper.

4. It is worth noting that a number of deniers specifically mentioned the victim's alleged interest in oral sex. Since our interview questions about sexual history indicated that the rapists themselves found oral sex marginally acceptable, the frequent mention is probably another attempt to discredit the victim. However, since a tape recorder could not be used for the interviews and the importance of these claims didn't emerge until the data was being coded and analyzed, it is possible that it was mentioned even more frequently but not recorded.

5. Research shows clearly that women do not enjoy rape. Holmstrom and Burgess (1978) asked 93 adult rape victims, "How did it feel sexually?" Not one said they enjoyed it. Further, the trauma of rape is so great that it disrupts sexual functioning (both frequency and satisfaction) for the overwhelming majority of victims, at least during the period immediately following the rape and, in fewer cases, for an extended period of time (Burgess and Holmstrom, 1979; Feldman-Summers et al., 1979). In addition, a number of studies have shown that rape victims experience adverse consequences prompting some to move, change jobs, or drop out of school (Burgess and Holmstrom, 1974; Kilpatnck et al., 1979; Ruch et al., 1980; Shore, 1979).

REFERENCES

Abel, Gene, Judith Becker, and Linda Skinner. (1980). "Aggressive behavior and sex." *Psychiatric Clinics of North America* 3(2): 133–151.

Abrahamsen, David. (1960). *The Psychology of Crime*. New York: John Wiley.

Albin, Rochelle. (1977). "Psychological studies of rape." *Signs* 3(2): 423–435.

Burgess, Ann Wolbert, and Lynda Lytle Holmstrom. (1974). *Rape: Victims of Crisis*. Bowie: Robert J. Brady.

———. (1979). "Rape: Sexual disruption and recovery." *American Journal of Orthopsychiatry* 49(4): 648–657.

Feldman-Summers, Shirley, Patricia E. Gordon, and Jeanette R. Meagher. (1979). "The impact of rape on sexual satisfaction." *Journal of Abnormal Psychology* 88(1): 101–105.

Glueck, Sheldon. (1925). *Mental Disorders and the Criminal Law*. New York: Little, Brown.

Groth, Nicholas A. (1979). *Men Who Rape*. New York: Plenum Press.

Hall, Peter M., and John P. Hewitt. (1970). "The quasi-theory of communication and the management of dissent." *Social Problems* 18(1): 17–27.

Hewitt, John P., and Peter M. Hall. (1973). "Social problems, problematic situations, and quasi-theories." *American Journal of Sociology* 38(3): 367–374.

Hewitt, John P., and Randall Stokes. (1975). "Disclaimers." *American Sociological Review* 40(1): 1–11.

Hollander, Bernard. (1924). *The Psychology of Misconduct, Vice and Crime*. New York: Macmillan.

Holmstrom, Lynda Lytle, and Ann Wolbert Burgess. (1978). "Sexual behavior of assailant and victim during rape." Paper presented at the annual meetings of the American Sociological Association, San Francisco, September 2–8.

Kilpatrick, Dean G., Lois Veronen, and Patricia A. Resnick. (1979). "The aftermath of rape: Recent empirical findings." *American Journal of Orthopsychiatry* 49(4): 658–669.

Ladouceur, Patricia. (1983). "The relative impact of drugs and alcohol on serious felons." Paper presented at the annual meetings of the American Society of Criminology, Denver, November 9–12.

Marolla, Joseph, and Diana Scully. (1979). "Rape and psychiatric vocabularies of motive." In Edith S. Gomberg and Violet Franks (eds.), *Gender and Disordered Behavior: Sex Differences in Psychopathology* (pp. 301–318). New York: Brunner/Mazel.

McCaghy, Charles. (1968). "Drinking and deviance disavowal: The case of child molesters." *Social Problems* 16(1): 43–49.

Mills, C. Wright. (1940). "Situated actions and vocabularies of motive." *American Sociological Review* 5(6): 904–913.

Nelson, Steve, and Menachem Amir. (1975). "The hitchhike victim of rape: A research report." In Israel Drapkin and Emilio Viano (eds.), *Victimology: A New Focus* (pp. 47–65). Lexington, KY: Lexington Books.

Queen's Bench Foundation. (1976). *Rape: Prevention and Resistance*. San Francisco: Queen's Bench Foundation.

Ruch, Libby O., Susan Meyers Chandler, and Richard A. Harter. (1980). "Life change and rape impact." *Journal of Health and Social Behavior* 21(3): 248–260.

Schlenker, Barry R., and Bruce W. Darby. (1981). "The use of apologies in social predicaments." *Social Psychology Quarterly* 44(3): 271–278.

Scott, Marvin, and Stanford Lyman. (1968). "Accounts." *American Sociological Review* 33(1): 46–62.

Scully, Diana, and Joseph Marolla. (1984). "Rape and psychiatric vocabularies of motive: Alternative perspectives." In Ann Wolbert Burgess (ed.), *Handbook on Rape and Sexual Assault*. New York: Garland Publishing.

Shore, Barbara K. (1979). "An examination of critical process and outcome factors in rape." Rockville, MD: National Institute of Mental Health.

Stokes, Randall, and John P. Hewitt. (1976). "Aligning actions." *American Sociological Review* 41(5): 837–849.

Sykes, Gresham M., and David Matza. (1957). "Techniques of neutralization." *American Sociological Review* 22(6): 664–670.

28

The Devil Made Me Do It: Use of Neutralizations by Shoplifters

PAUL CROMWELL AND QUINT THURMAN

Cromwell and Thurman offer a discussion of shoplifters' rationalizations that many people will find familiar. Stealing from stores has long been widespread among American youths, and practitioners have found it convenient and easy to rationalize pilfering from large companies that do not have a visible or identifiable local owner. This chapter compliments the Scully and Marolla analysis by drawing on Sykes and Matza's (1957) "techniques of neutralization," a conceptualization of accounts that predates the distinction between excuses and justifications. Cromwell and Thurman show how shoplifters make ample use of the latter existing categories and invent a few new ones of their own. These accounts, like all others, help individuals deflect the labeling process and the deviant identity.

How would you divide the techniques of neutralization discussed in this chapter into the categories of excuses and justifications laid out in Scully and Marolla's chapter? Which of the two chapters do you find offers the most compelling argument, and why?

"You know that cartoon where the guy has a little devil sitting on one shoulder and a little angel on the other? And one is telling him 'Go ahead on, do it,' and the angel is saying 'No, don't do it.' You know?... Sometimes when I'm thinking about boosting something, my angel don't show up." (30-year-old male shoplifter)

Nearly five decades ago Gresham Sykes and David Matza (1957) introduced neutralization theory as an explanation for juvenile delinquency. Sykes and Matza's (1957) theory is an elaboration of Edwin Sutherland's (1947) proposition that individuals can learn criminal techniques, and the "motives, drives, rationalizations, and attitudes favorable to violations of the law." Sykes and Matza argued that these justifications or rationalizations protect the individual from self-blame and the blame of others. Thus, the individual may remain committed to the value system of the dominant culture while committing criminal acts without experiencing the cognitive dissonance that might be otherwise expected. He or she deflects or "neutralizes" guilt in advance, clearing the way to blame-free crime. These neutralizations also protect the individual from any residual guilt following the crime. It is this ability to use neutralizations that differentiates delinquents from nondelinquents (Thurman, 1984).

While Sykes and Matza (1957) do not specifically maintain that only offenders who are committed to the dominant value system make use of these techniques of neutralizations, they appear to contend that delinquents maintain a commitment to the moral order and are able to drift into delinquency through the use of techniques of neutralization. This approach assumes that should delinquents fail to internalize conventional morality, neutralization would be unnecessary since there would be no guilt to neutralize.

One issue that has not been satisfactorily settled is when neutralization occurs. Sykes and Matza (1957) contend that deviants must neutralize moral prescriptions prior to committing a crime. However, most research is incapable of determining whether the stated neutralization is a before-the-fact neutralization or an after-the-fact rationalization.

TECHNIQUES OF NEUTRALIZATION

Sykes and Matza (1957) identified five techniques of neutralization commonly offered to justify deviant behavior—denial of responsibility, denial of injury, denial of the victim, condemning the condemners, and appeal to higher loyalties.

Five additional neutralization techniques have since been identified. These include defense of necessity (Klockars, 1974), metaphor of the ledger (Minor, 1981), denial of the necessity of the law, the claim that everybody else is doing it, and the claim of entitlement (Coleman, 1994).

This article examines neutralization theory as it might apply to a specific form of criminal activity that is highly prevalent across a wide range of the population. The purpose of this study is to determine the extent to which adult shoplifters use techniques of neutralizations and to analyze the various neutralizations available to them. We examine offenders who shoplift and explore the justifications that they say they rely upon to excuse behavior they also acknowledge as morally wrong.

STATEMENT OF THE PROBLEM

Shoplifting may be the most serious crime with which the most people have some personal familiarity. Research has shown that one in every 10–15 persons who shops has shoplifted at one time or another. Further, losses attributable to shoplifting are considerable, with estimates ranging from 12 to 30 billion dollars lost annually. Shoplifting also represents one of the most prevalent forms of larceny, accounting for approximately 15 percent of all larcenies, according to data maintained by the Federal Bureau of Investigation (1996).

Unlike many other forms of crime, people who shoplift do not ordinarily require any special expertise or tools to engage in this crime. Consequently, those persons who shoplift do not necessarily conform to most people's perception of what a criminal offender is like. Instead, shoplifters tend to be demographically similar to the "average person." In a large study of nondelinquents, Klemke (1982) reported that as many of 63 percent of those persons he interviewed had shoplifted at some point in their lives. Students, housewives, business and professional persons, as well as professional thieves constitute the population of shoplifters. Loss prevention experts routinely counsel retail merchants that there is no particular profile of a potential shoplifter. Turner and Cashdan (1988) conclude, "While clearly a criminal activity, shoplifting borders on what might be considered a 'folkcrime.'" In her classic study, Mary Cameron (1964: xi) wrote:

> Most people have been tempted to steal from stores, and many have been guilty (at least as children) of "snitching" an item or two from counter tops. With merchandise so attractively displayed in department stores and supermarkets, and much of it apparently there for the taking, one may ask why everyone isn't a thief.

Neutralization theory argues that ordinary individuals who engage in deviant or criminal behavior may use techniques that permit them to recognize extenuating circumstances that enable them to explain away delinquent behavior. Without worrying about guilty feelings that would stand in their way of committing a criminal act, the theory asserts that those persons are free to participate in delinquent acts that they would otherwise believe to be wrong.

METHOD

The data presented here were obtained in 1997 and 1998 in Wichita, Kansas. We obtained access to a court-ordered diversion program for adult "first-offenders" charged with theft. Of these, the majority of offenders were charged with misdemeanor shoplifting and [were] required to attend an eight-hour therapeutic/education program as a condition of having their record expunged. A new group met each Saturday. The average group size was 18–20 participants. Participants were encouraged by the program facilitator to discuss with the group the offense that brought them to the diversion program, why they did what they did, and how they felt about it. We obtained interviews with 137 subjects from approximately 350 subjects who were approached. Ethnicity and gender of the sample are shown in Table 28.1. The mean age of the sample was 26. The age range was 18 to 66 years of age. Although the diversion program was designed for first offenders, over one-half of the participants had been apprehended for shoplifting in the past.

TABLE 28.1   Neutralizations by Shoplifter Respondents by Gender and Ethnicity (N = 137)

	White	Hispanic	Black	Total
Male	48	11	29	88
Female	30	6	13	49
Total	78	17	42	137

FINDINGS

The informants appeared to readily use neutralization techniques. We identified nine categories of neutralizations; the **five Sykes and Matza (1957) categories**, the **Defense of Necessity** and **Everybody Does It**, identified by Coleman (1994) and two additional, which we labeled **Justification by Comparison** and **Postponement**. Only 5 of the 137 informants failed to express a rationalization or neutralization when asked how they felt about their illegal behavior. Three of these subjects responded by admitting their guilt and expressing remorse. Two others simply stated that they had nothing to say on that issue. In many cases, the respondents offered more than one neutralization for the same offense. For example, one female respondent stated, "I don't know what comes over me. It's like, you know, is somebody else doing it, not me[?]" (Denial of Responsibility). "I'm really a good person. I wouldn't ever do something like that, stealing, you know, but I have to take things sometimes for my kids. They need stuff and I don't have any money to get it" (Defense of Necessity). They frequently responded with a motivation ("I wanted the item and could not afford it") followed by a neutralization ("Stores charge too much for stuff. They could sell things for half what they do and still make a profit. They're just too greedy"). Thus, in many cases, the motivation was linked to [an] excuse in such a way as to make the excuse a part of the motivation. The subjects were in effect explaining the reason the deviant act occurred and justifying it at the same time. The following section illustrates the neutralizations we discovered in use by the informants.

Denial of Responsibility ("I Didn't Mean It")

Denial of responsibility frees subject[s] from experiencing culpability for deviance by allowing [them] to perceive themselves as victims of their environment. The offender views him- or herself as being acted upon rather than acting. Thus, attributing behavior to poor parenting, bad companions, or internal forces (the devil made me do it) allows the offender to avoid disapproval of self or others, which in turn, diminishes those influences as mechanisms of social control. Sykes and Matza (1957: 666) describe the individual resorting to this neutralization as having a "billiard ball conception of himself in which he see himself as helplessly propelled into new situations."

I admit that I lift. I do. But, you know, it's not really me—I mean, I don't believe in stealing. I'm a church-going person. It's just that sometimes something takes over me and I can't seem to not do it. It's like those TV shows where the person is dying and he goes out of his body and watches them trying to save him. That's sorta how I feel sometimes when I'm lifting. (26-year-old female)

I wasn't raised right. You know what I mean? Wasn't nobody to teach me right from wrong. I just ran with a bad group and my mamma didn't ever say nothin' about it. That's how I turned out this way—stealin' and stuff. (22-year-old female)

If it wasn't for the bunch I ran with at school I never would have started taking things. We used to go the mall after school and everybody would have to steal something. If you didn't get anything, everybody called you names—chicken-shit and stuff like that. (20-year-old male)

Many of the shoplifter informants neutralized their activities [by] citing loss of self-control due to alcohol or drug use. This is a common form of denial of responsibility. If not for the loss of inhibition due to drug or alcohol use, they argue, they would not commit criminal acts.

I was drinking with my buddies and we decided to go across the street to the [convenience store] and steal some beer. I was pretty wasted or I wouldn't done it. (19-year-old white male)

I never boost when I'm straight. It's the pills, you know? (30-year-old white female)

Denial of Injury ("I Didn't Really Hurt Anybody")

Denial of injury allows the offender to perceive of his or her behavior as having no direct harmful consequences to the victim. The victim may be seen as easily able to afford the loss (big store, insurance company, wealthy person) or the crime may be semantically recast, as when auto theft is referred to as joyriding, or vandalism as a prank.

They [stores] big. Make lotsa money. They don't even miss the little bit I get. (19-year-old male)

They write it off their taxes. Probably make a profit off it. So, nobody gets hurt. I get what I need and they come out O.K. too. (28-year-old male)

Them stores make billions. Did you ever hear of Sears going out of business from boosters? (34-year-old female)

Denial of the Victim ("They Had It Coming")

Denial of the victim facilitates deviance when it can be justified as retaliation upon a deserving victim. In the present study, informants frequently reported that the large stores from which they stole were deserving victims because of high prices and the perception that they made excessive profits at the expense of ordinary people. The shoplifters frequently asserted that the business establishments from which they stole overcharged consumers and thus deserved the payback from shoplifting losses.

> Stores deserve it. It don't matter if I boost $10,000 from one, they've made 10,000 times that much ripping off people. You could never steal enough to get even... I don't really think I'm doing anything wrong. Just getting my share. (48-year-old female)
>
> Dillons [food store chain] are totally bogus. A little plastic bag of groceries is $30, $25. Probably cost them $5.... Whatta they care about me? Why should I care about them? I take what I want. Don't feel guilty a bit. No sir. Not a bit. (29-year-old female)
>
> I have a lot of anger about stores and the way they rip people off. Sometimes I think the consumer has to take things into their own hands. (49-year-old female)

Condemning the Condemners ("The System Is Corrupt")

Condemning the condemners projects blame on lawmakers and lawenforcers. It shifts the focus from the offender to those who disapprove of his or her acts. This neutralization views the "system" as crooked and thus unable to justify making and enforcing rules it does not itself live by. Those who condemn [the offenders'] behavior are viewed as hypocritical since many of them engage in deviant behavior themselves.

> I've heard of cops and lawyers and judges and all kind of rich dudes boosting. They no better than me. You know what I'm saying. (18-year-old male)
>
> Big stores like J.C. Penneys—when they catch me with something—like two pairs of pants, they tell the police you had like 5 pairs of pants and 2 shirts or something like that. You know what I'm saying? What they do with the other 3 pairs of pants and shirts? Insurance company pays them off and they get richer—they's bigger crooks than me. (35-year-old female)
>
> They thieves too. Just take it a different way. They may be smarter than me—use a computer or something like that—but they just as much a thief as me. Fuck'em. Cops too. They all thieves. Least, I'm honest about it. (22-year-old male)

Appeal to Higher Loyalties ("I Didn't Do It For Myself")

Appeal to higher loyalties functions to legitimize deviant behavior when a nonconventional social bond creates more immediate and pressing demands than one consistent with conventional society. The most common use of this technique among the shoplifters was pressure from delinquent peers to shoplift and the perceived needs of one's family for items that the informant could not afford to buy.

This was especially common with mothers shoplifting for items for their children.

> I never do it 'cept when I'm with my friends. Everybody be taking stuff and so I do too. You know—to be part of the group. Not to seem like I'm too good for 'em. (17-year-old female)
> I like to get nice stuff for my kids, you know. I know it's not O.K., you know what I mean? But, I want my kids to dress nice and stuff. (28-year-old female)

The Defense of Necessity ("I Had No Other Choice")

The defense of necessity (Coleman, 1998) serves to reduce guilt through the argument that the offender had no choice under the circumstances but to engage in a criminal act. In the case of shoplifting, the defense of necessity is most often used when the offender states that the crime was necessary to help one's family.

> I had to take care of three children without help. I'd be willing to steal to give them what they wanted. (32-year-old female)
> I got laid off at Boeing last year and got behind on all my bills and couldn't get credit anywhere. My kids needed school clothes and money for supplies and stuff. We didn't have anything and I don't believe in going on welfare, you know. The first time I took some lunch meat at Dillons (grocery chain) so we'd have supper one night. After that I just started to take whatever we needed that day. I knew it was wrong, but I just didn't have any other choice. My family comes first. (42-year-old male)

Everybody Does It

Here the individual attempts to reduce his or her guilt feelings or to justify his or her behavior by arguing that the behavior in question is common (Coleman, 1998). A better label for this neutralization might be "diffusion of guilt." The behavior is justified or the guilt is diffused because of widespread similar acts.

> Everybody I know do it. All my friends. My mother and her boyfriend are boosters and my sister is a big-time booster. (19-year-old female)
> All my friends do it. When I'm with them it seems crazy not to take something too. (17-year-old male)
> I bet you done did it too... when you was coming up. Like 12–13 years old. Everybody boosts. (35-year-old female)

Justification by Comparison ("If I Wasn't Shoplifting I Would Be Doing Something More Serious")

This newly identified neutralization involves offenders justifying their actions by comparing their crimes [with] more serious offenses. While it might be argued that Justification by Comparison is not a neutralization in the strict Sykes and Matza (1957) sense in that these offenders are not committed to conventional norms, they are nonetheless attempting to maintain their sense of self-worth by arguing that they could be worse or are not as bad as some others. Even persons with deviant lifestyles may experience guilt over their behavior and/or feel the necessity to justify their actions to others. The gist of the argument is that "I may be bad, but I could be worse."

> I gotta have $200 every day—day in and day out. I gotta boost a thousand, fifteen-hundred dollars worth to get it. I just do what I gotta do…. Do I feel bad about what I do? Not really. If I wasn't boosting, I'd be robbing people and maybe somebody would get hurt or killed. (40-year-old male)
>
> Looka here. Shoplifting be a little thing. Not a crime really. I do it' stead of robbing folks or breaking in they house. [Society] oughta be glad I boost, stead of them other things. (37-year-old male)
>
> It's nothing. Not like its "jacking" people or something. It's just a little lifting. (19-year-old male)

Postponement ("I Just Don't Think About It")

In a previous study one of the present authors (Thurman, 1984) suggested that further research should consider the excuse strategy of Postponement, by which the offender suppresses his or her guilt feelings—momentarily putting them out of mind to be dealt with at a later time. We found this strategy to be a common occurrence among our informants. They made frequent statements that indicated that they simply put the incident out of their mind. Some stated that they would deal with it later when they were not under so much stress.

> I just don't think about it. I mean, if you think about it, it seems wrong, but you can ignore that feeling sometimes. Put it aside and go on about what you gotta do. (18-year old male)
>
> Dude, I just don't deal with those kinda things when I'm boosting. I might feel bad about it later, you know, but by then it's already over and I can't do anything about it then, you know? (18-year-old male)
>
> I worry about things like that later. (30-year-old female)

DISCUSSION AND CONCLUSION

We found widespread use of neutralizations among the shoplifters in our study. We identified two new neutralizations: *Justification by Comparison* and *Postponement*. Even those who did not appear to be committed to the conventional moral order used neutralizations to justify or excuse their behavior. Their use of neutralizations was not so much to assuage guilt but to provide them with the necessary justifications for their acts to others. Simply because one is not committed to conventional norms does not preclude their understanding that most members of society do accept those values and expect others to do so as well. They may also use neutralizations and rationalizations to provide them with a convincing defense for their crimes that they can tell to more conventionally oriented others if the need arises.

As stated earlier, our research approach could not determine whether the informants neutralized before committing the crime or rationalized afterward. Pogrebin, Poole, and Martinez (1992: 233) suggest that postevent reasons given for deviant behavior are not neutralizations but accounts, or "socially approved vocabularies that serve as explanatory mechanisms for deviance." No one, however, has yet been able to empirically verify the existence of preevent neutralizations. In fact, neutralization theory depends upon analysis of postevent accounts by the offender. We suggest that accounts, neutralizations, and rationalizations are essentially the same behavior at different stages in the criminal event. We argue that Hirschi (1969) was correct in stating that a postcrime rationalization may serve as a precrime neutralization the next time a crime is contemplated. Whether neutralization allows the offender to mitigate guilt feelings before the crime is committed or afterward, the process still occurs. Once an actor has reduced his or her guilt feelings through the use of techniques of neutralization, he or she can continue to offend, assuaging guilt feelings and cognitive dissonance both before and after each offense. It would follow that continued utilization of neutralization and rationalization habitually over time might serve to weaken the social bond, reducing the need to neutralize at all.

Our exploratory study of shoplifters' use of neutralization techniques also suggests that neutralization (theory) may not be a theory of crime but rather a description of a process that represents an adaptation to morality that leads to criminal persistence. Neutralization focuses on how crime is possible, rather than why people might choose to engage in it in the first place. In a sense, neutralization serves as a form of situational morality. While the offender knows an act is morally wrong (either in his or her eyes or in the eyes of society), he or she makes an adaptation to convention that permits deviation under certain circumstances (the various neutralizations discussed). Whether the adaptation is truly neutralizing (before the act) or rationalizing (after the act) the result is same— crime without guilt.

REFERENCES

Cameron, Mary. (1964). *The Booster and the Snitch*. New York: Free Press.

Coleman, James W. (1994). "Neutralization Theory: An Empirical Application and Assessment." Ph.D. Dissertation, Oklahoma State University, Department of Sociology, Stillwater.

Coleman, James W. (1998). *Criminal Elite: Understanding White Collar Crime*. New York: St. Martin's Press.

Federal Bureau of Investigation. (1996). *Crime in the United States—1995*. Washington, DC: U.S. Department of Justice.

Hirschi, Travis. (1969). *Causes of Delin-quency*. Berkeley, CA: University of California Press.

Klemke, Lloyd. (1982). "Exploring Juvenile Shoplifting." *Sociology and Social Research* 67: 59–75.

Klockars, Carl B. (1974). *The Professional Fence*. New York: Free Press.

Minor, William W. (1981). "Techniques of Neutralization: A Reconceptualization and Empirical Examination." *Journal of Research in Crime and Delinquency* (July): 295–318.

Pogrebin, M., E. Poole, and A. Martinez. (1992). "Accounts of Professional Misdeeds: The Sexual Exploitation of Clients by Psychotherapists." *Deviant Behavior* 13: 229–52.

Sutherland, Edwin H. (1947). *Principles of Criminology*. Philadelphia: Lippincott.

Sykes, Gresham M., and David Matza. (1957). "Techniques of Neutralization: A Theory of Delinquency." *American Sociological Review* 22(6): 664–70.

Thurman, Quint C. (1984). "Deviance and Neutralization of Moral Commitment: An Empirical Analysis." *Deviant Behavior* 5: 291–304.

Turner, C. T., and S. Cashdan. (1988). "Perceptions of College Students' Motivations for Shoplifting." *Psychological Reports* 62: 855–62.

29

Contesting Stigma in Sport: The Case of Men Who Cheer

MICHELLE BEMILLER

Just as the lesbian athletes discussed by Blinde and Taub violated gender roles by being intercollegiate athletes, Bemiller's male cheerleaders encounter gender stigma from venturing into a female-dominated activity. Female labels of lesbianism in sport correspond to male labels of homosexuality in cheerleading. The men Bemiller studied know that their descent into a girls' realm will lead to masculinity challenges of various sorts from the people they encounter, and they take a variety of measures intended to forestall their deviant labeling. Their face-saving strategies are fairly aggressive, as they attempt to invoke hypermasculine demeanors to counter their taint of female association. This chapter, like many in the book, reveals the hierarchy of gender stratification that positions hypermasculine men at the top, soft/gentle/androgynous men in a lower position, gay men below them, and all women at the bottom. In attempting to elevate themselves to a higher rung on this ladder, the male cheerleaders demean the role of their female squad mates in order to distance themselves from them and to enhance their own position. Earlier, we noted this behavior in Tuggle and Holmes's article on the antismoking campaign, where the authors described how nonsmokers gained status and power by stigmatizing and diminishing smokers. Similarly, male cheerleaders draw on high-status attributes of their gender role by emphasizing hypermasculine features such as toughness and the sexual objectification of women. Demeaning women is revealed as more than an innocuous form of male jocularity; it is a powerful strategy for maintaining differential access to status, opportunity, and power in society.

How would you compare the stigma faced by the women athletes with that encountered by the male cheerleaders? How would you compare the two groups' adaptations? To what do you attribute these differences?

Reprinted with permission from the author, Michelle Bemiller.

The institution of sports has long been associated with the construction and maintenance of masculinity among boys and men. As an institution, sport reinforces the patriarchal superstructure in which masculinity is valued over femininity. The devaluation of femininity is reflected in the subordination of women, as well as in men who participate in nonmasculine activities or who exhibit nonmasculine characteristics or mannerisms. Participation in competitive sports that emphasize physical size, strength, and power reinforces and reaffirms the masculinity of men who participate as viewers or players. Thus, sports such as football, basketball, ice hockey, and baseball, which emphasize mental toughness, competitiveness, and domination, are viewed as the domain of men.

Athletics provides young men with status among their peers, increasing their popularity and acceptance, assuming, of course, that these men are participating in appropriate "male" sports. Men who do not participate in masculine sports are stigmatized, leading to negative appraisals regarding their gender and sexuality.

Despite the possibility of negative appraisals, men have become more visible within female-dominated sports such as cheerleading, leading to a unique opportunity for research on gender presentation, relations, and identity. Yet this area has not received a lot of attention to date. In contrast, notable work has been done on men's entry into female-dominated *occupations*. Williams (1989, 1995), for example, indicates that, when men do women's work, gender differences are reproduced. Men are viewed as highly competent at their work and rise quickly through the ranks. The same is not true for women in male-identified occupations.

Men in female-dominated occupations do encounter questions regarding their sexuality. This questioning, however, does little to impede their progress in the organization. To reaffirm their masculinity, these men seek out male-identified specialties, emphasize masculine aspects of the job, and pursue administrative positions (Williams, 1989, 1995). Similarly, in her work on men working in "safe" and "embattled" organizations, Dellinger (2004) found that, when men work in an organization dominated by feminist ideals, they construct their masculinity by separating themselves from women in the work context and aligning with other males. In contrast, when men work in an environment which is supportive of masculinity, they have better relationships with their female coworkers.

Although these findings are useful in helping us understand men who do women's work, we still know little about men who do women's *sports*. Do these same patterns and outcomes emerge when men participate in women's sports? To further our understanding of men in female-dominated arenas, this paper will use cheerleaders at one northeastern Ohio university to examine the maintenance of gender and sexuality in a female-dominated sport.

METHODS

Sample and Procedures

Participants for this study were selected from a cheerleading squad at a northeastern Ohio public university. To recruit participants, I attended a cheerleading practice and asked for volunteers to participate in a study about men who cheer. I was introduced by the cheerleading coach as a sociologist interested in doing research on cheerleaders. The coach's introduction provided me with status as a legitimate researcher, a standing that may have influenced the cheerleaders' decision to participate in the study. Upon introduction, I simply told the cheerleaders that I was interested in learning about men's participation as cheerleaders and that I would appreciate their assistance in learning more about the men who cheer. At the time of recruitment, the men and women had just finished practice and were talking among themselves about their plans after practice. Most of them were planning on going out to the local bars. A few men and women agreed to participate in the study. Once these individuals agreed, their friends decided to participate as well. My assumption was that they agreed to participate together because they were all going out after the practice.

I used convenience sampling—sampling cases that were available at the time of the study. Out of 25 possible participants, 17 volunteered: 8 men and 9 women between the ages of 18 and 25. Four men and four women declined to participate. Because both men and women were equally willing to participate, there does not seem to be anything unique about the individuals who chose to participate. The individuals who chose not to participate did not provide me with a reason for declining. Of the 17 participants in the study, no racial minorities were present. Had any minorities participated, some of the information provided by the respondents might have been different, because experiences may have differed on the basis of not only gender, but also race.

The data provided are meant to present a glimpse into the lives of men who cheer at one university, not to be generalized to all cheerleaders at all universities. University and regional culture may certainly play a role in an individual's experiences, with the size of the university and region and the cultural diversity of the campus and community affecting individual attitudes regarding men's and women's participation in gendered sports. The community in which the university is located is a large urban area with over 200,000 residents. The residents of the city are predominantly White, middle-class individuals with a high school education. The university is a large, urban campus with roughly 20,000 undergraduate students; approximately 11,000 are women and 9,000 are men. Non-Whites make up about 20 percent of the overall student population.

Data collection consisted of two stages. Focus group discussions were held in the summer of 1999. Reviews of earlier versions of the focus group research indicated that men may not be willing to discuss issues related to sexuality in focus group settings, so in-depth, one-on-one interviews were conducted in 2001 to probe for a deeper understanding of issues related to sexuality and stigmatization. One-on-one interviews were conducted with both new participants ($n = 7$) and participants from the 1999 focus groups ($n = 4$).

FINDINGS

Two main themes emerged from the cheerleaders' narratives. The first was the stigma that coincides with being a male cheerleader. Stigmatization included the negative experiences men had because they are cheerleaders, as well as accusations regarding their sexuality. The second theme analyzes face-saving strategies that men use to protect their masculinity in a female-dominated sport. These strategies include emphasizing the masculine qualities of cheerleading (e.g., injury and risk) and acting in hyperheterosexual ways (e.g., claiming ownership over the sport and objectifying the female cheerleaders).

Stigma

Goffman (1963) argued that individuals are stigmatized when they possess an undesired differentness from what is anticipated. In the present study, the differentness under investigation is men participating in cheerleading. Goffman pointed to the need to determine whether one's stigma is evident (e.g., as in the case of race) or less visible and difficult to discern. Men who cheer possess a potentially discreditable identity due to their deviation from gendered proscriptions regarding participation in sports. This identity is "discreditable" because it is not a status that everyone necessarily knows about, yet, if known or found out, it can be stigmatizing.

Participation in a Feminine Sport   Throughout the focus group discussions and one-on-one interviews, the male and female cheerleaders talked about the stigmatization and resulting undesirable outcomes associated with being a male cheerleader. This talk is consistent with prior research showing that men are stigmatized because of their participation in what is considered a feminine sport. Thus, male cheerleaders are labeled as nonmasculine and homosexual for "crossing over" into a female domain. The quotes that follow capture some of these comments. According to Betty in her one-on-one interview,

> Cheerleading is known more as a girl's sport than a guy's sport, so the guys get a lot of teasing. I give them a lot of credit for putting up with people talking about them. The comments bother them, but they really like doing it, so they put up with them.

Annie elucidates this point further: "I think it is more acceptable to be a girl cheerleader than a guy cheerleader on campus. It is hard for people to adjust to the idea that men are cheerleaders too." In the all-male focus group discussion, one participant asserted that his fraternity brothers give him a hard time about being a cheerleader, teasing and taunting him about participation in a female-dominated sport. Another stated that he works in construction and that the guys at work "rib him pretty good" about being a male cheerleader.

The stigma of participation is further magnified because cheerleading is in a category of activities that many people would not define as a sport at all. John discusses other men's views of cheerleading by specifically focusing on the teasing that goes along with cheerleading. "Men from other sports," he claims, "say 'you can't participate in a real sport.'" The suggestion that cheerleading is not a "real" sport implies that men should be playing football or some other contact sport that demonstrates their manliness and that the majority of society labels as a sport.

Brett provides an interesting look at the importance of playing a "real" sport as opposed to a feminine sport. Although Brett participates in cheerleading practices, he has not yet performed at a football or basketball game. In other words, he has not made a "public" appearance as a male cheerleader. Brett tells me that he does not intend to cheer at the games. Although he enjoys cheerleading, he fully understands and attempts to avoid the stigma that he faces as a male cheerleader. "I'd rather be playing football than cheering for the football team. It would hurt my self-esteem, because guys don't cheer. I mean what percentage of the male population cheers? I mean, I like cheerleading, but it's a girl's sport." By participating in the cheerleading practices but not the public events, Brett attempts to have his cake and eat it too. In other words, Brett enjoys participating in the practices, but has avoided being publicly labeled as "a man who cheers." Thus, he has attempted to avoid the stigma associated with participation in a female activity. The emphasis on "real" sports not only stigmatizes males who choose to be cheerleaders but also marginalizes cheerleading as a sport altogether.

Sexuality Besides the stigma that occurs with participation in a female-dominated sport, men's participation in cheerleading calls into question their sexual identity, interpreted through the lens of gender. Men who cheer are perceived and labeled as homosexual. As a case in point, prior to becoming a male cheerleader, Adam assumed that men who cheered were homosexual: "I was hesitant to become a cheerleader because of the idea that you have to be gay to be a cheerleader." After discussing this issue with several of the men and women who cheer, Adam decided to participate in the sport despite these stereotypes. Even after discussing the issue with his teammates, however, Adam still faces stereotyping regarding his and his teammates' sexuality: "A lot of people think that the men are gay and ask me if they are. People make fun of them. People say they're all a bunch of fags." Similarly, Brett is teased by his close friends: "People definitely perceive that the male cheerleaders are fags. All of my friends give me a hard time about being a cheerleader." Brett's words demonstrate that strangers react not only to male cheerleaders, but also to the people who know them. Because significant others (e.g., family and friends) often have a strong impact, Brett's friends' reactions affected how he saw himself as a cheerleader. In fact, his friends' reactions probably had a stronger impact on how he saw himself as a cheerleader than strangers' comments would.

Perceptions about male cheerleaders' sexuality, however, do not just come from external forces. As in the cases of Adam and Brett, men who cheer bring internalized stereotypes with them into the sport. Specifically, many of the respondents claimed that, prior to their own participation in the sport, they thought that male cheerleaders were gay, and some current participants indicated that some of the men on the squad might be gay. Brett is one example of someone who holds that belief:

> The men here have more feminine characteristics. I would assume that
> they are gay. I would say they seem to be gay because they are cheerleaders,
> but then I put myself in that group. I think that I perceive them as gay

because they gossip a lot and have more girl friends than guy friends. Something is weird about a guy who can have all girls for friends.

By claiming that men who exhibit these qualities are gay, Brett buys into the stereotypes associated with nonmasculine men or men who participate in non-masculine activities. Brett acknowledges that he is a male cheerleader, but he does not belong "in that group" (i.e., gay cheerleaders). Both Adam and Brett distance themselves from the homosexual stereotype of male cheerleaders by referring to other male cheerleaders as "the men" and "they." By using the third person, they verbally and mentally remove themselves from an association with men whom they perceive as gay. In other words, they might be cheerlea-ders, but they are certainly not gay.

Brett and Adam articulate the gendered belief system which limits men and women to certain activities and which claims that deviation must relate to sexuality and must be a departure from heterosexual normativeness. For example, Brett's comment regarding the strangeness of males with female friends coincides with per-ceptions that all male–female relationships are inherently sexual. In other words, if a male has a female friend, he must have a sexual interest in her, and if he does not, he must be gay. Brett's comments reaffirm what is normatively assumed and accepted: that women and men cannot be friends unless a sexual relationship exists. Brett's assumptions regarding cross-gender friendships are not necessarily supported by research on this topic. Although some research on cross-gender friendships does indicate that male–female friends may experience sexual feelings toward one another, other research finds the opposite to be true. In sum, it is possible for men and women to be friends without having a sexual interest or dynamic intervene.

In Annie's discussion of male cheerleaders, she draws attention to their ten-dency to believe in the stereotypical image of men who cheer:

> We had a guy come in the last week of tryouts, and he fit the media's image of a gay person, and the guys made fun of him behind his back. It's almost like the guy cheerleaders believe the perception that male cheerleaders are gay. So, they participate, but they still believe the stereotype.

This adherence to stereotypical images of male cheerleaders, however, does not rest solely with men who cheer. The female cheerleaders also indicated that they had thought all male cheerleaders were gay until they participated in college cheerleading. In her one-on-one interview, Cindy asserted,

> Before I started cheering, I thought it was different for guys to cheer because, when I was in high school, we only had girls on the squad. People think the guys are gay and they say they wouldn't want to cheer. They say "they're gay," and I tell them that I know all of them and they're not gay.

Cindy's views are not far from the other women's perceptions. In the focus group discussions, one woman said, "I thought they were gay," and another woman stated, "Yeah, I thought they were freaky."

Throughout both the focus group discussions and the one-on-one interviews, the male and female cheerleaders acknowledged that, on the basis of perceptions about sexuality, it is more acceptable for females than males to cheer. One woman explains why participation is less acceptable for the men: "The people that don't know the men still have the idea that the guys are gay and the girls are okay." This response was followed by nods of agreement by the rest of the females in the focus group. In his one-on-one interview, Adam states, "I've been made fun of by people; they've said, like, 'What are you doing? You're in a girl's sport, you're gay, blah, blah, blah.'" All these responses capture a central issue: If the image of male cheerleading can be heterosexually validated, then any man should be able to cheer without the assumption that he is homosexual. Unfortunately, the link that was established by the female participants between being gay and being "freaky" reflects the power of the labeling and stigmatization process of people believed to be sexually deviant. Homophobia limits choices by labeling anyone who deviates from gender-appropriate norms as sexually deviant or gay.

SAVING FACE

In response to stigma and stereotyping, the male cheerleaders participate in strategies to help them save face as they participate in a female-dominated activity. According to Goffman (1959), face-saving behavior involves attempts to salvage an interactional performance that hasn't gone as planned. By participating in a female-dominated sport, men who cheer fail to "do gender" appropriately because of their failure to participate in masculine sports. They therefore must use face-saving techniques to be viewed as acceptable in interactions. The men in this study saved face by acting in hypermasculine ways. According to Connell (1992), hypermasculinity becomes manifest when hostility exists toward homosexual men and heterosexual men attempt to create social distance from homosexuality by emphasizing their heterosexuality. For the men who cheer, this hypermasculinity was demonstrated as they claimed territoriality over the sport, stressed the masculine nature of the sport, and sexually objectified the female cheerleaders. Through these actions, the men attempted to deflect accusations of homosexuality.

Territoriality   Both the male and female cheerleaders discussed the fact that cheerleading is viewed as a female-dominated sport. However, many of the male cheerleaders insisted that cheerleading is becoming a male-dominated sport and that the female cheerleaders would not be able to participate in it without the help of the men. Female cheerleaders have been cheering without men at the high school and professional level for quite some time. Yet, these statements coincide with attitudes of entitlement, superiority, and solidarity that exist in male athletics, a realm that encourages homophobia and sexism.

This sense of entitlement, superiority, and solidarity was apparent during the focus group discussions. The men were adamant about the importance of men in cheerleading, marginalizing female participation. The tenor of the conversation became tense and defensive as the men proclaimed their dominance in the sport. One male stated, "Males started the sport of cheerleading in the first place." This comment was followed by agreement from other respondents, "Yeah, they wouldn't let women do it, so we started it. We're just coming back." Males' reclaiming the sport requires marginalizing the women's status. Women's importance in the sport has been relegated to the margins because "men started the sport." Once this territoriality was established, the atmosphere became much more comfortable. The male cheerleaders had asserted their authority and defended their masculinity.

In the all-female focus group, the female cheerleaders acknowledged that cheerleading started as a male-dominated sport. One female said, "Cheerleading started with men. I think the media has turned it into a sexualized female sport. That's what people like to see." In so saying, the women recognize the secondary, sexualized status of women in cheering. At the same time, the women are invested in helping the men to protect and maintain their masculine guise. The women's willingness to elevate men into a superior position within the sport demonstrates that gender maintenance occurs in social interactions, with both men and women cooperating. By subordinating their role as cheerleaders, the women "do femininity" as they demonstrate submissiveness to the men who cheer. The women also help the males "do masculinity" by assisting in the construction of cheerleading as a masculine sport through the emphasis placed on aggression and injury.

Masculine Aspects of the Sport: Toughness and Aggression   The male cheerleaders used imagery and examples from their cheerleading lives to demonstrate their heterosexuality and establish their masculinity in much the same way that people working in other gendered or sexualized arenas might draw on aspects of those scenarios to validate their claims to membership. Through these masculine examples, the men saved face, or protected their masculine identities. Specifically, some of the men referred to fighting, or the possibility of fighting, with individuals who made derogatory comments to them about participation in cheerleading. These men said that no one would have the nerve to say anything offensive for fear of being injured for their remarks. One male stated, "I don't think that anyone's ever had the balls to say anything negative to me," implying that there would be consequences for anyone who voiced negative comments. Another respondent demonstrated his masculinity by stating, "Me and Bob fought the football team one year. I don't think that had anything to do with cheerleading though. We kicked their asses." Demonstrating physical prowess and the ability to defeat the most masculine of sports teams, the two cheerleaders not only saved face, but also were elevated in status. In a one-on-one interview, Adam alluded to this incident with the football team, saying, "I've heard of cheerleaders taking on the football team and winning before. I wasn't a part of that, but I did hear about it." Again, he uses the incident to demonstrate that male cheerleaders should not be messed with and that they are able to protect themselves through violence if necessary. Because he had heard of the incident, it is apparent that the story had been talked about within the cheerleading group, serving to reinforce the idea that the men who cheer can take care of themselves physically. This storytelling demonstrates how the men collectively maintain their masculinity in a female-dominated activity.

During the female focus group discussion, this masculine facade was also discussed. One female commented, "They always say, 'They won't mess with me, I'm a cheerleader.'" This comment and the aforementioned ones illustrate how the men are creating and exuding a collective identity in which male cheerleaders are defined as dominant and aggressive. The statements cited demonstrate how the male cheerleaders maintain their masculinity by emphasizing their ability to defend themselves in physical altercations with other men. The use of violence and aggression are common characteristics utilized in maintaining masculinity for men.

The emphasis on toughness and aggression is also apparent when the male and female cheerleaders discuss the men's role as cheerleaders. Male-dominated sports, such as football, wrestling, and hockey, are often labeled as "real" sports because they contain elements of violence and an increased likelihood of injury. In order to equate cheerleading with these "real" sports, Leeann talks about the men's injuries and the need for strength to masculinize cheerleading, showing that cheerleading is not just a "girl's sport":

> You're always going to have people that think it's just a girl's sport. But, as people start to see more men doing it and once people see what they do and see all of these bloody noses and broken mouths, then they realize that it's not a girl's sport. It's tough for them. They do a lot of lifting and stuff that requires a lot of strength.

Similarly, Mary states, "I think guys like cheerleading because they get the satisfaction of knowing that they can lift two girls at once. It takes a lot of strength." Cindy also acknowledges the difficulty of the sport: "It's a lot more physical than you would think." Adam agrees with the women, stating, "I don't think that people realize how hard it is." Brett also emphasizes the difficulty of cheerleading for guys. He says, "It's physical. I'm tired [when I cheer]."

The male and female cheerleaders define cheerleading as a sport because it requires strength, manual labor, and physical exertion and it involves competition, much like sports defined as masculine (e.g., football, wrestling). The male cheerleaders act tough and aggressive to prove their masculinity, which then confirms their heterosexuality. The link between toughness, masculinity, and heterosexuality for these men is woven throughout their face-saving strategies.

Sexual Objectification of Women   The use of sexual objectification of women to "save face" is one of the most important findings that demonstrate how gender inequalities on a macrolevel are replicated on a microlevel through denigration of the subordinate sex. We see this happen in other areas besides sports. There are many arenas of gendered work in which men advance more quickly than women in female-dominated occupations. Male cheerleading demonstrates all aspects of gender: gender on the individual level (i.e., socialization into gender roles through various agents of socialization), gender on the interactional level (i.e., "doing gender" through male–female interactions that demonstrate power dynamics), and gender on the structural level (i.e., replication of gendered relationships on a microlevel that are seen on a macrolevel in patriarchal societies). Patriarchal societies are built upon the premise of compulsory heterosexuality: Everyone is straight until proven gay. As an illustration, this attitude can be seen when lesbian women attend sex toy parties: The assumption is that they are there to please their man, that they are heterosexual when, in fact, they are homosexual and are purchasing toys to enjoy with their female companions or are there simply to support their friend who is throwing the party. "Straight until proven gay" holds true until someone enters a nontraditional gender realm such as cheerleading. Men who cheer are faced with immediate accusations of homosexuality because of their participation in a female-dominated sport. Given the emphasis placed upon compulsory heterosexuality in our society, they act in a hypermasculine manner to counter such attacks, even when they may in fact be homosexual.

Connell (1992) asserts that hypermasculinity becomes manifest when hostility exists toward homosexual men and heterosexual men attempt to create social distance. In cheerleading, the men do this in large part by sexualizing the female cheerleaders. The women indicated this sexual objectification in both the focus group discussions and the one-on-one interviews. One participant stated, "People say the guys are girly. Not at all … they are probably more manly than guys who don't participate in the sport. They're perverted and sexual sometimes." In agreement with this female, another participant stated, "The male cheerleaders are the most heterosexual males I have ever met, so people have the wrong perception when they say they are gay. It makes me laugh. They are the most perverted guys that I have ever met." By "perverted," the women mean that the men talk about the women's bodies among themselves and to the women. For example, one female focus group respondent stated, "They say, like, 'Oh, people think I'm gay, but I get to grab your butt.' They're perverts." Another cheerleader in the female focus group said, "They always think all the girls want them. They'll say, 'She wants me.'"

The female focus group participants agreed that, when a group of men and women work closely, sexual innuendos occur and sexual tensions build. Cindy provides an explanation of why the male cheerleaders sexualize the females:

> When the guys are accused of being gay, I think they try to overcompensate for these accusations by being all about the girls. They act very sexual toward girls. I think that the guys look at us as sex objects, but all guys would have that attraction to the women. But they've never made me or any of the other girls feel uncomfortable. I'm sure they find us attractive, but it's kind of more like a sisterly–brotherly thing.

Cindy's statement is, of course, contradictory, asserting that the men are sexually attracted to the women but that the relationship is sisterly–brotherly. This confusion continues as another participant also comments on the attraction between the men and women, as well as the familylike relationship between the male and female cheerleaders:

> At first, I think that the girls and guys are attracted to each other. It's only natural when you have guys and girls together in a sport. You are constantly touching and it is sexual at first, but you get over that. It becomes like kissing your grandma.

The contradiction between sexual and familial relationships demonstrated by the females reveals the limited frame of reference available for mixed gender settings. The women construct the men as both "brotherly" and "perverted." Both instances provide defenses for the men's sexualization of the women. In the first, the women argue that the men's comments are often misconstrued as perverted when, in actuality, they are simply jokes and that their touching is more like that of a brother than a boyfriend. With this claim, the women become accepting of the men's antics. In the second instance, the women do claim that the men are perverted. However, the women defend the men's comments and actions as "boys being boys." In essence, the women normalize the perverse comments and actions by inadvertently stating that this sexual banter is part of the men's attempts to demonstrate their masculinity. By defending the men's actions, the women allow the men to display their gender, thereby negating perceptions that male cheerleaders are gay.

Both the men and the women see the objectification of the women as a component of being a man or being masculine within society. While the women demonstrate contradictions regarding their relationships with the male cheerleaders, the men emphasize a sexualized relationship with the women as opposed to a familial relationship. When asked why they chose to cheer, many of the men asserted that it was because of the presence of the beautiful girls and because they get to have intimate physical contact with the women. Brett said,

> I participate in cheerleading because of the girls. Most of the cheerleaders are hot all around the board. I would love to go to a cheerleading competition. There are tons of girls there. That's the only reason that I'm a cheerleader.

George sees the female cheerleaders as sex objects because all of the women are in good physical shape. George states, "If they were ugly cheerleaders, I probably wouldn't cheer."

In the male focus group discussion, the men continuously talked about the fact that they get to touch the female cheerleaders. The males asserted that one of the best perks of being a male cheerleader is the closeness to the females' bodies. The "hot chicks" that stand beside them are seen as prizes and a major reason for participating in the sport of cheerleading. One male even stated that people who might make allegations of homosexuality do not realize that they get to "touch the girls' butts and stuff." Another respondent stated, "I worked at a trucking company for 3 years, and all the guys loved it [that he was a male cheerleader]. They would come over and ask me stories and stuff, like, what do you do with these girls?"

The construction of masculinity through the objectification of women is applied when men are in the minority, engaged in a female-dominated activity. The route to power in this situation is the use of sex and sexuality to realize gender relations. By talking about these women as sex objects, they are subordinating women within the sport. The men were so excited about touching the women that they even suggested that more men might be interested in cheering if they were aware of how beautiful the women are and how much time they would get to spend with them, talking to them and touching them. One male specifically suggested that the athletic department needs to "sell the female cheerleaders" in order to recruit more men.

The objectification of women provides the male cheerleaders with a mechanism for asserting their heterosexuality and masculinity to the noncheering world in the context of cheering. In so doing, the male cheerleaders are maintaining a socially constructed ideal of manliness through the objectification of women—an ideal that confirms their heterosexuality and masculinity to themselves while also reinforcing hegemonic, dominating, sexualized masculinity for public consumption.

CONCLUSION: CROSSING THE GREAT GENDER DIVIDE

The aim of this research was to explore male participation in a female-dominated sport by focusing on men who cheer. The findings highlight the methods used by both male and female cheerleaders to redefine men's roles and activities within cheerleading in ways that stabilize and reassert hegemonic masculinity. Both the male and female cheerleaders articulated the stigma that goes hand in hand with participation as a male cheerleader.

In an attempt to masculinize cheerleading, the male and female cheerleaders focused attention on the men's strength and toughness. The male cheerleaders asserted their dominance through claims that cheerleading started as a male sport and that the women "need" the men in order to perform stunts. In addition, the males used objectification of the women to pronounce their heterosexuality and counter the stigma attached to men engaged in "women's" activities.

Claiming territoriality over a female domain happens in other female-dominated areas besides sports, because in a patriarchal society maleness is valued over femaleness, allowing men to take over organizations that were previously identified as female oriented. For example, Williams (1989, 1995) found that men in female-dominated occupations were paid better and promoted more quickly than females in the organizations in which they worked. Some of these men, however, were perceived as homosexuals when outsiders observed their participation in a female-dominated occupation. Because of these perceptions, the men adhered to a macho image and distanced themselves from the women in the organization by participating in male-identified specialties, emphasizing the masculine nature of their jobs, and by pursuing higher administration jobs.

Similar to Williams' findings is the finding here that men who cheer also experience accusations of homosexuality because of their participation in a female-dominated sport. Contrary to Williams' findings, however, the men who cheer did not distance themselves from their female counterparts. Instead, to prove their masculinity, these men physically objectified the female cheerleaders. This sexual objectification may be unique to the sport of cheerleading because of the close physical contact that is essential within the sport. Men and women are expected to perform stunts that require physical contact, and this physical contact is used by the men to demonstrate their heterosexuality.

The findings presented in this paper demonstrate that both male and female cheerleaders redefine cheerleading to emphasize masculinity and subordinate femininity. The tactics used to re-create and maintain masculinity for the male cheerleaders demonstrate the importance of maintaining male dominance over women and gay men. The cheerleaders' adherence to male dominance and male power within cheerleading reinforces the maintenance of a masculinity that emphasizes strength, skill, aggression, and competition, as well as heterosexuality.

REFERENCES

Connell, Robert W. (1992). "A Very Straight Gay: Masculinity, Homosexual Experience, and the Dynamics of Gender." *American Sociological Review* 57: 735–751.

Dellinger, Kirsten. (2004). "Masculinities in 'Safe' and 'Embattled' Organizations: Accounting for Pornographic and Feminist Magazines." *Gender & Society* 18: 545–566.

Goffman, Erving. (1963). *Stigma: Notes on the Management of Spoiled Identity.* Englewood Cliffs, NJ: Prentice Hall.

———— (1959). *The Presentation of Self in Everyday Life.* Garden City: Doubleday.

Williams, Christine. (1995). *Still a Man's World: Men Who Do 'Women's Work.'"* Berkeley, CA: University of California Press.

———— (1989). *Gender Differences at Work: Women and Men in Nontraditional Occupations.* Berkeley, CA: University of California Press.

30

Moral Stigma Management
Among the Transabled

JENNY L. DAVIS

Some forms of deviance are so different that they challenge even those who customarily think of themselves as nonjudgmental. The Internet has enabled even extreme groups of people to find others like themselves who can help each other find legitimacy and acceptance. Davis's research on people with body integrity identity disorder (BIID) documents the stigma cast upon those who see themselves as having a disjuncture between how they see themselves and how they appear externally. Although they are able bodied, they want and/or need to live in a body that is physically impaired. This chapter touches on the ways people deal with this disjuncture and how they manage the stigma of wanting or claiming a deviant physical status that they do not physically have.

To what do you attribute the extreme reactions voiced by readers of the blog Transabled.org to the transabled people who post on it? How would you assess their claims? To what extent are they trying to manage the impressions others have about them by downplaying or concealing the sexual dimension of their disorder? By lying or omitting significant other details of their conditions? How would you assess these people's claims to legitimacy in terms of the three S's of stigma attribution? How do people in our society generally feel about physical versus mental disabilities? Voluntary versus involuntary physical impairment? The claims transabled people have for assistance as individuals with special needs? Should we as a society feel responsible for accommodating them? How do you feel about this condition compared with other mind–body disjunctures, such as gender identity disorder or body dysmorphic disorder?

Do you find the stigma management claims of the transabled effective? Why or why not? What are the most successful aspects of their claims, and what are the least successful? How do those claims relate to the claims made in the accounts of rapists and shoplifters presented in Chapters 27 and 28?

I just know that ever since I was a little kid, the most comforting sensation I am capable of experiencing involves having my neck encased

Jenny L. Davis (2014). Morality Work among the Transabled, *Deviant Behavior*, 35:6, 433–455.

and immobilized. Some people have a favorite singer, or a movie, or a place they like to go when they need to be comforted... . I need to be unable to move my head (and preferably the rest of my body as well) (Sarah—blogger on Transabled.org).

Transabled.org is a website and interaction forum for people with body integrity identity disorder (BIID). BIID is a condition of incorrectly abled embodiment. Individuals with BIID are born able bodied, but want or need to live in a body that is physically impaired.[1] They experience a painful schism between a physically able body and a disabled self-image (e.g., amputated limbs, paraplegia, blindness, deafness, etc.) (First and Fisher, 2012; First, 2005). Perhaps being transabled is best described on the site's home page by *Sean*, the founder of *Transabled.org*:

> So you'll ask: "*That 'thing', transabled, just exactly what is it?*" It is hard to define in just a few words, the best way to learn is by going through the site, but in a nutshell, someone who is transabled "wants" to be disabled. But it is not so much a "*want*" as much as a "*need.*" Our "*desire*" is more a reflection of the fact that our self-image is that of a paraplegic (or amputee, or blind, or any number of other disabilities) [rather] than that of an able bodied man or woman (*Sean* http://transabled.org/) (emphasis in original).

I demonstrate stigma management by using data from an extensive qualitative analysis of 17 years (1994–2011) of blog posts, archived content, and links to and from *Transabled.org*.

OVERVIEW AND HISTORY OF BIID

In 1977, John Money and colleagues wrote the first article about the need for physical impairment (Money, Jobaris, and Furth, 1977). From two related case studies, Money and colleagues developed two key diagnostic terms: apotemnophilia and acrotomophilia. The former refers to those who wish to have their own limbs amputated and who fantasize sexually about themselves as amputees. The latter refers to those who require amputee partners (real or imagined) in order to experience sexual satisfaction.

Money and colleagues' germinal article did two key things: First, it defined the need for physical impairment as amputee specific; second, it defined the need for physical impairment as a sexual pathology. The focus on amputation remains strong within the literature. Despite a large and long-standing presence of nonamputee BIID sufferers online, only recently have nonamputee manifestations been professionally acknowledged. Sexuality, however, has received more critical attention from the onset, with scholars and doctors offering opposing perspectives.

Most recently, the condition has been defined primarily as one of identity incongruence. Michael First (2005) introduced the identity model, using data from his 2005 survey of people with the professed need for amputation.

His was the first piece of research on the topic to go beyond individual case studies ($n = 52$). The majority of participants rooted their need for impairment in a quest to "restore their bodies to be in line with their 'true-self' identities" (2005:22). First coined the term "body integrity identity disorder" and explicitly connected it to gender identity disorder (GID)—a condition in which an individual feels that she or he was born in an incorrectly sexed body (First, 2005). Today, this terminology and the meanings embedded within it dominate the literature on the need for impairment. However, debates still remain as scholars and doctors work to make sense of this emergent condition.

In addition to the formal terminology discussed in connection with Money and colleagues and with First, members of the BIID community use a self-created set of lay terminology. Those who profess a need for physical impairment are often called "wannabes" or "need-to-bes." Those who enact their desired embodiment (e.g., by using crutches, a wheelchair, braces, opaque contacts, etc.) are called "pretenders." Those who experience fetishistic attraction toward the physically impaired bodies of others are called "devotees." The boundaries between these three designations (wannabe, pretender, and devotee) are quite permeable, and none of the terms are mutually exclusive with any other of them. Many individuals who fall into one category also fall into at least one, if not both, of the others. Note, however, that wannabes and pretenders are more likely to overlap with each other than with devotees (Elliott, 2003; First, 2005).

Encompassing wannabes, pretenders, and, sometimes, devotees, is the term "transabled." *Sean,* the founder of *Transabled.org,* was the first to coin the term. He uses it as an informal alternative to BIID, one that is personal and inclusive. Inclusivity is of particular importance, in that being transabled necessarily includes the need for *all kinds* of physical impairments, combating the heretofore emphasis on amputation. The term has caught on and is used not only on *Transabled.org,* but also in other BIID-related interaction forums online.

As with many marginalized groups, members of the transabled community hotly debate the language with which they describe themselves. In particular, debates surround the use of the terms "wannabe" and "pretender." Those who reject these terms argue that they connote a lack of authenticity and thereby stand in direct opposition to the notion that impaired embodiment is, for them, *the* authentic way to live. Through these debates, members of *Transabled.org* came up with the term "persons with BIID" (PWBs) to describe themselves. This terminology seems to avoid the divisive connotations of "wannabe" and "pretender." Accordingly, throughout the remainder of the work, I will refer to people with the need for physical impairment as either transabled or PWBs.

MORALITY WORK: A THEORETICAL FRAMEWORK

Goffman (1963: 3) canonically defined stigmatization as that which is "deeply discrediting" and reduces the subject "from a whole and usual person to a tainted, discounted one." Stigmatization can be defined more broadly as incorporating labeling, stereotyping, separation, loss of status, and discrimination.

This encompassing conceptualization of stigma entails not only difference, but devaluation, tangible consequences, and power deficits of the stigmatized subject relative to stigmatizing agents.

PWBs fit the status of stigmatized subjects. They desire, and sometimes enact or even pursue, physical impairment, a socially devalued bodily state. They are not recognized by medical authorities, and they risk losing jobs, friends, families, and even freedom (through institutionalization) if their desires are exposed. Moreover, their condition represents a *moral* stigma, as those with BIID are often accused of "not trying hard enough" to rid themselves of their abnormal desires.

Moral stigmatization is charged against those who seem to hold a degree of control over their stigmatizing attribute(s). For example, physical disabilities and mental illness are both stigmatized statuses, but only the latter is a *moral* stigma. Although all stigmatized persons experience negative effects from their stigmatization, those who hold moral stigmas are most at risk of internalizing negative self-evaluations and suffering emotional consequences.

Recent empirical research, however, tempers this claim, showing that, under some conditions, those with moral stigmas can actively resist negative moralizing judgments and maintain a positive sense of self. I refer to this process of remoralization as *morality work*. *Transabled.org* is a fruitful site for studying the remoralization process.

METHODS

Data for the present work come from a qualitative analysis of 17 years of blog posts, comments, content, and links to and from *Transabled.org*, a publicly accessible, nonpassword-protected website centered around the experiences of bloggers with BIID. The site was founded by *Sean* in 1994 as a personal blog, on which he talked about his own need for below-the-waist paralysis. In 2005, *Sean* reconstructed the site into its current form, with an expanded list of blog authors. The site remained active until 2013, although my formal analysis for the present work ended in 2011. Blog authors write about various experiences with BIID, offer tips and suggestions, and ask for advice. All posts are open for public comment, and all site content was archived, making the space particularly fruitful for research.

The location of this community in an online space is theoretically and empirically significant. The potential for anonymity and geographic transcendence makes computer-mediated communication ideal for the exploration and enactment of marginalized and secretly held identities. Further, the status of the site as a public arena and its links to and from other sites (e.g., www.biid-info.org, the Wikipedia entry for BIID, and several Yahoo chat groups) speak to the purpose of the community: *Transabled.org* provided a space for people to talk about their experiences with BIID and for outsiders to learn of, become educated about, and engage with, being transabled. In short, the site not only connected transabled people, but also shared their individual and collective voice(s), offering a safe space in which transability could be personally and publicly negotiated as a way to be in the world, even while community members remain anonymous.

In collecting data, I read each of the blog posts and all of the comments written throughout the existence of the site (1994–2011), a total of approximately 2,900 pages of text. Although this is an extensive data set in its own right, Internet research is necessarily multisited. Consequently, I followed all hyperlinks to outside sites included in the archived content. I also Googled the terms "transabled," "*Transabled.org*," "BIID," "body integrity identity disorder," "apotemnophilia," and "amputee identity disorder," reading and analyzing all sites that linked back, directly or indirectly, to *Transabled.org*.

One of the major themes that emerged was that of morality. The content from this category is the focus of the present work. Specifically, I look at four types of interaction: (1) the moral accusations made by those who come to the site expressing disagreement with and/or hostility toward BIID and PWBs. (2) The moral accusations made against BIID and PWBs on outside sites, linked either directly or indirectly to *Transabled.org*. (3) The responses by PWBs to the aforementioned moral accusations. (4) Preemptive or hypothetical responses by PWBs to potential moral accusations (e.g., "some people say that we are being dishonest when we use a wheelchair….").

STIGMA MANAGEMENT STRATEGIES

I begin with descriptive accounts of moral stigmatization and remoralization practices among PWBs. The first part of this section lays the empirical groundwork for the second part, in which participants rebut accusations and offer "accounts" to legitimate their beliefs and actions.

Moral Accusations

Moral stigmas are as idiosyncratic as they are numerous. As such, each moral stigma brings with it specific accusations of moral failings. Detractors aimed the following attributions of immorality at PWBs: (1) sexual perversion, (2) emotional weakness, (3) dishonesty, and (4) greed. Detractors articulate these broad failings via specific moral accusations. I discuss and illustrate each of these (im)morality claims in turn, pulling from comments written by visitors to *Transabled.org*, as well as from outside websites and articles linked via *Transabled.org*. Although, in practice, these moral claims often overlap and exist in conjunction with one another, I separate them here for purposes of clarity.

The first moral accusation is that of *sexual perversion*. As mentioned earlier, Money and colleagues (1977) were the first to introduce the notion of the need for impairment and so were the first to coin the term "apotemnophilia," categorizing such needs into the class of sexual disorders known as "paraphilia." The need for physical impairment was therefore grouped with disorders as disparate as necrophilia (the desire for sex with dead bodies), pedophilia (the desire for sex with children), and others commonly categorized as perversions.

This sexual pathology model continues to have an effect and is a means by which detractors deem PWBs immoral. *Miska* (2009), the moderator of a feminist website (*fabmatters.com*), writes a critical post about transability, pondering its fetishistic qualities and providing a link to *Transabled.org*. The following are comments written by *fabmatters* readers in response to the article:

> I followed your link to *Transabled.org* … maybe after I stop emotionally puking I will come back and respond to it more in-depth. For now, I would point to the very disturbing way … this person [is] essentially describing a BDSM [bondage, discipline, sadism, and masochism]…" consensual encounter" with a surgeon, in order … to "achieve" her desired deafness. (oh crap, I just emotionally vomited again) *(Factcheckme)*.
>
> This is beyond fucked up. I'm not going to apologize for being extremely pissed off at the so-called "transabled" folk here…. That you fetishize the suffering of the disabled (and yes, this is fetishization) is just as fucked up beyond all regard *(Calliope)*.

A second moral accusation is that of **emotional weakness**—in particular, the need for attention. This accusation (along with that of sexual perversion) complicates the identity model. For example *Penny*, a nontransabled visitor to *Transabled.org*, writes the following:

> Sean, I have been reading your comments on a variety of forums including this site. I wanted to share some thoughts. I think it is about attention…. Disabled people get more attention, and people expect less of them and consider small achievements great successes when compared to the gauge associated with AB[able-bodied] persons … you want to be a sufferer … you want to have attention … you want to be special… you want to be a paraplegic as it is the easy way to ensure [that] you are always seen as special and different and a battler….

A third moral accusation is that of **dishonesty**. Outsiders often reprimand PWBs for falsely representing themselves. This accusation is aimed particularly at those who enact their desired embodiments by using wheelchairs, crutches, braces, etc. We can see it in the following comment left by a visitor to *Transabled.org*:

> Ugh. You are NOT disabled, and it is offensive that you think it is OK to pose as someone who needs a wheelchair (*michiru*—emphasis in original).

Similarly, when a D/deaf blogger brings up BIID on a deafness forum (*alldeaf.com*), commentators by and large decry the dishonest performances of those who enact impaired embodiment. They particularly express empathetic anger for those whom PWBs "deceive" with their performances. This anger is seen in the following exchange between poster *RedFox* (2007) and commenter *Morbid-Mongoose*: *RedFox* says, "I found another deaf forum with a thread about…a hoh [hard of hearing] pretender who made friends with a real hoh person. What would happen if the real one found that the other one is a pretender?" *Morbid-Mongoose* reposts this segment of *RedFox's* text and provides the following response: "If someone BSed [bullshitted] me like that, I'd be pissed. 🐭" So not only is dishonesty itself considered morally reprehensible, but the way in which this dishonesty affects an already disadvantaged group (i.e., the disability community) makes it doubly so.

The fourth moral charge against PWBs is that of **greed**. People with BIID are thought to desire a disproportionate amount of resources—be it physical care, government benefits, medical treatment, or convenient parking. Further, detractors accuse PWBs of unfairly using or wanting to use a set of *scarce* resources—resources that those with "real" disabilities rely upon.

For example, when *Sean* contributes to the disability forum *disaboom* (to which he provides a link on *Transabled.org*), he is met with the following comment:

> How about you tell this new group how fond you are of taking handicapped spaces/stalls that people who really are handicapped need to have?... be in a wheelchair all you want, but don't take advantage of services that are actually needed for handicapped people, because regardless of what you say, you are a horribly selfish person! *(Sharon).*

Through moral accusations of greed, detractors paint PWBs as consumers of more than their fair share of resources. Not only is this egregious consumption depicted as generally "unfair" but it purportedly undermines the rights and resources available to those with "real" need (i.e., people with nonvoluntary physical impairments).

Discursive Techniques of Remoralization

PWBs engage in their own morality work, responding to their accusers by relying on (1) medicalization/biologization and (2) claims to self-actualization/authenticity. The former works to neutralize (im)morality claims. The latter goes beyond neutralization and locates PWBs on a moral high ground.

Medicalization and Biologization   Moral stigmatization is characterized by stigmatized subjects' perceived control over their stigmatizing attribute(s). Thus, one way for stigmatized individuals and groups to neutralize claims of immorality is to redefine their condition as existing outside of personal control.

Medicalization and biologization, or the "biological drives" (Scott and Lyman, 1968) technique, are related methods by which morally stigmatized subjects can neutralize their moral denigration. These terms refer to the practice of defining a condition as biologically rooted and/or medically treatable. By reformulating BIID into a condition with genetic and/or neurological roots, one legitimized by the mental health profession, PWBs relocate their stigmatizing attribute outside of themselves—absolving themselves from blame.

The importance of medicalization is shown in many PWBs' explicit support for BIID's inclusion in the *Diagnostic and Statistical Manual of Mental Disorders* (*DSM*). As noted in footnote 1, at the time of this publication, BIID was not included in the *DSM-IV*, but was under consideration for inclusion in the *DSM-V*. The significance of inclusion is both symbolic and material, as it would provide particular identity meanings, but also potential resources and treatment protocols. *Sean* articulates his support: "Having BIID documented would…ensure us a certain level of legitimacy. Too many people are saying we're "just sick." Being able to point to the *DSM* and say "yes, we are…it is a REAL condition" would be very helpful."

In this vein, many bloggers and participants at *Transabled.org* define BIID as "just another disability." This phrase was first used by *Sean* to articulate the debilitating effects of BIID—linking it to the debilitating effects of physical impairment—and has become a common discursive tool among PWBs.

As sufferers of a contested illness, PWBs struggle to legitimate their medical claims. In an era of biomedical advancement, society expects illnesses to have visible and measurable effects. When symptoms are not connected with bodily abnormality, the experience is largely negated—or contested. People with contested illnesses therefore have a stake in finding biomedical abnormalities associated with their symptoms.

By decoupling PWBs from control over their condition, medicalization and biologization allow PWBs to confront and neutralize the explicit claims of immorality charged against them. In combating claims of dishonesty, transabled bloggers reject the notion that their use of assistive devices is deceitful, arguing instead that it is simply a means of treating their disabilities with the available technology. *Chloe* articulates this clearly when she justifies her use of a wheelchair and braces by saying, "I am treating BIID, not paraplegia." Using this same logic, *Elizabeth* says,

> I am a transabled wheelchair user. I never pretend to be a paraplegic, I never told anybody I was one....The use of a wheelchair depends on my state of mind. Sometimes I need a high dosage, I need to wheel all the time. Other times, walking is fine with me.... But I am the one who makes the medical choice.... Because after all wheeling is my medical choice, it's not a lifestyle choice.

Bloggers often discuss the issue of honesty when pondering the dilemma of what to say when asked by others about their use of assistive devices. They discursively maintain a sense of honesty by explaining their use of these devices as treatment for a "neurological condition" and leaving it at that. *Sasha,* for example, gives the following anecdote from an experience in which she was asked about her wheelchair:

> I told...somebody that I have a "very rare neurological disorder that is not progressive but is incurable, can be extremely painful, and do you have any more questions?" It was actually the most truthful statement about my BIID.

In combating the moral accusation of greed, PWBs' professed membership in the "disability community" allows them to argue that they do not use a disproportionate share of resources, but rather utilize only that which their medical condition requires. *Sean* makes the following argument in defense of his disability parking placard:

> I use a wheelchair all day, every day... I require a wider parking space so I can get my wheelchair in and out of the car. I need the disability space because without it, I cannot use my wheelchair. And I need my wheelchair to function....What it really comes down to is a question of validity: Is BIID a valid condition or not?... How do we make people understand that BIID is indeed a valid condition?

Finally, in combating claims of sexual perversion, transabled bloggers use not only the legitimacy of medicalization, but also the particular meanings associated with the medical diagnosis of having a neurological condition that results in mind–body incongruence. Specifically, they emphasize the language and meanings of "body integrity identity disorder" over that of "apotemnophilia."

PWBs often point to the use of BIID in the literature and as a potential diagnostic category in the *DSM* to promote the need for impairment as an identity-based condition while rejecting the etiology of sexual pathology. Exemplifying the importance of this distinction for PWBs is *Kyla's* response to a visitor on *Transabled.org* who used the term "apotemnophilia":

> While I see nothing inherently wrong with someone who has peculiar sexual fetishes, and will defend their right to have them, I don't see clouding the issue of transability with fetishism as doing any good for anyone....The point is that when there is a condition associated with sexual fetish ('paraphilia'), it gives society justification (rightly or not) to marginalize us and discriminate against us—and to ignore the issue that needs to be addressed. The condition of being transabled with the need to have an amputation was long ago called 'apotemnophilia,' which is classified as an unhealthy paraphilic sexual fetish...the term BIID has been coined to distinguish us from those for whom the desire is sexual in nature.

Kyla's comment not only articulates the "correct" way to make sense of transability (i.e., as a disorder of identity, *not* sexuality) but also points to the importance of doing so in terms of remoralization and resistance to stigma. Specifically, she points out the danger of the "paraphilia" label (i.e., *when...a condition is associated with sexual fetish...it gives society justification...to marginalize and discriminate against us*).

By broadly placing their desires (and related actions) in a "natural" rather than a "social" frame, PWBs are able to combat a host of specific moral accusations. Their use of assistive devices is not dishonest, because they use these devices to treat BIID—a real disability. Their desire for others to treat them as persons with disabilities does not signify emotional weakness, but is instead a form of treatment for mind–body incongruence. They take no more resources than their disability requires of them; and they have BIID, a condition of incorrect embodiment, *not* apotemnophilia, a condition of sexual pathology.

Authenticity and Self Actualization: The Moral High Ground In the last 150–200 years, the Western world has come to view authenticity as a sacred moral value. As bioethicist Carl Elliott eloquently states,

> The ideal of authenticity says that if you are not living life as *yourself* ... you are squandering your short time on this earth...this is not simply the sense that an authentic life is a happier life; it is the sense that an authentic life is a *higher* life" (2003:39 emphasis in original).

Therefore, to live a "good" and "righteous" life is to live a fulfilling life. To live a fulfilling life is to be true to the self. Not surprisingly, research has long shown the use of "self-fulfillment" (Scott and Lyman, 1968) as a discursive remoralization technique. Following this tradition, transabled bloggers claim not only that they have a *right* to inhabit physically impaired bodies, but have a moral *duty* to do so. The following excerpts from transabled bloggers exemplify this claim:

> When you embrace your BIID and start living it, you disable your body some, you get physical limitations but you free your soul and spirit *(Elisabeth)*.
>
> If we believe that our true selves are disabled, then is it not the bigger lie, and therefore the bigger sin, to present ourselves as able bodied? *(Chloe)*.

This discourse of authenticity renders claims of sexual perversion irrelevant, turns the greedy use of resources into the essential use of tools for self-fulfillment, reworks the need for attention into a need for self-verification, and justifies lying as a necessary evil in the sacred quest for an authentic and fulfilling life.

As noted previously, PWBs actively police the use of medical terminology, rejecting "apotemnophilia" in favor of "BIID." Such language effectively relocates the discussion out of the realm of sexuality entirely and into the realm of identity and, in turn, authenticity. The language of authenticity paints the need for physical impairment not as a sexual fetish to be gratified, but as a destiny to be fulfilled. This sentiment is exemplified by *Chloe's* response to a visitor who claims that BIID is "clearly" sexual in nature. *Chloe* dismisses the claim as laughable and cites the physiological sexual limitations that she will encounter as a paraplegic: "Oh yes, REALLY clear. That's exactly why I need to have no genital sensation and be anorgasmic. It's for the sex!!! Thanks for giving me a good laugh though" (emphasis in original). As *Chloe* (sarcastically) points out, it is about the self. It is not about sex.

It is also, as discussed earlier, not about greed. By medicalizing the need for physical impairment, the use of assistive devices and societal resources is defined as just that: a morally neutral use of medical equipment and services for the treatment of a (neuropsychological) disability. Upping the morality ante, the language of authenticity redefines these resources as essential tools for achieving self-fulfillment. *Sean* and *Claire,* who both wheel (almost) full time and have obtained legal disability parking placards, describe the existential strife they would feel if they were unable to utilize this particular resource:

> Here is the deal with my placard. Without it, I am a quivering mess of fear and anxiety… With it I am much more calm and stable… *(Claire)*.
>
> It has been suggested by some people with disabilities that as I am physically able to walk, I should park in a regular space, and walk to the trunk where I should stow my wheelchair, then sit in the chair…The thing is, I could physically do that, yes. Emotionally, it would rip me apart. The mere idea of doing that makes me shaky *(Sean)*.

Finally, the language of authenticity works to remoralize PWBs in the face of accusations of dishonesty. Bloggers on the site recognize that, at times, they tell only partial truths and that sometimes, they must straight-out lie. They utilize the moral ethic of authenticity, however, to reframe lying as a small sacrifice to make in fulfilling the larger moral purpose of living a good and true life within a hostile society. In short, bloggers morally privilege being true to the self over being truthful with others. In the following passages, *Sean* articulates this complex and nuanced treatment of honesty:

> I say often enough that I don't pretend to use a wheelchair. I am a wheelchair user. Period. I spend nearly 100% of my public life as a wheelchair user. I do, however, pretend. I pretend to have a physical impairment. The reason I pretend is that I have long felt the need to have that physical impairment. I am pretending (lying) because society at large is not ready to accept me as an individual with BIID. I am telling a lie to live my own truth *(Sean)*.

In sum, the language of authenticity more than neutralizes claims of immorality. It places PWBs on a higher path of self-discovery, self-fulfillment, and the "good" life. This morally stigmatizing trait is not something to be fought against (such a fight is self-destructive) but is to be embraced and accepted by oneself and others. Indeed, under the Western ethic of authenticity, those who *prevent* PWBs from realizing their desires could very well be the true culprits of immorality. In a time and place in which morality is found by looking inward and morality is achieved by following what one finds inside, transabled individuals can ethically do no better than to enact, or even pursue surgically, impaired embodiment.

CONCLUSION

Moral identity is not a binary state (moral/immoral) but instead operates along a continuum (more moral/less moral). The morally accused begin at a relative moral deficit. Morality work is a means by which to rectify this deficit, relocating the morally accused higher on the moral continuum and potentially relocating the accuser to a lower moral status as well. In this vein, PWBs neutralize, rise above, and flip the moral script in response to moral accusations.

PWBs employ medicalization and biologization discourses to pull themselves out of a moral deficit. By locating impairment desires within the material of the body, they externalize the blame for these desires and divorce themselves from control. In doing so, they decouple the stigmatizing trait from moralizing judgments, moving their need for impairment from a morally reprehensible to a morally neutral way of being in the world.

Claims of authenticity effectively move PWBs further along the morality continuum, placing bloggers on a moral high ground. By appealing to the value of inner truth, PWBs make abnormal embodiment not only acceptable, but a righteous pursuit. With this logic, PWBs not only heighten their own moral standing, but also flip the moral script on those who stand in the way of such a pursuit and relocate their accusers into a position of relative moral deficit, to become objects of moral derision.

In addition, morality work operates simultaneously at the interpersonal and institutional levels. Interpersonally, the morally accused come together to validate experiential claims, construct a narrative with which to talk about themselves and their collective condition, and combat those who rail against them. At the institutional level, the morally accused work toward official recognition, inclusion, protection, and/or entitlements.

We see interpersonal moral labor as PWBs comment supportively on each other's posts, enveloping each other in shared experiences and providing a safe space in which to articulate these experiences. This support is illustrated further as newcomers adopt the moral discourse of established bloggers, who validate the newcomers' embodied experiences and welcome the newcomers into the community through mutually verifying exchanges. Indeed, new bloggers often report "finding themselves" in the words of others, and established bloggers reaffirm identity meanings through newcomers' fresh articulations. In this way, bloggers simultaneously produce and consume their individual and collective selves into being.

Morality work, however, goes beyond the interpersonal level, as the morally accused strive for material resources. Indeed, this striving is the drive behind the fight for the inclusion of BIID in the *DSM* and the related access to insurance coverage, disability services, and protocols for ability reassignment surgery. As Giddens (1991) points out, the contemporary era is characterized by a tension between an increase in self-advocacy and a continued reliance on institutional structures. Stigmatization necessarily involves power differentials, and the morally stigmatized, though agentic, depend upon institutional channels of support. Morality work functions as potential entrée into these channels. For instance, one of the major arguments against ability reassignment surgery is the Hippocratic Oath, which dictates that doctors are to "do no harm." By formulating BIID as a medical condition, surgery can be construed as a treatment, rendering voluntary impairment a viable ethical procedure. Moreover, to assert that impaired embodiment is the only route to existential fulfillment and that failure to achieve correct embodiment will forever be a source of existential strife is to assert that a refusal to treat this condition of incorrect embodiment is, in fact, doing far greater harm than that caused by, say, a spinal cord transection.

NOTE

1. at the time of this publication, BIID was not included in the *DSM-IV*, but was under consideration for inclusion in the *DSM-V* (First and Fisher 2012).

REFERENCES

Elliott, Carl. (2003). *Better than Well: American Medicine Meets the American Dream.* New York: W.W. Norton and Company, Inc.

First, Michael. (2005). "Desire for Amputation of a Limb: Paraphilia, Psychosis, or a New Type of Identity

Disorder." *Psychological Medicine* 35(6): 919–928.

First, Michael, and Carl E. Fisher. (2012). "Body Integrity Identity Disorder: The Persistent Desire to Acquire a Physical Disability." *Psychopathology* 45(1): 3–14.

Giddens, Anthony. (1991). *Modernity and Self-Identity: Self and Society in the Late Modern Age*. Stanford, CA. Stanford University Press.

Goffman, E. (1963). *Stigma: Notes on the management of spoiled identity*. New York: Simon and Schuster, Inc.

Miska. (2009). "The Transabled." *Fabmatters*. Retrieved January 2010 (http://fabmatters.wordpress.com/2009/11/13/the-transabled/).

Money, John, Russell Jobaris, and Gregg Furth. (1977). "Apotemnophilia: Two Cases of Self-Demand Amputation as a Paraphilia." *The Journal of Sex Research* 13(1): 115–125.

RedFox. (2007). "Something Shocking and Creepy, Deaf Wannabes, Pretenders and Others." *Alldeaf.com*. Retrieved December 2011 (http://www.alldeaf.com/topic-debates/45323-something-shocking-creepy-deaf-wannabes-pretenders-others.html).

Scott, Marvin B., and Stanford M. Lyman. (1968). "Accounts." *American Sociological Review* 33(1): 42–62.

31

Passing as Black: Identity Work Among Biracial Americans

NIKKI KHANNA AND CATHRYN JOHNSON

Over the last decade, we have become more aware of the multidimensional aspects of people's racial identities, highlighted by such celebrities as President Obama, Tiger Woods, Halle Berry, and others. With this new awareness, and reinforced by the 2010 U.S. Census finally incorporating a multiracial category, U.S. society has seen an increase in the number of multi- and biracial people living among us.

For many years, multi- and biracial people were classified according to the "one drop" rule, wherein those who had even one Black ancestor were viewed as Black. But more recently, as Khanna and Johnson document, the trend is changing, opening a wider range of options. This chapter discusses the conditions under which biracial Americans cast themselves as either White or Black and the reasons they do so, noting in particular the rise in optional Black self-labeling. The identity strategies that the people the authors studied employ are useful to consider beyond the notion of racial identity self-presentation. Like Bemiller, Khanna and Johnson go beyond the idea of simply denying a deviant identity (by passing for White) to examine the way people engage in covering up or diminishing the relevance of a known identity to accent an alternative, more acceptable identity by highlighting its existence and relevance to the situation and group. The authors' discussion of the various motivations for such "identity work" are likely to resonate with readers beyond the one context they present.

How would you assess the prevalence of the "not-me" identity strategy in relation to other deviant statuses? Can you think of other multiple identities in which people shift back and forth in their self-presentation? Do you think their strategies are effective? Are they moral?

My father has sixteen brothers and sisters and … . a lot of them used to pass as white … I mean it's easier if you can go to any movie theater you want. [A] few of my aunts told me about a place they used to go to and eat all the time that was "whites only" … they did it as a joke … they did it because they wanted to show how stupid [segregation] was. –Olivia, age 45

From Nikki Khanna and Cathryn Johnson, "Passing as Black Racial Identity Work among Biracial Americans", *Social Psychology Quarterly*, 73(4), 380–397. Copyright © 2010 by Sage Publication. Reprinted with permission by the publisher.

Until relatively recently, few racial options have been available to multiracial people—especially those with black ancestry. The one-drop rule, rooted in slavery and Jim Crow segregation, defined multiracial people with any drop of black blood as black. Just like their monoracial black counterparts, they had few, if any, rights (e.g., they were enslaved, they could not vote, they were restricted from many public facilities). According to Daniel (1992), "Multiracial individuals for the most part have accepted the racial status quo, and have identified themselves as Black. A significant number of individuals, however, have chosen the path of resistance … . Individual resistance has taken the form primarily of 'passing'" as white (91). Like Olivia's aunts (described above), many Americans passed as white to resist the racially restrictive one-drop rule and the racial status quo of the Jim Crow era.

Racial passing has generally been understood as a phenomenon in which people of one race identify and present themselves as another (usually white). According to Ramona Douglass, a multiracial activist and cofounder of the Association of Multiethnic Americans, however, the *concept* of passing (not the act itself) is racist in origin (Russell, Wilson, and Hall, 1992) because it is entwined with the racist one-drop rule. Even if people have white ancestry and look white, they are considered "really" black because of their black ancestry (no matter how distant); white identity is perceived as somehow "fraudulent." Kennedy (2003) provides a more precise definition and defines passing as "a deception that enables a person to adopt specific roles or identities from which he or she would otherwise be barred by *prevailing social standards*" (283; emphasis added). Thus, if people were "really" black, as defined by the social standards of the Jim Crow era (e.g., the one-drop rule) and presented themselves as white, they were perceived as deceiving the public with a false identity.

Passing as white was especially attractive during the Jim Crow era when blacks had few rights and opportunities, yet little is known about racial passing today. Some scholars argue that given the increase in opportunities to black Americans, passing is a relic of the past. While the driving force behind passing may have faded, we ask: *Are biracial people still passing today? If yes, how so and why?* This study also investigates the ways in which black–white people manage their racial identities in day-to-day interactions. *How much individual strategy is involved in racial identity today? And what types of strategies, other than passing, are used?*

Symbolic interactionists suggest that identity is process—society influences identities, yet individuals are also active agents in shaping their identities. We find that respondents use various strategies—verbal identification/disidentification, selective disclosure, manipulation of phenotype, highlighting/downplaying cultural symbols, and selective association. We delineate a typology that distinguishes passing from other types of identity work in order to create a more nuanced and fluid analysis of identity and the ways in which people manage their identities—respondents use these strategies to conceal aspects of their racial ancestry (i.e., to **pass** as monoracial), but when passing is not feasible or desired, they may **cover** (i.e., downplay) (Goffman, 1963) or **accent** (i.e., draw attention to) particular ancestries. We further extend research on identity work by identifying structural-level factors, such as social class and social networks, which limit the accessibility and/or effectiveness of some strategies of identity work (in addition to individual-level factors like phenotype).

Moreover, we find that the majority of these biracial respondents identify as biracial or multiracial, but occasionally pass as monoracial. While passing during the Jim Crow era involved passing as white, we find a striking reverse pattern of passing today—only a few respondents situationally pass as white, while the majority of respondents describe situations in which they pass as black. After describing the identity strategies that respondents use, we explore their motivations for identity work—with a focus on passing as black.

DATA AND METHODS

This paper is part of a larger study examining racial identity among black–white biracial adults. In 2005 and 2006, Khanna conducted semi-structured interviews with 40 black–white biracial adults living in a large urban area in the South. To participate in the study, respondents must have had one black and one white parent (as identified by respondents). Respondents were asked open-ended questions on a range of topics such as their racial identities, how others have influenced their identities, how their identities have changed over time and situation, and if and how they assert particular identities to others. Interviews were audio-taped and respondents' names were replaced with pseudonyms.

Because locating biracial individuals within the general population is often difficult, we primarily relied on convenience sampling. Khanna began recruiting respondents by placing flyers in a variety of places, including local colleges, universities, and places of worship. Flyers read, "Do you have one black parent and one white parent?" We omitted terms such as "biracial" or "multiracial" from the flyers, aware that individuals who did not consider themselves biracial or multiracial may not have responded. Khanna also asked interviewees to pass along her information to others with similar backgrounds.

Our data collection efforts resulted in a sample of 40 black–white biracial individuals. The ages ranged from 18 to 45, with the average age a little over 24 years of age. More than half of the respondents fell between the ages of 18 and 22, which is typical college age; this is not surprising considering that our recruitment efforts began at local colleges and universities. Of the remaining respondents, 27.5 percent fell between the ages of 23 and 30, and 15 percent were over the age of 30. Regarding gender, 22.5 percent are men and 77.5 percent are women.

In terms of socioeconomic background, the majority of respondents have a middle-to upper-middle class background as measured by their educational backgrounds and that of their parents. All respondents are currently enrolled in college or are college-educated—67.5 percent are current college students and 32.5 percent had completed a bachelor's degree; 15 percent of respondents are pursuing advanced degrees. While respondents often had limited information about their parents' incomes, they frequently described parents who were highly educated. Most had at least one parent with a bachelor's degree (75 percent) or some college (87.5 percent), and 47.5 percent had at least one parent who held an advanced degree.

Finally, regarding racial identification, the majority of respondents (33 of 40) label themselves [by] using multiracial descriptors (e.g., as biracial, multiracial, mixed-race). In comparison, only six respondents labeled themselves as black, and one respondent as white. The fact that so few respondents labeled themselves as black mirrors recent studies, which similarly show a weakening of the one-drop rule and widening of racial options. Furthermore, respondents' identifications are situational; they generally identified themselves to others as biracial, but in some contexts, passed as black or white.

RACIAL IDENTITY WORK: STRATEGIES AND MOTIVATIONS

We find that respondents regularly do racial "identity work" and employ a variety of strategies to present their preferred racial identities to others. In this section, we first explore the strategies that respondents use to manage their identities, and we identify factors which influence the accessibility and efficacy of these strategies. After outlining the various identity strategies and limitations, we then examine the motivations of identity work with a focus on passing as black. Of those presenting monoracial identities to others, we find that they more often situationally pass as black (31), rather than white (3).

STRATEGIES OF IDENTITY WORK

Respondents use a variety of strategies to *pass* or, when passing is not desirable or feasible, to *cover* an identity (i.e., downplay its obtrusiveness [Goffman, 1963]). Further, identity work is not just about concealing or covering a stigmatized identity, but highlighting a nonstigmatized or preferred identity, or what we term *accenting*. While covering involves downplaying an attribute, accenting involves emphasizing or accentuating it. Further, accenting differs from passing; not all people can pass as black or white (e.g., their ancestry may be well known; their phenotype may prevent it), but they may be able to accent their black or white ancestry as a form of identity work. To conceal (i.e., pass), cover, or accent particular aspects of their racial ancestries, respondents use five strategies: (1) verbal identification/disidentification, (2) selective disclosure, (3) manipulation of phenotype, (4) highlighting/downplaying cultural symbols, and (5) selective association.

First, respondents do "identity talk" (Snow and Anderson, 1987) via *verbal identification/disidentification*. In short, they claim or disclaim identities by verbally saying, "I'm this" or "I'm not that." Anthony presents himself as black through verbal identification, and says, "I guess I just always make sure people know I'm black. Like even when I went to an all-white school, I used to say, 'I'm black … Even though they knew I had a white father, if they ask, 'I'm black. That's it." By saying "I'm black," Anthony invokes a "me" identity (McCall, 2003).

According to McCall (2003), identity processes must be studied in terms of the "Me" (*identifications*), but also the "Not-Me" (*disidentifications*). Caroline, for example, verbally resists being classified as black and says:

> In my [graduate] program, I think we maybe have like four black peo-
> ple, not including myself. And the other day, one of the [black] guys
> said to me, "Oh, in our class, there are only three of us." And I said,
> "Three of who?" And I didn't know what he was talking about and he
> looked at me. And I was like, "Don't do that. Don't lump me in [with
> being black] because I don't see myself that way and I don't like it when
> you just assume that." And then one time, I was taking an African
> Cultural Studies class and our teacher was black and she made reference
> to the black students in the class and lumped me in there with them.
> And I raised my hand and I was like, "I'm not black." And she almost
> wanted to argue with me like, "Yes you are." And no, no I'm not.

A second, and related identity strategy is ***selective disclosure***—selectively revealing and/or concealing particular racial identities to others. Unlike Anthony and Caroline, where peers were aware of their biracial backgrounds, some respondents manage the racial information they give to others to pass as mono-racial (often as black). In school, Samantha intentionally conceals her biracial background and says, "There was a time in middle school [that] I never told anyone what I was. A lot of times they never asked. They just kind of assumed, Well she's black. They just assumed … I [was] like, Okay, what's the reason for bringing it up?" Similarly, Natasha, who currently attends an HBCU (an Histor-ically Black College/University) selectively reveals only her black background to her black peers. She says, "Since I've been at college, I don't even mention [that I'm biracial]. I don't bring it up unless it's brought up to me. …I would just rather say 'I'm black' and that be the end of it. It's definitely not something that I advertise." Both Samantha and Natasha conceal their white/biracial ances-try to pass as black.

In addition to selectively disclosing particular identities in face-to-face social interactions, respondents strategically reveal and conceal particular ancestries when filling out race questions commonly found on school, job, and scholarship applications. While Natasha reveals only her black identity to black peers, she consciously manipulates her identity in different ways on forms. Highlighting the situational nature of her identity, she says, "[W]hen I fill out the little ques-tion things, I used to always check 'other.' Now I just check 'black' … . I've also learned to manipulate the situation that I'm in. I know that if I say I'm 'biracial,' I will get certain things, and if I say I'm 'black' I will get certain things. So I know I probably play with that a little bit." Like Natasha, the majority of respondents use selective disclosure on applications.

Third, respondents ***manipulate their phenotypes*** (e.g., hair, skin) to manage their identities; this parallels Snow and Anderson's (1987) "cosmetic facework or arrangement of personal appearance" in their study of the homeless. Most respondents cannot alter their phenotypes in ways to present themselves as white, but they often describe modifying their phenotypes to pass as black or to accent their black ancestry. Others alter their phenotypes to cover or downplay their white ancestry. When growing up, Olivia's peers knew she was biracial, yet she manipulated her hair to downplay her white ancestry and says:

> When I was younger ... I had very long hair and I identified more with
> African Americans ... So I usually kept my hair pulled black or kept it
> up or tried to do different things to blend in more ... . [Other black
> girls] used to call me "white girl" because my hair was very long and it
> would blow in the wind ... . Back then I used to get up in the morning
> for school and leave my hair down and run out. And then when they
> started saying I was a white girl ... I would never leave my hair down.

Likewise, Anthony modifies his hair to pass as black and says, "I used to have really long hair and sometimes I would pick it up into a "fro." Michelle, who claims that she looks white, covers her white background by manipulating her skin color (e.g., tanning) because she does not "want to be seen as a white person."

Fourth, respondents manage their identities by highlighting and/or downplaying **cultural symbols** they perceive associated with whiteness and blackness (e.g., clothing/dress, food, language). While altering phenotype is not an option for everyone (e.g., not everyone can pick their hair into an afro), invoking cultural symbols is frequently employed in racial identity work. For example, to pass as black, Anthony draws on cultural symbols of clothing and language. Describing how he presented himself as black in school, he says, "You know, pants sagging ... I used to kind of slur my speech a little bit because I used to talk very properly and I used to force myself to sound different. Sound like I was more black." Denise, too, describes passing as black especially when "trying to get into a step team or... choir...or... something where people in the organization are black, like a fraternity or sorority." When asked how she presents herself as black, she responds, "Probably the way I style my clothes ... looking like I dress like I'm a black person ... I have to change how I appear."

While Anthony and Denise highlight black cultural symbols (via clothing and language) to manage their black identities, Stephanie managed her black identity in school by distancing herself from cultural symbols of whiteness:

> [I attended] an all black school and so all my friends were black then
> ....I remember NSync being out ... and my friends listened to them
> and I hated that. I hated any music that wasn't black. I hated any clothes
> that black people didn't wear ...I felt like I had to stress to people that I
> was black ... . So I felt like "I hate NSync. I hate this white music."

By distancing herself from these so-called white symbols (e.g., white music, clothing), Stephanie works to downplay or cover her white ancestry.

A final identity strategy is **selective association**. Respondents selectively associate with a particular racial group (via peers, friends, and romantic partners) and organizations/institutions (e.g., clubs, colleges, churches), which mirrors Snow and Anderson's (1987) strategy of "selective association with other individuals or groups." This strategy is often used by respondents to pass as black or to accent their black identity. For example, Stephanie says, "When I got to high school, all the white people were so nice ....And I hated them. I didn't want to be friends with them. I didn't want to sit with them. I didn't want them to talk to me. I wanted to sit at the black table. I felt like I had to stress to people that I was black." While Stephanie associated only with black peers, Olivia dated only dark black men as a strategy to emphasize her black identity: "I used to only date very dark-skinned black men because I didn't want people to think I was trying to be white ....So I stayed with dark-skinned men because it's like I want to prove that I was black. Yeah, that I'm this black woman. 'See, I've got this very dark man.' It sounds stupid now, but back then it was important."

Other respondents manage black identities by joining organizations that reflect their preferred black identities. Alicia limits her peer network to black people and dates only black men, and she also describes being drawn to black organizations:

> I'm pretty black ... . Maybe I'm just more concerned about being black right now... . And I want to have kids that are part of Jack and Jill and I'm infatuated with my [black] boyfriend. I want to marry him and have children with him. I can't imagine a life where I wasn't part of Jack and Jill and I wasn't in AKA ... things that are exclusively black ... I feel like I'm pretty segregated. I kind of segregate myself and I pretty much just hang out with black people.

Alicia, who is "concerned about being black," consciously controls the racial makeup of her social circles and purposefully participates in organizations (Jack and Jill of America, Alpha Kappa Alpha[1]) that reflect her preferred black identity.

MOTIVATIONS FOR PASSING AS BLACK

Motivations for passing as white, especially during the Jim Crow era, are well-documented. Less is known, however, about the motivations for passing as black. While we find [that] a few respondents have passed as white in rare situations, the majority of respondents have, at one time or another, passed as black and they do this for several reasons—to fit in with black peers, to avoid a (white) stigmatized identity, and/or for some perceived advantage or benefit.

To fit in   Not wanting to stand out, especially in adolescence, respondents often describe working to "blend in" to feel accepted by peers. In some cases, they try to fit in with both black and white peers. Kristen grew up attending a predominantly white school and a predominantly black gymnastics program, and says, "Going to school and going to the gym were just two totally different things for me. So it's like I had to switch. I was like Superman. I was kind of like Clark Kent—take off my glasses going to the gym and then put them back on when I was in school ....I would just kind of change. I would just do little things that I very well knew what I was doing." The "little things" included changing her clothing and speech depending upon the race of her audience (i.e., drawing on black and white cultural symbols). When asked why she altered her appearance and behavior between friends, she says, "To fit in probably. Because I wanted friends in both areas."

While some respondents employ identity strategies to "fit in" with their black *and* white peers, the majority claim that their black characteristics (e.g., dark skin) prevent their full acceptance by whites (unlike Kristen, above, who claims she looks white). Feeling thwarted by whites, many respondents pass as black to find a place with their black peers. Stephanie describes her experiences in school and says, "First grade through eighth grade I was in the same school and it was an all black private school. So everybody there was black... . And all the kids ... basically told me I was white ....And I got so frustrated because I wanted to fit in and they kind of made me feel like I wasn't going to fit in if I didn't go along with being totally black." Stephanie uses several strategies (e.g., downplaying white cultural symbols, selective association with black peers) to present a black identity.

Fitting in with black peers also appeared more important for women than men in the sample; [women] more often describe situations in which they were discredited as black if their biracial background was revealed. Rockquemore and Brunsma (2002) find that biracial women often encounter negative experiences with black women because of their looks and/or biracial ancestry, and we also find that they, at times, find their blackness challenged. Describing her experiences with black women, Natasha says:

> For some [black] people, [a biracial background] is a strike against you ...with girls, I can't escape [my white] side. It's constantly being brought up ... they always seem to make sure to tell me I'm not really black. If I would tell someone I'm black, they would say, "No, you're mixed"... when people are always reminding you, "You're mixed"... trying to discredit you, it's hard.

Natasha is constantly reminded that she is biracial and "not really black." Olivia, too, describes how some black women do not see her as black: "I think when I was growing up, [black girls] just did not accept me as being a black girl ... with [black] women, I still think there are some instances where they don't see me as an authentic black woman... ." Thus, wanting to fit in, not have their blackness discredited, nor feel contention with black peers, some respondents consciously concealed their white/biracial ancestries.

To avoid a stigmatized identity   In the Jim Crow era, blackness was stigmatized (e.g., as inferior, backward) and is arguably stigmatized today. In describing an experience as an undergraduate, Caroline notes the stigma and says:

> I can remember when I was an undergrad, one time I got braids in my hair ... that were down my back. And it wasn't anything dramatic and I thought it looked really nice and I liked it. And as soon as I went back to school in the city ... I was immediately on guard when I was walking down the street. And I was like, "Oh gosh, I don't want people to think I'm black because I have these braids in my hair." ... I was so nervous ... that was all that went through my mind, "I don't want people to think that I'm black." ... I know it sounds awful, but I don't want

people to think that I'm stupid or that I'm bitchy or anything like that. So I didn't keep them in for very long.

Here, Caroline manipulates her phenotype (removes her braids) to avoid negative stereotypes she associates with blackness (e.g., stupidity, bitchiness).

For advantage   Finally, whiteness in the slave and Jim Crow eras conferred many advantages and privileges, and three respondents describe occasionally passing as white, even today, for some perceived benefit. Beth describes a context when she passed as white via selective disclosure: "I used to be a caseworker. Some of [my white coworkers] assumed I was white and I just rolled with it … yeah, you're just sitting there like, 'You really don't have a clue. I'll just continue to be white, if that's what you're going to insist on.' … I just left it as 'I'm going to let you assume [I'm white]. And I'll go along with it.'" When asked why she allowed others to assume she is white, she describes this as a protective strategy to avoid prejudice from coworkers. Similarly, Michelle uses selective disclosure to pass as white at work, and says:

> I [identify as white] more so when it's convenient to me in corporate America. I've witnessed where white people get further than the black people … . And I just think in my whole experience, not just with this job but other jobs, I have to … put forth that I'm white. Then they're more likely to trust me ….I think I use it to my advantage when I need to. [In the work setting?] Yes, because I'm trying to get ahead.

While these respondents pass as white for some perceived workplace advantage, the majority (29 respondents) pass as black in other contexts for perceived advantage—in particular, on college, scholarship, financial aid, and job applications. Frequently unaware that being biracial is often sufficient for affirmative action purposes, they presented themselves exclusively as black. While Michelle describes passing as white at work to "get ahead," she also describes passing as black on college applications. Explaining why she checks the "black box," she says, "I thought maybe if I chose black, especially in college, I'd get more financial aid. I'd get more opportunities, and so I kind of thought it was to my best advantage to just say I was black."

DISCUSSION AND CONCLUSION

That people perform race is not new; during the Jim Crow era some "blacks" passed as white. With the implementation of civil rights legislation, however, many argue that the strategy of passing is a relic of the past. We surprisingly find, however, that passing still occurs today and quite frequently, although it looks different: a few respondents occasionally pass as white, but the majority describe situations in which they pass as black.

That so few respondents passed as white is not surprising given that this option is unavailable to most (unless they have white skin and appearance) and because passing as white is often viewed with disdain by other blacks today. Also, not surprising were the motivations for the few individuals who did pass as white—in all three cases, respondents passed as white to avoid prejudice/discrimination and/or for advantage in the workplace. Further, we find that passing as white today is temporary and situational, not the continuous type of passing that marked the Jim Crow era.

Most interesting, however, are not the few respondents who passed as white, but the many that passed as black. Scholars understand the motivations of passing as white in a society dominated by whites, but less is known about motivations for passing as black. We find that biracial people pass as black for several reasons. Most notably, we argue, because they can. While passing as white is difficult for most, passing as black is less difficult given the wide range of phenotypes in the black community regarding skin color and other physical features. With generations of interracial mixing between blacks and whites and the broad definition of blackness as defined by the one-drop rule, most Americans cannot tell the difference between biracial and black. Hence, there is little difficulty when many biracial people conceal their biracial background; this is because many "blacks" also have white phenotypic characteristics (because they, too, often have white ancestry). Further, we find that biracial respondents pass as black for additional reasons—to fit in with black peers in adolescence (especially since many claim that whites reject them), to avoid a white stigmatized identity, and, in the postcivil rights era of affirmative action, to obtain advantages and opportunities sometimes available to them if they are black (e.g., educational and employment opportunities, college financial aid/scholarships).

Passing as black is an interesting concept in and of itself given the unique history of race in this country, and it further illuminates changes in race and politics in the United States. In previous decades, the notion of passing as black was impossible given the one-drop rule—if people had black ancestry, they did not pass as black, they *were* black. People could only pass as white based on a concept that was inherently racist and asymmetrical (i.e., one drop of black blood made one black, but one drop of white blood did not make one white). As the one-drop rule weakens, what it means [for a person] to pass is arguably undergoing modification, especially in an era where blackness (at least in some contexts) confers tangible benefits. While the notion of passing has historically conjured up images of black–white people (who were perceived as really black) passing as white, shifting definitions of blackness may change this and draw new attention to the concept of passing as black.

NOTE

1. Jack and Jill of America is one of the oldest black social organizations in the United States; Alpha Kappa Alpha was the first sorority established by black women.

REFERENCES

Daniel, Reginald G. (1992). "Passers and Pluralists: Subverting the Racial Divide." Pp. 91–107 in *Racially Mixed People in America*, edited by Maria P. P. Root. Newbury Park, CA: Sage Publications.

Goffman, Erving. (1963). *Stigma: Notes on the Management of Spoiled Identity*. Englewood Cliffs, NJ: Prentice Hall.

Kennedy, Randall. (2003). *Interracial Intimacies*. New York: Vintage Books.

McCall, George J. (2003). "The Me and the Not-Me: Positive and Negative Poles of Identity." Pp. 11–25 in *Advances in Identity Theory and Research*, edited by Peter J. Burke, Timothy J. Owens, Richard T. Serpe, and Peggy A. Thoits. New York: Plenum.

Rockquemore, Kerry Ann, and David L. Brunsma. (2002). *Beyond Black: Biracial Identity in America*. Thousand Oaks, CA: Sage Publications.

Russell, Kathy, Midge Wilson, and Ronald E. Hall. (1992). *The Color Complex: The Policies of Skin Color Among African Americans*. New York: Harcourt Brace Jovanovich.

Snow, David A., and Leon Anderson. (1987). "Identity Work among the Homeless: The Verbal Construction and Avowal of Personal Identities." *American Journal of Sociology* 92(6):1336–71.

32

Fitting In and Fighting Back: Homeless Kids' Stigma Management Strategies

ANNE R. ROSCHELLE AND PETER KAUFMAN

The homeless occupy a particularly stigmatized position in society, serving as general pariahs. People go out of their way to avoid seeing them or interacting with them, and as Reinarman notes, the American "vocabulary of attribution" assigns blame to them based on their individual failures rather than to the unequal opportunity structure of society. Nowhere is this blame less worthy than it is for homeless kids, who have inherited their homeless status from parents unable to house them. On the basis of research conducted in moderate-term housing shelters for (usually) single mothers with children, Roschelle and Kaufman articulate some of the strategies these youngsters use to disguise or manage their deviant status.

They divide their techniques into strategies of inclusion and exclusion. In the former, children attempt to fit into society and attain the appearance, friendships, and acceptance they see among their "homed" peers. Their attempts to pass, cover, and forge relationships are poignant, revealing their pain and insecurity. When these fail, however, and they cannot escape their stigma, they muster a bravado designed to achieve a tough and intimidating persona. Like the cheerleaders and the bankrupt, they seek to raise their status by comparing themselves with others whom they deem lower than themselves, although in so doing, they denigrate others in similarly unfortunate situations whom they can construct as more socially and morally abject than themselves.

Why do you think these homeless kids pursue such different approaches in dealing with in-groups and out-groups? How effective do you think these two approaches are at protecting them from stigma?

Beginning in the 1980s the gap between the rich and the poor widened significantly, and there was a concomitant increase in homelessness. Social scientists responded to this crisis by studying the rates and causes of homelessness. However, with the exception of a few notable studies (Liebow, 1993; Snow and Anderson, 1993), there has been little ethnographic research on the daily struggles this population encounters. Moreover, research has been slow to address the

From Anne R. Roschelle and Peter Kauffman, "Fitting In and Fighting Back: Homeless Kids' Stigma Management Strategies." *Symbolic Integration*, Vol. 27, No. 1, 2004. All rights reserved. Reprinted by permission of the University of California Press Journals.

growing feminization of the homeless. The proportion of women and children constituting the urban homeless population increased from 9 percent in 1987 to 34 percent in 1997 to a staggering 40 percent in 2001. Although there has been some recent work on familial homelessness, most accounts have disproportionately examined street homelessness among single men. More research is needed on women and children who are homeless and, in particular, on the strategies they use to construct meaning, participate in social interactions, and negotiate the boundaries with the nonhomeless world....

In our analysis, stigma and homelessness are both construed as structural locations. Like stigma, homelessness must be recognized as a structural component characterizing the individual's relationship to the social world. Individuals may find themselves delegitimized socially, politically, or economically if they cannot accumulate normative resources, are unable to receive positive appraisals, and are incapable of following expected rules of behavior. Such is the case with homeless individuals and other stigmatized populations. In daily life, these populations often find themselves disadvantaged because they do not have the means to engage in successful social interactions. Implicitly then, stigma, like homelessness, is about power—or the lack thereof. As Link and Phelan (2001, p. 367) argue, "[S]tigmatization is entirely contingent on access to social, economic, and political power that allows the identification of differentness, the construction of stereotypes, the separation of labeled persons into distinct categories, and the full execution of disapproval, rejection, exclusion, and discrimination."

Recognizing stigma as a condition deriving from macro-level forces does not negate the micro-level processes that characterize an individual's stigma management. Although stigma should be understood as a consequence of structural relationships, the ways in which individuals manage their stigma is microinteractional. Acknowledging the macro-origins of stigma forces researchers to incorporate an analysis of the structural while studying the ways in which individuals manage their stigma interpersonally. Because stigmatized individuals lack power, their ability to protect their sense of self may be severely compromised. As a result, their actions may unintentionally augment their stigmatization and perpetuate the structural relationships that generate their stigmatized status (Link and Phelan, 2001). Much like impoverished individuals, therefore, acquiring the capital (social, cultural, economic, political, etc.) that is needed to overcome stigma may be difficult because of the spoiled identity. In short, stigma may be a chronic status until the individual accumulates the resources necessary to break the cycle.

Much of the existing literature concentrates on strategies that presumably lessen stigma and neglects to consider the extent to which stigma management strategies may perpetuate the individual's stigmatized status. Since Goffman's 1963 work, the assumption that stigma management strategies result in positive outcomes has prevailed. Researchers generally explore the ways in which stigmatized individuals protect their sense of self and attempt to gain social acceptance. The underlying assumption is that through these strategies, deviant individuals will achieve some degree of normalcy. Clearly, some actions benefit the individual's social standing. However, more attention should be given to the process through which stigma management strategies might result in the acquisition of characteristics that further spoil one's identity. Recognizing stigma management strategies as micro-level reactions to macro-level conditions elucidates how such strategies may lead to unintended and potentially negative consequences.

In the case of kids who are homeless, the structural disadvantages are significant. [These kids] come from poor families and are also marginalized because of their age, race, ethnicity, and gender. Unlike some of their privileged peers who may go to great lengths to distinguish themselves through dress, taste, and behavior, kids who are homeless experience what Goffman (1963, p. 5) calls "undesired differentness." Homeless kids are social outcasts because of the situation into which they were born or because of their parents' current predicament. Consequently, when we attempt to understand the homeless kids' strategies to manage their stigma, we must examine how structural disparities influence these strategies and how the nonhomeless interpret them. As we illustrate, although some of their stigma management strategies produced positive results with regard to their identities and social standing, others led to unanticipated and negative consequences. In effect, it is through the use of these strategies that kids who are homeless, like some stigmatized populations, unwittingly reproduce their stigmatization and cement their status loss.

RESEARCH SITE

The research site is an organization in San Francisco called A Home Away From Homelessness. Home Away serves homeless families living in shelters, residential motels, foster homes, halfway houses, transitional housing facilities, and low-income housing. The program operates a house in Marin County (called the Beach House), provides shelter support services, a crisis hotline, a family drop-in center (the Club House), a mentorship program, and, in conjunction with the San Francisco Unified School District, an afterschool educational program (the School House). Home Away is neither a typical homeless service agency nor a shelter.

Home Away serves a population of approximately one hundred five- to eighteen-year-old children annually. Fifty percent of the participants are boys and 50 percent are girls. The racial–ethnic breakdown of homeless families participating in Home Away programs is 40 percent African American, 30 percent white, 20 percent Latino, and 10 percent multiracial. All the kids participating in Home Away programs lived with at least one parent (none are runaways or homeless youth[s] living independently), although some were essentially on their own as a result of parental neglect. Many of the kids suffered from a variety of physical, emotional, and developmental deficits, which is not surprising given the harsh conditions under which they live (e.g., Bassuk and Gallagher, 1990; Rafferty, 1991).

A majority of Home Away's families come from northern California; some families are from southern California and the Pacific Northwest. These families, often mother-only, typically come from chronically poor communities and usually become homeless after losing a job or stable housing. When extended kinship networks deteriorate as a result of poverty (Menjivar, 2000; Roschelle, 1997) and violence, these transient families are forced to rely on institutional forms of social support. There are primarily two types of families using the services of Home Away: short-term participants who experience brief spells of homelessness and the chronically homeless who use programs over several years....

Chronically homeless families who cycled in and out of Home Away were the most frequent users of Home Away programs. Because of the severe housing shortage and exorbitant rents in San Francisco, many homeless families had their shelter stays extended past the three-month deadline. Some families eventually moved to motels, others doubled up with family members, and some shuffled between substandard apartments in violent neighborhoods. Many of the chronically homeless kids went to the Beach House on a weekly basis over the course of several years. As they got older, some of the teenagers who no longer wanted to go to the Beach House went to the Club House a few times a week. Some of these kids attended the afterschool educational program several times a week, but their transience made it difficult for many to stay in the program. During the past six years, nearly 1000 children and their parents have participated in various programs provided by Home Away.

METHODS AND DATA

The ethnographic component of our research included participant observation at the Beach House, the School House, and the family drop-in center where Anne collected the data and did volunteer work for four years (1995–99). During this time, she also attended meetings of relevant social service agencies at which she took copious notes and conducted observational research at residential motels, transitional housing facilities, and homeless shelters throughout the San Francisco Bay area. In addition, she conducted thousands of hours of informal interviews, sometimes asking leading questions (Gubrium and Holstein, 1997) and sometimes just listening. Many of the informal interviews were taped and transcribed. In addition, she conducted and transcribed verbatim formal taped interviews with 97 kids and parents who were currently or previously homeless. While in the field, when something that seemed profound about the homeless experience was uttered or occurred, Anne would go to the bathroom and quickly record it. This strategy was necessary because sitting in a room full of kids and simultaneously taking notes creates an artificial atmosphere that inspires skepticism and mistrust among kids who may already be wary of adults. Anne typically spoke her field notes and critical interpretations of the days' happenings into a tape recorder immediately after working with the kids. These tapes were eventually transcribed....

Gathering ethnographic data among kids who are homeless presents a number of methodological challenges. The first is the difficulty of gaining entry, although this proved easier than expected. Anne developed rapport with the kids by spending considerable time with them at a variety of locations. Entry was further facilitated by the House Mother (a formerly homeless woman) who regularly conveyed to the kids her support for and fondness of Anne.

This legitimacy was crucial because it allowed kids to be interviewed and observed in the context of their social group interactions....

The second, related challenge is being attentive to their devalued social position. This awareness is particularly necessary because of our interest in how homeless kids manage their stigma. As noted, because these kids are poor, young, and predominantly racial–ethnic minorities, their behavior is often interpreted pejoratively even when it mirrors that of their nonhomeless peers. For this reason, it is important to record and respect the ways in which kids construct meanings and to not assume "the existence of a unidimensional external reality" (Charmaz, 2000, p. 522)....

The third strategy, and challenge, was participating in impromptu conversations, which captured the ways kids managed their devalued social status. These gatherings, which Anne tape recorded and later transcribed verbatim, were not focus groups because they had neither an identifiable agenda nor a formal mediator. Rather, these were freewheeling conversations that allowed Anne to "remain as close as possible to accounts of everyday life while trying to minimize the distance between [herself] and [the] research participants" (Madriz, 2000, p. 838). These conversations were particularly appropriate because they allowed for the expression of ideas in a forum where individuals felt comfortable speaking up. Furthermore, talking with kids in relaxed group settings in which the kids outnumber the adults minimizes the inherent power differential between the adult researcher and the respondents and gives voice to those who have been subjugated.

STIGMA MANAGEMENT STRATEGIES

Goffman (1963, p. 3) suggests that stigma should be understood as "a language of relationships" as opposed to the attributes a person possesses. Moreover, he notes that "the stigmatized are not persons but rather perspectives" (p. 138). These ideas are particularly relevant for understanding the stigmatization of kids who are homeless. While these kids do not necessarily exhibit physical manifestations of their stigma, they are stigmatized because of their marginalized status in society. The discourse surrounding poverty and homelessness has a long history of blaming individuals for their predicament. Kids who may have been defined in the past as deserving of societal sympathy have more recently been recast as part of the legions of the undeserving poor.

Home Away kids were aware of their relationship to normative society and understood that others generally saw them as undesirable. Because of the large numbers of homeless in San Francisco and the visibility of the street homeless, there were numerous negative public discussions about the issue. The local newspapers constantly featured articles that were contemptuous of the homeless, and mayoral races often focused on ridding the city of its unsightly homeless population. The kids in the sample were attuned to the negative discourse surrounding homelessness and often expressed their anxiety about being demonized by the public. In fact, the kids spent many hours sitting in front of the fireplace discussing how hard it was for them to be routinely disparaged. As Dominic, an articulate

16-year-old, said, "Everyone hates the homeless because we represent what sucks in society. If this country was really so great there wouldn't be kids like us."...

Kids who are homeless manage their stigma in a variety of ways. One common typology of stigma management strategies assumes an in-group/out-group dichotomy (Anderson, Snow, and Cress, 1994; Blum, 1991). Although this typology is useful for understanding some stigmatized populations, the stigma management strategies of Home Away kids were more fluid and could not be so neatly categorized. As kids who are homeless make their way through the world, they transgress—yet simultaneously create—boundaries and often use similar stigma management strategies with both peers and strangers.

Although the in-group/out-group dichotomy did not fit our data, we developed an alternative schema that enabled us to better categorize the stigma management strategies Home Away kids use. We observed that the kids displayed two sets of strategies. The first set conformed to societal norms of appropriate behavior and aimed at creating a harmonious environment with both peers and strangers. In other words, these strategies represented attempts at establishing self-legitimation in both hostile and supportive environments. We refer to these as *strategies of inclusion* because they reflected the kids' desire to eradicate the boundary between a homeless and a nonhomeless identity. Through these inclusive strategies, Home Away kids hoped to be recognized simply as kids—without the stigmatizing label and discredited status of being homeless. The most common strategies of inclusion among Home Away kids were forging friendships, passing, and covering.

The second set of strategies ... also [consisted of] attempts at gaining social acceptance but were not necessarily aimed at creating a harmonious atmosphere. We refer to these as *strategies of exclusion* because Home Away kids use them to distinguish themselves from both peers and strangers. With these strategies, kids who are homeless attempted to redress their spoiled identity by declaring themselves tougher, more mature, and better than others. These exclusive stigma management strategies included verbal denigration and physical and sexual posturing. Unlike strategies of inclusion in which kids used conciliatory tactics to be accepted, strategies of exclusion were aggressive and forceful attempts by the kids to blend in. These strategies largely reflected the kids' interpretations of socially acceptable behavior. However, given their disenfranchised social position, these behaviors were often perceived as maladaptive and threatening. As a result, when kids engaged in strategies of exclusion, they provided members of the dominant culture with a seemingly legitimate justification to further disqualify and disparage them....

STRATEGIES OF INCLUSION

Forging Friendships

The language of relationships that identifies kids who are homeless as a stigmatized group is so embedded in society that it is implicit in all their social relations. Even when these kids are in situations in which their homelessness is unknown or unimportant, they still feel the need to manage their spoiled identities. For example, one would expect that in the consonant social environment of the Beach House they would not have to engage in stigma management strategies. However, because their stigmatization is defined by their relationship to mainstream society, their stigma is always part of their identity....

At the Beach House, the kids attempted to construct a positive identity. Children were treated with dignity and respect by volunteers and were included in important decision-making processes. The services provided by Home Away gave kids an opportunity to experience childhood in a way most individuals take for granted. In addition to providing a break from the difficulties of living in shelters, transitional housing, residential motels, and so on, Home Away gave the kids a place to forge friendships with caring adults. Many of these relationships offered the only opportunity for homeless kids to obtain positive self-appraisals from nondisenfranchised adults. This supportive environment allowed for greater self-legitimation, gave the kids a respite from their stigmatization, and provided them with a sense of belonging (Brooks, 1994). As a result of this inclusive strategy, many of the kids developed more favorable self-images. Silvia's discussion of her friendship with Tami, a volunteer, illustrates the importance of forging friendships as a way to manage stigma.

SILVIA: I call her whenever I feel really bad. She is so nice. She makes me feel better when I'm depressed and she never makes me feel like a freak because I'm homeless.

ANNE: Is that important to you?

SILVIA: Yeah. Most of the time I feel pretty bad about my life. I mean my mom has a new boyfriend every few weeks, we live in this nasty ass hotel, and I feel like everyone knows I'm a loser. Tami is always trying to make me feel better about my life. She tells me how smart and pretty I am and that I should feel good about being such a great older sister. Tami really cares about me and makes me feel better about myself. I don't know what I'd do without her.

Home Away also provided these kids with the opportunity to befriend other impoverished youths. By embracing other disenfranchised children, Home Away kids created a safe space where they could construct a positive identity and manage their stigmatized status. For example, when new kids were incorporated into Beach House activities, instead of rejecting them, the "old-timers" served as mentors and attempted to create a harmonious climate. Old-timers initiated the new kids into the program by teaching them the rules, showing them around, and introducing them to secret hiding places. Much like the traditional African American role of "old heads" and "other mothers," this mentor relationship allowed the old-timers to gain self-esteem. By initiating newcomers, they became experts at something and shared their wisdom with other impoverished kids. In addition, as the following conversation suggests, old-timers took great pride in their knowledge of the surrounding environment.

CARLOS: Hey, Hernán, don't go in the water without your life jacket.

HERNÁN: *No hay problema*—I'm a great swimmer.

CARLOS: It doesn't matter, man, the undertow is really strong and you can get swept under really easily.

HERNÁN: Don't be such a pussy.

CARLOS: Seriously man, people drown here all the time. Last week some kid almost died.

HERNÁN: Really?

CARLOS: Yeah, man. Listen I been coming here for two years and I've seen a lot of shit—you gotta believe me.

HERNÁN: *Gracias*, Carlos—thanks for watching my back. You really are a cool dude. I guess I better to stick close to you so I don't get into any trouble.

CARLOS: Hey after we finish swimming Marcus and I can show you the secret hiding place.

HERNÁN: Cool.

Forging friendships with newcomers is an important stigma management strategy because it [gives] kids who are homeless an opportunity to transcend their discredited status and assume a role invested with interactional legitimacy. Through forging friendships, Carlos is able to negate feelings of worthlessness bestowed on him by society and feel like a valuable member of the kids' community. Similarly, Hernán's stigma is mitigated because he now feels like a legitimate member of Home Away. This strategy of inclusion anchors kids in a particular social group and subsequently provides them with a desired social identity that is conferred by significant others.

Passing

Goffman (1963) suggests that visibility is a crucial factor in attempts at passing. To pass successfully, an individual must make his or her stigma invisible so that it is known only to himself or herself and to other similarly situated individuals. Unlike their nonhomeless peers who have legitimate access to the public domain, kids who are homeless must often pass as nonhomeless as a means of appropriating heretofore unavailable and legitimate public space. One "passing" strategy kids in our study used entailed adopting the dress and demeanor of non-homeless kids. Whenever clothing was donated to Home Away, the kids selected outfits based on style rather than function. It was extremely important that clothing and shoes looked new and were "hip." On numerous occasions, kids refused to take donated coats during the winter because they were ugly and out of style. The importance of fitting in is evidenced by the following conversation between Anne, Jamie, and Cynthia (Jamie's mom) while they were hanging out at Stonestown Mall.

JAMIE: Hey, do you think these people can tell we are homeless?

ANNE: No, how could they possibly know?

JAMIE: I don't know. I always feel like people are looking at me because they know I am poor and they think I am a loser.

CYNTHIA: I feel like that a lot too—it makes me feel so bad—like I'm a bad mother and somehow being homeless is my fault. I feel so ashamed.

JAMIE: Me too.

ANNE: Jamie, what are some of the ways you keep people from knowing you are homeless?

JAMIE: I try and dress like the other kids in my school. When we get clothes from Home Away, I always pick stuff that is stylin' and keep it clean so kids won't know I'm poor. Sometimes it's hard though because all the kids try and get the cool stuff and there isn't always enough for everyone. I really like it when we get donations from people who shop at the Gap and Old Navy. I got one of those cool vests and it made me feel really great.

ANNE: Is it important for you to keep your homelessness a secret?

JAMIE: Yeah, I would die if the kids at school knew.

Home Away kids also attempt to pass by using code words. Their use of code words illustrates the importance of language as a symbolic indicator of membership in a social group. As Herman (1993) argues, individuals selectively withhold or disclose information as a way to maintain secrecy about their negative attributes. Knowing the rules of social interaction and the symbolic importance of words, Home Away kids resist language that reveals their homelessness to others. They purposefully choose words they associate with middle-class life to articulate their social reality and simultaneously conceal their marginalized social existence. The following impromptu conversation illustrates this point.

ROSINA: I hate that kid Jamal that lives with us in Hamilton [Family Shelter], you know his cot is three over from mine.

SHELLEY: Sssh, be quiet, someone will hear you and then people will know we are homeless.

ROSINA: I don't care.

SHELLEY: I do.

LINDA: So do I. You should say that you don't like Jamal who lives three houses down—that way people will think you are talking about a kid in your neighborhood.

ANNE: Is that how you guys talk in school?

SHELLEY: Yeah, we say things like that and we make up other stuff so people don't know we live in the shelter.

LINDA: Or in the motels.

SHELLEY: When we talk about our caseworkers we say our aunts and when we talk about shelter staff we call them our friends.

ANNE: That is a really clever way of keeping people from knowing you're homeless.

ROSINA: It totally is, but then you can't invite friends over after school because they think you live in a house.

LINDA: We can't hang out with kids who are not poor. There is no way I would invite a kid from school over to my room in that skanky motel we live in, I'd be so embarrassed.

SHELLEY: Yeah, we can't make friends with a lot of kids because how can we bring them to the shelter after school?

This example illustrates how homeless kids use words to construct a positive identity, protect their sense of self, and feel integrated with the larger society. Moreover, it shows that language reflects and reproduces power relations in society. Interestingly, although these kids use code words to mask their stigmatized status and construct a positive identity, their language choices disqualify them from participating in normative social interactions. These verbal gestures reflect the extent to which homeless kids are in danger of having their homelessness revealed. When they do not use code words, their stigmatized identity will be evident, and they will be unlikely to befriend middle-class children. Alternatively, when they do use code words, they reduce their chances of befriending nonhomeless kids because they risk exposing their stigma. In managing stigma through passing, the threat of discovery is ever present. Although the kids desire acceptance in mainstream society, the threat of discovery leads them to monitor and curtail their social involvement.

Covering

Individuals engage in covering when they attempt to minimize the prominence of their spoiled identity. Covering allows individuals to participate in more normative social interactions by reducing the effects their stigma elicits (Goffman, 1963). Unlike passing, the point of covering is not to deny one's stigma but rather to make it less obtrusive and thereby reduce social tension (Anderson, Snow, and Cress, 1994). Home Away kids often engaged in this stigma management strategy when they became friends with nonhomeless kids who knew about their predicament. Home Away kids would especially use this strategy around the parents of their nonhomeless friends, and they would often ask the staff to advise them on appropriate dress or behavior when they were going out with nonhomeless kids and their families.

One Home Away kid, Ellie, became a close friend of an affluent schoolmate, Carol, who lived in the Marina district of San Francisco. Ellie was honest with Carol and her family about where she lived. Still, when Ellie went to Carol's house for dinner, she wore the least tattered, cleanest clothes she owned and would eliminate her ghetto swagger and jargon. As she said to Anne, "They know I'm homeless and all, but I don't want to act like I'm homeless. I don't want to embarrass them. You know my clothes aren't that fancy and I'm used to eating at the shelter where everyone is talking really loud and eating with their hands and being kinda sloppy." To fit in and be "normal," Ellie decided to minimize the obvious manifestations of her homelessness.... For the most part, these strategies had the intended effect of protecting the kids' identities by offering them a degree of social legitimacy and by aligning them with nonhomeless individuals. In spite of the chaotic nature of their lives and their awareness of their stigmatization, it is noteworthy that these kids attempt to conform to society through socially acceptable means. In this sense, Home Away kids do not fit the traditional conception of the homeless as disaffiliated and socially isolated individuals. They seem surprisingly resilient in their attempts to develop relationships with nonhomeless people in spite of being denigrated by them....

STRATEGIES OF EXCLUSION

Verbal Denigration

When individuals face a social world that labels them deviant, they are likely to fight back by maligning others as a way to augment their self-esteem. Termed "defensive othering" by Schwalbe et al. (2000), this type of stigma management strategy was common among Home Away kids. Many kids in the sample protected their sense of self by verbally denigrating other stigmatized groups such as homosexuals and homeless street people. This was a form of identity work that allowed homeless kids to distance themselves from the stigmatized "other" (Snow and Anderson, 1987) and proclaim their superiority over these similarly disparaged groups. Because homeless kids recognize that they are problematic in the eyes of society, their denigration of other stigmatized groups is a "largely verbal effort to restore or assure meaningful interaction" and align themselves with the dominant culture (Stokes and Hewitt, 1976, p. 838).

Along these lines, many of the kids were homophobic and freely expressed their disgust for homosexuality. This attitude can be attributed to the kids' (arguably accurate) interpretation of societal norms and values with regard to homosexuality. By portraying gays and lesbians as "freakin' faggots" and "child molesters," the kids were placing others lower than themselves in the social hierarchy. The kids' stigmatized status was deflected onto others, thereby bolstering their own sense of self. This exclusionary strategy was especially interesting in the context of San Francisco—a city in which there is a large population of gay and lesbian activists and citizens, many of whom have achieved positions of power and prestige and who themselves often denigrate the homeless. The following conversation illustrates the pejorative discourse Home Away kids often used when talking about gays and lesbians:

ANTOINE: Look at those homos, they make me sick.

FRED: Yeah, they be all trying to grab your ass when you walk by.

ANTOINE: I know man, they'd love a piece of us but I'd make 'em suck on a bullet before I'd let 'em suck on my thang.

FRED: People think we are nasty because we live at the Franciscan, but at least we don't do little boys. I mean those guys are freaks.

ANTOINE: I hear ya, I'd rather be a dope fiend livin' in the Tenderloin than a fuckin' faggot! [Loud laughter]

FRED: I'd rather be a dope fiend, livin' in the Tenderloin, selling crack to ho's than be a fuckin' faggot! [Louder laughter]

Homeless kids also spoke disparagingly about homeless street people. Homeless kids in our sample lived in shelters, residential motels, transitional housing, foster homes, and other institutional settings. Though many of them have spent a night or two sleeping in parked cars, in a park, or on the street with their parents, their homeless experience has taken place primarily in some type of sheltered environment. The main reason they have not spent the majority of their lives on the streets is because there are many more programs for homeless families in San Francisco than there are for childless homeless adults. Essentially, the only factor preventing their parents from becoming part of the homeless street population is them. Despite this, Home Away kids often made fun of homeless street people.

ROSITA: Man look at those smelly street people, they are so disgusting, why don't they take a shower.

JALESA: Yeah, I'm glad they don't let them into Hamilton with us.

ROSITA: Really, they would steal our stuff and stink up the place! [Laughter]

JALESA: Probably be drunk all the time too.

ROSITA: Yeah, and smokin' crack all night long!

Ironically, the mothers of both these girls have struggled with drug and alcohol addiction and have had several episodes of homelessness. By distancing themselves from the "true" social pariahs of San Francisco, Jalesa and Rosita mitigate their own stigmatized status and maintain some semblance of a positive self. Further, identifying a group as more discredited than themselves allows the girls to feel superior to "those losers on the street."…

Physical Posturing

Physical posturing is another form of identity work that grants homeless and nonhomeless kids a momentary degree of empowerment. Kids who are homeless use physical posturing to feel powerful in their interaction with nondisenfranchised kids, a feeling that is rarely available to them. It has long been recognized that many adolescents are filled with insecurity as they attempt to establish their identities. However, middle- and upper-class kids have more socially acceptable means than low-income ones to demonstrate to others (and to themselves) that they are important. As Anderson (1999, p. 68) notes, privileged kids "tend to be more verbal in a way unlike those of more limited resources.

In poor inner-city neighborhoods verbal prowess is important for establishing identity, but physicality is a fairly common way of asserting oneself. Physical assertiveness is also unambiguous." In this sense, Home Away kids' use of physical demeanor is a stigma management technique that emerges from their social structural location and protects them from a hostile world.

One common manifestation of this physical posturing was the use of body language. When encountering nonhomeless children, kids in our study often adopted threatening postures by altering their walk, speech, and clothing to mimic "gangsta" bravado. In short, these kids adopted what Majors (1986) refers to as the "cool pose." For example, to get from the parking lot to the beach, we had to walk over a very small, narrow wooden bridge. As we crossed the bridge, we often passed kids coming from the opposite direction. As soon as the Beach House kids spotted other kids coming toward them, they would immediately change their demeanor. They metamorphosed from sweet cuddly kids to ghetto gangstas. The kids grabbed their pants and pulled them down several inches to mimic the large baggy low-riding pants of the gangbanger. They turned their baseball caps around so that the brim was in the back and swaggered in exaggerated ways. Their speech became filled with ghetto jargon, and they spoke louder than usual. When encountering other kids, the topic of conversation on the bridge would also suddenly shift from how much fun we just had to the fate of a gang member they knew who had just been arrested.

These self-presentations mirror what Anderson (1990, p. 175) calls "going for bad." Anderson argues that this intimidation tactic is clearly intended to "keep other youths at bay" and allow disempowered kids to feel tougher, stronger, and superior. Like verbal denigration, this strategy of exclusion also helped Home Away kids to lessen their stigma. Interestingly, these behaviors clearly imitate images of adolescence that pervade popular culture. In music videos, movies, television shows, video games, and so on, images of baggy pants, baseball caps, and swaggering walks abound. Although this strategy had the intended effect of empowering Home Away kids by intimidating their nonhomeless peers, it also reinforced their stigmatization. As Nonn (1998, p. 322) notes, "While the coping mechanism of cool pose weakens the stigma of failure, it undermines identity … and contribute[s] to their own alienation from other groups in society." Though some kids recognized that their behavior may stigmatize them further, they still engaged in such posturing because it was one of the few areas in their lives over which they had some control….

Sexual Posturing

Some Home Away kids used sexuality to validate themselves. Although nonhomeless youth[s] also engage in sexual posturing to establish a sense of self in their social groups, Home Away kids engage in it more explicitly and overtly than their nonhomeless peers. For Home Away kids, sexual posturing and promiscuity articulate the sexual exploitation and violence they experienced both within and outside the family. As researchers have documented, victims of child sexual abuse often resort to promiscuity in adolescence and adulthood.

Molestation by older men was not uncommon, and many of these kids learned at an early age to use their sexuality to gain status among their peers. For example, several young women in the sample dressed and behaved in overtly sexualized ways that surpassed what we expect from the budding sexuality of "normal" adolescents. In one poignant incident, a 14-year-old girl was teaching a younger girl how to perform oral sex on an eleven-year-old boy. By bragging to his friends about his newfound sexual prowess and achieving prominence among his peers, this boy exhibited an exaggerated masculine alternative commonly found among lower-income racial–ethnic males (Oliver, 1984). The girls involved increased their status by demonstrating a level of sexual maturity that is typically associated with older women.

In another example, Patti, a 13-year-old, discussed the self-esteem she gained through her sexual posturing.

> I mean not everyone is pretty enough to get a man to pay for a room for the night. At least I know that I can get anything I need from a man because I am totally hot. I get lots of attention and some of my friends are jealous of me because they know I got it going on. Some girls are so butt ugly no man would ever even want to fuck them anyway—but not me—they think I'm a hotty—you know what they say—if you got it flaunt it!

As these examples illustrate, sexual posturing is an exclusionary tactic kids use to distinguish themselves from their peers and to lessen their stigmatization. Research has shown that some poor racial–ethnic girls invoke a discourse of sexuality in order to negotiate and construct a more empowering identity (Emihovich, 1998), and others may attempt to achieve similar results through motherhood (Luker, 1996). In a society bombarded with images of hyper-sexuality, it is not surprising that Home Away kids imitate these sexual scripts in an attempt to strengthen their social standing....

CONCLUSION

In this article, we examine the stigma management strategies of kids who are homeless. Although researchers of stigma management have studied a variety of populations, our work contributes to existing knowledge by including this previously neglected group. By examining the stigma management strategies homeless kids use, we address a number of gaps in the literature. First, our work emphasizes the need to acknowledge stigma and homelessness as structural locations. Much like homelessness, we must posit stigma not as an individual attribute but as a relationship to the social structure (Goffman, 1963; Link and Phelan, 2001). This perspective is noteworthy because it attends to how individuals' behaviors are both informed by and interpreted through their social structural location. This insight is particularly true of homeless kids who are oppressed socially because of their age, race, ethnicity, and social class. Kids who are homeless encounter and interpret a world that is characterized by hunger, uncertainty, chaos, pain, drug abuse, violence, sexual abuse, degradation, and social rejection, among other things. This social reality epitomizes the violence of poverty and ultimately results in a lack of consistency, stability, and safety in their lives. Furthermore, these kids exist in a constrained public domain and are forced to carve out their own space in a limited urban environment that is generally hostile to them. It is in this environment that kids who are homeless engage in social interactions that aim to construct positive identities and overcome their discredited status.

By positing stigma and homelessness as social structural phenomena, our work also illustrates the processes whereby stigma may become a chronic status. In contrast to much of the literature on stigma management, we found that the tactics used by Home Away kids sometimes had the unintended effect of perpetuating their spoiled identities. Although their actions were attempts at protecting their sense of self and gaining a degree of social acceptance, strategies such as sexual posturing and verbal denigration substantiated societal stereotypes of kids who are homeless as violent, disrespectful, and dangerous delinquents to be avoided. Obviously, it is not these kids' intention to engage in behaviors that perpetuate their stigmatized identities and their disenfranchised positions. Rather, it is their social structural location that offers them limited opportunities to exert their agency in a socially acceptable way.

In attempting to negate their stigmatization, Home Away kids engaged in strategies that often imitated the behaviors of mainstream youth[s]. By interpreting popular culture and the social interactions of their nonhomeless peers, homeless kids behaved in ways that are often exhibited, condoned, and even rewarded when enacted by their more privileged peers. In mimicking this behavior, Home Away kids naively expected to obtain a modicum of social acceptance. Unfortunately, they failed to understand that the parameters of acceptable and unacceptable behavior are mediated by one's social location. In other words, the deviant behaviors middle-class and affluent peers engage in have different consequences than when they are perpetrated by homeless kids (Chambliss, 1973).

REFERENCES

Anderson, Elijah. (1990). *Streetwise: Race, Class, and Change in an Urban Community.* Chicago: Chicago University Press.

———. (1999). *Code of the Street: Decency, Violence and the Moral Life of the Inner City.* New York: Norton.

Anderson, Leon, David A. Snow, and Daniel Cress. (1994). "Negotiating the Public Realm: Stigma Management and Collective Action among the Homeless." *Research in Community Sociology* (Supplement 1): 121–43.

Bassuk, Ellen L., and Ellen M. Gallagher. (1990). "The Impact of Homelessness on Children." *Child and Youth Services* 14(1): 19–33.

Blum, Nancy. (1991). "The Management of Stigma by Alzheimer Family Caregivers." *Journal of Contemporary Ethnography* 20(3): 263–84.

Brooks, Robert B. (1994). "Children at Risk: Fostering Resilience and Hope." *American Journal of Ortho-psychiatry* 64(4): 545–53.

Chambliss, William J. (1973). "The Saints and the Roughnecks." *Society* 2(1): 24–31.

Charmaz, Kathy. (2000). "Grounded Theory: Objectivist and Constructivist

Methods." pp. 509–35 in *Handbook of Qualitative Research*. N. K. Denzin and Y. S. Lincoln, eds. Thousand oaks, CA: Sage.

Emihovich, Catherine. (1998). "BodyTalk: Discourses of Sexuality among Adolescent African American Girls." pp. 113–33 in *Kids Talk: Strategic Language Use in Later Childhood*. S. M. Hoyle and C. Temple Adger, eds. Oxford: Oxford University Press.

Goffman, Erving. (1963). *Stigma: Notes on the Management of Spoiled Identity*. New York: Simon and Schuster.

Gubrium, Jaber F., and James A. Holstein. (1997). *The New Language of Qualitative Method*. New York: Oxford University Press.

Herman, Nancy J. (1993). "Return to Sender: Reintegrative Stigma-Management Strategies of Ex-Psychiatric Patients." *Journal of Contemporary Ethnography* 22(3): 295–330.

Liebow, Elliot. (1993). *Tell Them Who I Am: The Lives of Homeless Women*. New York: Penguin.

Link, Bruce G., and Jo C. Phelan. (2001). "Conceptualizing Stigma." *Annual Review of Sociology* 27: 363–85.

Luker, Kristine. (1996). *Dubious Conceptions: The Politics of Teenage Pregnancy*. Cambridge, MA: Harvard University Press.

Madriz, Esther. (2000). "Focus Groups in Feminist Research." pp. 835–50 in *Handbook of Qualitative Research*. N. K. Denzin and Y. S. Lincoln, eds. Thousand Oaks, CA: Sage.

Majors, Richard. (1986). "Cool Pose: The Proud Signature of Black Survival." *Changing Men: Issues in Gender, Sex and Politics* 17: 5–6.

Menjivar, Cecilia. (2000). *Fragmented Ties: Salvadoran Immigrant Networks in America*. Berkeley: University of California Press.

Nonn, Timothy. (1998). "Hitting Bottom: Homelessness, Poverty, and Masculinity." pp. 318–27 in *Men's Lives*, edited by M. S. Kimmel and M. S. Messner. Boston: Allyn and Bacon.

Oliver, W. (1984). "Black Males and the Tough Guy Image: A Dysfunctional Compensatory Adaptation." *Western Journal of Black Studies* 8: 201–2.

Rafferty, Yvonne. (1991). "Developmental and Educational Consequences of Homelessness on Children and Youth." pp. 105–39 in *Homeless Children and Youth*. J. H. Kryder-Coe, L. M. Salamon, and J. M. Molnar, eds. New Brunswick, NJ: Transaction Publishers.

Roschelle, Anne R. (1997). *No More Kin: Exploring Race, Class, and Gender in Family Networks*. Thousand Oaks, CA: Sage.

Schwalbe, Michael, Sandra Godwin, Daphne Holden, Douglas Schrock, Shealy Thompson, and Michele Wolkomir. (2000). "Generic Processes in the Reproduction of Inequality: An Interactionist Analysis." *Social Forces* 79(2): 419–52.

Snow, David A., and Leon Anderson. (1987). "Identity Work among the Homeless: The Verbal Construction and Avowal of Personal Identities." *American Journal of Sociology* 92: 1336–71.

———. (1993). *Down on Their Luck: A Study of Homeless Street People*. Berkeley: University of California Press.

Stokes, Randall, and John R. Hewitt. (1976). "Aligning Actions." *American Sociological Review* 41: 838–49.

33

Dark Secrets and the Collective Management of Inflammatory Bowel Disease

ALEX I. THOMPSON

The way organizations help people collectively manage their stigma is the focus of Thompson's study of a group that supports sufferers of inflammatory bowel diseases (IBDs). IBDs comprise a set of conditions of the small intestine and bowel and are found mostly in people with Crohn's disease and ulcerative colitis. Thompson discusses his participant observation of two groups, a project that was driven by his own recent diagnosis and his transition from a membership to a research role.

IBDs, Thompson found, are especially stigmatized because they deal with issues normally considered "dirty," such as flatulence, diarrhea, incontinence, and other shame-generating symptoms. People customarily seek out support groups of like others to find a safe haven where they can surround themselves with those who will understand and not judge them and, at the same time, give them advice on how to manage the symptoms and stigma of their condition. Yet, surprisingly, Thompson found that even members of the two groups he studied censored themselves in this backstage realm, using group-approved euphemisms to replace certain topics and language.

Thompson's study is poignantly troubling because he lays bare the emotional struggles he and others navigated in handling their medical and social selves as they learned how to manage this stigmatized disability.

How would you compare the stigma management strategies members used in these groups with those discussed in the individual stigma management chapters preceding this one? At what point is it useful for them to pass over, cover up, or accent their deviant conditions? Why do you think the shame associated with these diseases was so strong for participants that they imposed such strict normative guidelines for conversing with each other? What do their stigma management strategies mean for the way they likely present themselves to outsiders?

Reprinted with permission from the author, Alex I. Thompson.

As the women and men around me, in turn, told of their recent triumphs over and tragedies caused by IBD, I became increasingly anxious. It wasn't what they said but what they didn't say that hit me in wave after distressing wave. They seemed to be discussing the dirty details of the disease without actually discussing them. As my turn approached, I felt like an actor about to be thrust onto a stage without having seen the script. Nevertheless, the director, Corey, gave my cue: "Alex, why don't you tell us a little bit about yourself." [Fieldnote excerpt]

Our bodies are made meaningful in the unfolding drama of social life in the settings in which we find ourselves, the props we find scattered about in them, the clothes we wear, and the words we choose. Bodies play a key role in shaping the selves that we present and *are* as the curtain rises and falls on stages we inhabit. Therefore, those living within chronically defiant "failed bodies" face an unending threat to their selves. When this bodily defiance is also entangled in stigma, people often encounter an additional barrier against maintaining a moral embodied self. By exploring the ways in which individuals cope with this dilemma, we have the opportunity to clarify the relationship between body and self, the contemporary boundaries of the private body, and the role of the unmentionable aspects of our physical bodies in the ongoing assembly of our dramaturgical bodies.

This discussion is born out of an ethnographic study broadly focusing on the "dramaturgical body." Goffman (1959) observed that, just as every successful theater production relies on a dirty, chaotic backstage, so does the presentation of self. Self-presentation depends on the "secretive creaturely body." In focusing on the dramaturgical embodiment of inflammatory bowel disease (IBD), I ask what happens when one is unable to keep this creature secretly tamed. I explore IBD support group dynamics and the disparate ways in which group members collectively harness language, and its absence, to protect their embodied selves from the stigma of fecal matters.

I intend this research to contribute to our understanding of the unique genre of interaction taking place in support groups. I aim to demonstrate, specifically in the case of those organized around chronic illness, that the alternative label of "self-help group" is true in more ways than one. A chronic illness is a chronic role, persisting across time, place, and audience. When support group participants come together to define their physiological and behavioral deviations, they are defining a permanent and pervasive part of their bodies and selves. I present a view into another setting focusing around the social significance and individual experience of feces and defecation—of the private body. This focus uncovers the pressures we feel and efforts we go to as we labor tirelessly to sanitize a relentlessly unsanitary body.

This paper explores the interactive processes by which the body afflicted with IBD comes to be. Life for the over 1.4 million Americans living within such a body is characterized by frequent and unpredictable diarrhea, flatulence, abdominal cramping, and incontinence. Therefore, IBD ties together both our understanding of the social and individual experience of chronic illness and that of fecal matters. IBD implies a life at the intersection of the unmentioned and the unmentionable, where the private body is constructed and maintained. This aspect of the human body, as all others, comes about dramaturgically.

METHODS

Data were drawn from a yearlong ethnographic study of a chapter of the Crohn's and Colitis Foundation of America (CCFA). I attended support groups, seminars, social outings, and fund-raising drives in a CCFA chapter located in the northeastern United States. Ultimately, my attendance was an attempt to better understand the particularities of the lives of those around me. However, finding the connection between the personal and the professional in qualitative work, I entered this setting not as a social scientist but as a young man.

In the spring of 2009, I was diagnosed with Crohn's disease, one of the two conditions classified as IBD. In the months that followed, I was overcome with embarrassment, shame, confusion, anger, and resentment. Although I had never kept a personal journal before, I began writing in one daily. Feeling that I needed to find an additional way to deal with these issues, I joined the CCFA and began attending one of the support groups sponsored by my chapter.

During my first support group meeting, I was fascinated and troubled by the ways that the other IBD sufferers around me chose to refer to their condition and its consequences. While I had initially expected to be able to discuss my experiences with IBD to whatever extent I chose, I soon realized that this was not the case. Nevertheless, because being in the group made me feel that I was not alone in trying to cope with the emotions and dilemmas I had recently been dealing with, I decided to return the next month. In the subsequent months, as I kept attending the support group meetings, my journal entries took on the appearance of field notes. I found myself paying attention to the specific words the support group members used and encouraged, those they avoided and discouraged, and the way that stigma shaped them. Curious about whether I would find the same phenomenon elsewhere, I increased my involvement in my CCFA chapter and gradually came to view my participation as an ethnographic endeavor. Over the next year, I took on the role of participant observer in three monthly, 90-minute, CCFA-sponsored support groups. My status as a Crohn's disease patient and CCFA member enhanced my ability to establish myself in that role. I also found that this personal connection allowed me to gain rapport with my respondents, a bond that aided in the interview process.

Following a year of participant observation, I conducted semistructured interviews with 12 support group members with IBD. These interviews ranged from 25 minutes to 90 minutes in length. Their primary focus was on interviewees' perception of, and experiences with, the stigma of IBD. During the interview process, I drew significantly on my own experiences with IBD, both to structure my interviews and to comfort my respondents during highly emotional conversations. This intimate personal connection did, however, pose a dilemma.

As my research progressed, my health declined. I was progressively over-whelmed by an aversion to the research. I found it suffocating to fight my own battle against IBD while I simultaneously sought to immerse myself in the battles fought by other sufferers. This conflict came to a head when I was hospitalized a year after my diagnosis. One incident, in particular, typified my bewilderment, as I recalled in the following field note excerpt:

> At just past 2:00 a.m. last night, I found myself impatiently trying to untangle my IV tube so that I could write a few jottings on the back of a "get well soon" card. After beginning to write, I paused and let a small ink splotch seep slowly from the tip of my pen onto the card's glossy surface. I was recalling the argument for ethnographers working where we live. What I was doing seemed like something else entirely, how-ever. It felt like debasing self-exploitation. I violently capped my pen and threw it across the room, yelling: "Stop! You're in the fucking hospital!"

An incident two days later, however, significantly changed my perspective:

> In preparation for today's colonoscopy and endoscopy, I had to take two different laxatives last night. In the midst of my countless trips to the bathroom, my stomach pain increased dramatically. So, as I tried, in vain, to sleep, my nurse gave me a considerable dose of morphine. While the morphine masked my pain, it also made me semi-comatose. I woke this morning to find that I had soiled myself in the night. I called for a nurse, who came in and left for clean bedding. As she stepped into the hallway, a nurse at the nurse's station called to her: "What's up?" She replied: "The guy in 10 shit all over himself." [Fieldnote excerpt]

Although I was initially mortified by the episode, I came to see it differently later. Having experiences like this allowed me to be intimately attuned to the implications of IBD in a way that a nonafflicted researcher might never have been. If I were not a sufferer myself, perhaps I never would have picked up on the way that the stigma and interactional consequences of this illness shaped IBD support group discourse.

IT'S A "BATHROOM DISEASE"

Living with IBD is no more or less than living with a chronic illness entrenched in the stigma of "fecal matters." The patients I spoke with expressed immense distress over the associated stigma. Of greatest concern was the strong association of IBD with restroom behavior, as Julian attempted to convey:

> I feel that it's kind of embarrassing just because it involves, you know, the whole bathroom situation so it's embarrassing so [long pause] I don't know [long pause] it's hard for people to understand it. It's just very

involved. The bathroom is like one of those forbidden areas that people just don't like to talk about. It's like sex and a few other things [long pause] actually sex is probably more acceptable. Everyone gets a little squeamish when you start talking about your bowel habits [emphasis in original].

As it was for Julian, all of the IBD patients I spoke with reported feeling shame and humiliation at the thought of, and pressure to avoid, discussing their illness publicly. IBD sufferers often explained it in the same guarded manner as Karl: "I'm ashamed about it. Don't ask me why I am; I just am." Elaborating on this attitude, Nancy confided in me: "It's a very difficult disease to communicate about and feel okay about having. No one wants to admit they have it because it's [pause] a 'bathroom disease.' No one wants to hear about that." In the following field note excerpt, I wrote of Shane's reflecting, during one of his support group's monthly meetings, on what this feeling meant for his visits to the pharmacy:

With his elbows on the table and resting his head in his cupped hands, Shane stared with glazed eyes at the laminate tabletop. Whispering almost imperceptibly, he said: "Buying enemas is humiliating, just humiliating. Because, they make you say what you want to refill. Then, they hand you that big bag that can only be one thing. Everyone knows it. I wish I could go through the drive-up window but that goddamn bag won't fit through the window!" As he wiped a tear from the outside corner of his right eye with his middle finger, he said, "I just want a normal life. I just want a normal life and to do normal activities and not have to worry they'll all mock me [emphasis in original]."

Most told of keeping their condition a secret from both friends and family. While some seemed to have taken it upon themselves to do so, others spoke of feeling pressured to remain silent. Nate recounted being overtly muzzled: "After I told her about my diagnosis, my mom gave me a look I'll always remember and said to me, 'Nobody wants to hear about it. We all have our issues.'" In contrast, others realized the importance of remaining silent among family members when they learned of a family history of gastrointestinal issues that those members refused to talk about.

Social castigation and tacit accusation of being physically unacceptable often pushed the IBD sufferers I spoke with to use an array of "normalizing tactics" throughout their daily lives. As Karen described,

I just have no one. No one else knows that I have this problem. I don't want them to know. I mean, I'll sit there and hold it for 8 hours if I have to. If I'm at a friend's house I'll eat anything they give me because I don't want them to know anything is wrong. That's what I do. I really *really* pay for it later, but I'd rather do that than be humiliated.

Echoing Karen's fear of humiliation, others reported traveling with a personal supply of toilet paper to mask the frequency of their bathroom visits, carrying a change of clothes in case of incontinence, limiting or eliminating their food intake for as much as a full day before and/or during social engagements, and mapping out the location of public restrooms along travel routes.

Although Turner (1984, 112) observed that "all illness is social illness," not all illnesses are created equal. The social sentiments attached to fecal matters stand as a significant hurdle in the lives of IBD sufferers above and beyond the hurdles created by the physiological challenges of the disease. The negative social construction of feces and defecation has made discussion of these matters largely taboo. The inability to openly discuss their illness with others places IBD sufferers in a precarious position as they attempt to go about dealing with the day-to-day implications of, reconciling themselves to a lifelong affliction with, and embodying this bodily abomination. Turning to the IBD support group, they find that the affliction that pulls the members together is also the dark secret threatening to push them apart. For, as Jill imparted to me, "Sometimes, I think I might say too much during the support group [meeting]."

SOILED WORDS

In a dramaturgical sense, support groups are teams working toward, and performing a working consensus on, the definition of the situation, a definition of the commonality that brings them together. As the curtain rises and the facilitator begins the support group meeting, the actors are ultimately responsible for presenting their ideal selves and for creating the opportunity for the others to do the same. This interdependence between actors, documented in Alcoholics Anonymous, Codependents Anonymous, Gamblers Anonymous, posttraumatic stress disorder group therapy sessions, and all kinds of support groups (for parents of children with disabilities, separation, divorce, and bereavement; parents of troubled teens; caregivers; organ transplant recipients; infertile women; and gay and ex-gay Christians), entails limitations on individual performances.

Across these settings, members arrive with a uniquely shaped narrative that is often forcefully recast to fit into the group's rigid framework. The actors are given a relatively fixed script containing the words they may choose from and the ways in which they may combine them. This script provides structure, answers, and a sense of normalcy for those who are often seeking all three.

In the IBD support group, the commonality structuring the performance was an unspeakable failed body. As IBD sufferers defined their condition, they were defining a permanently pervasive condition of the body and self. Although they were solely amongst those already "in the know" about IBD, support group members expressed statements about their bodies in such a way as to implicitly deny the entrenchment of their illness in fecal matters. By disregarding fecal matters in their discussions, support group members were able to maintain a self unsoiled by them; they were able to avoid internalizing a soiled self. There were several strategies they used to do so.

Sidestepping

As Inglis (2000, 164) suggests, "[W]ords are dirty if they refer to dirty things, and to avoid the shame carried by the latter, one must censor the former." Across the attempts at explaining the silences surrounding us is a consensus that silence acts as a barrier against embarrassment, shame, and attributions of immorality. In the IBD support group, members disregarded the potentially damaging influence of deviant bodies caked with fecal matters and reaffirmed the order of interaction by participating in "a conspiracy of silence," sidestepping soiled words.

In order to overcome the immorality often attributed to individuals who are incapable of stringently maintaining the boundary between their public and private bodies, IBD support group participants actively fostered silence through avoidance and euphemisms. Early on in my observations, Gary's attempt to describe, to the New Holland support group facilitator Ben, the recent improvements he had seen in his health encouraged me to explore this process. I documented the following in my field notes:

BEN: And how about you, Gary? How have things been goin' lately?

GARY: I'm good. It's much better.

BEN: Great! How so?

> Gary smiled, though it was small and tight-lipped, and fidgeted in his chair. He then furtively let his eyes and head dip towards his abdomen as he began:

GARY: Digestively it's [pause] things down there are [pause] I'm going less. I'm having more solid [pause] you know [pause] it just feels like things down there are [pause] better than they are otherwise."

Here, two common patterns in IBD support group discourse are evident. Members of each group made frequent references to things "down there," accompanied by eye and/or hand gestures to indicate the lower half of their body. "You know" was also a frequent catchphrase, indicating that support group members were relying on others to mentally fill in a blank in their narration.

Individuals also frequently made statements intimating that they relied on others to understand them solely upon the basis of those others' own intimate understanding of IBD. For example, Sheila informed the Mercer support group, "It's all, well [pause] not good. It's just a guess but I'm probably anemic [pause] had to buy a box of lotion Kleenex." It was clear to me, and likely to the others, from her statement that she had recently been dealing with bloody diarrhea and anal irritation but neither she nor anyone else said anything further on the matter. In a similar incident in the New Holland group, Rhonda simply shared "Well, lately I've had to start carrying a book of matches with me wherever I go." Again, this statement likely meant that she lately had had to manage excessive flatulence, although she did not choose to elaborate. Other examples included "Last week, all hell broke loose for me," "I'm a straight pipe," "I'm having symptoms," "Are you anemic?" "I'm stuck eating absolutely no fiber," and

> I did a naughty thing. I had a piece of banana cream pie. Bananas don't like me. I definitely paid for it. If I'm not careful, I always pay for it. I suffered from it for almost two weeks. I felt like I was going to die!

In addition to appealing to general avoidance and euphemisms, the Mercer support group had established a unique substitutive vocabulary that I have labeled "system speak." During their meetings, when attendees wished to refer to their bowel, their digestive tract, or its processes or by-products, they instead made reference to their "system." Moreover, if members wished to highlight the degree of severity of their diarrhea, they spoke instead in terms of the speed of their system. For example, instead of Allen stating that he had lately been suffering from frequent diarrhea, he said, "My system has been moving very quickly lately." Instead of describing oatmeal as having the ability to solidify his stools, Seth described it as "great for slowing down your system, really gums things up." In a related example, Olivia asked Nate, "Before your surgery, did you have a really fast system?" He replied, "Nope. I was actually always really slow." In my experience with the three support groups, the Mercer group had the steepest learning curve for new members. I credit this to the unifying utility of "system speak."

Soaking Up the Vocabulary

According to Goffman (1959), when preventative practices employed by a team fail and the definition of the situation is threatened by members acting "out of character," actors frequently turn to corrective practices. Often, these practices take the form of "staging cues"—"techniques for saving the show" (Goffman, 1959)—to reorient their performance. In the case of the IBD support group, when such preventative practices as avoidance, euphemism, and/or system speak failed, actors zealously gave staging cues. For example, new members to the group would make mistakes, speaking openly of diarrhea, flatulence, laxatives, and the like. But soon the old members told them that they were using soiled words and taught them to do otherwise.

Here, facilitators were the most vocal and active in their attempts to socialize new members. They acted as the directors, working to ensure that the show went on. They usually did so in one of two ways. At times, facilitators attempted to teach new members the proper script, or "sayings," through what I term "vocabulary lessons." These cues took the form of interruption and rephrasing directed toward relatively new members struggling to learn their role. Facilitators subtly rephrased individuals' statements in an acceptable form. For example, the following is an example from an evening at the Mercer support group:

> Tonight was Lora's first time attending the group. She was silent for about the first thirty minutes, sitting on her hands on the front few inches of her chair, avoiding eye contact with all of us. When it came time for her to share, she began by only answering Corey's probing questions with short, quiet, simple answers. After a couple of minutes, though, she began to speak more freely, slid back into her chair, and

began to make brief eye contact with each of us. In discussing a medication she'd recently began taking, she reflected: "I feel better and it
helps to keep the diarrhea under con…" Corey's hand shot up as if he
were a traffic cop urgently signaling her to stop. He interrupted her
with: "So, your system is moving pretty fast, right?" Lora's posture
stiffed slightly and her eyebrows came together as her mouth seemed to
fumble for an answer: "I [pause] um [pause] sure. I guess you could say
that. But I've just been having a lot of diarrhea." His hand still in the air,
Corey replied, "Yeah, when your system refuses to slow down like that,
it's tough. We've all been there." [fieldnote excerpt]

More often, however, facilitators preemptively interrupted new members. In
a closely related happening at a New Holland support group meeting, Ben used
the form of interruption common to the group facilitators:

BRYAN: The cramps are the worst and [pause] I can't take any laxatives and I
[pause]

BEN: Oh, ok. Yeah. Yeah. You can't. I know. I hear ya.

In these instances of definitional disruption of new members, the other group
members played a smaller role in correcting the situation. Their cues were largely
passive and included downcast eyes, fidgeting, and "conspiracies of noise," such
as nervous laughter and audible sighs. Although most new members, apparently
picking up on these cues, adopted the discursive practices imposed upon them by
the group, a few seemed unable to internalize the need to do so and stopped
attending.

I asked the Mercer support group facilitator what type of influence he perceived such seemingly unaware group members as having on his support group.
He reflected on the challenges that frankness and visibility can create:

Well [pause] you have people like Greg. You remember him, right? He
showed up for a few months but hasn't been around for more than a
while. He was just sooo [pause] *outright*. You know what I mean? He'd
just sit there and tell you his *entire* life story. It's like when you've got
those people who come in here with an IV 'cause they can't keep
anything down or like that guy that came last month who's got no
treatment options left. It all just makes me, and I'm positive everybody
else too, *really* uncomfortable. This is a safe place, but not *that* safe.

As the facilitator highlighted, the structural integrity of the situated reality fostered in the group is akin to a house of cards. When members' words or bodies
betray this reality, their presence is often akin to the slight breeze capable of top-
pling the house and burying everyone inside.

Suspending the Rules

There were exceptions to the restrictions placed on using soiled words in the
IBD support groups. When group discussions were either explicitly clinical or

joking in nature, the members' usually tight lips were noticeably looser. In these circumstances, support group members were protected from attributions of fault or shame either by the power of technical medical jargon to sterilize their spoiled bodies or by the utility of satire to relieve the tension created by their deviance. Group members demarcated two acceptable extremes within which fecal matters were appropriate elements of discussion, on either side of an "unmarked" region where such discussions were profane.

In avoiding the midrange of the continuum between "sterilization" and "satirization," a range over which public discussion of fecal matters is discrediting, support group members engaged in a form of "the ceremonial labor of person production." At the extremes, only when in the presence of either medical professionals or those choosing to make light of IBD did support group members allow one another to discuss the dirty details of their illness. The "marked" categories of "sterilization" and "satirization" constitute a reprieve from the moral necessity of shying away from soiled words and from the acknowledgment of the body and self that they imply.

Sterilization

Numerous scholars have argued that a biomedical diagnosis carries with it legitimation and forgiveness for an ill individual's physiological and social limitations. Goffman (1963) pointed to the utility of being defined as "physically sick," for when individuals are so defined, their deviance no longer threatens the moral evaluation of their group.

Each support group I attended hosted a physician at least once during the year that I spent with them. On each of these occasions, when the IBD support group setting could be framed as clinical, the members were open about the details of IBD that were otherwise unspeakable. As I wrote in my field notes one evening after a meeting of the Joliet support group,

> We had a guest speaker at the meeting this evening. She was a well-known gastroenterologist from the [a nearby city] IBD specialty clinic. I knew before the meeting that she would be in attendance but I suppose I didn't think much of it until a few minutes into the meeting, at which point I was quite astonished. The physician turned to Chad, who had just discussed the bowel resection he underwent as a child, and asked, "Have you been dealing with considerable diarrhea? When a patient has the type of bowel resection that it sounds like you had, it's not uncommon for them to end up having to manage a high level of motility." I felt my eyes widen as I listened to the conversation that followed:

CHAD: Yeah. They say the body adjusts and I've been with it for seventeen years now. So, there's a little adjustment but it's kind of like [pause]

DR.: Loose?

CHAD: Still a little loose, yep.

DR.: Not watery?

CHAD: Not watery. I don't really get the diarrhea unless [pause] well, caffeine is a big killer. I remember when I first had the surgery; a can of soda would come out of me within minutes because of the caffeine.

JOAN: Do you eat any kind of natural foods, organic, that type of thing?

CHAD: I stay away from fresh vegetables because of the gas and the fiber. I don't need the fiber. I don't need anything to be pushing anything out any quicker than it normally does.

CELIA: I had two bowel resections. Now no matter what I eat, it doesn't stay in there.

SARAH: See, I'm the opposite. A lot of people have more diarrhea than me.

Upon hearing and later recalling this conversation, I noted how incongruous it was relative to what I had come to expect from this group and these members. Chad, Joan, and Celia each had been attending the Joliet group for at least 5 years, and although in meetings before this one they had stuck strictly to the sidestepping, avoidance, and euphemism tactics described earlier, on this evening they lapsed into openness, without pause, at the addition of medical professionals. Throughout the meeting, members brought up and freely discussed bowel movement frequency, consistency, odor, and incontinence.

Two months after the aforementioned exchange in Joliet, the members of the Mercer support group showed trepidation at allowing physicians into their group because they feared that they would be unable to oppose the "sterile" clinical approach that the physicians' conversation might foster. They resolved their mixed feelings by inviting only one doctor instead of the four who offered to come because they were reluctant to open their dynamics up to the kind of medical sterility and frankness the doctors' discourse would foster.

CONCLUSIONS: SELF-HELP AND FECAL MATTERS

The self is less a finished product than an ongoing work in progress found within the discrete scenes in which we act. We are the actions we undertake, the statements we make, and the narratives we tell. However, for the stigmatized, finding a place to make these declarations and tell these stories can be a thorny proposition. Moreover, even once they have seemingly found such a place in a support group, they often find themselves both the censored and the censors within it.

Goffman (1963, 36) argued that the veteran stigmatized go to great lengths to "instruct him in how to manage himself physically and psychically." This management is aimed not only at helping the new member but also at helping other members of the support group. Those in attendance view themselves and their bodies through new members' gazes, gestures, and expressions, through the scene that they help to stage. As attendees share their stories with each other, they help each other amend them so as to not offend one another. For, support groups are far less settings for people to talk about their selves and far more settings that they *bring about* by learning how to talk about their selves.

In short, with no need or utility to be found in passing, support group members use their gatherings not as performances but as rehearsals. They come together and rehearse the lines making up the narratives that allow for the emergence of the most admirable self, given their shared hardship. Recognizing this tendency among the chronically ill, Charmaz (1991, 4) noted that "most people live with their illnesses rather than for them. Often they try to keep illness at the margins of their lives and outside the boundaries of their self-concepts." In this regard, Charmaz champions the liberating power of private gatherings of those kept silent by fear and humiliation. In breaking the silence, group members are capable of forming favorable feelings about their selves through the eyes of those around them.

The IBD support group offers a unique window into these support group processes and their interplay with members' body perceptions and self-perceptions. Participants in IBD support groups practice misdirection and use euphemisms to hide their befouled private bodies from their ideally immaculate public selves. Even in what is advertised to be a setting of honesty and openness, in which a frequent facilitator refrain is "Don't worry; we've heard it all," they do not allow themselves or those around them to verbally acknowledge that IBD is a bowel condition, a life spent on a toilet. In so doing, individuals with IBD alter and fashion the body and self that their stories exhibit. They discursively remold an immoral failed body steeped in feces into a vaguely challenged one.

REFERENCES

Charmaz, Kathy. 1991. *Good Days, Bad Days: The Self in Chronic Illness and Time*. New Brunswick, NJ: Rutgers University Press.

Goffman, Erving. 1959. *The Presentation of Self in Everyday Life*. New York: Doubleday.

———. 1963. *Stigma: Notes on the Management of Spoiled Identity*. Englewood Cliffs, NJ: Prentice Hall.

Inglis, David. 2000. *A Sociological History of Excretory Experience: Defecatory Manners and Toiletry Technologies*. Lewiston, NY: The Edwin Mellen Press.

Turner, Bryan S. 1984. *The Body and Social Theory*. Thousand Oaks, CA: Sage.

The Social Organization
of Deviance

In Part VI, we turn to a closer examination of the lives and activities of deviants. Once they get past dealing with outsiders, they must deal with other members of their deviant communities and with the specifics of accomplishing their deviance. There are several ways of looking at how deviants organize their lives. We start by examining the relationships among groups of deviants, focusing on the character, structure, and consequences of different types of organizations. This typology encompasses the structure or patterns of relationships in which individuals engage when they enter the pursuit of deviance.

As Best and Luckenbill (1980) have noted in their analysis of the social organization of deviants, relationships among deviants can follow many models. All of the models vary along a dimension of sophistication involving complexity, coordination, and purpose. Deviant associations differ in their numbers of members, the task specialization among members, the stratification within the group, and the amount of authority concentrated in the hands of a leader or leaders. Some groups of deviants are loose and flexible, with members entering or leaving at their own will, uncounted and unmonitored by anybody. Others maintain more rigid boundaries, with access granted only by the consent of one or more insiders. Membership rituals may vary from none to highly specific acts that must be performed by prospective inductees, thereby granting them not only membership, but also a place in the pecking order once they are inside. In some ways, rigidity inside deviant groups is related to their insulation from conventional society: The more a group's members withdraw into a social and economic world of their own, the more they will develop norms and rules to guide them, replacing those of the outside order.

401

Groups of deviants also vary in their organizational sophistication, with the more organized groups capable of more complex activities. Such organized groups provide greater resources and services to their members; pass on the norms, values, and lore of their deviant subculture; teach novices specific skills and techniques when necessary; and help one another out when they get in trouble. As a result, individuals who join more tightly knit deviant scenes tend to be better protected from the efforts of social control agents and more deeply committed to a deviant identity.

Best and Luckenbill outlined a range of ways by which deviants may organize socially. **Loners** are the most solitary, interacting with people, but keeping their deviant attitudes, behaviors, or conditions secret. They lack the company of other, similar deviants with whom they can share their interests, troubles, and strategies. Serial rapists or murderers and embezzlers commit their acts without the benefit of the camaraderie of like others. However, through the rise of Internet communities, many individuals who hold themselves as loner deviants in the real world are now finding that they are not alone. Without jeopardizing their identities and social relations with conventional people, they have found ways to connect anonymously with people who share their deviance and from whom they seek company and advice. People such as sexual asphyxiates, self-injurers, anorectics and bulimics, computer hackers, depressives, pedophiles, and others now have the opportunity to go online and find international cybercommunities populated 24/7 by a host of like others. These Websites offer chat rooms, Usenet groups, email discussion lists, and bulletin boards or message boards for individuals to post where they can provide or seek the advice and cybercompany of others. Some, such as the "proana" (anorexia) and "promia" (bulimia) sites explicitly state that they reinforce and support the deviant behavior, regarding it as a lifestyle choice (Force, 2005). Others, such as many self-injury sites, purport to help users desist from their deviance but may actually end up reinforcing it by providing a supportive and accepting community to whom individuals can go when they feel misunderstood and rejected by the outside world, as we note in our own Chapter 35 on the cybercommunities of self-injury.

Whether the sites aim to reinforce or discourage the deviance, nearly all tend to serve several unintended functions that have significant consequences for participants. First, they transmit knowledge of a practical and ideological sort among people, enabling them to engage in and legitimate the behavior more effectively. This knowledge helps people learn new variants of their activities, how to carry them out, how to obtain medical or legal services, and how to deal with outsiders. Second, the sites tend to be leveling, bringing people together into a common discourse, regardless of their age, gender, marital status, ethnicity, or socioeconomic status (although users do need a computer, and most have high-speed Internet access). Third, they bridge huge spans of geographic distance, putting Americans in contact with English-speaking people from the United Kingdom, Australia, New Zealand, Canada, and, in fact, all over the world.

These interactions, regularly conducted among a range of heavy and moderate users, as well as periodic posters and "lurkers" (those who read but do not post), forge deviant communities. Participants develop ties to those communities by virtue of the support and acceptance they offer, especially to individuals who are lonely or semi-isolated. People who are unable to find "real" friends "FTF" (face-to-face) may come to rely on these cybercommunities and cyberrelationships, interacting with members for years and even traveling large distances to meet each other. Such Internet relationships may come to take the place of core friendships. In this way, if in no other, they reinforce continuing participation in the deviance as a way of maintaining membership. The stronger and more frequent the bonds, the greater effect they have on strengthening members' deviant identities. Deviant cybercommunities thus provide a space and a mechanism for deviance to grow and thrive in a way that it has not previously had.

Colleagues represent the next most organizationally sophisticated associational form. Participants have face-to-face relationships with other deviants like themselves but do not need the cooperation of fellow deviants to perform their deviant acts. The jump from loner to colleague is the greatest leap in the spectrum of deviant individuals, as mutual association brings the possibility of membership in a deviant subculture or counterculture from which people can learn specific norms, values, and rationalizations; helpful information; specialized terms or vocabulary; and gossip about people like them. From others, they can gain social support, as we see with the expressive groups discussed earlier, and a sense of their position in the status hierarchy of their kind. Deviants organizing as colleagues include the homeless, recreational drug users, and con artists. Colleagues may interact and perform their deviance with nondeviants, such as prostitutes' clients, or johns.

Deviants socially organized as **peers** engage in their deviance with others like themselves, but have no more than a minimal division of labor. Members of neighborhood gangs who congregate with their friends generally engage in all of the same types of activities and see role specialization only when it comes to the leader versus the followers. Most peers traffic in a black market of illegal goods and services, such as drugs, guns, endangered species, stolen art, and exotic forms of sex.

Especially fascinating to the media and the public is the **crew** form, in which groups of anywhere from three to a dozen individuals band together to engage in more sophisticated deviant capers than less organized deviants can accomplish. Crews fascinate observers because their more sophisticated division of labor usually requires specialized training and socialization, giving them a more professional edge. Bounded by their lack of affiliation with other crews, they are dependent on a leader who organizes and recruits them, sets and enforces the group rules, plans their activities, and organizes travel and lodging if they go on the road. Crews usually commit intricate forms of theft, but they may also engage in smuggling and in hustling at cards and dice.

At the top of this organizational continuum are deviant **formal organizations**, which are much larger than crews and extend over time and space. They may stand alone or be connected to other, similar organizations domestically or even internationally, as we see with the Cosa Nostra Mafia families and the Colombian drug cartels. Their affiliations may take the form of what Godson and Olson (1995) call "transnational links," whereby they have regular connections to do business or exchange services with other criminal organizations, or they may have a "global scope," by which they conduct extensive operations in various continents through franchised branches of their own organization located in different places. In this regard, they are akin to multinational corporations. Much larger than crews, deviant formal organizations may have 100 or more members, so even if their leaders are killed, the group endures. Ethnically homogeneous, these organizations trade in a currency of violence, are vertically and horizontally stratified, and have the resources to corrupt law enforcement. They are the most organizationally sophisticated of the deviant associations formed for purely deviant purposes.

Legitimate individuals and organizations also engage in deviant activities, although these activities may be their side, rather than primary, purpose. It is worth noting that, although most studies of crime and deviance focus primarily on crime in the streets, a more socially injurious amount of deviance occurs at the top, in the suites, through **white-collar crime**. But the FBI's *Uniform Crime Reports* publishes statistics only on "street crime in the United States," so, in this document, you will read about burglary, robbery, and theft, but not about price-fixing, corporate fraud, pollution, or public corruption.

In the Introduction to Part III, we noted Cloward and Ohlin's belief, advanced in their differential opportunity theory, that access to illegitimate opportunity is unequal. They talked about people from distinct neighborhoods, ethnic groups, and criminal ladders having better criminal opportunities, but their concept of differential opportunity can apply to the privileged as well. White-collar crime is directly related to opportunities to abuse positions of financial, organizational, and political power. We often tend to associate white-collar crimes with strictly financial activities, but they extend to bodily injury and death as well. The Federal Bureau of Investigation (FBI) estimates that 19,000 Americans are murdered every year. Compare this figure with the 56,000 Americans who die every year on the job or from occupational diseases such as black lung and asbestosis and with the tens of thousands of other Americans who fall victim to the silent violence of pollution, contaminated foods, hazardous consumer products, and hospital malpractice. These deaths are often the result of criminal recklessness. They are sometimes prosecuted as homicides or as criminal violations of federal laws. And environmental crimes often result in death, disease, and injury. In 1998, for example, a Tampa, Florida, company and the company's plant manager were found guilty of violating a federal hazardous-waste law. Those illegal acts resulted in the deaths of two 9-year-old boys who were playing in a dumpster at the company's facility.

Perhaps the most telling statistic in the field of deviance/criminality is that the cost of white-collar deviance to the average U.S. citizen is much greater than that of the so-called street crime. The FBI estimates that the United States loses $3.8 billion a year through burglary and robbery; Compare this figure with the loss of $300 to $500 billion a year for health care fraud alone. Yet the perception of many is that we are at a greater risk from street crime than from crime that emanates from executive suites. This perception is reflected in the justice system: The average sentence for bank robbery is 7.8 years; it is 2.4 years for embezzling money from the same institution.

White-collar crime can be divided into two main sections: occupational crime and organizational crime. **Occupational crime** is pursued by individuals acting on their own behalf. Employees at all levels of organizations may steal from their companies, and we have also seen the rise of embezzlement and computer crime. Corporate executives at such firms as Enron, Tyco, and WorldCom looted their companies, shareholders, and employee retirement plans of millions of dollars through fraudulent accounting, offshore and dummy corporations, and the manipulation of information so that they could live in high style. Individuals in charge of purchasing for their firms also frequently accepted bribes to give business to vendors. Ponzi schemes, such as the ones run by Bernard L. Madoff, Allen Stanford, Michael Kelly, the Villalobos brothers, and others defrauded investors by allegedly generating high rates of returns on investments that were secretly paid for only by the influx of funds from new subscribers. These schemes are named after Charles Ponzi, the "king of get rich quick," who became a millionaire in 6 months by promising investors a 50 percent return in 45 days on international postal coupon investments.

In the government sector, we see people evading taxes through offshore companies and fraudulent tax shelters, often sold to them by top accounting and brokerage firms that charged millions of dollars for these services. Politicians—especially those in charge of awarding government contracts—have been caught selling power. A case in point is now-jailed U.S. Republican Congressman Randy "Duke" Cunningham, the former Air Force flying ace (whom Tom Cruise portrayed in the film *Top Gun*), who had a price list in the tens of thousands of dollars for military appropriations. Politicians also sell business to companies (sometimes in no-bid contracts) whose products and services are inferior or if they have a stake in those companies. Politicians and police officers may also receive individual remuneration for selling immunity from prosecution to criminals or companies, in either direct cash payments or campaign contributions. People connected to the government may also sell their influence, as we saw in the scandals surrounding former U.S. Republican Congressman Tom DeLay and lobbyist Jack Abramoff, both of whom collected millions from Indian tribes to secure their gambling interests.

Nor are professionals above collecting money for their individual benefit, as we see with doctors who accept gifts from pharmaceutical companies to steer their patients toward certain drugs, who overcharge and overservice their patients, and who commit Medicare and Medicaid fraud, some of which is discussed in Liederbach's Chapter 22. Top stockbrokers and their clients make money through insider trading, such as what we saw in the Martha Stewart scandal.

Organizational crime, committed with the support and encouragement of a legitimate formal organization, is intended to advance the goals of the firm or agency. Looking at the corporate world, we see that environmental crimes top the list, with double the dollar amount of the next-closest offenders. We also see many instances of false advertising, with products (e.g., air and water purifiers, fire retardants) misleadingly alleged to do something and either failing to do so or having the reverse effect, and fraud, with companies misrepresenting themselves to investors and the general public. Antitrust violations are the second most common types of corporate offenses. In these situations, companies (e.g., Microsoft, Comcast) engage in monopolistic practices to control the market, artificially subsidizing or cheapening their products or services (microchips, airlines, entertainment products) to drive competitors out of business or conspiring (Samsung, General Electric) with other companies to set minimum threshold prices for consumers. Corruption among companies with government service contracts is rife, with $5 billion to $7 billion lost annually during the Iraq war alone. Because of their immense political power, big corporations have the resources to defend themselves in courts of law and in the court of public opinion.

Injury and loss of life may result from unsafe working conditions, as we saw in 2005 in a number of mining deaths. In these tragedies, governmental regulators let enforcement slide among several drilling organizations so that they could strengthen their business. Widespread illness and fatalities have also arisen because of the working conditions found in nuclear power plants, oil and chemical companies, and pesticide manufacturers. But few Americans realize that, when they buy Exxon stock or when they fill up at an Exxon gas station, they are in fact supporting a criminal recidivist corporation. For every company convicted of polluting the nation's waterways, many others are not prosecuted because their corporate defense lawyers are able to offer up a low-level employee to go to jail in exchange for a promise from prosecutors not to touch the company or any of its high-level executives. Unsafe products represent another case in which companies put their balance sheet above the lives of consumers, figuring that it is cheaper to settle lawsuits against them than to fix the company's goods. Notable offenders are the pharmaceutical companies, the automobile and tire industry, medical manufacturers, the cattle industry, and even peanut butter companies. This system continues to thrive because corporations define the laws under which they live. For example, the automobile industry has worked its will on Congress over the past 30 years to block legislation that would impose criminal sanctions on knowing and willful violations of the federal auto safety laws. Now, if an auto company is caught violating the law and if the cops are not asleep at the wheel, only a civil fine is imposed.

A disturbing amount of government activity also falls into this category, with politicians abusing the public trust, manipulating information, and breaking laws to advance their administrations. In fact, campaign finance represents the third most common area of corporate crime. Yet, for every corporation convicted of bribery or of giving money directly to a public official in violation of federal law, there are thousands who give money legally to candidates and political parties through political action committees. These companies profit from a system that has effectively legalized bribery. American international policy is often clearly tied to the interests of the corporate sector, most notably recently with the oil industry in Middle Eastern diplomatic and military activities. The K Street project was designed by Republican strategist Karl Rove to engineer a Republican takeover of the lobbying industry, bringing corporate and governmental interests and financial obligations closer together, to enrich politicians and favor their contributors. Numerous domestic governmental scandals, such as the Watergate break-in, the Iran-Contra scandal, and the National Security Agency (NSA) warrantless surveillance controversy, have erupted in violation of the law. The spying of the CIA, the military, the FBI, and private burglars into the telephone records, bank accounts, Internet logs, library records, and credit card transactions of alleged terrorists may be acceptable to the American public, but when these things are done in violation of law or in the interest of political parties and against reporters, political opponents, chaplains or lawyers counseling political prisoners, or antiwar activists, they are regarded by a large number of Americans as very grave offenses indeed. International violations have also been common, with the secret CIA "black site" torture prisons in unknown Eastern European countries, Iraq's Abu Ghraib prisoner abuse violations, Guantanamo Bay prison in Cuba, and numerous political dirty tricks and secret assassination attempts. All of these white-collar crimes have led to greater cost, more loss of life, forfeit of international prestige, and the violation of conventional norms and values than the sum total of conventionally recognized crime and deviance. Yet big companies that are criminally prosecuted represent only the tip of a very large iceberg of corporate wrongdoing.

34

Drug Use and Disordered Eating Among College Women

KATHERINE SIRLES VECITIS

The intersection between eating disorders and drug use represents a growing form of contemporary deviance. Sirles examines this hidden and highly stigmatized behavior, drawing stark contrasts between her two dimensions: priority (Which came first, eating disorder or drug use?) and the legality of the substances used (pharmaceuticals versus street drugs). Her resulting fourfold table illuminates several pathways that college women take in adapting to the enormous pressures found in contemporary society—especially among the population she interviewed— to conform to feminine standards of beauty. You may be alternately drawn to the easy solutions her subjects offer and repulsed by them, but their solutions will not be easily forgotten. This behavior compares to other loner forms of deviance, including sexual asphyxia, anorexia and bulimia, embezzling, rape, and physician and pharmacist drug addiction.

How would you assess the relative stigma of each of Sirles's four types? How would you assess them in terms of Heckert and Heckert's categories? How do Sirles's types make you feel about these categories?

When I was an undergraduate sociology student, I read the story of a young woman's long-time struggles with disordered eating. Her narrative had an impact on me, particularly because of her descriptions of street amphetamine use for appetite control. While I was certainly familiar with eating disorders generally, I was awestruck that this young woman reported using powerful, potentially dependence-producing substances in order to control her weight. Entering graduate school, I turned to this topic avidly, and was not surprised to find that sociological research had not previously been used to analyze the relationship between drug use and eating disorders, most prior studies coming from the medical establishment. Sociological scholarship on substance use seemed to focus on

From Katherine Anne Sirles, *Drug Use and Disorderly Eating Among College Women*. Reprinted by permission of the author.

either recreational or medicinal manifestations, of which this specific type was neither. Instead, the use of drugs for weight control struck me as utilitarian, purposeful, and extreme. In addition, the use of substances in this manner was largely the pursuit of women, a population that I was most suited to research. This research reports original research on college women who used licit pharmaceutical drugs or illicit street drugs in an ongoing effort to manage their body weight. My goal was to analyze women's use of these substances as a tool in their attempts to modify their bodies.

Extreme weight control practices, which I term eating disorders or disordered eating, are a serious problem. Although there is evidence that eating disorders have existed since ancient times, before the late 1960s they were virtually unknown to the general public. Diagnosis of anorexia nervosa, bulimia nervosa, and other eating related medical syndromes skyrocketed during the 1970s. As rates of disordered eating increased among young women, the public took notice. Mass media fueled interest in the subject as women began starving themselves, or bingeing and purging. This increased interest led to not only popular reporting of the phenomenon, but also academic scholarship in the field. Since the 1980s, researchers have analyzed problematic eating behaviors from numerous different perspectives.

The majority of research focusing on women's efforts to control and manage their weight has failed to recognize the potentially extreme measures that young women may take in an effort to conform to an American ideal of feminine beauty. Increasingly, women who are or wish to be a part of the "cult of thinness" are turning to street drugs and illegally or unethically obtained prescription drugs in an effort to be ultra-slim. Despite the substantial amount of research in the area of risky weight management, this drug use, both licit and illicit, has not been documented. Existing literature on the relationship between substance use and weight loss has understood drug use (e.g., alcohol, depressants, or antidepressants) mainly as a coping mechanism for the trials and tribulations associated with extreme weight-controlling behaviors. While the exact extent of drug use and abuse is not known, the lack of scholarly inquiry into the relationship between substance use and intentional weight management has left the impression that they [are] two completely distinct behavioral patterns.

METHODS

In this paper I draw on in-depth life-history interviews I conducted with 57 college age women at a large, public university. This was an appropriate place to study drug use and eating disorders because body image distortions and disturbances in eating were common among college women, and illicit drug use was widespread at university campuses. As women transitioned from high school to college, changes occurred socially, psychologically, and academically. Women's eating patterns, usually set by family in the past, were subject to change, and the pressures at college of academics, dating, and peer expectations were precursors to developing eating disorders. Finding subjects for this study was difficult, as this behavior is hidden by people, even from their closest friends. I gathered a convenience sample of anyone I could find to interview, culling research participants through the posters with which I blanketed the campus and from visits I conducted to many classes (upper and lower division), where I announced my intentions for this project. Participants contacted me via email or phone after my research solicitations, whether they heard about it directly or through others on campus. I prescreened research candidates before scheduling an interview to insure that individuals currently or historically used drugs, either licit or illicit, for the primary purpose of weight management.

Class rank among participants ranged from freshman to seniors, and also included three recent college graduates who remained engaged in the college environment. Women lived both on and off campus. All participants were between the ages of 18–25, with most identifying as white. As college students, women in this sample represented a relatively privileged, well-educated group, predominantly reporting growing up in either middle or upper-middle class homes. As such, participants fell squarely within the population that has already been identified by researchers as most prevalent among women with disordered eating. I conducted semi-structured interviews that were very open and conversational in the privacy of a campus faculty office or empty classroom, and these usually lasted between one and two hours. After assuring my participants of complete confidentiality, I found that they were more than willing to talk to me about their drug use and histories of disordered eating.

TYPOLOGY OF INSTRUMENTAL DRUG USERS

Participants reported using two distinct types of substances: licit (pharmaceutical) and illicit (street) drugs, all falling into the category of stimulants. The packaging and distribution of these substances varied, but they all shared the side effect of appetite control. Women's goal-orientation (as opposed to recreation or experimentation) in their drug use defined them as *instrumental* users, motivated by the substances' specific effects. Not unlike bodybuilders who used steroids instrumentally, women in this research conceived of their drug use as essentially performance enhancing.

At the same time, instrumental drug using women varied according to the *temporal* nature of their disturbed eating and drug use. Some women reported disordered eating behaviors *before* the onset of their instrumental substance use. For these women, the discovery that there were drugs that could assist them in achieving their body and weight-control goals followed months or years of experience with significant and deliberate weight management. Other participants reported the development of nonnormative weight managing behaviors *after* a period of drug use. Generally, these women transformed their personal drug usage patterns from recreational or medicinal into instrumental use. This usually followed positive social reinforcement and personal satisfaction about

TABLE 34.1   Typology of Instrumental Drug Users

	Licit Drug Users	Illicit Drug Users
Disordered Eating Foundation	Conventional Over-Conformist	Scroungers
Drug Use Foundation	Journeyers	Opportunists

changes in their bodily appearance. Women who then identified drug use as a desirable means to maintain or continue weight loss, and perceived the results of their efforts as positive, were compelled to further develop their behaviors. To be clear, I termed the behavior women engaged in first, whether it was drug use or problematic eating, as women's instrumental drug using *foundation*. Table 34.1 introduces the four types of instrumental drug users analyzed in this research:

The first category of women consisted of the "conventional over-conformists." These women reported a history (foundation) of disordered eating prior to their instrumental prescription drug use for weight loss. They were conventional in their choice to use the more socially accepted prescription drugs instead of street drugs. Although they used drugs instrumentally, their overall motivation centered on achieving the cultural ideal of thinness; their goal was conformist. Most of these women's weight management goals evolved from conforming to over-conforming, with an acute fixation on weight that tended to exceed average social expectations for personal body modification. Participants' excessive adherence to ideals of beauty, along with their nonnormative means used to accomplish this ideal, distinguished them as deviants (although their deviant means were largely unknown to others in everyday life). Most conventional over-conformists presented as thin, but not too thin, and constituted the largest category in my typology of users ($n = 24$).

The second largest category of instrumental drug users ($n = 13$) [was] the "scroungers." This group consisted of women who reported a foundation of disordered eating, only later (after the onset of problematic weight control) turning to street drugs for weight control. I call them scroungers because their choice of street drugs represented a much less socially accepted form of substance use. Many of the women conceived of these drugs as "dirty," "unacceptable," or "inappropriate," highlighting their conception of illicit drugs. Their access to illicit substances was customarily not as consistent or reliable as it was for those using pharmaceuticals, leaving them, at times, forced to forage or scrounge for their supply. Generally speaking, however, women reported scarcely more difficulty in obtaining street drugs than was typically expected for illicit substances.

Third, women who used prescription drugs recreationally or medicinally prior to their instrumental use for weight control made up the category of "journeyers" ($n = 11$). This term depicts the journey, or evolution, through which their drug use patterns evolved. While some women in this category reported using pharmaceuticals instrumentally for academic purposes, intentionally gearing their drug use toward weight management shifted aspects of their deviant career in ways that were specific to their body modification goals. For example, journeyers tended to ration their supply of pharmaceuticals in a different manner once weight management became a priority.

The last, and smallest, category was the "opportunists" ($n = 9$). Opportunists were women who initially used street drugs recreationally, and later transformed their drug use into instrumental patterns for weight control purposes. I call them opportunists for the way they recognized the positive social feedback about their drug-assisted weight loss and defined it as an opportunity for the transformation of self. Like journeyers, opportunists engaged in substance use before they turned into instrumental drug users. Opportunists, like journeyers, shifted a wide array of behavioral practices in order to accommodate their instrumental drug use. Opportunists did not necessarily stop using drugs recreationally during their periods of instrumental use; in fact, many participants (not just journeyers or opportunists) reported recreational drug use, some on a fairly regular basis.

SOLITARY DEVIANCE

Some deviance is practiced by individuals alone, away from the company of others. Scholars have conceptualized forms of solitary deviance in a number of ways. Best and Luckenbill (1980) described individuals engaged in deviance alone as "loners." These people did not associate with others "for the purpose of sociability, the performance of deviant activities, or the exchange of supplies and information" (p. 15). Prus and Grills (2003) further elaborated categories of individual deviance through their work on "solitary operators" and "subcultural participants." Solitary operators, like loners, acted alone, and did not associate with deviant others. Subcultural participants also acted alone, but their deviant behaviors were heavily influenced by particular group memberships. Other examples of loner deviants include sexual asphyxiates, self-injurers, substance abusing pharmacists, embezzlers, and anorectics and bulimics. While loners deviants lack contact with the company of fellow similar deviants but may victimize others, as is the case with obscene phone callers or violent individuals, the women in this research were completely solo actors.

In Adler and Adler's study of self-injurers (2005: 352), they described loners as "on their own," noting that they "must find the deviance, decide to engage in it, and figure out how to do it themselves." Unlike other forms of deviant associations, where individuals learn from and support each other, loners evolve from conforming behavior to nonconforming behavior without any aid. Such an evolution demanded that individuals singularly create the means and rationalizations for any given behavior. This, in a large way, differentiated loners from other varieties of deviants. They lacked others to introduce them to specific behavior and guide them through the learning process, to offer support and guidance throughout their deviant careers, and to provide them with alternative values and ideals, helping them define and rationalize their behavior. They also lacked help dealing with the practical hurdles or problems of the deviant lifestyle. Instrumental users were unable or unwilling to seek out deviant others, and thus lived without the support that potentially accompanied deviant associations.

Generally, loners subscribed to normative social values, ideas, and behaviors. They were "entrenched" in the dominant culture and "likely, then, to view their deviant acts through the value system of conventionality" (Adler and Adler 2005: 347). This contradiction of normative values and deviant actions was ever-present for women in this research, causing a strain that many felt forced them to keep nonnormative behaviors private.

Loners and Secrecy

For the most part, women in this research reported employing lies and secrecy during the course of their deviant behaviors. They were fully capable of engaging in, or at least sharing information about, their deviance with trusted others, yet they consistently chose to keep these behaviors hidden. In an effort to lessen the potential negative consequences (such as labeling) that might result from others knowing, individuals chose the loner lifestyle. Their fear of stigmatization was not the only motivation for keeping behaviors private; many women feared that if others knew about their instrumental drug use, they would be forced to stop.

Women who were motivated to adopt extreme weight control practices needed to look no further than an Internet search on "pro-anorexia" or "pro-bulimia" to find other like-minded individuals gathered in an online community. They could have thus easily sought out the company of like others. Online, individuals were willing to trade tips and tricks for weight loss, explaining and neutralizing their behaviors as a "lifestyle choice." However, outside of the anonymous support and camaraderie an Internet community might have offered, women practicing extreme weight control were often embarrassed about their methods. As a result, they went to great lengths to maintain secrecy. This consistent drive to conceal their chosen weight control method highlights the extreme deviance of these behaviors.

In addition to drug use, the specific weight controlling behaviors my subjects practiced at one time or another included severe caloric restriction, episodes of bingeing and purging, laxative abuse, cigarette smoking, dishonesty in the course of medical care, and obsessive thoughts about weight and body management. Most of these behaviors were likely to be viewed as nonnormative by the women themselves. Although their means were deviant, these behaviors were specifically aimed at achieving an appearance that would be praised and admired by others. Thus women's secrecy did not surround the results of their drug use (i.e., their thinner bodies); rather it centered on their pathways.

Generally, my research participants were high achievers or perfectionists, fulfilling or exceeding societal expectations in a variety of arenas. These privileged and socially accepted individuals included honors students, college athletes, social leaders, award winners, and future professionals. Through their prosocial bonds, ideals, and entrenchment, these women had high levels of attachment, and thus a high "stake" in conventional society. With much to lose, many perceived the consequences of being caught and labeled as deviant to be severe, even unacceptable.

Secrecy among Licit Substance Users   Women who engaged in pharmaceutical substance use displayed differences from those using illicit substances in their level and intensity of secrecy about their weight control methods. Many college campuses reported high rates of prescription stimulant use, but such behaviors were more socially acceptable when aimed at academic performance (McCabe et al. 2005; McCabe, Teter, and Boyd 2005). Women in this study feared that their instrumental drug use would be too readily associated with disordered eating, which was generally negatively stigmatized by others. For example, Corina, a 22-year-old college senior studying sociology, reported that:

> Um, I don't think anyone knows. I never said anything and um.... I don't really want to. I've definitely, uh, thought about it, like who knows? Who knows? ... But I think mostly I'm being paranoid 'cause I don't know why they would.... Sometimes I talk about using Adderall to study, um, but that's, you know, different. No one cares about that; they'll just ask you if they can have some.

Given that it was relatively easy to hide both pharmaceutical drugs and their ingestion, many licit substance users suspected that no one knew about their use at all.

Pharmaceutical users employed a variety of means for hiding their behavior. Women often employed secrets and lies to obtain their medications. Getting their drugs required them to construct "covers" for their frequent visits to the doctor. Students, who routinely had to visit doctors on campus, did so during the weekdays, spending an hour at the campus health center instead of eating lunch with friends. They created a host of alibis for these visits. Cynthia, a 21-year-old young woman studying biology, spoke directly to this point:

> What would I tell people? Oh I dunno. Ha, I mean it's not that much of a stretch.... Once a month.... I say something like uh, um, say, "I have to get my teeth cleaned." Or "I have a group project meeting in [the dorms]." ...No one, no one thinks, you know, no one thinks too much of it.

Secrecy among Illicit Substance Users   When compared to licit substance users, women using illicit drugs reported a different kind of secrecy. The use of street drugs was not only illegal, but it was also largely frowned upon socially. Regardless, subcultures surrounding black market drugs have existed throughout history, populated by individuals who, despite dominant cultural proscriptions, engaged in deviant drug use. As a result, illicit drug users had access to others who valued and used, to one degree or another, psychoactive substances. For example, Opal, 20-year-old college sophomore and cocaine user, told me:

> A lot it's by myself but, uh, you know, I always do, when my dealer is around, uh, not too much, you know, but always some.... I have friends who I sniffed up with but, uh, also people who did not by any means, you know, blow coke, like ever.

Study participants were generally discrete about using illicit drugs in casual everyday conversations with their peers, but were able to let loose in the company of others using substances. Their secrecy surrounded their motivations for using drugs and the regularity of their consumption. Sasha, a 21-year-old college senior, reported extreme reluctance to talk with friends or family who did not use illegal drugs, noting that her family was particularly unfamiliar with drugs:

> Absolutely not, no way, uh–uh.... They would shit. I mean, look, let me tell you that my family is completely against drugs so I don't know why they even let me come to [this school].... Really they just have no clue about anything. They'd be like that woman who thought that a bong was, um, a horn.... If they knew about any of this, well I bet my dad'd say that I was going to end up jumping out of a window.

SOCIAL ISOLATION

As a result of keeping their deviance hidden from others, loners kept an important dimension of their lives secret. Unlike the self-injurers described by Adler and Adler (2005) who wanted to concentrate fully on themselves during the act, these instrumental drug users preferred to be alone primarily so others would not find out what they were doing. Some women reported that they had rituals surrounding drug use which they enjoyed, but isolating themselves during episodes of drug use was a choice. Ophelia, a 21-year-old junior, told me:

> I did coke and stuff at parties; it's not like I have a problem with drugs at all. I mean, I don't really care what people do, whatever. I'd do stuff with my friends and drink. But mostly since I was using all the time, I ended up keeping it to myself because, uh, I don't know. I was embarrassed how much I cared about being skinny.

Many instrumental users feared that if others found out how they were controlling their weight, the benefits others associated with their own weight control would be diminished. Women enjoyed the idea that others thought they were "naturally thin." In order to maintain that appearance, they had to keep secrets from their friends and sneak around, creating a wall between others and themselves.

PRACTICAL HURDLES

To manage the impressions others developed about them, the women in my study were forced to navigate the practical problems they encountered in maintaining their deviance alone. Many women identified with the common narratives of eating disorders, but a piece of the puzzle was missing. They lacked access to information regarding substance use for weight control.

Health Consequences of Instrumental Drug Use   Very little was known or discussed about the behaviors described in this research. Consequently, women were left to create their own drug-using systems. These included assessing their health risks, such as how much drug use was physiologically safe. They were forced to rely on self-created systems of interpreting signs and signals regarding changes in their bodies. This began, but did not end, with rapid or excessive weight loss. Carla, a 20-year-old college junior, discussed some of the other symptoms she noticed:

> Sometimes I can, uh, feel like, my heart beating a little bit faster. Like a little bit.... It definitely raced sometimes, but not a lot.... And it sometimes gives you this taste, um, "metally," in your mouth. That happens when you really don't eat.

When physical signs like these presented, women dealt with them on their own.

While pharmaceutical drug users could mention any physical symptoms to their doctors, many of my subjects felt constrained, often withholding negative information from their doctors in order to protect their prescriptions. Cheyenne, a 20-year-old junior and former cross-country runner, said:

> These appointments are short, ya know? So I just kinda, um, say things are good, you know, whatever. It's working.... Sometimes in the beginning he would ask me how, uh, or why I thought it was working and I'd give some bullshit story ... But it's not like a physical exam or anything.... He doesn't ask too much about my health besides asking about how it's going with my dose.

Since many women went to psychiatrists for their medications, appointments did not generally include any physical exams. As such, any health problems that arose from women's drug use often went undocumented and untreated by the medical establishment.

Illicit substance users, like licit users, most often dealt with health-related concerns that resulted from drugs, on their own. The types of health questions or problems often varied according to the amount of drugs they were consuming. Higher doses or prolonged use often led to greater health consequences. Sammy, a college sophomore, recalled an incident that was related to her drug use and health:

> Well, I passed out once, full on.... I was standing by my closet folding some laundry and I sort of, uh, fell into the clothes that were hanging there. I hit the wall with my head.... Everything went totally black.... But it only lasted like, um, five seconds. It was weird, but it didn't, like, scare me or anything.

Like Sammy, a few participants reported dizzy spells, which were sometimes blamed on caloric restriction instead of drug use. In addition, many women explained that consistent low calorie diets caused feelings of weakness or disconnection.

All participants reported that using drugs had effects on their moods and energy levels. For many, this aspect of drug use was a two-sided coin. First, stimulants gave an initial boost, a general feeling of well-being for a period of time following consumption. However, after time, many women reported that these substances made them irritable or cranky. Savannah, a college senior, told me:

> I'd be gritting my teeth, just so irritated with the world.... You know, especially in the morning, I'd sometimes just feel like shit.... You'd want to stay away from me, trust me on that one.

Reports of mood swings were quite common, and were generally regarded as an annoyance by participants. Stimulants also interfered with normal sleeping and eating patterns.

Financing Drug Use   Another practical problem these women encountered involved financing their drug use. The associated costs varied according to the regularity and dosages of drug use, varying also by specific substances. For example, illicit drug users, who tended to prefer cocaine, faced relatively high street costs for their supplies. On average, women using cocaine reported spending between fifty and one hundred dollars per gram. Pharmaceutical users reported variations in cost as well, but these were generally reported as a per-month estimation, whereas illicit users reported their costs by weight. Prescriptions tended to cost women between forty and two hundred dollars a month.

Supporting a cocaine habit placed different levels of strain on women's finances depending on how much and how often they used, as well as [on] their general financial situations. For some women, money flowed from their families, [so these women were] endowed with hefty enough allowances to support drug use. I asked Odessa, a 21-year-old college student who received such an allowance, if she told her parents what she spent money on. She said:

> They don't really know. I mean, I think clothes and going out or whatever, they don't really ask.... Once my dad found shit in my bathroom though, and he asked me about it. Not money, I mean; he asked about if I was using drugs.... I don't think they know.

Some women, on the other hand, did not have parents willing or able to support them while they were in school. Even those who received moderate support were sometimes forced to finance drugs on their own.

Specific means for securing monies ranged greatly among participants. Many women held jobs while they went to school. Outside employment was not generally full-time, but for many, it was enough to supplement their income. Sally, a junior studying biology, reported a scheme she used throughout college:

> I sign up for eighteen credits, then my parents pay my tuition.... Then I wait 'til school starts and drop classes, so I've got like, uh, twelve credits.... I have it set up so that the refund from that tuition, you know, goes into my bank.... I've done it a few times.... I get a chunk of money.

Contrary to my expectations, participants did not report engaging in illicit drug sales in order to fund their drug use. While a few women had acted at one point or another, as middle-men, "hooking" friends up with drugs, regular systematic dealing was not reported.

Although women did not become sellers themselves, many reported that the more time they spent immersed in drug subcultures, hanging out with dealers or fellow users, the less money their drugs cost them. However, because most illicit users were relatively secretive about the regularity of their use, their dealers were privy to more information than were their fellow drug users. Close relationships with dealers were useful for some women, but price structures for cocaine were still set by the local black market. Accordingly, even though frequent contact with the drug world may have sometimes yielded lower prices, drugs ended up being a major financial cost in illicit users' lives.

Pharmaceutical users' reports of financial strain were very different from street drug users' experiences. Since women did not obtain their substances from the black market, their costs remained relatively stable, depending on how many prescriptions they were filling and what types of pills they received. In addition, many women were able to at least partially, if not wholly, cover the costs of their medications with health insurance. A 30-day supply of Adderall cost one woman roughly forty dollars. That same supply, same brand, same dose, cost another participant around one hundred dollars. The former had health insurance; the latter paid out-of-pocket.

Financing drug use certainly put constraints on many women. At times, use patterns would be affected by individuals' pocketbooks. Times of plenty enabled more frequent or heavier use. When money was scarce, many women were forced to use less. Plenty of participants reported a regular system for funding drug use. Women used certain amounts regularly and figured out how to pay for their supplies. While some women reported shifts in their use patterns, many participants maintained fairly stable habits for significant chunks of time. The relative privilege of women who participated in this research may have accounted for their ability to continue using. In addition, these types of weight managing behaviors may have been largely self-selected by those who were able to support consistent drug use. For example, there could have been women who employed extreme weight control techniques who would have adopted drug use as well, if not for the cost. Instrumental users' reports of financial strain were common during interviews. However, most did not report money as the cause of significant concern. While one might think that, especially for illicit drug users, finances would eventually force them out of their deviant career, this was generally not the case.

CONCLUSION

The deviant associations reported by the women discussed in this study varied by the legal status of their drug of choice. Although a few pharmaceutical users obtained their pills on the black market, most purchased these drugs through conventional medical routes. As a result, licit drug users were not forced to inter-act with other drug users. Street drug users were more likely, however, to inter-act with other drug users, and reported many nonnormative associations. While they were not comfortable talking with these people about their drug use for weight control, illicit drug users were at least likely to engage in sociability with other drug users. Yet all of the women discussed in this study hid their instrumental drug use and its relation to their feelings about their bodies and weight. The goals of their drug use, the regularity of their drug use, and the combination of their drug use with their eating disorders shamed them into maintaining a degree of secrecy that kept them from opening up to even their closest friends and family members about this behavior. As such, they were alone with their deviance.

REFERENCES

Adler, Patricia, and Peter Adler. 2005. "Self-Injurers as Loners: the Social Organization of Solitary Deviance." *Deviant Behavior* 26: 345–378.

Best, Joel and David F. Luckenbill. 1980. "The Social Organization of Devi-ants." *Social Problems* 28: 14–31.

McCabe, Sean E., John R. Knight, Christian J. Teter, and Henry Wechsler. 2005. "Non-Medical Use of Stimulants among U.S. College Students: Prevalence and Correlates

from a National Survey." *Addictive Behaviors* 30: 78–805.

McCabe, Sean E., Christian J. Teter, and Carol J. Boyd. 2005. "Illicit Use of Prescription Pain Medication among College Students." *Drug and Alcohol Dependence* 77: 34–47.

Prus, Robert, and Scott Grills. 2003. *The Deviant Mystique: Involvements, Reali-ties, and Regulation*. Westport, CN: Praeger.

35

Cybercommunities of Self-Injury

PATRICIA A. ADLER AND PETER ADLER

Self-injurers—those who cut, burn, brand, pick at, or otherwise injure themselves in a deliberate, but nonsuicidal, attempt to achieve relief by harming themselves— grew from a relatively small and unknown population into a burgeoning, but largely secretive, group in the late 1990s. Although these people are no longer regarded as mentally ill or suicidal, as they once were, a strong stigma remains attached to the behavior. Consequently, they are still, for the strongest part, loners in the solid (or real-life) world, hiding or giving legitimate accounts for their scars. Yet the early 2000s saw the rise of online communities of self-injurers, first just as places where individuals could find each other and gain nonjudgmental accep-tance, but later as support groups composed of like-minded others. Self-injurers thus represent a hybrid associational form, behaving as loners in the solid world and colleagues in the cyberworld. Based on extensive online research and more than 135 in-depth, life-history interviews, this contribution by the Adlers gives us a rich portrayal of these deviant communities and the relationships that form within them.

How do you think these cyber support groups compare and contrast with those of other deviants? With those who meet in person, like Thompson's in Chapter 33, or with those who meet online? How do they compare with those people mentioned in Davis's discussion of the transabled? Do you think that this hybrid associational form is unusual or more common? What does it suggest about the types of interactions people have online versus in person? Do you think that cyberselves and cyberrelationships serve as a staging ground, helping to prepare people to interact more successfully in the real world, or do they take people away from the real world, replacing it with a community that is more important to them?

The cyberworld represents a new frontier, one that extends what has often been colloquially referred to as the fourth and fifth dimensions: time and space. The cyberworld occurs in a new form of space that is both "out there," and "in here," simultaneously public and social, while remaining private and solitary. It is created by technology and populated by disembodied people in a virtual universe detached from any physical location. This space is a fertile location for the rise of virtual communities, which challenge traditional notions of identity and associations. In this paper, we focus on the way self-injury—the deliberate, nonsuicidal destruction of one's own body tissue, incorporating practices such as self-cutting, burning, branding, scratching, picking at skin or reopening wounds, biting, head banging, hair pulling (trichotillomania), hitting (with a hammer or other object), and bone breaking, has been affected by the Internet.

In an earlier work (Adler and Adler, 2005), we described the way self-injurers hid their behavior from others and embodied the deviant associational form (Best and Luckenbill, 1982) of "loners," bereft of the subcultural support, knowledge, and interaction with others who, like themselves, live on the margins. Yet, through the computer-mediated communication of the World Wide Web, these individuals, unconnected in what they (and almost everybody else) call real life, have constructed myriad cyberforums and cybercommunities. We describe some outgrowths of this cybercommunication—specifically, the rise of cyber subcultures that transform face-to-face loner deviants into cyber "colleagues," and we analyze the implications of this development for the concept of deviant associations in the postmodern world.

Drawing on data gathered through analysis of self-injury bulletin boards and Usenet groups, in-depth life history interviews with self-injurers, and email communications and relationships formed over a period of 5 years, we begin by discussing the way people discovered the existence of self-injury on the Internet and some of the ways they engaged it. We then examine self-injury cybercommunities and their characteristics. Beyond community, the virtual networks of self-injurers reveal complex forms of social relationships, as people engage and disengage from contact with others they meet online, forging bonds that bypass and transcend the corporeality of real life. We examine the differences between face-to-face relationships and those developed in the anonymity, intimacy, and invisibility of cyberspace. We conclude by analyzing the implications of the development of cybercommunities and cyber subcultures in the postmodern world and reflect on the effect of this postmodern medium on self-injurers' cyberselves and their lives in the solid world.

METHODS

The nascent idea for this research began in 1982 when a student of Peter's spoke to him about her cutting. Over the next few years, we continued to hear reports from friends and students about cutting, burning, branding, and bone breaking, further piquing our curiosity about this behavior and its spread. We began our formal research in 2000 and since that time have conducted 125 in-depth interviews, in person and on the telephone, generating the largest existing qualitative data set with a nonclinical self-injuring population. Participants ranged in age from 16 to their mid-fifties, with more women (100) than men (25), nearly all

Caucasian. We initially found participants through a convenience sample of individuals who heard, usually on one of our two campuses, that we were looking to talk with people who self-injured. Interested parties requested interviews via email.

In addition, beginning in early 2002 we began to explore the Websites, message/bulletin boards, and public postings of self-injurers, joining a growing group of Internet ethnographers. Since 2002, we have collected tens of thousands of Internet messages and emails, including those posted publicly and those written to and by us. This computer-mediated communication offered insight into the subcultures of self-injurers and their naturally occurring conversations. Further, like other cyberresearchers, we used the Web as a means of recruitment, going to various sites, lists, groups, boards, and chat rooms to solicit subjects. We posted copies of our informed consent form and the complete range of interview topics on Patti's university Website, directing potential participants to view and sign these before agreeing to be interviewed.

Because of the vulnerable nature of the self-injuring population, we specifically told people that we were not interested in interviewing minors over the telephone and we asked people who agreed to talk to us to print, sign, and FAX, or mail us their consent form along with proof of their age. The resulting telephone interviews ranged from locations all over the United States to Canada, the United Kingdom, Sweden, Australia, New Zealand, Bulgaria, and Germany. Of the interviews we conducted, nearly two-thirds of respondents said that used the Internet in connection with their self-injury. Finally, over the course of these years, we developed deeper cyberrelationships with people we "met" on message boards and in groups, through interviews, through side instant messages and email conversations, and through friends of these friends, and these dozen or so individuals helped us as we struggled to conceptualize our data, sharing their life stories with us and regularly responding to questions we posed out of the sociological literature.

The extremely sensitive nature of this topic and the gender patterning of the people who came forward dictated that Patti conduct the interviews. These followed a natural-history approach and then moved to specific sociological concerns that evolved creatively over the years. This paper relies upon data from both Internet postings and in-depth interviews. All the postings are presented unedited, with the exception of replacing real screen names with pseudonyms.

SELF-INJURY IN CYBERSPACE

During our early face-to-face interviews, most of the people we encountered worked hard to hide their self-injury and felt the sting of social condemnation and shame. They were primarily loner deviants, unconnected to other self-injurers and lacking the social support and information diffusal prevalent in deviant subcultures.

Beginning in the late 1990s and early 2000s, people began going to the Internet as a resource. Most felt confused and alone, unable to find counterparts in the solid world, and sought help for themselves. Paula, a 38-year-old holistic massage therapist who had picked open wounds for years, described the frustration that led her, in 2000, to search the Web:

> Sometimes the picking episodes would be like three or four hours long and when I would use the needle, this wasn't a hugely bloody thing, but it was a little bloody. And I'd be in a position like this [*leaning the top of her head forward toward the mirror, but with her eyes peeking up*], in kind of a grimace, because you can't be in a position like this for three hours without being really physically just pshhh. And you know, I'm emotionally disconnected, so there'd be this sort of like insane look in my eyes, and I'm looking in the mirror and I'm not really seeing my reflection because I'm focusing on this. Blood is gathered on my hands, so I have caked blood all over my fingers, maybe some moments where more blood comes, and it really starts to drip. So I had one of these, "I'm here for hours," and then all of a sudden the veil comes up and I see myself like this, and I see the look in my face and I see the blood on my hands, and that's when I went to my computer and I said, "I need fuckin' help." And I know there's gotta be something out there, and I don't know what the hell it is, but I need help. And that's when I got online and just put in words.

We describe here the ways that self-injurers' lives were dramatically changed by their cybercommunication, compared with their status as isolates in the solid world. Self-injurers participated in three of the four common modes of Internet engagement. They participated passively by going to Websites, reading others' postings and poetry, and viewing their images, some of which were rather graphic. They participated interactively in message/bulletin boards, newsgroups, Usenet groups, or listservs that offered supportive communities. They found real-time communication in chat rooms, populated around the clock all over the world. We found no virtual spaces—the interactive, multiuser, online cybergames, cyberpubs, cybercafes, or other forums that go beyond mere words to offer visual representations of characters in "textual virtual reality"—however, marked by or for self-injurers.

Cybercommunities

Critical to becoming a regular participant in self-injury cyberspace was establishing membership in one of the many cybercommunities.

Finding a Community   People searched until they found a community that seemed to fit their specialized needs. In this endeavor, they considered the size, level of activity, demographics, and orientation of the group. People with the greatest communication needs gravitated toward busier communities. Some of those communities marketed themselves specifically as teen oriented (the largest numbers of people), others were for older people (twenty somethings were still numerous, people in their thirties were steady but fewer, in their forties still a group, and the smallest numbers were in their fifties), and many invited a mixture of ages.

Individuals came to rely on their cybercommunities to help them. They found it important to talk to people who knew what they were going through, given that most people they knew in real life did not really understand self-injury. When they found a community that fit them well, the experience was rewarding. Paula, the holistic massage therapist, expressed the sense of community she got from her group:

> It's a good feeling to find a community that can accept your darkest shadows, but it's also a really scary thing to see those shadows. So it was double-edged. But it got to where, the same way I would look forward to coming home to pick, I would look forward to getting home and getting on the computer and reading all the emails, and I would go on the chats. It's a world, it is definitely a world.

Yet many others found that no one community completely satisfied their needs. Self-injurers searched around, moving through different sites and groups, finding some that were partially, but not completely, satisfying.

Others used memberships in multiple communities to express different identities or different aspects of their identities. Tim, a 21-year-old part-time college student who held various part-time jobs and who moderated his own self-injury group, felt that he had to offer a hopeful self-presentation to the people he was helping on his site, so he proclaimed himself free of self-injury for 2 years when, in fact, he had lapsed in and out of cutting. Tim also used his membership in another group to discuss his ongoing problems, presenting a different self and identity there, although he used the same screen name.

Nature of the Community Communities differed in their policies, their norms, and their orientations toward self-injury. Some were highly regulated, while others were fairly open; some were highly focused, while others were diffuse; some were stringently anti-self-injury, while others were more accepting of people's continuing practice. A moderator on one group regularly posted the following policy:

> The reason that this group exists is to help people in recovery. All members are asked to identify the alternatives s/he tried to use to avoid using [self-injury] as a coping mechanism. For those who are not ready to embrace recovery, this is the wrong group.

Linda, a 40-year-old former medical transcriptionist, discussed how she felt frustrated by her membership in a group that was highly regulated and that prohibited any actual discussion of self-injury. She had to supplement this site with visits to another that was not as rigidly recovery oriented, one in which the absence of a "no trigger" policy enabled her to vent her feelings about suicide and injuring.

Finally, a few sites were more avowedly pro-self-injury. These sites approached self-injury in much the same way as the pro-anorexia and pro-suicide movements. They treated it as a voluntary lifestyle choice and a long-term coping mechanism. Considering individuals' decision to injure themselves, rather than injuring others, constructive, they encouraged people to help themselves embrace their self-injury and, like others in a tertiary deviant stage (Kitsuse, 1980), to shed the stigma. Along with this encouragement, they offered practical suggestions for engaging in self-injury and dealing with the physical problems that it generally engendered. Zoe, a member of an unregulated Website, posted the following view of self-injury:

> I honestly dont see what is so wrong with cutting. I think Im kinda looking to see if anyone agrees. I mean, instead of punching a pillow, you just take it out on yourself. As long as you dont do it too deep, whats the big deal??? Its better than abusing the people around you. The real problem with it is the emotions and the depression BEHIND the cutting, right? If it isnt "adversely affecting one's life," as is required for anything to be a legal disorder, then why does everyone else think it is wrong…

> Am I making any sense to anyone???

She received the following response from Angie:

> Hi there!! Nice to hear from you, welcome! As I was reading your posts, I couldn't help but feel as tho I was reading something that I had written!!! I don't see too much wrong with it either, it doesn't hurt anyone but myself.

These deviant cyber subcultures may be thought of as "back places," where people of similar preferences felt no need to conceal their pathology and openly sought out one another for support and advice. As Deshotels and Forsyth (2007:212) noted, "Socially proscribed and severely sanctioned behavior that was once relegated largely to secrecy among isolated individuals is now at the center of a cybercommunity in which all manner of support is readily available." Jenkins (2001) discussed the feeling of freedom and safety that child pornographers felt whose participants were loner deviants in real life but members of an online subculture.

Identification with the Community   Although people belonged to various sites and sometimes went for long periods between postings, when they found a community that fit well, it gave them a sense of identity. They experienced this sense whether they were actively self-injuring or not. Jones (1997) noted that our sense of identity is derived not only from identification with the group, but also from our understanding of the group identity. As Erica, an 18-year-old college freshman, noted,

> You've been there; you know what it's like. I have traits in common with other members of the community: being sexually abused, being a perfectionist, having an ED [eating disorder]. Always like, trying to help

other people, doing community service, volunteer work, I'm really into that. Like everything they say on those Websites is completely me. I don't think it's all cutters; I think it's the majority of cutters. I just happen to fit. So it makes me feel more connected to the community as a member.

Identifying with members of the community was vitally important to most people we encountered, whether they had fully functioning work and social lives and hid their self-injury or whether they were trapped in their houses or bedrooms, unable to make contacts with people in the solid world. McKenna and Bargh (1998) suggested that people with concealable stigma identify strongly with these Internet support groups and consider them important to their identities. As a result, they are also more likely to achieve greater self-acceptance, decreased estrangement from society, and decreased social isolation. Deshotels and Forsyth (2007) proposed that identities forged with the aid of Internet groups may help people disengage themselves from normative social control. Yet, although people found these sites helpful, their identification with the community might also reinforce their self-injurious behavior, as Amber, a 20-year-old college junior, noted:

> If you go to, like, the same chat room and stay there, you kind of get this group of friends, maybe. I guess you could get a sense of belonging or something. It's like you need to cut to stay in that group, you know? Because that's what the chat rooms are for. It's a cutting chat room I guess, even though it says it's a no-cutting chat room. And so I think it just escalates people because we're kind of co-dependent in a way because, like say someone tells their friends the experience of it in that group, everyone will try it and they'll just keep on doing it and it'll just keep on escalating because, like, that's what's expected in that group and it just gets worse because there's no outside force preventing you from doing that, I guess.

When people affiliated with a community, that identity often transferred to them. Lemert (1967) discussed how primary deviants, who keep their deviance hidden from others, have the luxury of denying self-identification with their behavior. Becker (1963) echoed this theme, arguing that "secret deviants" are unlikely to conceive of themselves through the deviant lens. Erica, the college freshman, explained what it was about membership in her site that changed her identity:

> Just the fact that there were other people doing it. Maybe like it really is, there's a group of people. I *am* part of this group, obviously. That helped me connect my identity to a self-abuser. Whereas before I was just, like, one of two people doing it so it wasn't really an identity, it was more of a problem. I didn't really think it was a problem, just a habit. Whereas on the Internet it's a lifestyle almost, the way you are, instead of just a habit. They were connected to it in a more long-term way. It was a more central focus of people's lives. It was the central focus of mine for quite a while.

Oscillating In and Out of Communities   As many scholars have noted, a common feature of cybercommunities, much more than communities in the physical world, was members' transience. People moved fluidly through groups, looking for one that felt right. As their lives evolved and changed, what had once been a good fit might no longer suit them. People's support needs were dependent on the stage of their self-injury career. Bonnie, a 33-year-old bankruptcy coordinator for a student loan service, commented on how she was ready to move on to a different kind of group:

> And I think [site name] is a good group and I've met a lot of really nice people there. But there's also a lot of that constant crisis, like help me *right now, right now, right now* kind of thing. And I'm trying to avoid being in that situation again. Because I was in that situation so I have a little harder time with that group now.

Communities differed in their composition, with some having "regulars" who posted a lot and others characterized by people who posted sporadically and then disappeared. Regular posters were often moderators or people likely to be tapped as moderators in the future when current moderators burned out and moved on to something else. It was also common to see people announce that they were going to either leave or take a break from the group. Sometimes people indicated what was causing their departure, but other times they just said goodbye. Bob, a regular on one list, posted this message, in which he clearly was upset:

> I must excuse myself for a while while i cope with this situation.. i will look in on mail daily - please contact me directly. i will not open group mail for a while until i feel better about things. I sign off wishing everyone a safe weekend while i mull over whether or not i wish to remain a member of this group or not.

Eventually, Bob returned.

Oscillations in and out of participation were an expected characteristic of these communities. People dropped in when they were having trouble and left when they were excessively triggered or when they felt able to cope without the group support.

Eventually, some people left the community for good. Many held on long after they had (allegedly) desisted from self-injuring for months or years, enjoying the outreach they provided to those still in the throes of the behavior. Cindy, a 19-year-old retail sales clerk, found a better job, got into therapy, and met a boyfriend, and her life improved significantly. She no longer felt the need to self-injure, and although her group had been a huge part of her life for 3 years, she gradually faded out of the picture. At first she did not write as often, but did read some of the posts; eventually, however, she found people's stories depressing and self-absorbed. She stayed with the group for as long as she felt strong urges to self-injure, but as these weakened, she was able to leave. Reflecting on her life after self-injury, she noted that if she had a problem or got upset, she was likely to turn to her boyfriend or find some other way of dealing with it.

CONCLUSION

Embarking on the journey into cyberspace is a simultaneously communal and lonely experience. Although people seek out online interaction in order to bond into relationships and communities, they do so in a physical space of separateness. The technology of the Internet thus offers both separation and connection, a time to be alone yet be with others, leading people to spend their days alone at their computers trying to retribalize.

The cyberworld represents an ephemeral space of creation and destruction. It offers people who are dispossessed by mainstream society a reservoir of cultural hiding places where they can form their own cultures and communities, even though normative standards and assumptions are not totally absent. "Space" refers not only to physical, but to social, proximity—to how far or close we feel with others, the connections between people. It includes everything with the character of "beside-each-otherness."

The cyberworlds of self-injurers have altered classic typologies about the social organization of deviance, transforming individuals who were isolated loner deviants in their face-to-face life into deviant cybercolleagues. While self-injurers with no Internet connections, as well as those who regularly frequented the bulletin boards, email groups, and chat rooms, remained loner deviants in everyday life, the latter lived Goffman's classic enactment of a "double life:" They were closeted to people around them yet open about their behavior in cyberspace. Accordingly, they reaped the benefits of membership in a global cyber subculture without the associated risk of exposure, stigma, or rejection, gaining practical advice, legitimation for their behavior, social support, and a set of nonjudgmental cyberfriends and cyberacquaintances.

Only a few studies have emerged that discuss loner deviants who found community in cyberspace. (e.g., studies of pedophiles, men who seek castration, eating disorders, depressives, and female-to-male transsexuals). This research suggests that we may have to change our conception of the social organization of deviance because loner deviants may be becoming obsolete in the era of the cyberworld. Deshotels and Forsyth (2007) noted that the Internet is an especially effective environment for spawning and supporting communities formed around extreme behaviors, which characterize much loner deviance.

REFERENCES

Adler, Patricia A., and Peter Adler. 2005. "Self-Injurers as Loners: The Social Organization of Solitary Deviance." *Deviant Behavior* 26(4): 345–378.

Becker, Howard S. 1963. *Outsiders*. New York: Free Press.

Best, Joel, and David F. Luckenbill. 1982. *Organizing Deviance*. Englewood Cliffs, NJ: Prentice-Hall.

Deshotels, Tina H., and Craig J. Forsyth. 2007. "Postmodern Masculinities and the Eunuch." *Deviant Behavior* 28: 201–218.

Jenkins, Philip. 2001. *Beyond Tolerance: Child Pornography on the Internet.* New York: New York University Press.

Jones, Steven G. 1997. "The Internet and its Social Landscape." In S. Jones, (Ed.), *Virtual Culture: Identity and Communication in Cybersociety* (pp. 5–35). Thousand Oaks, CA: Sage.

Kitsuse, John. 1980. "Coming Out All Over: Deviants and the Politics of Social Problems." *Social Problems* 28: 1–13.

Lemert, Edwin M. 1967. *Human Deviance, Social Problems, and Social Control.* Englewood Cliffs, NJ: Prentice Hall.

McKenna, Katelyn Y. A., and John A. Bargh. 1998. "Coming Out in the Age of the Internet: Identity 'Demarginalization' through Virtual Group Participation." *Journal of Personality and Social Psychology* 75:681–694.

36

Subcultural Evolution: The Influence of On- and Off-line Hacker Subcultures

THOMAS J. HOLT

The influence of subcultural norms is the focus of this research on subcultures of Internet hackers formed both online and through face-to-face association. Holt's research is especially interesting because he integrates and compares these two realms and asks how each realm influences members and why. Incorporating strings of thread postings, cybergroups, and a convention of cyberhackers, Holt integrates his interview data with group presentations and with posted Web blogs and accounts. From these, he distills five "normative orders," or subcultural norms, that guide hacker behavior, stratification, and social status: technology, knowledge, commitment, categorization, and law. These orders frame the way individuals account for their behavior, legitimate legal vs. illegal hacks, and generate identity within the subculture.

Reading this chapter, be sensitive to the importance of the members' presentations in encouraging or discouraging various types of behavior. How important do you think subcultural feedback is to people who hack? Harking back to the chapters on accounts, how would you compare their excuses and justifications and their techniques of neutralization with those of rapists and shoplifters? Do you think that such accounts work merely for self-legitimation, or do they neutralize deviant acts for others as well? How important is a deviant subculture to Internet hackers compared with self-injurers or people with IBD? What functions does such a subculture serve?

A great deal of criminological research has explored the impact of subcultures in a variety of contexts, including street gangs, drug sellers, professional thieves, and other deviant groups. This research has provided insight into the relationships and knowledge shared between individuals in the real world, and has important ramifications for the explanation of crime. At the same time,

From Thomas J. Holt (2007). Subcultural Evolution: the Influence of On- and Off-Line Hacker Subcultures. *Deviant Behavior*, 28(2), 171–198.

numerous deviant and criminal subcultures have developed in cyberspace, including hate groups and pedophiles. The Internet and computer-mediated communication methods, such as newsgroups and Web forums, allow individuals to exchange all sorts of information almost instantaneously. Deviant and criminal peers can communicate online across great distances, facilitating the global transmission of subcultural knowledge without the need for physical contact with other members of the subculture.

However, most research couches deviant or criminal subcultures in either cyberspace or real-world social situations. A limited number of studies have considered the overlap and effects of on- and off-line experiences in the development of gender, race, and political identity off-line. Recent research by Wilson and Atkinson (2005) explored the structure of subcultural norms, values, and beliefs as a result of on- and off-line experiences in Rave and straightedge youth subcultures. Yet there have been no real considerations of the role of virtual and real experiences in the development and structure of deviant subcultures, despite the growing number of online deviant subcultures. As a result, there is a need to consider how deviant subcultures may be structured by people's experiences in social environments in the real world and in cyberspace.

This research attempts to address this gap by examining the influence of on- and off-line experiences on the subculture of one of the most easily recognized computer criminals: computer hackers. Hackers are individuals with a profound interest in computers and technology [who] have used their knowledge to access computer systems, often with malicious illegal consequences. For example, unauthorized access of computer systems cost U.S. businesses $31 million dollars in 2005 alone. Hackers have also been linked with the creation of malicious viruses that affect all computer users, such as the Melissa virus, which infected computers worldwide causing at least $80 million in damages.

However, computer hackers do not use computers as a means of attack only; they use the Internet as a way to communicate with others around the globe. At the same time, they often have connections with individuals in the real world through various regional groups and conventions. As a result, this is an ideal criminal behavior to explore the contours of subculture stemming from experiences in the real world and cyberspace.

THE PRESENT STUDY

This study uses three qualitative data sets to examine hacker subculture across both social settings, including a series of 365 strings from 6 hacker Web forums, interviews with active hackers, and observations from Defcon, a hacker convention held annually in Las Vegas, Nevada. These multiple data sources are triangulated and used to explore the normative orders of hacker subculture. This is followed by an exploration of the ways that experiences in virtual and real social settings impact the socialization process and normative structure of computer hackers, and deviant groups generally.

The first data set consists of a series of 365 posts to 6 different public Web forums run by and for hackers. Web forums are online discussion groups where individuals can discuss a variety of problems or issues. Web forums demonstrate relationships between individuals and provide information on the quality and strength of ties between hackers. They also include a variety of users with different skill levels and knowledge of hacker subculture. For this study I included Web forums with both large and small user populations, high-traffic forums, and public forums. A snowball sampling procedure was used to develop the sample of six forums used in this analysis. The six forums that compose this data set include a total of 365 strings, providing copious amounts of data to analyze. These strings span two and a half years, from August 2001 to January 2004. Moreover, they represent a range of user populations, from only 20 to 400 users.

The second data set is a series of 13 in-depth face to face and e-mail interviews with active hackers. These interviews probed individuals' experiences and impressions of the normative orders of hacker subculture on- and off-line. They were asked to describe their experiences with hacking, interactions with others in on- and off-line environments, and their direct opinions on what constitutes hacker subculture. Interviewees were identified through the use of a fieldworker, word of mouth solicitations at a Midwestern university, [the] Defcon 12 hacker convention, and IT listservs. Hackers who could be met in person were asked to participate in an open-ended semi-structured interview, lasting between two and three hours. These interviews (n = 5) were taped and transcribed verbatim.

The third data source consists of first-hand observations of hackers and data collected at the 2004 Defcon, the largest hacker convention held in the United States. This three-day convention draws participants from around the world, including law enforcement agents, attorneys, and hackers of all skill levels. Written and tape-recorded field notes were made during scheduled events at the convention, including panels and games. Observations and discussions with attendees were also documented in unscheduled social situations, such as gatherings in the halls, pools, and in room parties. These observations provide ample insight into hacker subculture as attendees were continuously exchanging ideas and engaging in social intercourse off-line.

FINDINGS

Subcultural values and norms are measured using the concept of "normative order." This is a "set of generalized rules and common practices oriented around a common value." An order "provide[s] guidelines and justifications" for behavior, demonstrating how subcultural membership impacts actions (Herbert 1998:347). This gives a dynamic view of culture, recognizing that individual behavior can stem from individual decisions as well as through adherence to subcultural values. Normative orders also provide for the identification of informal rules considered important by members of the subculture because of the values they uphold. Furthermore, this frame allows the researcher to recognize conflicts in the subculture based on the presence of contradicting orders.

The social world of computer hackers is shaped by five normative orders including technology, knowledge, commitment, categorization, and law.

The orders are used to generate justifications for behavior, affect attitudes toward hacking, and structure identity and status within the subculture.

Technology

One of the most significant normative orders in the hacker subculture is the relationship between hackers and technology. Hackers across the data sets clearly possessed a deep connection to computers and technology, which played an important role in structuring the interests and activities of hackers. For example, all of the interviewees reported developing an interest in technology before or during adolescence. Spuds wrote, "my Grandparents saw my aptitude for all things technical at an early age. At the age of seven, my grandparents decided to nurture that interest and aptitude by purchasing me my first PC." Mutha Canucker wrote, "I got a computer when I was 12, and my interest grew from there. The more I played with it, the more I realized what I could do."

Once hackers were given access to a computer, they spent their time becoming acquainted with its functions in a variety of ways. They played video games and developed interests in the many different facets of computer technology. This gave them an appreciation for a variety of technical skills such as programming, software, hardware, and computer security. For example, Spuds "learned how to program the machine to make my own programs to do things for which there were no programs readily available to do. I learned how to fix the machine, upgrade the machine, and so much more."

The more time hackers spent familiarizing themselves with technology, the more their skill level increased. Whether on- or off-line, hackers discussed the need to understand the interrelated elements of computer systems, as a hacker's knowledge level directly relates to [his or her] ability and skill. To meet the intense internal desire to understand computer technology, hackers sought out a variety of online resources, especially Web forums. For instance, the poster MorGnweB wrote,

> You might want to remember that this forum is designed for people to
> ask questions, despite the fact that you can find almost anything on
> google. Soif [sic] we all should just searched [sic] for things ourselves
> thered [sic] be no forums.

Defcon also illustrated hackers' fascination with computers and technology. Most of the panels held during the course of the convention related to technology. A wide range of topics [was] discussed, including hardware hacking, phreaking, computer security, exploits, cryptography, privacy protections, and the legal issues surrounding hacking and piracy. Technology structured many of the competitions held during the convention as well, including the IP Appliance showcase where participants integrated fully functional computer hardware into common household appliances. Contestants also demonstrated their technological know-how in wardriving, WiFi, and robotics challenges. All these elements demonstrated the importance of technology in the activities and interests of hackers in the real world and cyberspace.

Knowledge

Another important order identified in hacker subculture is knowledge. Hackers across the data sets demonstrated that hacker identity is built on a devotion to learn and understand technology. As one forum poster suggested, "if you want to be a hacker, then you should start to learn ... hacking is all about learning new stuff and exploring." Forum users and interviewed hackers stressed the notion of curiosity and a desire to learn. For example, the interviewee MG defined a hacker as "any person with a sincere desire for knowledge about all things and is constantly trying to find it."

As a result, hackers spent a great deal of time learning and applying their knowledge on- and off-line. Most hackers stressed that the learning process began with the basic components of computer technology. An understanding of the rudimentary functions of computers provided hackers with an appreciation for the interrelated nature of computer systems. Dark Oz explained his own learning experience with computers:

> You do this long enough, with many different technical projects, and you begin to really learn a lot. ... Once you learn the logic and how to "think" like a computer or programmer would, you can just guess at how things are working, based on existing knowledge.

Forum users echoed this sentiment as well, such as the poster dBones who wrote, "you will become more knowledgeable if you were to find out information by yourself and then teach yourself that information." As a result the notion of learning on one's own was addressed in a different fashion online. When an individual asked a question, users would almost always respond by giving a Web link that would help answer the question. Users would have to actively open the link and read in order to find their answer. These links provided specific information about an issue or topic discussed in the string without repetition or wasted time for the other posters. Tutorials were also provided, giving detailed explanations on topics from programming to the use of hack tools. In some instances, users made actual programs available for download to help individuals learn. However, hackers did not rely solely on their online social connections to learn about hacking. Cyberspace relationships existed in tandem with real world social networks to provide hackers with new information and techniques. Vile Syn explained, "we [a small group of hackers] would constantly spend time trading pirated software, and discussing the next find. Here my interest in electronic engineering, cryptography and the lack of respect for software copyrights developed."

Learning and knowledge were also intimately tied to status within hacker subculture on- and off-line. When an individual shared useful information with others, [he or she was] able to gain status and respect. For example, those who successfully applied their knowledge in the unique challenges at Defcon were acknowledged during an award ceremony at the end of the convention. Contest winners were announced, brought on stage, and distinguished for their achievements. Individuals were given a black conference admission badge and, in some cases, a black leather Defcon logo jacket. These items could not be purchased, and the black badge provided the recipient with free convention admission for life. Since the convention badge design changed each year, the black badge stood out as a symbol of achievement and ability.

Hackers' levels of knowledge affected how they were viewed and labeled by others within the subculture. Several different terms were used by interviewees, forum users, and even at Defcon, to differentiate between hackers based on their skill level. Those with a deep understanding of technology were referred to as hackers. The extremely skilled hacker was also called elite, spelled "1337," or "leet." A hacker with little skill who used tools and scripts was referred to as a script kiddie. New hackers or people with little knowledge were labeled noobs. A unique term was used at Defcon for those who knew nothing of hacker culture but attended the convention. These individuals were referred to as scene whores, which had an extremely negative connotation. The terms script kiddie and noob also had negative associations as individuals often applied these terms to the unskilled or uninitiated. Thus, knowledge plays a significant role in hacker subculture on- and off-line.

Commitment

Commitment structured individual behavior on- and off-line through [the] continued study and practice of hacking techniques. The poster Ashy Larrie suggested:

> If you are just starting you might not have a clue what to learn, or what you should know. As for me, I just started reading texts for a long time … after awhile things become more clear and you get the idea of what hacking is all about.

Although these comments echo elements of the order knowledge, the poster makes very distinct statements about the importance of commitment to learning. A commitment to learning and understanding computers and technology was needed to discover what topics [hackers] truly find interesting. In addition, continuous changes and improvements in technology compounded the length of time required to learn. Thus, hackers must be committed to the continuous identification and acquisition of new information. Mack Diesel emphasized the importance of commitment, saying, "the minute you feel you've learned everything is the minute you're out. There's always something new to learn." Hence, hacker subculture placed tremendous value on constant learning over time.

Commitment also reflects the significant amount of effort applied to learn the tradecraft of hacking. It was apparent across the data that the time people spent hacking improved their skills. Dark Oz reflected on this, writing, "You do this long enough, with many technical projects, and you begin to really learn a lot, and then it becomes quicker to pick up more things faster then you did before." Although this statement references the importance of learning, it clearly indicates the value of expending constant and consistent effort in the process of learning.

Commitment also refers to the hours or days required to complete some hacks. For example, Defcon competitors demonstrated a high level of commitment to performing and completing complicated lengthy hacks. Individuals spent months developing and testing equipment for use in the games. Participants also spent many hours in competition to win the contests. The Root Fu hacking challenge lasted for 36 straight hours with no scheduled breaks for the competitors. Thus, these games demonstrated the skill of each competitor and emphasized the value of commitment to hacking as an important way to gain status within the subculture off-line.

Commitment to hacking has a significant impact on the activities and interests of hackers. The importance of this order makes it clear why the forum poster WisdomCub3 wrote, "hacking is a lifestyle. Spend all your time on it and you will get better and better." Many forum users and interviewees echoed this sentiment, especially when defining the term hacker. For instance, MG indicated in his interview that "to be a hacker, you must live the life, not just play the part. You must be hacker in everything you do."

Categorization

The ways individuals create and define the hacker identity constitutes the fourth normative order of hacker subculture: categorization. There was significant discussion over how to define hackers and their motives in the forums. Posters spent considerable time explicating who and what is a hacker. Disputes over the nature of hackers and hacking allowed posters to define and differentiate themselves from others within the subculture. One such discussion began because an individual asked "When did you start thinking you were a 'hacker'?" This post was a survey giving users options including when they "used a port scanner [a tool that identifies the programs running on a target computer]," "used a lamer program with 'hacker tools,'" or "tried to download malicious scripts [programs including viruses and worms] and only ended up hurting yourself." The options available to posters accentuated behavioral measures or benchmarks in a hacker's development. Many users validated the use of such measures, arguing that once they performed a certain task or understood a complicated process they could consider themselves a hacker.

At the same time, many posters suggested there were attitudinal components of their definition of "hacker." This included a certain state of mind or spirit, such as [that referenced by] Brainiackk who wrote, "the hacker seeks for knowledge, the unknown and tries to reach his own goals. That's the spirit." Curiosity and a desire to learn was an important part of most definitions of hacker.

Individual conceptions also generated much of the discussion about what different types of hackers do and how this relates to their label or title. This was especially true of the ideology or behaviors associated with each type.

For example, there were many disagreements and discussions surrounding two of the main subtypes of hackers: white hats and black hats. Both were very skilled types of hackers who engaged in different behaviors because of different ideologies. As the forum user j@ck0 indicated, "the black hats use their knowledge to destroy things. The whitehats use it to build things." However, there was some disagreement over the malicious nature of black hat hackers. For example, kFowl3r responded to j@ck0's comments, suggesting,

> One thing about blackhats, its [sic] totally wrong that blackhats only use their knowledge to destroy ... blackhats just hack ... not like whitehats which arent [sic] really hackers since they work against hackers, they build tools to stop people from breaking into systems etc.

These comments indicate [that] white hats were active in the computer security industry, securing systems from hacks. Black hats were more prevalent in the hacking community identifying weaknesses and exploits for later attack.

Law

The final normative order identified in this analysis is law. This was reflected in discussions on the legality of hacking and information sharing in the real world and in cyberspace. Hackers in the forums often discussed whether some hacks or related activities were legal, and if they should be performed. There was a split between hackers who felt [that] no illegal hacks were appropriate and those who viewed hacking in any form as acceptable. Such competing perspectives were addressed in the following exchange. An individual asked for information on a password cracking tool and how to use it. Pilferer answered the poster's question and gave an admonition that was quickly contradicted:

PILFERER: You do understand that using these password crackers on machines which you don't own or have no permission to access is ILLEGAL?

LEETER: Illegal ... So is masturbation in a public place, but we don't get reminded of that every time anyone thinks about it do we?

Legal matters were also addressed off-line during multiple different presentations at Defcon. For example, a panel of attorneys from the Electronic Freedom Foundation, a legal foundation supporting digital free speech rights and hacker interests, spoke on the current state of law relative to computer hacking. A similar talk was given by an attorney addressing the ways that the Digital Millenium Copyright Act could be used to deal with civil and criminal hacking cases. There was also a panel titled "Meet The Fed," where attendees could ask a number of different law enforcement agents questions on the law and hacking.

However, concern over potential law violations appeared to have little effect on hacker behavior. Individuals across the data sets provided information that could be used to perform a hack regardless of their attitude toward the law. This led to a contradiction in the process of information sharing. If hackers shared knowledge with possible illegal applications, they justified its necessity.

Individuals on- and off-line stated [that] they provided information in the hopes of educating others, as in this statement from a tutorial posted in one of the forums on macro-virus construction:

> This is an educational document, I take no responsibility for what use the information in this document is used for. I am unable to blamed for any troubles you get into with the police, FBI, or any other department. ... It is not illegal to write viruses, but it is illegal to spread them—something I do not condone and take responsibility for.

Similar justifications were used at Defcon, especially when a presenter's content had rather obvious or serious illegal applications. An excellent example of this was a presentation titled "Weakness in Satellite Television Protection Schemes or 'How I learned to Love The Dish.'" The presenter, A, indicated, "I will not be teaching you how to steal [satellite Internet connectivity] service, but I will give you the background and information to understand how it could be done." However, the second slide in his presentation included the message "Many topics covered may be illegal!" as well as the potential laws they [might] break by stealing service.

This legal warning reduced the presenter's accountability for how individuals used the information he provided. He simply shared his knowledge on satellite systems and television service. If someone used the information to break the law, A had clearly described the laws they could violate by engaging in these actions. Just as with the warning in the macrovirus tutorial, he justified sharing information that could be used to engage in illegal behavior, stealing satellite service, as part of the pursuit of knowledge.

Nevertheless, forum users and the Defcon staff did not condone the exchange or supply of overtly illegal information. In the forums, hackers eschewed posting blatantly illegal content and forcefully explained this idea. For example, an individual proclaimed himself, "the kind of hacker police really hate" and posted someone's credit card information. One of the senior users posted the following comments in reply:

> No [sic] only do the police hate you Regardless if this is a honeycard and regardless if its good or bad to card (btw its bad), someone should delete it because it IS illegal and this is an open forum Go away

However, seven interviewees, including those in the IT field, felt [that] some types of hacks were acceptable. For example, Vile Syn said, "when dealing with hacking that isn't violating privacy, it shouldn't be illegal at all." Hacks that did no damage or left no trace of entry were also deemed acceptable. For example, Dark Oz felt these sorts of hacks were a measure of one's skill, writing, "It takes little skill to get into a system and cause damage, destroy, or make it unavailable. The true skill is in getting in, looking around, doing whatever you want, but no one ever knows you were there." Thus, hackers created boundaries within the subculture based on the ability to hack, as well as their beliefs about hacking.

DISCUSSION

Hacker subculture places significant value on concealing blatantly illegal behavior from law enforcement on- and off-line, but justifies open involvement in certain activities. Hence, this study provides support for the contradictory role of secrecy found in previous examinations of hacker subculture.

Researchers must also consider how virtual and real experiences can differentially impact the normative structure of deviant subcultures. This study revealed a lack of uniformity in hackers' subcultural experiences on- and off-line. Specifically, the normative order categorization was largely present in hackers' online experiences. There was vigorous debate between forum posters on the various terms used to define hackers and the act of hacking. Users discussed the meaning of labels, like white and black hat hackers, and the various activities of these hacker subcategories. The meaning of these terms differed across posters, demonstrating the significance of personal opinion in the development of an individual's beliefs about [who] and what constituted a hacker online.

Yet there was very little [off-line] discussion on how to define hackers and what constitutes a hack. The interviewed hackers used Web forums and online resources with some frequency, though there was little disparity in their definitions for different hacker terms. There were also no protracted discussions [at Defcon] on the meaning of hacking. This suggests that active debate over hacker identity may be common online, but does not play a significant role in hacker subculture off-line. However, this finding is contradictory to other studies that report identity is shaped by both virtual and real experiences. Hence, further research is needed examining the way hackers and other deviants and criminals form and accept a deviant identity as a consequence of on- and off-line experiences. Such information is necessary to specify the dynamics between cyberspace and real-world experiences in the creation and acceptance of deviant identity.

REFERENCES

Herbert, Steve. 1998. "Police Subculture Reconsidered." *Criminology* 36: 343–369.

Wilson, Brian, and Michael Atkinson. 2005. "Rave and Straightedge, the Virtual and the Real: Exploring Online and Offline Experiences in Canadian Youth Subcultures." *Youth and Society* 36(3):276–311.

37

Gender and Victimization Risk among Young Women in Gangs

JODY MILLER

Miller offers a glimpse into the contemporary urban world of street gangs in this analysis of the role and dangers faced by female gang members. Gang members not only associate together, but need each other's participation in the deviant act in order to function. (No man or woman is a gang unto him- or herself.) Once nearly faded to obscurity, gangs made a rebound in U.S. society in the late 1980s, fueled by the drug economy and the increasing economic plight of urban areas. Since that time, they have evolved considerably, adding sophisticated nuances and female members. Miller finds that, although women gain status, social life, and some protection from the hazards of street life by joining gangs, they assume a new set of dangers: By entering the gang world, they are exposing themselves to violence, from both rival gang members and their own homeboys. Miller discusses the particularly gendered status dilemmas and risks for these young women, and how these vary depending on their activities, stance, and associations within the group.

An underdeveloped area in the gang literature is the relationship between gang participation and victimization risk. There are notable reasons to consider the issue significant. We now have strong evidence that delinquent lifestyles are associated with increased risk of victimization (Lauritsen, Sampson, and Laub 1991). Gangs are social groups that are organized around delinquency (see Klein 1995), and participation in gangs has been shown to escalate youth's involvement in crime, including violent crime (Esbensen and Huizinga 1993; Esbensen, Huizinga, and Weiher 1993; Fagan 1989, 1990; Thornberry et al. 1993). Moreover, research on gang violence indicates that the primary targets of this violence are other gang members (Block and Block 1993; Decker 1996; Klein and

From Jody Miller, "Gender and Victimization Risk Among Young Women in Gangs." *Journal of Research in Crime & Delinquency* (Vol. 35, Issue 4). Copyright © 1998. Reprinted by permission of Sage Publications, Inc.

Maxson 1989; Sanders 1993). As such, gang participation can be recognized as a delinquent lifestyle that is likely to involve high risks of victimization (see Huff 1996: 97). Although research on female gang involvement has expanded in recent years and includes the examination of issues such as violence and victimization, the oversight regarding the relationship between gang participation and violent victimization extends to this work as well.

The coalescence of attention to the proliferation of gangs and gang violence (Block and Block 1993; Curry, Ball, and Decker 1996; Decker 1996; Klein 1995; Klein and Maxson 1989; Sanders 1993), and a possible disproportionate rise in female participation in violent crimes more generally (Baskin, Sommers, and Fagan 1993; but see Chesney-Lind, Shelden, and Joe 1996), has led to a specific concern with examining female gang members' violent activities. As a result, some recent research on girls in gangs has examined these young women's participation in violence and other crimes as offenders (Bjerregaard and Smith 1993; Brotherton 1996; Fagan 1990; Lauderback, Hansen, and Waldorf 1992; Taylor, 1993). However, an additional question worth investigation is what relationships exist between young women's gang involvement and their experiences and risk of victimization. Based on in-depth interviews with female gang members, this article examines the ways in which gender shapes victimization risk within street gangs....

METHODOLOGY

Data presented in this article come from survey and semistructured in-depth interviews with 20 female members of mixed-gender gangs in Columbus, Ohio. The interviewees ranged in age from 12 to 17; just over three-quarters were African American or multiracial (16 of 20), and the rest (4 of 20) were White. The sample was drawn primarily from several local agencies in Columbus working with at-risk youths, including the county juvenile detention center, a shelter care facility for adolescent girls, a day school within the same institution, and a local community agency.[1] The project was structured as a gang/nongang comparison, and I interviewed a total of 46 girls. Gang membership was determined during the survey interview by self-definition: About one-quarter of the way through the 50+ page interview, young women were asked a series of questions about the friends they spent time with. They then were asked whether these friends were gang involved and whether they themselves were gang members. Of the 46 girls interviewed, 21 reported that they were gang members[2] and an additional 3 reported being gang involved (hanging out primarily with gangs or gang members) but not gang members. The rest reported no gang involvement.

The survey interview was a variation of several instruments currently being used in research in a number of cities across the United States and included a broad range of questions and scales measuring factors that may be related to gang membership.[3] On issues related to violence, it included questions about

peer activities and delinquency, individual delinquent involvement, family violence and abuse, and victimization. When young women responded affirmatively
to being gang members, I followed with a series of questions about the nature of
their gang, including its size, leadership, activities, symbols, and so on. Girls who
admitted gang involvement during the survey participated in a follow-up interview to talk in more depth about their gangs and gang activities. The goal of the
in-depth interview was to gain a greater understanding of the nature and meanings of gang life from the point of view of its female members. A strength of
qualitative interviewing is its ability to shed light on this aspect of the social
world, highlighting the meanings individuals attribute to their experiences
(Adler and Adler 1987; Glassner and Loughlin 1987; Miller and Glassner 1997).
In addition, using multiple methods, including follow-up interviews, provided
me with a means of detecting inconsistencies in young women's accounts of
their experiences. Fortunately, no serious contradictions arose. However, a limitation of the data is that only young women were interviewed. Thus, I make
inferences about gender dynamics, and young men's behavior, based only on
young women's perspectives.

GENDER, GANGS, AND VIOLENCE

Gangs as Protection and Risk

An irony of gang involvement is that although many members suggest [that] one
thing they get out of the gang is a sense of protection (see also Decker 1996; Joe
and Chesney-Lind 1995; Lauderback et al. 1992), gang membership itself means
exposure to victimization risk and even a willingness to be victimized. These
contradictions are apparent when girls talk about what they get out of the
gang, and what being in the gang means in terms of other members' expectations
of their behavior. In general, a number of girls suggested that being a gang member is a source of protection around the neighborhood. Erica,[4] a 17-year-old
African American, explained, "It's like people look at us and that's exactly what
they think, there's a gang, and they respect us for that. They won't bother us....
It's like you put that intimidation in somebody." Likewise, Lisa, a 14-year-old
White girl, described being in the gang as empowering: "You just feel like, oh
my God, you know, they got my back. I don't need to worry about it." Given
the violence endemic in many inner-city communities, these beliefs are understandable, and to a certain extent, accurate.

In addition, some young women articulated a specifically gendered sense of
protection that they felt as a result of being a member of a group that was predominantly male. Gangs operate within larger social milieus that are characterized by gender inequality and sexual exploitation. Being in a gang with young
men means at least the semblance of protection from, and retaliation against,
predatory men in the social environment. Heather, a 15-year-old White girl,
noted, "You feel more secure when, you know, a guy's around protectin' you,
you know, than you would a girl." She explained that as a gang member,

because "you get protected by guys ... not as many people mess with you." Other young women concurred and also described that male gang members could retaliate against specific acts of violence against girls in the gang. Nikkie, a 13-year-old African American girl, had a friend who was raped by a rival gang member, and she said, "It was a Crab [Crip, the name of the rival gang] that raped my girl in Miller Ales, and um, they was ready to kill him." Keisha, an African American 14-year-old, explained, "If I got beat up by a guy, all I gotta do is go tell one of the niggers, you know what I'm sayin'? Or one of the guys, they'd take care of it."

At the same time, members recognized that they [might] be targets of rival gang members and were expected to "be down" for their gang at those times even when it meant being physically hurt. In addition, initiation rites and internal rules were structured in ways that required individuals to submit to, and be exposed to, violence. For example, young women's descriptions of the qualities they valued in members revealed the extent to which exposure to violence was an expected element of gang involvement. Potential members, they explained, should be tough, able to fight and to engage in criminal activities, and also should be loyal to the group and willing to put themselves at risk for it. Erica explained that they didn't want "punks" in her gang: "When you join something like that, you might as well expect that there's gonna be fights.... And, if you're a punk, or if you're scared of stuff like that, then don't join." Likewise, the following dialogue with Cathy, a white 16-year-old, reveals similar themes. I asked her what her gang expected out of members and she responded, "to be true to our gang and to have our backs." When I asked her to elaborate, she explained,

CATHY: Like, uh, if you say you're a Blood, you be a Blood. You wear your rag even when you're by yourself. You know, don't let anybody intimidate you and be like, "Take that rag off." You know, "You better get with our set." Or something like that.

JM: OK. Anything else that being true to the set means?

CATHY: Um. Yeah, I mean, just, just, you know, I mean it's, you got a whole bunch of people comin' up in your face and if you're by yourself they ask you what's your claimin', you tell 'em. Don't say, "Nothin'."

JM: Even if it means getting beat up or something?

CATHY: Mmhmm.

One measure of these qualities came through the initiation process, which involved the individual submitting to victimization at the hands of the gang's members. Typically this entailed either taking a fixed number of "blows" to the head and/or chest or being "beat in" by members for a given duration (e.g., 60 seconds). Heather described the initiation as an important event for determining whether someone would make a good member:

When you get beat in if you don't fight back and if you just like stop and
you start cryin' or somethin' or beggin' 'em to stop and stuff like that,
then, they ain't gonna, they'll just stop and they'll say that you're not gang
material because you gotta be hard, gotta be able to fight, take punches.

In addition to the initiation, and threats from rival gangs, members were expected to adhere to the gang's internal rules (which included such things as not fighting with one another, being "true" to the gang, respecting the leader, not spreading gang business outside the gang, and not dating members of rival gangs). Breaking the rules was grounds for physical punishment, either in the form of a spontaneous assault or a formal "violation," which involved taking a specified number of blows to the head. For example, Keisha reported that she talked back to the leader of her set and "got slapped pretty hard" for doing so. Likewise, Veronica, an African American 15-year-old, described her leader as "crazy, but we gotta listen to 'im. He's just the type that if you don't listen to 'im, he gonna blow your head off. He's just crazy."

It is clear that regardless of members' perceptions of the gang as a form of "protection," being a gang member also involves a willingness to open oneself up to the possibility of victimization. Gang victimization is governed by rules and expectations, however, and thus does not involve the random vulnerability that being out on the streets without a gang might entail in high-crime neighbor-hoods. Because of its structured nature, this victimization risk may be perceived as more palatable by gang members. For young women in particular, the gen-dered nature of the streets may make the empowerment available through gang involvement an appealing alternative to the individualized vulnerability they oth-erwise would face. However, as the next sections highlight, girls' victimization risks continue to be shaped by gender, even within their gangs because these groups are structured around gender hierarchies as well.

Gender and Status, Crime and Victimization

Status hierarchies within Columbus gangs, like elsewhere, were male dominated (Bowker, Gross, and Klein 1980; Campbell 1990). Again, it is important to high-light that the structure of the gangs these young women belonged to—that is, male-dominated, integrated mixed-gender gangs—likely shaped the particular ways in which gender dynamics played themselves out. Autonomous female gangs, as well as gangs in which girls are in auxiliary subgroups, may be shaped by different gender relations, as well as differences in orientations toward status, and criminal involvement.

All the young women reported having established leaders in their gang, and this leadership was almost exclusively male. While LaShawna, a 17-year-old Afri-can American, reported being the leader of her set (which had a membership that is two-thirds girls, many of whom resided in the same residential facility as her), all the other girls in mixed-gender gangs reported that their Original Gangster [OG] was male. In fact, a number of young women stated explicitly that only male gang members could be leaders. Leadership qualities, and qualities attributed to high-status members of the gang—being tough, able to fight, and willing to "do dirt" (e.g., commit crime, engage in violence) for the gang—were perceived as characteristically masculine. Keisha noted, "The guys, they just harder." She explained, "Guys is more rougher. We have our G's back but, it ain't gonna be like the guys, they just don't give a fuck. They gonna shoot you in a minute." For the most part, status in the gang was related to traits such as the willingness to use serious violence and commit dangerous crimes and, though not exclusively, these traits were viewed primarily as qualities more likely and more intensely located among male gang members.

Because these respected traits were characterized specifically as masculine, young women actually may have had greater flexibility in their gang involvement than young men. Young women had fewer expectations placed on them—by both their male and female peers—in regard to involvement in criminal activities such as fighting, using weapons, and committing other crimes. This tended to decrease girls' exposure to victimization risk comparable to male members, because they were able to avoid activities likely to place them in danger. Girls *could* gain status in the gang by being particularly hard and true to the set. Heather, for example, described the most influential girl in her set as "the hardest girl, the one that don't take no crap, will stand up to anybody." Likewise, Diane, a white 15-year-old, described a highly respected female member in her set as follows:

> People look up to Janeen just 'cause she's so crazy. People just look up to her 'cause she don't care about nothin'. She don't even care about makin' money. Her, her thing is, "Oh, you're a Slob [Blood]? You're a Slob? You talkin' to me? You talkin' shit to me?" Pow, pow! And that's it. That's it.

However, young women also had a second route to status that was less available to young men. This came via their connections—as sisters, girlfriends, cousins—to influential, high-status young men.[5] In Veronica's set, for example, the girl with the most power was the OG's "sister or his cousin, one of 'em." His girlfriend also had status, although Veronica noted that "most of us just look up to our OG." Monica, a 16-year-old African American, and Tamika, a 15-year-old African American, both had older brothers in their gangs, and both reported getting respect, recognition, and protection because of this connection. This route to status and the masculinization of high-status traits functioned to maintain gender inequality within gangs, but they also could put young women at less risk of victimization than young men. This was [so] both because young women were perceived as less threatening and thus were less likely to be targeted by rivals, and because they were not expected to prove themselves in the ways that young men were, thus decreasing their participation in those delinquent activities likely to increase exposure to violence. Thus, gender inequality could have a protective edge for young women.

Young men's perceptions of girls as lesser members typically functioned to keep girls from being targets of serious violence at the hands of rival young men, who instead left routine confrontations with rival female gang members to the girls in their own gang. Diane said that young men in her gang "don't wanna waste their time hittin' on some little girls. They're gonna go get their little cats [females] to go get 'em." Lisa remarked,

> Girls don't face [as] much violence as [guys]. They see a girl, they say, "We'll just smack her and send her on." They see a guy—'cause guys are like a lot more into it than girls are, I've noticed that—and they like, well, "we'll shoot him."

In addition, the girls I interviewed suggested that, in comparison with young men, young women were less likely to resort to serious violence, such as that involving a weapon, when confronting rivals. Thus, when girls' routine confrontations were more likely to be female on female than male on female, girls' risk of serious victimization was lessened further.

Also, because participation in serious and violent crime was defined primarily as a masculine endeavor, young women could use gender as a means of avoiding participation in those aspects of gang life they found risky, threatening, or morally troubling. Of the young women I interviewed, about one-fifth were involved in serious gang violence: A few had been involved in aggravated assaults on rival gang members, and one admitted to having killed a rival gang member, but they were by far the exception. Most girls tended not to be involved in serious gang crime, and some reported that they chose to exclude themselves because they felt ambivalent about this aspect of gang life. Angie, an African American 15-year-old, explained,

> I don't get involved like that, be out there goin' and just beat up people like that or go stealin', things like that. That's not me. The boys, mostly the boys do all that, the girls we just sit back and chill, you know.

Likewise, Diane noted,

> For maybe a drive-by they might wanna have a bunch of dudes. They might not put the females in that. Maybe the females might be weak inside, not strong enough to do something like that, just on the insides.... If a female wants to go forward and doin' that, and she wants to risk her whole life for doin' that, then she can. But the majority of the time, that job is given to a man.

Diane was not just alluding to the idea that young men were stronger than young women. She also inferred that young women were able to get out of committing serious crime, more so than young men, because a girl shouldn't have to "risk her whole life" for the gang. In accepting that young men were more central members of the gang, young women could more easily participate in gangs without putting themselves in jeopardy—they could engage in the more routine, everyday activities of the gang, like hanging out, listening to music, and smoking bud (marijuana). These male-dominated mixed-gender gangs thus appeared to provide young women with flexibility in their involvement in gang activities. As a result, it is likely that their risk of victimization at the hands of rivals was less than that of young men in gangs who were engaged in greater amounts of crime.

Girls' Devaluation and Victimization

In addition to girls choosing not to participate in serious gang crimes, they also faced exclusion at the hands of young men or the gang as a whole (see also Bowker et al. 1980). In particular, the two types of crime mentioned most frequently as "off-limits" for girls were drug sales and drive-by shootings. LaShawna explained, "We don't really let our females [sell drugs] unless they really wanna and they know how to do it and not to get caught and everything." Veronica described a drive-by that her gang participated in and said, "They wouldn't let us [females] go. But we wanted to go, but they wouldn't let us." Often, the exclusion was couched in terms of protection. When I asked Veronica why the girls couldn't go, she said, "so we won't go to jail if they was to get caught. Or if one of 'em was to get shot, they wouldn't want it to happen to us." Likewise, Sonita, a 13-year-old African American, noted, "If they gonna do somethin' bad and they think one of the females gonna get hurt they don't let 'em do it with them.... Like if they involved with shooting or whatever, [girls] can't go."

Although girls' exclusion from some gang crime may be framed as protective (and may reduce their victimization risk vis-à-vis rival gangs), it also served to perpetuate the devaluation of female members as less significant to the gang— not as tough, true, or "down" for the gang as male members. When LaShawna said her gang blocked girls' involvement in serious crime, I pointed out that she was actively involved herself. She explained, "Yeah, I do a lot of stuff 'cause I'm tough. I likes, I likes messin' with boys. I fight boys. Girls ain't nothin' to me." Similarly, Tamika said, "girls, they little peons."

Some young women found the perception of them as weak a frustrating one. Brandi, an African American 13-year-old, explained, "Sometimes I dislike that the boys, sometimes, always gotta take charge and they think, sometimes, that the girls don't know how to take charge 'cause we're like girls, we're females, and like that. " And Chantell, an African American 14-year-old, noted that rival gang members "think that you're more of a punk." Beliefs that girls were weaker than boys meant that young women had a harder time proving that they were serious about their commitment to the gang. Diane explained,

> A female has to show that she's tough. A guy can just, you can just look
> at him. But a female, she's gotta show. She's gotta go out and do some
> dirt. She's gotta go whip some girl's ass, shoot somebody, rob somebody
> or something. To show that she is tough.

In terms of gender-specific victimization risk, the devaluation of young women suggests several things. It could lead to the mistreatment and victimization of girls by members of their own gang when they didn't have specific male protection (i.e., a brother, boyfriend) in the gang or when they weren't able to stand up for themselves to male members. This was exacerbated by activities that led young women to be viewed as sexually available. In addition, since young women typically were not seen as a threat by young men, when they did pose one, they could be punished even more harshly than young men, not only for having challenged a rival gang or gang member but also for having overstepped "appropriate" gender boundaries.

Monica had status and respect in her gang, both because she had proven herself through fights and criminal activities, and because her older brothers were members of her set. She contrasted her own treatment with that of other young women in the gang:

> They just be puttin' the other girls off. Like Andrea, man. Oh my God, they dog Andrea so bad. They like, "Bitch, go to the store." She like, "All right, I be right back." She will go to the store and go and get them whatever they want and come back with it. If she don't get it right, they be like, "Why you do that bitch?" I mean, and one dude even smacked her. And, I mean, and, I don't, I told my brother once. I was like, "Man, it ain't even like that. If you ever see someone tryin' to disrespect me like that or hit me, if you do not hit them or at least say somethin' to them...." So my brothers, they kinda watch out for me.

However, Monica put the responsibility for Andrea's treatment squarely on the young woman: "I put that on her. They ain't gotta do her like that, but she don't gotta let them do her like that either." Andrea was seen as "weak" because she did not stand up to the male members in the gang; thus, her mistreatment was framed as partially deserved because she did not exhibit the valued traits of toughness and willingness to fight which would allow her to defend herself.

An additional but related problem was when the devaluation of young women within gangs was sexual in nature. Girls, but not boys, could be initiated into the gang by being "sexed in"—having sexual relations with multiple male members of the gang. Other members viewed the young women initiated in this way as sexually available and promiscuous, thus increasing their subsequent mistreatment. In addition, the stigma could extend to female members in general, creating a sexual devaluation that all girls had to contend with.

The dynamics of "sexing in" as a form of gang initiation placed young women in a position that increased their risk of ongoing mistreatment at the hands of their gang peers. According to Keisha, "If you get sexed in, you have no respect. That means you gotta go ho'in' for 'em; when they say you give 'em the pussy, you gotta give it to 'em. If you don't, you gonna get your ass beat. I ain't down for that." One girl in her set was sexed in and Keisha said the girl "just do everything they tell her to do, like a dummy." Nikkie reported that two girls who were sexed into her set eventually quit hanging around with the gang because they were harassed so much. In fact, Veronica said the young men in her set purposely tricked girls into believing they were being sexed into the gang and targeted girls they did not like:

> If some girls wanted to get in, if they don't like the girl they have sex with 'em. They run trains on 'em or either have the girl suck their thang. And then they used to, the girls used to think they was in. So, then the girls used to just come try to hang around us and all this little bull, just 'cause, 'cause they thinkin' they in.

Young women who were sexed into the gang were viewed as sexually promiscuous, weak, and not "true" members. They were subject to revictimization and mistreatment, and were viewed as deserving of abuse by other members, both male and female. Veronica continued, "They [girls who are sexed in] gotta do whatever, whatever the boys tell 'em to do when they want 'em to do it, right then and there, in front of whoever. And, I think, that's just sick. That's nasty, that's dumb." Keisha concurred, "She brought that on herself, by bein' the fact, bein' sexed in." There was evidence, however, that girls could overcome the stigma of having been sexed in through their subsequent behavior, by challenging members that disrespect them and being willing to fight. Tamika described a girl in her set who was sexed in, and stigmatized as a result, but successfully fought to rebuild her reputation:

> Some people, at first, they call her "little ho" and all that. But then, now she startin' to get bold.... Like, they be like, "Ooh, look at the little ho. She fucked me and my boy." She be like, "Man, forget y'all. Man, what? What?" She be ready to squat [fight] with 'em. I be like, "Ah, look at her!" Uh huh.... At first we looked at her like, "Ooh, man, she a ho, man." But now we look at her like she just our kickin'-it partner. You know, however she got in that's her business.

The fact that there was such an option as "sexing in" served to keep girls disempowered, because they always faced the question of how they got in and of whether they were "true" members. In addition, it contributed to a milieu in which young women's sexuality was seen as exploitable. This may help explain why young women were so harshly judgmental of those girls who were sexed in. Young women who were privy to male gang members' conversations reported that male members routinely disrespect girls in the gang by disparaging them sexually. Monica explained,

> I mean the guys, they have their little comments about 'em [girls in the gang] because, I hear more because my brothers are all up there with the guys and everything and I hear more just sittin' around, just listenin'. And they'll have their little jokes about "Well, ha I had her," and then and everybody else will jump in and say, "Well, I had her, too." And then they'll laugh about it.

In general, because gender constructions defined young women as weaker than young men, young women were often seen as lesser members of the gang. In addition to the mistreatment these perceptions entailed, young women also faced particularly harsh sanctions for crossing gender boundaries—causing harm to rival male members when they had been viewed as nonthreatening. One young woman[6] participated in the assault of a rival female gang member, who had set up a member of the girl's gang. She explained, "The female was supposingly goin' out with one of ours, went back and told a bunch of [rivals] what was goin' on and got the [rivals] to jump my boy. And he ended up in the hospital." The story she told was unique but nonetheless significant for what it indicates about the gendered nature of gang violence and victimization. Several young men in her set saw the girl walking down the street, kidnapped her, then brought her to a member's house. The young woman I interviewed, along with several other girls in her set, viciously beat the girl, then to their surprise the young men took over the beating, ripped off the girl's clothes, brutally gang-raped her, then dumped her in a park. The interviewee noted, "I don't know what happened to her. Maybe she died. Maybe, maybe someone came and helped her. I mean, I don't know." The experience scared the young woman who told me about it. She explained,

> I don't never want anythin' like that to happen to me. And I pray to God that it doesn't. 'Cause God said that whatever you sow you're gonna reap. And like, you know, beatin' a girl up and then sittin' there watchin' somethin' like that happen, well, Jesus that could come back on me. I mean, I felt, I really did feel sorry for her even though my boy was in the hospital and was really hurt. I mean, we coulda just shot her. You know, and it coulda been just over. We coulda just taken her life. But they went farther than that.

This young woman described the gang rape she witnessed as "the most brutal thing I've ever seen in my life." While the gang rape itself was an unusual event, it remained a specifically gendered act that could take place precisely because young women were not perceived as equals. Had the victim been an "equal," the attack would have remained a physical one. As the interviewee herself noted, "we coulda just shot her." Instead, the young men who gang-raped the girl were not just enacting revenge on a rival but on a *young woman* who had dared to treat a young man in this way. The issue is not the question of which is worse—to be shot and killed, or gang-raped and left for dead. Rather, this particular act sheds light on how gender may function to structure victimization risk within gangs.

DISCUSSION

Gender dynamics in mixed-gender gangs are complex and thus may have multiple and contradictory effects on young women's risk of victimization and repeat victimization. My findings suggest that participation in the delinquent lifestyles associated with gangs clearly places young women at risk for victimization. The act of joining a gang involves the initiate's submission to victimization at the hands of her gang peers. In addition, the rules governing gang members' activities place [the initiates] in situations in which they are vulnerable to assaults that are specifically gang related. Many acts of violence that girls described would not have occurred had they not been in gangs.

It seems, though, that young women in gangs believed [that] they … traded unknown risks for known ones—that victimization at the hands of friends, or at least under specified conditions, was an alternative preferable to the potential of random, unknown victimization by strangers. Moreover, the gang offered both a semblance of protection from others on the streets, especially young men, and a means of achieving retaliation when victimization did occur.…

Girls' gender, as an individual attribute, can function to lessen their exposure to victimization risk by defining them as inappropriate targets of rival male gang members' assaults. The young women I interviewed repeatedly commented that young men were typically not as violent in their routine confrontations with rival young women as with rival young men. On the other hand, when young women are targets of serious assault, they may face brutality that is particularly harsh and sexual in nature because they are female—thus, particular types of assault, such as rape, are deemed more appropriate when young women are the victims.

Gender can also function as a state-dependent factor, because constructions of gender and the enactment of gender identities are fluid. On the one hand, young women can call upon gender as a means of avoiding exposure to activities they find risky, threatening, or morally troubling. Doing so does not expose them to the sanctions likely faced by male gang members who attempt to avoid participation in violence. Although these choices may insulate young women from the risk of assault at the hands of rival gang members, perceptions of female gang members—and of women in general—as weak may contribute to more routinized victimization at the hands of the male members of their gangs. Moreover, sexual exploitation in the form of "sexing in" as an initiation ritual may define young women as sexually available, contributing to a likelihood of repeat victimization unless the young woman can stand up for herself and fight to gain other members' respect.

Finally, given constructions of gender that define young women as non-threatening, when young women do pose a threat to male gang members, the sanctions they face may be particularly harsh because they not only have caused harm to rival gang members but also have crossed appropriate gender boundaries in doing so. In sum, my findings suggest that gender may function to insulate young women from some types of physical assault and lessen their exposure to risks from rival gang members, but also to make them vulnerable to particular types of violence, including routine victimization by their male peers, sexual exploitation, and sexual assault.

NOTES

1. I contacted numerous additional agency personnel in an effort to draw the sample from a larger population base, but many efforts remained unsuccessful despite repeated attempts and promises of assistance. These [agency personnel] included persons at the probation department, a shelter and outreach agency for runaways, police personnel, a private residential facility for juveniles, and three additional community agencies. None of the agencies I contacted openly denied me permission to interview young women; they simply chose not to follow up. I do not believe that much bias resulted from the nonparticipation of these agencies. Each has a client base of "at-risk" youths, and the young women I interviewed report overlap with some of these same agencies. For example, a number had been or were on probation, and several reported staying at the shelter for runaways.

2. One young woman was a member of an all-female gang. Because the focus of this article is gender dynamics in

mixed-gender gangs, her interview is not included in the analysis.

3. These [surveys] include the Gang Membership Resistance Surveys in Long Beach and San Diego, the Denver Youth Survey, and the Rochester Youth Development Study.

4. All names are fictitious.

5. This is not to suggest that male members cannot gain status via their connections to high-status men, but that to maintain status, they will have to successfully exhibit masculine traits such as toughness. Young women appear to be held to more flexible standards.

6. Because this excerpt provides a detailed description of a specific serious crime, and because demographic information on respondents is available, I have chosen to conceal both the pseudonym and gang affiliation of the young woman who told me the story.

REFERENCES

Adler, Patricia A. and Peter Adler. 1987. *Membership Roles in Field Research.* Newbury Park, CA: Sage.

Baskin, Deborah, Ira Sommers, and Jeffrey Fagan. 1993. "The Political Economy of Violent Female Street Crime." *Fordham Urban Law Journal* 20: 401–17.

Bjerregaard, Beth and Carolyn Smith. 1993. "Gender Differences in Gang Participation, Delinquency, and Substance Use." *Journal of Quantitative Criminology* 4: 329–55.

Block, Carolyn Rebecca and Richard Block. 1993. "Street Gang Crime in Chicago." Research in Brief, Washington, DC: National Institute of Justice.

Bowker, Lee H., Helen Shimota Gross, and Malcolm W. Klein. 1980. "Female Participation in Delinquent Gang Activities." *Adolescence* 15(59): 509–19.

Brotherton, David C. 1996. "'Smartness,' 'Toughness,' and 'Autonomy': Drug Use in the Context of Gang Female Delinquency." *Journal of Drug Issues* 26 (1): 261–77.

Campbell, Anne. 1990. "Female Participation in Gangs." pp. 163–82 in *Gangs in America.* G. Ronald Huff, ed. Beverly Hills, CA: Sage.

Chesney-Lind, Meda, Randall G. Shelden, and Karen A. Joe. 1996. "Girls, Delinquency, and Gang Membership." pp. 185–204 in *Gangs in America,* 2nd ed. C. Ronald Huff, ed. Thousand Oaks, CA: Sage.

Curry, G. David, Richard A. Ball, and Scott H. Decker. 1996. Estimating the National Scope of Gang Crime from Law Enforcement Data Research in Brief. Washington, DC: National Institute of Justice.

Decker, Scott H. 1996. "Collective and Normative Features of Gang Violence." *Justice Quarterly* 13(2): 243–64.

Esbensen, Finn-Aage and David Huizinga. 1993. "Gangs, Drugs, and Delinquency in a Survey of Urban Youth." *Criminology* 31(4): 565–89.

Esbensen, Finn-Aage, David Huizinga, and Anne W. Weiher. 1993. "Gang and Non-Gang Youth: Differences in Explanatory Factors." *Journal of Contemporary Criminal Justice* 9(2): 94–116.

Fagan, Jeffrey. 1989. "The Social Organization of Drug Use and Drug Dealing among Urban Gangs." *Criminology* 27(4): 633–67.

____. 1990. "Social Processes of Delinquency and Drug Use among Urban Gangs." pp. 183–219 in *Gangs in America.* C. Ronald Huff, ed. Newbury Park, CA: Sage.

Glassner, Barry and Julia Loughlin. 1987. *Drugs in Adolescent Worlds: Burnouts to Straights*. New York: St. Martin's.

Huff, C. Ronald. 1996. "The Criminal Behavior of Gang Members and Nongang At-Risk Youth." pp. 75–102 in *Gangs in America*, 2d ed., edited by C. Ronald Huff Thousand Oaks, CA: Sage.

Joe, Karen A. and Meda Chesney-Lind. 1995. "Just Every Mother's Angel: An Analysis of Gender and Ethnic Variations in Youth Gang Membership." *Gender & Society* 9(4): 408–30.

Klein, Malcolm W. 1995. *The American Street Gang: Its Nature, Prevalence, and Control*. New York: Oxford University Press.

Klein, Malcolm W. and Cheryl L. Maxson. 1989. "Street Gang Violence." pp. 198–231 in *Violent Crime, Violent Criminals*. Neil Weiner and Marvin Wolfgang, eds. Newbury Park, CA: Sage.

Lauderback, David, Joy Hansen, and Dan Waldorf. 1992. " 'Sisters Are Doin' It for Themselves': A Black Female Gang in San Francisco." *The Gang Journal* 1(1): 57–70.

Lauritsen, Janet L., Robert J. Sampson, and John H. Laub. 1991. "The Link between Offending and Victimization among Adolescents." *Criminology* 29(2): 265–92.

Miller, Jody and Barry Glassner. 1997. "The 'Inside' and the 'Outside': Finding Realities in Interviews." pp. 99–112 in *Qualitative Research*, edited by David Silverman. London: Sage.

Sanders, William. 1993. *Drive-Bys and Gang Bangs: Gangs and Grounded Culture*. Chicago: Aldine.

Taylor, Carl. 1993. *Girls, Gangs, Women and Drugs*. East Lansing: Michigan State University Press.

Thornberry, Terence P., Marvin D. Krohn, Alan J. Lizotie, and Deborah Chard-Wierschem. 1993. "The Role of Juvenile Gangs in Facilitating Delinquent Behavior." *Journal of Research in Crime and Delinquency* 30(1): 75–85.

38

Hezbollah's Global Criminal Operations

MICHAEL P. ARENA

One of the international terrorist organizations most germane to world peace is the Arab group Hezbollah, a Shi'a Islamic militant group and political party based in Lebanon yet active in other countries. Hezbollah, which began as only a small militia, has grown into a large pyramid-shaped hierarchal organization that receives military training, weapons, and financial support from Iran and political support from Syria. It wields political power, holding seats in the Lebanese government and sponsoring a radio station and a satellite television station as well as programs for social development. At the same time, it is involved in terrorism, sometimes under the name of the Islamic Jihad Organization.

Less well known about Hezbollah are the criminal operations that support it. Uncertain of continuing financing by its sponsors in the Arab world, Hezbollah has followed a path to financial independence taken by many international criminal organizations. In this chapter, Arena discusses the diversity of Hezbollah's financial dealings that support its international political mission.

How would you assess the financial basis of an organization such as Hezbollah compared with other, traditional international criminal groups, such as the Italian mafia, the Colombian cartels, the Chinese triads and dyads, and the Russian mobs? How does Hezbollah compare against them in terms of ethnic and family homogeneity, long-term commitment, specialization or diversification, vertical and/or horizontal differentiation, and loose or tight structuring? How does the political mission of Hezbollah compare and contrast with these other groups? What kind of influence can governments and law enforcement agencies have over such groups?

On July 12, 2006, Lebanese Hezbollah militants attacked Israeli military outposts in the Shebaa Farms, kidnapping two Israeli soldiers and leaving at least eight dead. The attack sparked a month long volley of Israeli air strikes in southern Lebanon and a barrage of Hezbollah Katyusha rocket attacks on northern Israel.

From Michael P. Arena (2006). Hezbollah's Global Criminal Operations. *Global Crime*, 7(3–4), 454–470.

The crisis peaked with an Israeli ground assault before a ceasefire was finally reached in mid-August. The fighting caused hundreds of deaths and injuries, extensive damage to civilian infrastructure, and hundreds to become homeless.

The Lebanese Hezbollah (Party of God) first came onto the world stage in April 1983 when one of its operatives drove a van packed with explosives into the U.S. Embassy in Beirut, killing 63 people. A second attack followed in October 1983 when an explosive-laden pickup was driven into the U.S. Marine barracks also in Beirut and detonated, killing 241 Marines. Moments later, 58 French paratroopers were killed in the Beirut suburb of Jnah when another explosive-filled vehicle detonated in their barracks. Initially, an organization called the Islamic Jihad claimed responsibility for the attacks but it was soon determined that it was a front for Hezbollah [which] was intent on expelling Western influence from the region. These attacks made Hezbollah responsible for more American deaths than any other terrorist organization prior to September 11, 2001.

Hezbollah evolved from a ragtag group of militants into a sophisticated organization with several thousand supporters and members and a multi-dimensional infrastructure complete with military, political, social welfare, and finance wings requiring a substantial amount of money to operate. While Hezbollah is often quoted as being allied with Syria, the organization's chief financial sponsor is Iran which is believed to provide somewhere between $60 [million and] $100 million dollars a year to the group. Iran has been a steady sponsor of Hezbollah since its inception[;]however, its funding has fluctuated. Under the presidencies of Rafsanjani (1989–1997) and Khatami (1997–2005), Iran's financial support to Hezbollah was cut by almost 70 percent. This reduction may have been a key factor in Hezbollah's reliance upon funding from business interests and external sources such as charitable donations from the Lebanese diaspora community. Recently, it has also come to light that Hezbollah receives a great deal of funding from a worldwide network of criminal enterprises.

Hezbollah appears to be profiting from a variety of criminal enterprises, most notably, Intellectual Property Crime (IPC), drug smuggling, cigarette smuggling, and the exploitation of the African diamond trade. This essay aims to provide a brief description of these enterprises, [an account of] instances in which Hezbollah's members and supporters have been found to be involved in such crimes, and some indication as to how much money has been funneled to the Hezbollah organization. However, it begins with a brief history of the Hezbollah organization.

A BRIEF HISTORY OF HEZBOLLAH

Lebanese Hezbollah is a radical Shia organization which was formed in 1982 as a result of an interaction of several disparate forces. One of these forces was the political mobilization of the southern Shia community during the 1960s. With Shia, Sunni, and Druze Muslims, Christian Maronites, and Greek Orthodox Christians, Lebanon utilizes a confessional system in which political offices and bureaucratic appointments are distributed along sectarian lines to govern its religiously divided population. Inhabiting the rural regions of south Lebanon and the Biq'a Valley, Shia Muslims tended to be politically and economically marginalized until a charismatic Imam by the name of Musa al-Sadr began organizing the community and fighting for official representation in the late 1950s.

The political awakening of the Shia community was also spurred by the growing power of the Palestinian refugee community [that] had escaped to Lebanon after the declaration of Israel in 1948 and whose population was growing after the violence of the Six Day War in 1967. By the early 1970s, the Palestinian Liberation Organization (PLO) [had] relocated its base of operations to Lebanon after King Hussein evicted the militants from Jordan. Once the PLO arrived, [it] proceeded to launch attacks on Israel's northern border and, subsequently, invite periodic incursions by the Israel Defense Force (IDF) into Lebanon. With tensions in the region at a peak, civil war erupted in 1975. Intent on defending itself from the various armed forces in the region, Musa al-Sadr founded the Lebanese Resistance or what would come to be known as the Amal militia.

A second force to influence the creation of Hezbollah was the Iranian Revolution. The revolution had what Ahmad Hamzeh called a "demonstration effect" on the Lebanese Shia by showing them [that] it was possible to create an Islamic regime. Heavily influenced by Khomeini, Hezbollah became fiercely anti-Western, dedicated to destroying Israel, and committed to establishing an Islamic state in Lebanon in accordance with Islamic law. The Iranian regime and its supreme leader Ayatollah Khamenei continue to be a source of inspiration for Hezbollah.

The third and perhaps the most influential force to contribute to the creation of Hezbollah was the 1982 Israeli invasion of Lebanon dubbed Operation Peace for Galilee. While Israel had made periodic incursions into Lebanon in the past, this large-scale invasion was aimed at permanently eliminating the PLO's threat to Israel's northern border. Israeli forces stormed the southern region with relative ease before moving north to Beirut and forcing the PLO leadership to flee to surrounding Arab states. At first, the Shia community welcomed the IDF hoping that, once the PLO was gone, Amal would be free to reign. However, the Shia community soon realized [that] Israeli officials had no intention of leaving Lebanon before implementing [their] own security plan. In response, small groups of Shia militants began attacking the Israeli troops who[m] they viewed as occupiers. Anxious to spread its brand of Islamic revolution, Iran dispatched 1,500 members of its Revolutionary Guard to the Biq'a Valley to infiltrate organizations such as Amal and train Lebanese Shia militants. Eventually, these militants gathered under one umbrella to form Hezbollah. The IDF withdrew from Lebanon in 1985 but continued to occupy a 15 kilometer wide security zone on the Lebanese border to help prevent further cross-border attacks. This strip of land would ultimately become Hezbollah's proving ground in its resistance to Israel which finally made a complete withdrawal from the zone in 2000.

HEZBOLLAH'S PYRAMID STRUCTURE

Since evolving from an underground movement into a bona fide political party, Hezbollah's once guarded organizational structure has become more open to public view. Hezbollah is organized into a hierarchical pyramid whose territorial divisions parallel Lebanon's governorates, more specifically, the regions of Beirut,

Biq'a, and south Lebanon as they have the highest concentration of Shia Muslims. The first administrative tier of this pyramid is Hezbollah's seven-member council called the Majlis al-Shura or Consultative Council. It consists primarily of Shia clergy but also has some positions for laypersons. The Shura council's primary function is to oversee the second tier of the organization, the Executive Administrative Apparatus (Shura Tanfiz). This administrative body consists of the Politburo which oversees political activities, the Parliamentary Council which is made up of Hezbollah members who have won seats in the Lebanese Parliament, the Judicial council which resolves conflict in the Hezbollah controlled Shia community, and the Executive Council which handles the day-to-day activities of the organization through its eight units.

The eight units of the Executive Council can be viewed as the organization's third tier and [are] further subdivided into the Social Unit[,] which oversees welfare services to members, supporters, and the families of martyrs; the Islamic Health Unit[,] which provides health care; the Education Unit, which provides financial aid and scholarships; the Information Unit, which controls the organization's media outlets and propaganda campaigns; the Syndicate Unit, which oversees professional associations; the External Relations Unit, which acts as a liaison between Hezbollah and the government; the Finance Unit, which is responsible for fundraising, expenditures, and budgeting; and the Engagement and Coordination Unit, which is responsible for general security and addressing threats against the organization. There are two additional components that fall outside the primary hierarchy and report solely to the Shura Council[:] the Islamic Resistance and the Security Apparatus. The Islamic Resistance is the closest entity Hezbollah has to a standing army. The Security Apparatus is broken into two parts[:] the internal security, which polices the Hezbollah membership, and the external security[,] which is responsible for countering foreign intelligence services. The final components organized under the Executive Council and Military and Security Apparatus are the various geographic regions broken into Beirut, Biq'a, and South Lebanon and corresponding sectors, branches and groups. In whole, it is estimated that Hezbollah has several thousand supporters and members and a few hundred terrorist operatives.

Hezbollah was designated as a Foreign Terrorist Organization by the U.S. Department of State in 1997 based on the terror campaign it began in the 1980s. The bombings of the U.S. Embassy and the Marine barracks were soon followed by the 1984 kidnapping and detention of seventeen Americans in Lebanon. The last hostage was finally released in December 1991. In June 1985, hijackers overtook TWA Flight 847 in route from Athens to Rome. One hundred and forty-three passengers and crew members were held hostage as they were shuttled back and forth from Beirut to Algiers over a four-day period. During the ordeal, Robert D. Stethem, a Navy diver on board the aircraft, was murdered and dumped on the tarmac at Beirut International Airport prior to all of the remaining passengers eventually being released. During the 1990s, Hezbollah was focused primarily on expelling Israel from southern Lebanon, however[.][W]orking under a front organization, Hezbollah is believed to have detonated a car bomb outside the Israeli embassy in Buenos Aires, Argentina, killing 32 people in 1992. Two years later, in 1994, a second attack in the form of another vehicle bomb was detonated outside the Israeli–Argentine Mutual Association (AMIA) in Buenos Aires, killing 86 people. As mentioned earlier, prior to the September 11, 2001 attacks Hezbollah was responsible for killing more Americans than any other terrorist organization.

Intellectual Property Crime

One criminal enterprise Hezbollah has profited from is the IPC. According to Ronald K. Noble, Secretary General for the International Criminal Police Organization, intellectual property is defined as:

> [T]he legal rights that correspond to intellectual activity in the industrial, scientific, and artistic fields. These legal rights, most commonly in the form of patents, trademarks, and copyright, protect the moral and economic rights of the creators, in addition to the creativity and dissemination of their work[.] Industrial property, which is part of intellectual property, extends protection to inventions and industrial designs.[1]

Therefore, IPC refers to "…the counterfeiting or pirating of goods for sale where the consent of the rights holder has not been obtained".[2] The World Customs Organization reported [that] there were over 4,000 cases in 2004 involving the seizure of more than 166 million counterfeit or pirated articles by customs officials.

IPC is considered a low risk, high gain crime meaning [that] it has the potential to provide considerable monetary rewards but does not carry substantial penalties if [the perpetrator is] caught. As one suspect in an IPC investigation explained "It's better than the dope business, no one[']s going to prison for [pirating] DVDs".[3] This makes IPC an attractive enterprise to organized crime, street gangs, and international terrorists associated with groups such as Hezbollah. Terrorist groups typically take two types of roles in IPC. The first type of role is direct involvement[,] where the group is implicated in the production, distribution, or sale of counterfeit goods and uses the money to fund its activities. The second is indirect involvement[,] where sympathizers or militants are involved in IPC and remit some of the profits back to the group through couriers or money wires via third parties. It appears [that] Hezbollah is indirectly involved in IPC as its members and supporters are committing the act of pirating intellectual property and then funneling the proceeds back to the group via third parties.

Aside from their many operations abroad, Hezbollah profits from the commission of intellectual property theft in North America. During a recent hearing conducted by the U.S. Senate Committee on Homeland Security and Governmental Affairs, Los Angeles County Sheriff's Department Lieutenant John C. Stedman detailed two instances in which his investigative team encountered what he believed were indications that some associates of Hezbollah may be involved in IPC in Los Angeles, California. The first instance was experienced while [he was] executing an IPC related search warrant of a suspect's home. During the search he saw small Hezbollah flags displayed in the suspect's bedroom along with a photograph of Secretary General Hassan Nasrallah. The suspect's wife asked Lt. Stedman if he knew the individual in the portrait to which he responded in the affirmative. The wife explained, "We love him because he protects us from the Jews." The investigators also found dozens of audio tapes of Nasrallah's speeches and a locket which contained a picture of the male suspect on one side and a picture of Nasrallah on the other. The second instance involved the execution of a search warrant at a clothing store in which thousands of dollars in counterfeit clothing and two unregistered firearms were seized. The suspect revealed a tattoo of the Hezbollah flag on his arm while being booked into custody.

Stedman explained that one of the biggest issues for those who commit IPC is how to disperse the money generated by the enterprise. He identified one instance, not explicitly identified as being linked to Hezbollah, in which a suspect who was arrested at Los Angeles International airport by U.S. Customs officers had more than $230,000 strapped to her body. She told authorities that she was traveling to Lebanon for "vacation". The suspect was soon identified as the owner of a chain of cigarettes shops. Investigators seized more than 1,000 cartons of counterfeit cigarettes, an additional $70,000 in cash, and wire transfers to banks throughout the world.

Drug Trafficking

The illegal drug trade has proven to be another profitable enterprise for the Lebanese Hezbollah. While testifying before a U.S. Congressional Subcommittee, Ambassador-at-large for counterterrorism, Francis X. Taylor, stated that drug trafficking, in particular, is one illicit venture terrorist organizations have used to replace funding lost as a result of declining state sponsorship. He went on to say that the terrorist organizations' role in the drug trade has taken several forms, ranging from protecting production operations to directly trafficking illicit drugs.

Both drug traffickers and terrorist organizations have similar needs, such as covert funding sources, weapons, and operational security, and possess similar skill sets, such as the production of fraudulent documents, establishing front organizations, money laundering, and the clandestine movement of logistics and people. These similarities have made terrorist organizations a natural and effective fit for participation in the drug trade.

Hezbollah's involvement in the drug trade may have begun with the smuggling of opiates out of the Bi'qa Valley, a Hezbollah stronghold. Lebanese and Syrian farmers have traditionally raised opium poppies and cannabis in the region to make hashish for local consumption. However, the cultivation and processing of opium and cannabis has been largely suppressed since the Lebanese and Syrian governments joined international drug eradication initiatives in the early 1990s. According to the International Policy Institute for Counterterrorism, the success of the government-led eradication program prompted Hezbollah to find an alternate drug supply, mainly from the Latin America route (Colombia, Peru, Brazil, etc.) and the Far East route (Pakistan, Afghanistan, Iran, Turkey, and Syria). The Tri-Border area is believed to be a focal point for Hezbollah's drug trafficking operations in Latin America as evidenced by several recent, highly publicized cases.

Cigarette Smuggling

Another U.S.-based Hezbollah criminal operation was revealed in July 2000 when federal authorities levied charges against 25 suspects for their involvement in racketeering, money laundering, immigration fraud, credit card fraud, marriage fraud, visa fraud, bribery, and providing material support to a terrorist organization. Dubbed Operation Smokescreen, the investigation began in the mid-1990s after an off-duty sheriff, working as a security officer at a cigarette wholesaler in Statesville, North Carolina, watched three Arabic-speaking men enter the store, collect 299 cartons of cigarettes apiece, pay for the merchandise in cash, and load the goods into minivans before heading north. With the passing of the Contraband Trafficking Act of 1978, it is a federal offense to transport more than 60,000 cigarettes or 300 cartons from one state to another without proof [that] the appropriate state tax had been paid in each state. Realizing this, the off-duty deputy reported his suspicions to the Bureau of Alcohol, Tobacco, and Firearms (ATF) which initiated an investigation.

The ATF soon discovered a multimillion dollar cigarette smuggling ring which was using a classic organized crime scheme. The ring purchased cigarettes in a low tax state, in this case, North Carolina whose rate was 50 cents a carton, and transported them to Michigan, a high tax state which charged $7.50 a carton. The ring was also taking advantage of the fact that neither North Carolina nor Michigan marked cigarettes with a tax stamp making it difficult for authorities to determine if appropriate taxes were paid. By exploiting the discrepancy in tax rates, the smugglers stood to make roughly $13,000 per van load. In all, the ring is estimated to have purchased $8 million in cigarettes making roughly $2 million in profit. As authorities began to make preparations to dismantle the ring, agents from the Federal Bureau of Investigation (FBI) approached the federal prosecutor, informing him that they had stumbled onto a Hezbollah cell operating out of Charlotte, North Carolina, and explained that they already had two of the cell members under electronic surveillance.

Members of the cigarette smuggling ring became well versed in the art of criminal conspiracy. They opened multiple bank accounts, fraudulently obtained credit cards, and learned how to establish multiple identities by obtaining the names of departing international students by retrieving corresponding driver's licenses, social security numbers, and Immigration and Naturalization Service work authorizations and credit histories. Many of the credit card scam and identity theft skills were taught by Said Mohamad Harb, one of the cell members who was described by authorities as "a one-man crime wave". Through the assistance of the Canadian Security Intelligence Service, it was revealed that Harb was the key link in the cell's provision of material support to Hezbollah. Harb was childhood friends with Mohammud Dbouk who was sent to Canada from Lebanon by Hezbollah's chief procurement officer to lead its North American Procurement Program which was charged with obtaining dual use military equipment from retail outlets in Canada and the United States. Through Dbouk, Harb arranged for the Charlotte cell members to fulfill equipment orders from Hezbollah in Lebanon in exchange for a percentage of the purchase price. The Charlotte cell provided items such as night-vision goggles, cameras, global-positioning systems, metal-detection gear, video equipment, computers, and stun guns.

Exploiting the African Diamond Trade

The link between the diamond trade and international terrorism was publicized in November 2001 when the *Washington Post* published an article claiming international terrorist groups, namely al Qaeda and Hezbollah, had made millions of dollars from the illicit sale of diamonds mined by rebel groups in Sierra Leone.

The topic of international terrorists profiting from the illicit diamond trade received a more comprehensive treatment when the Global Witness, a British nongovernmental organization dedicated to exposing the relationship between natural resource exploitation and human rights abuses, published "For a few dollars more: How al Qaeda moved into the diamond trade". The report detailed how international terrorists had infiltrated West African diamond networks by taking advantage of illicit trading structures, weak governments, and lax trade regulations.

The U.S. General Accounting Office (GAO) identified the illegal trafficking of precious metals and stones as one means by which terrorist groups raise funds, explaining that illicit diamond dealing is an attractive criminal enterprise for several reasons. Diamonds serve as a form of currency that can be used for money laundering, exchanged for cash, or to purchase an array of illicit goods such as arms or drugs. The international diamond industry is fragmented and there is limited scrutiny of the flow of diamond[s] from mine to consumer. In addition, there are numerous small mining operations across Africa which has porous borders and little rule of law, making it less complicated for traders to acquire and smuggle diamonds. Areas such as Sierra Leone became notorious for the sale and export of conflict diamonds, the name given to diamonds mined in a war zone and used to finance armed conflicts. The Revolutionary United Front used the proceeds of the sales to drive [its] failed attempt to overthrow the government of Sierra Leone, during which [it] committed countless human rights abuses. Diamonds are conducive to smuggling because of their low weight, lack of odor, and untraceable origins. Additionally, they can be easily sold on the black market. Authorities report that in Antwerp, the world's largest diamond trading center, a substantial number of diamonds are sold on the black market with no transaction records.

Hezbollah has been profiting off the African diamond market since the 1980s when it first began establishing networks of operatives in the region. In the Congo, Hezbollah is reported to take a direct role in the market by purchasing diamonds from miners and local merchants at a fraction of the market value and selling them in Antwerp, Mumbai, and Dubai. In West Africa, Hezbollah is reported to be siphoning off profits from the region's diamond trade through the extortion of Lebanese merchants. As the Deputy Chief of mission for the U.S.

Embassy in Sierra Leone explained, "One thing [that] is incontrovertible is the financing of Hezbollah. ... It's not even an open secret[;]there is no secret. There's a lot of social pressure and extortionate pressure brought to bear: 'You had better support our cause, or we'll visit your people back home'".[4]

CONCLUSION

The gradual decrease in state sponsorship for terrorism has forced many terrorist organizations to seek alternative sources of funding. The Lebanese Hezbollah is no exception. It has come to rely upon a worldwide network of members and supporters who engage in various forms of criminal enterprise and remand a portion of its illicit profits back to the group. Regardless if the Hezbollah leadership is completely witting of the source, it is presumed that generating its own funding has enabled the organization to become less dependent upon Iran, whose funding has fluctuated over the years, and more self-sufficient. Furthermore, this source of funding may prove vital in the aftermath of recent fighting between Hezbollah and the IDF. The group will desperately need money to rebuild its weapon and equipment stockpiles and infrastructure. It will also need money to rebuild goodwill among the Lebanese Shia population. In mid-August, its Secretary General pledged to pay for rent and furniture to those Lebanese Shia whose homes were destroyed.

It is anticipated that Hezbollah's criminal networks will remain dynamic and adaptable in the face of law enforcement initiatives to cut off terrorism funding. According to the FBI, terrorists will switch to another commodity or industry once they realize authorities have become aware of their illegal activities. With this utilitarian aim, these networks may become active in any number of illegal enterprises in an effort to meet the organization's financial needs. The enterprise may have been around for a while, like cigarette smuggling and trafficking in stolen automobiles, or it may yet [remain] to be contrived. The only limit is that the enterprise [be] lucrative and relatively undetected. One final observation is warranted when assessing Hezbollah's global criminal operations. There is concern that those who are participating in illicit enterprises worldwide could be tasked with committing an act of terrorism on behalf of Hezbollah or at the behest of Iran. With an intimate knowledge of the inner workings of smuggling routes, clandestine money transfers, and illicit materials procurement, Hezbollah may be well positioned to perpetrate a terrorist attack almost anywhere in the world.

NOTES

1. Intellectual property crimes: Are proceeds from counterfeited goods funding terrorism?: Hearing before the Committee on International Relations, House of Representatives (2003) 108th cong, 1st sess, GPO, Washington, DC. Available at: http://commdocs.house. gov/committees/intlrel/

hfa88392.000/hfa88392_0f.htm
(retrieved 11 November 2005).

2. *Ibid*.

3. Counterfeit goods: Easy cash for
criminals and terrorists: Hearing before
the Committee on Homeland Security
and Governmental Affairs, United
States Senate (2005) 109th cong, 1st
sess, GPO, Washington, DC. Available at: http://www.gpo.gov/fdsys/
pkg/CHRG-109shrg21823/html/
CHRG-109shrg21823.htm (retrieved
16 July 2006).

4. 'Hezbollah extorts diamond profits
from Lebanese, U.S. says', *Rapaport
News*, June 30, 2004, Available at:
http://www.diamonds.net/News/
NewsItem.aspx?ArticleID=9852&
ArticleTitle=Hizballah+Extorts+
Diamond+Profits+From+Lebanese%
2C+U.S.+Says (retrieved 11
November 2005).

39

State–Corporate Crime in the Offshore Oil Industry: The BP Oil Spill

ELIZABETH A. BRADSHAW

The complementary interplay between governmental and corporate interests has given rise to a rash of state-sponsored and supported crime. Over the last decade or two, we have seen in the political and economic realms the ascendance of a "neoliberal" philosophy that, drawing on neoclassical economic theory, advocates support for great economic liberalization, privatization, free trade, open markets, deregulation, and reductions in government spending in order to enhance the role of the private sector in the economy. During this period, we have also witnessed an increase in governmental functions contracted out to private industry. At the same time, government agencies have lost funding and employees, with an associated weakening in their regulatory and oversight capabilities. The original rationale underpinning this transfer was that private industry would operate more efficiently and profitably than government bureaucracy, because private industry is driven by competitive market forces. But what has happened, instead, is that corporate and governmental interests have become incestuously joined, and favoritism and no-bid contracts have replaced the free-market economy when it comes to bidding for government contracts. This climate has fostered one of the greatest environments for the growth of white-collar crime ever seen. The oil industry, the tobacco industry, and the pharmaceutical industry are three that notably come to mind as enmeshed in governmental connections and decision making.

The twenty-first century has seen one company after another taken down, their employees' pension funds and shareholders' investments robbed by the corporate fraud of top executives. Yet an even higher price has been less visibly paid by members of the ordinary public through state-supported corporate crime. Politicians have fostered this practice by decreasing business regulation and creating a revolving door, as Bradshaw aptly depicts, among business, lobbying, and government.

One of the most dramatic recent examples lies in the Deepwater Horizon oil spill, also known as the BP oil spill or the Gulf of Mexico spill. Here, we see the effects of the conflict between state policies aimed at protecting the environment and

the temptation to ignore environmental dangers in the face of the pressure for finan-
cial profit. Just as we saw in Liederbach's chapter on the doctors who regulated their
own behavior, we see here that, when industries are left to set their own policies and
enforce them, temptations trump restrictive guidelines and directives. We are faced,
again, with recognizing the ability of powerful groups and entities in society to
engage in deviance while resisting being labeled and treated as deviant. Thus, we
come face-to-face with the social reality that we live in a world characterized by
inequality—a world in which structural connections such as that which we see here
between business and government allow companies to take risks and evade their
consequences, leaving ordinary people to shoulder the ultimate cost.

To what extent do you think things have changed or stayed the same since
this major oil spill occurred? Since the economic meltdown of 2008 left the global
economy in shambles? What other kinds of organizations do you think might be
operating under these kinds of collusions?

On April 20, 2010, the Deepwater Horizon rig operated by British Petro-
leum (BP) exploded in the Gulf of Mexico, killing 11 workers and spilling
nearly 5 million barrels of oil over a period of 3 months. The cause of this environmental disaster was not rogue industry or governmental officials, but a systematic collusion between government policies and industry practices, caused
by a radical reshaping of the nature of federal and corporate relations in the offshore oil industry. Motivated to earn royalties from leasing the Outer Continental Shelf (OCS) of the Gulf of Mexico, the federal government granted the oil
industry greater access for expansive drilling. Yet, at the same time, it reduced
the royalties oil companies had to pay the government and curtailed the regulatory oversight by which it monitored offshore drilling.

Initially, these functions were carried out by different government agencies.
The U.S. Geological Survey studied and protected public lands and resources.
The Bureau of Land Management granted leases and collected royalties. However in 1982, in the context of increasing governmental deregulation, the
Minerals Management Service (MMS) was founded, and these two conflicting
missions were combined into one organization. From that point forward, the
operative goal of collecting royalties began to take precedence over regulation
of the industry. Within this context, a normalization of deviance developed not
only within the MMS, but between the MMS and the offshore oil industry. In
the years leading up to the Deepwater Horizon disaster, relations between the
MMS and the oil industry had become so close that at times it was impossible
to tell them apart. This normalized deviance within and between organizational
cultures paved the way for state–corporate criminality to flourish.

STATE–CORPORATE CRIME AS
ORGANIZATIONAL DEVIANCE

A major assumption underpinning the study of state–corporate deviance is that state
and corporate organizations are social actors in and of themselves that must be

understood as both connected to, yet distinct from, individual employees, managers, owners, and regulators. Although individuals occupy organizational positions, their thoughts, actions, and behaviors are fundamentally shaped by the goals, procedures, standards, and norms of an organization. Moreover, the structure of any organization is composed of positions occupied by replaceable people, an approach designed to ensure the longevity of the organization. Directing inquiry toward the goals, procedures, standards, and norms of organizations draws attention to the power and influence of organizations in society and helps to further the understanding of the socially injurious behaviors that result from such features. Applying an organizational perspective to the concept of white-collar crime turns the subfield away from focusing narrowly on the role of the individual, as with occupational crime, and reorients it toward the power of organizational structures.

The concept of state–corporate crime includes illegal or socially injurious actions that result from mutually reinforcing interactions between the policies and practices of one or more institutions of political governance and one or more institutions of economic production and distribution (Michalowski and Kramer, 2006:20). This definition of state–corporate crime encompasses both legal criteria and socially injurious actions while centering on the nexus between government and business.

Further refining the concept, state–corporate crime can be categorized into two distinct forms: *state-facilitated corporate crime* and *state-initiated corporate crime*. State facilitation of corporate crime occurs when government institutions of social control clearly fail to establish regulatory institutions that are capable of restraining deviant activities by business, because of either direct collusion or shared, common goals. State initiation of corporate crime occurs when a business employed by the government undertakes deviant or illegal actions at the direction, or with the tacit approval, of government.

The concept of state–corporate crime has three useful characteristics for understanding deviant interactions between government and business as organizational actors. First, by highlighting the relationships between social institutions, it refutes the notion that organizational deviance is a discreet act. Second, by embracing the relational character of the state, the concept of state–corporate crime demonstrates how the horizontal interactions between political and economic institutions contain the potential for illegal and social injurious actions to occur. Finally, adopting a relational approach to the state allows for not only a consideration of horizontal interactions, but also the vertical relationships between different levels of organizational action: political–economic, organizational, and interactional.

STATE FACILITATION OF CORPORATE CRIME
IN THE OFFSHORE OIL INDUSTRY

As guardian and administrator of the nation's offshore resources, the federal government has profited immensely from the private leasing of public offshore lands. Revenue from offshore leases has been the primary goal behind federal expansion of deepwater exploration and development. Regardless of political party, each presidential administration has played a key role in supporting legislation that paved the way for drilling in deeper waters within the Gulf of Mexico. Although offshore disasters, such as the 1969 Santa Barbara oil spill, have helped to raise awareness of the need for increased safety and environmental oversight in the oil industry, offshore development has consistently superseded environmental protection.

Particularly since the Reagan administration and the creation of the MMS, federal policy has allowed regulation and development of the OCS to become increasingly controlled by the oil industry itself. Although the relationship between the offshore industry and the Department of the Interior (DOI) was close from the beginning, it became even closer as drilling moved into deeper waters. However, federal legislation has provided the opportunity for corporations to take additional risks by reducing the royalties on OCS leases in deep water while simultaneously weakening regulatory oversight, especially in the Gulf of Mexico.

Government Pursuing Revenue over Regulation

Federal ownership over the OCS was established with the passage of the Outer Continental Shelf Lands Act (OCSLA) of 1953, which gave the federal government jurisdiction and control over the OCS, extending outward beyond the coastal states' 3-mile tidelands, and which furthermore stipulated a process for leasing the ocean lands involved. As authorized under the OCSLA, the secretary of the interior was charged with overseeing and administering the lease, a task that was to be coordinated by the Bureau of Land Management (BLM) and the U.S. Geological Survey (USGS). Under this arrangement, the BLM was responsible for reviewing nominations for leases and overseeing competitive bids on the basis of the highest cash bonus bid with fixed royalty or the highest percentage bid with fixed cash basis. After the sale, the USGS regulated OCS activities and collected royalties.

Established in 1982 by Ronald Reagan's Secretary of the Interior James Watt, the MMS combined the functions previously held by the USGS and the BLM. By placing the mandates of regulatory oversight of offshore development and revenue collection within the same agency, Watt set the seeds for the maximization of revenue to become the dominant mission of the MMS. Secretary Watt also introduced the practice of "areawide leasing" (AWL), which opened much larger sections of land to industry at one time, superseding the previous practice of offering only select areas specifically nominated by firms. Policy changes during this period, such as AWL, expanded the industry's access and choice of leasing areas while requiring less government oversight. These changes at the MMS mutually benefited both the federal government and the offshore oil industry. Income from offshore oil leases represents the second-largest source of revenue for the federal government, providing strong motivation for both government and industry to pursue revenue collection at the expense of regulation.

Diminished Government Oversight

By the mid-1990s, the offshore oil industry had expanded beyond the MMS's capacity to oversee it. Revenue generation had consumed the majority of the MMS's efforts at the expense of regulatory oversight, something that was openly acknowledged by former MMS directors for years. In November 1996, the MMS's **budget** had reached its lowest point, even further hindering its ability to oversee the industry effectively. With the lack of funding came fewer unannounced inspections. As highlighted in a report by the Department of Interior's inspector general, by 1999 MMS inspections had declined significantly and were no longer effective. Just as the need for regulatory oversight was at its greatest, the MMS's capacity for oversight had been undercut.

As the industry moved ever further offshore, the MMS struggled to keep pace with the evolving deepwater **technology**, and the little training inspectors did receive was inadequate. The MMS lacked a formal inspection certification program for the oil and gas industry, and no exam was required in order to become certified. So unfamiliar with the inspection process, some government inspectors depended on oil industry representatives to explain the technology at facilities. Federal salaries at the MMS stagnated, and the agency struggled to attract trained and qualified personnel, especially engineers. In the MMS Gulf of Mexico offices, for instance, the number of permits for offshore drilling increased 71 percent between 2005 and 2009, yet there were not enough qualified engineers to review them. As the agency became more and more overwhelmed with applications, operators began to "shop around" at different offices outside of the appropriate jurisdictional area to seek an engineer who would approve the permit (U.S. National Commission 2011:74).

A culture of complacency concerning federal environmental regulations also developed within the MMS owing to a lack of funding. Scientists at the agency experienced strong pressure from their managers to rapidly approve development plans without proper evaluation of the environmental effects. As the volume of lease applications increased, especially in the Gulf of Mexico, the capacity of MMS regulators to oversee implementation of federal environmental policy was reduced.

The means by which the MMS **collected revenues** from industry changed significantly in 1997, making it more difficult to verify royalty payments and giving the industry even greater control. Known as taking "royalties in kind" (RIK), accepting payment in this manner differed from the MMS's former policy of accepting cash payments based on the value of oil produced, known as "royalty in value" (RIV). This new method of royalty collection allocated payment to the MMS in the form of oil and gas. The agency could then transfer the commodities to other federal agencies or sell them to refineries. The switch from RIV to RIK advantaged the oil industry, because it reduced administrative costs and made it so that leases would not be subject to audit, despite being worth millions (and sometimes billions) of dollars. So lax was regulation under RIK, that the Government Accounting Office asserted that the program's management operated on an "honor system." Encouraged by exploration into ever deeper waters, lavish royalty relief programs facilitated an increasingly close relationship between the oil industry and the MMS.

Resulting from extensive lobbying by the oil industry, the RIK program became a central part of the Bush–Cheney administration's energy strategy. As RIK continued to blossom, so did the relationship between MMS and the oil industry. Exemplifying the *"**revolving door**"* between government and industry, there were multiple examples of high-ranking DOI and MMS officials serving during the Bush administration who left their regulatory appointments to work for companies they formerly oversaw. Likewise, the Obama administration favored a continuation and expansion of deepwater exploration and royalty relief through the RIK program.

Deviance Normalized: Government and Industry as One

Epitomizing the intimate relationship between the MMS and the offshore oil industry, in 2008 congressional reports revealed that up to a third of the MMS department employees involved in the RIK program had been engaged in serious misconduct over the past several years, including rigging oil contracts, taking money as oil consultants, and having sexual relationships and using drugs with oil and gas company representatives. The investigation into the MMS RIK program based in Denver, Colorado, uncovered a pattern of ethical failure that revealed "a pervasive culture of exclusivity, exempt from the rules that govern all other employees of the Federal Government" (U.S. Department of Interior, 2008, no page numbers in document).

The investigation revealed that MMS was no longer fulfilling its regulatory mandate and that employees had fully adopted a "business model" approach to the RIK program, embracing a private-sector perspective on almost everything they did. In an attempt to codify their unique relationship with industry and exempt themselves from the guidelines governing all other federal employees, MMS RIK employees formed a study group in June 2006 to consider altering the rules to commonly align government and business interests.

Allegations of inappropriate relations with the oil industry are not unique to the MMS RIK office in Denver. MMS officials at the Lake Charles, Louisiana, district office that oversaw drilling in the Gulf of Mexico were investigated in 2010 for accepting gifts, such as meals, tickets to sporting events, and hunting and fishing trips, from industry representatives. Following a 2007 investigation and termination of one regional MMS supervisor of the New Orleans office for accepting gifts from an offshore drilling contractor, employees at the Lake Charles office appeared to drastically decrease their participation in these illegal behaviors (U.S. Department of Interior, 2010). It seems that the MMS organizational culture was plagued by corruption at the highest levels, setting the tone for other members throughout the agency.

Without any oversight or regulation, employees of the MMS RIK program and the oil industry had melded to become one. Far from being perceived as "deviant" activity, intimate fraternization between the MMS and the industry had become the norm, enough to even consider legally codifying the relationship. This normalization of deviance had become so ingrained that employees of the RIK program sought to legalize their intimate relationships with industry that were prohibited by federal law. After the fallout from the RIK scandal, DOI Secretary Ken Salazar was forced to announce on September 16, 2009, that it was time to end the RIK program. Nevertheless, although the RIK program may have been terminated, the influence of the oil industry continued to pervade the MMS organizational culture.

As the offshore industry expanded, employees at the underfunded and inadequately staffed MMS turned to illegal means to perform their jobs. Fraternizing with oil industry representatives had become a normal part of the culture at the MMS despite federal ethics guidelines that prohibited such close interactions. By the time its employees were having sex and doing drugs with oil industry representatives, the regulatory mission of the MMS was overcome by the shared goal of profit for both the federal government and the offshore industry. Without regulatory controls, the disintegration of federal oversight further allowed the offshore industry to take additional risks in the pursuit of profit.

CONCLUSIONS

An Integrated Theoretical Model of State–Corporate Crime and the Deepwater Horizon Spill

In the decades leading up to the Deepwater Horizon explosion, the political–economic climate in the United States underwent substantive changes that paved the way for the spill to occur. Neoliberalsim (a trend characterized by the privatization of public goods and services) brought about important shifts in the institutional relationships between government and business. During the Reagan administration, government oversight and control were retracted while greater power and autonomy was granted to businesses and corporations. Within this institutional context, the reorganization of the MMS by Reagan's secretary of the interior combined the conflicting mandates of leasing offshore lands and royalty collection, on the one hand, with regulation, on the other, setting the seeds for disaster to occur.

Motivation Motivation, the first catalyst for action, concerns goal attainment, which in turn draws on the interactional level and on Sutherland's differential association theory. The federal government's pursuit of royalties from offshore leases aligned with the oil industry's goals of maximizing profit and the pressure for goal attainment between the two was amplified. It did not take long for the goal of royalty collection from offshore leases to supersede the agency's regulatory mission, and the motivation for criminality in the pursuit of profit developed. The MMS began to operate under a business model, offering industry more offshore leases of greater swaths of the Gulf of Mexico while simultaneously reducing supervision of offshore lands.

Opportunity Opportunity, the second catalyst for action, assumes that organizational deviance is more likely when legitimate means are scarce relative to goals. This approach draws on the organizational level and directs inquiry toward the goals, procedures, standards, and norms of organizations; draws attention to the power and influence of organizations in society; and helps to further the understanding of the socially injurious behaviors that result from such structures. Catering to the industry's interests became an implicit part of the MMS's mission, and corruption became a pervasive part of the organization in multiple sectors. Owing to the revolving door between government and industry, most of the employees at the MMS had at some point worked for the private sector and maintained deep bonds with friends in the industry—bonds rooted in the region and in the culture and personal histories of the players. As a result of these close interactions, the normalization of deviance between government officials and industry representatives provided the opportunity to pursue the goal of profit at the expense of safety.

Lack of Social Control   Finally, the third catalyst for action is the presence or absence of social control. Organizations subjected to a high operationality of social control are more likely to foster cultures that favor compliance with laws and regulations, and organizations that are not subject to such control are more likely to develop cultures of resistance. As the offshore oil industry expanded into deeper waters, the MMS experienced cuts in funding that hindered its ability to provide oversight. Furthermore, the MMS was unable to adapt its regulatory framework to address the new proliferation of specialized contractors relied on by the industry. The MMS and the DOI were unable to effectively regulate the rapidly evolving industry and the increasing reliance on outsourced contractors. Absent any government control and oversight from the MMS, the offshore oil industry as a whole was left to police its own behavior. By the time Secretary Salazar attempted to reform the MMS following the 2008 RIK scandal, the closeness between the MMS and the industry had become far too normalized to prevent the explosion of the Deepwater Horizon rig on April 20, 2010, causing loss of life and untold environmental damage.

REFERENCES

Michalowski, Raymond, and Ronald Kramer (Eds). 2006. *State–Corporate Crime: Wrongdoing at the Intersection of Business and Government.* New Brunswick, NJ: Rutgers University Press.

U.S. Department of the Interior, Office of the Inspector General, Minerals Management Service. 2010. *Investigative Report: Island Operating Company et al.* http://www.doioig.gov/images/stories/reports/pdf/IslandOperatingCo.pdf.

_____. 2008. *Investigative Report: MMS Oil Marketing Group—Lakewood.* September 9, 2008. http://media.washingtonpost.com/wp-srv/investigative/documents/mmsoil-081908.pdf.

U.S. National Commission on the BP Deepwater Horizon Oil Spill and Offshore Drilling. January 2011. *Deep Water: The Gulf Oil Disaster and the Future of Offshore Oil Drilling, Report to the President.* http://cybercemetery.unt.edu/archive/oilspill/20121211005728/http:/www.oilspillcommission.gov/sites/default/files/documents/DEEPWATER_ReporttothePresident_FINAL.pdf.

Structure of the Deviant Act

A lthough the structure of deviant associations is revealing, in Part VII we investigate the characteristics of the acts of deviance themselves. Deviant associations involve the social organization of the deviants as people, depicting the types of relationships they have with other deviants. Deviant acts focus on the particular instance of deviance, not the relationships surrounding those acts. When we look at the structure of deviant acts, we look at the nature of the transaction: its length, its goals, its stability, the degree of cooperation or conflict involved, the number of parties involved, and the way the participants interact (or do not interact) with one another.

Deviant acts involve one or more people aiming to accomplish a particular deviant goal. These acts vary widely in character, from those enduring over a period of months to the more fleeting encounters that last only a few minutes, from those conducted alone to those requiring the participation of several or many people, and from those in which the participants are face-to-face to those in which they are physically separated. At the same time, acts of deviance can be looked at in terms of what they have in common. All deviant acts consist of purposeful behavior intended to accomplish a gratifying end, require the coordination of participants (if there is more than one participant), and depend on individuals reacting flexibly to unexpected events that may arise in this relatively unstructured and unregulated arena. Like the relations among deviants, deviant acts fall along a continuum of sophistication and organizational complexity. Following Best and Luckenbill's (1981) typology, we arrange them here according to the minimum number of their participants and the intricacy of the relations among these participants, moving again from the least to the most organizationally sophisticated.

Some deviant acts can be accomplished by a lone **individual**, without recourse to the assistance or presence of other people. This does not mean that others cannot accompany the deviant, either before or during the deviant act, or even that two deviants cannot commit acts of individual deviance together. Rather, the defining characteristic of an individual act of deviance is that it can be committed by one person, to that person, on that person, and for that person. A teenager's suicide, a drug addiction, a skid row transient's alcoholism, and a self-induced abortion are examples of individual behavioral deviance. Nonbehavioral forms of individual deviance include obesity, minority group status, a physical disability, and a deviant belief system (such as alternative religious or political beliefs). Individual deviance has areas of overlap with loner deviance, discussed in Part VI, but there are also ways in which they diverge. Sexual asphyxia, eating disorders, and suicide all fall within both categories, able to be accomplished alone and usually done without the benefit of associations with other, like deviants. But individual deviants, unlike loners, can hang around and participate in subcultures and countercultures with other fellow deviants, as long as they can accomplish their deviant act alone. Individual deviants, such as illicit-drug users, stutterers, transvestites, the depressed, the obese, and the homeless, are colleagues. In contrast, loners, unlike individual deviants, need not rely exclusively on themselves to accomplish their acts of deviance; they may have victims. Rapists, murderers, embezzlers, physician drug addicts, and obscene telephone callers are loners, but not individual deviants.

A second type of deviant act involves the **cooperation** of at least two voluntary participants. This cooperation usually involves the transfer of illicit goods, such as pornography, arms, or drugs, or the provision of deviant services, such as those in the sexual or medical realm. Cooperative deviant acts may involve the exchange of money. When they do not, participants usually trade reciprocal acts. They both come to the interaction wanting to give and get something. In deviant sales, one participant supplies an illicit goods or service in exchange for money. One or more of the participants in such acts may be earning a living through this means.

Newmahr's Chapter 41, on the sadomasochistic scene of sexual play, illustrates the way sexual partners negotiate roles, communicate with other through these roles, and enact them with authenticity, even while in public and in front of spectators. Scull's chapter on male strippers (Chapter 42) takes us into another world where men and women enact sexual roles in a voluntary exchange, but one motivated by financial, rather than sexual, gain.

The final type of deviant act is one of **conflict** between the involved parties. In it, one or more perpetrators force the interaction on the unwilling other(s), or an act seemingly entered into through cooperation turns out with one party "setting up" the other. In either case, the core relationship between the interactants is one of the hostilities, with one person getting the more favorable outcome. Conflictual acts may be carried out through secrecy, trickery, or physical force, but they end up with one person giving up goods or services to the other, involuntarily and without adequate compensation. Conflictual acts may be highly volatile in character, with victims complaining to the authorities or enlisting the aid of outside parties if they have the chance. To be successful, therefore, perpetrators must control not only their victims' activities, but also the victims' perception of what is going on. Such acts can range from kidnapping and blackmail to theft, fraud, arson, pickpocketing, trespassing, and assault.

Various types of conflictual and exploitive deviance abound and are thriving, both domestically and internationally. We have witnessed the repopularization of kidnapping abroad for political, military, and financial purposes. Domestic rates of rape committed by strangers, friends, and family members have never been higher. The prevalence of fraud is also rising precipitously, aided by the Internet, through fake stock tips, travel scams, identity theft, "phishing" (a practice in which victims disclose account passwords and other data online in response to emails that seem to come from legitimate businesses), "pharming" (in which experienced hackers are able to redirect people from a legitimate site to a bogus site without the people even knowing it), mail-order bride schemes, and "advance fee" 419 scams (in which people are contacted by a solicitor from abroad offering fabulous riches if they will help the slicitor recover some lost fortune). Identity theft is also prevalent and, apparently, popular among methamphetamine addicts, because the skill set of meth users facilitates this kind of fraud.

40

Artificial Love: The Secret Worlds of iDollators

NANCY J. HERMAN-KINNEY, DAVID A. KINNEY, KARA TAYLOR, AND ASHLEY M. MILLER

Through documentaries, dramatic portrayals in movies and television shows, and some sociological studies, people have become increasingly aware in the 2000s of the existence of Real Dolls®: life-size silicone dolls between 4 and 5 feet tall that weigh up to 100 pounds and are quite realistic and customizable, especially in their genitalia. Men form meaningful relationships with these (male and female) dolls, substituting them for sexual and romantic partners. Yet, consorting with such artificial partners is highly stigmatized, and most people find themselves hiding these relationships.

In this study, Herman-Kinney and colleagues present data gathered through Internet chat rooms and interviews conducted over the phone and in person of men in this highly hidden world. Comprising both loners and individual deviants, they carry out these relationships completely on their own, inventing feelings, behaviors, and desires for their loved ones. Some people have more than one of these dolls. Why they turn to artificial girlfriends and boyfriends and the way they manage these relationships is the subject of this fascinating chapter.

In thinking about the social organization of this behavior, compare it with that of the loners and colleagues described earlier. What advantages do these people have compared with them? What disadvantages? Why do you think they seclude themselves as loners?

People either don't know us or [they] have this perception that we are freaks, nuts, sex addicts, sicko doll fuckers, women haters, or should just be plain locked up. They have a real narrow-minded, myopic view of iDollators. We're just like everyone else, except we choose to spend time with synthetic women instead of organic women... and it's not just about the sex! (John, a 32-year-old iDollator).

Reprinted with permission from the author, Nancy J. Herman-Kinney.

INTRODUCTION

Examination of the sociological literature indicates that researchers have categorized different types of deviance on the basis of the social organization of those participating and by the behaviors of the individuals involved. Sociologists have developed several categories to describe individuals who perform their deviance alone (Best and Luckenbill, 1980, 1982; Prus and Grills, 2003). Best and Luckenbill (1982) define "individual deviants" as those individuals who commit their acts completely by themselves, for themselves, to themselves, and on themselves. Similar to other individuals, such as the mentally ill (Herman, 1993, 1994), homeless people (Snow and Anderson, 1993), and drug users (Goode, 2005); iDollators, or doll lovers, do not need others in order to carry out their deviant actions. They may at times associate with like deviants within a deviant subculture, but the vast majority of the time they enact their deviance alone.

In a similar vein, Prus and Grills (2003) make this distinction when they speak of solitary deviance and explain the differences between solitary operators and subcultural participants. Solitary deviants operate alone, while subcultural participants may engage in deviance by themselves, but they are involved with a group by which they are shaped and influenced. The iDollators in this study fall within the domain of those who engage in individual or solitary deviance, because they engage in their actions alone. They act alone as free agents who perform their acts by themselves and for themselves on inanimate objects. Some do, however, associate with other iDollators in an online subculture.[1] In this chapter, we define and describe the social worlds of doll lovers, or iDollators—their "*definitions of the situation*" (Thomas and Thomas, 1928) and their "*constructions of reality*" (Berger and Luckmann, 1966). In particular, we begin by discussing what iDollatry is and the types of dolls that are available and purchased. We then discuss our methods of data collection. We subsequently address the motives underlying iDollators' decisions to purchase these dolls, the types of relationships they develop, and the functions that dolls serve in their lives.

iDollators, for the purpose of our chapter, are defined as persons who are aficionados of high-end love dolls and who use them *not only* as sex toys but also for companionship. Until recently, the world of iDollatry and iDollators was largely unknown, kept hidden or misunderstood. With their relatively new exposure on various reality television shows such as *Taboo* and *My Strange Addiction*; television sitcoms such as *Boston Legal*; a contemporary movie, "*Lars and the Real World*"; popular talk shows (such as Anderson Cooper); documentaries such as *Synthetic Dolls and the Men Who Love Them* (National Geographic Channel) and *Guys and Dolls* (BBC); and celebrities such as Howard Stern, Charlie Sheen, and Vince Neil of Mötley Crüe coming forward to speak about their dolls, society has become increasingly curious to find out about iDollators.

Ranging from $250 to $10,000, female synthetic dolls, or *Real Dolls*®, are often customized to individual preferences. These life-size silicone dolls are quite realistic, fluctuating in height from 4 to 5 feet tall and weighing up to 100 pounds. Purchasers can customize their dolls by choosing from a variety of head, face, skin, hair, eye, makeup, and fingernail color options. So too, can buttock, breast size, clitoris, and hymen be customized to the desires of the purchaser. For an additional cost, to heighten the realism of the doll, some owners add freckles, real eyebrows, custom wigs, artificial milk glands, pressure-released urination, and custom pubic hair (ranging from "trimmed" to "full"). In addition, the genitalia are fully customizable: a purchaser can order the "she-male" style, which means that the doll has an interchangeable penis and vagina, a permanently attached penis and vagina (without the testicles), or a permanently attached penis with no vagina.

METHODS

The idea for this project burgeoned out of a project that the third author (Taylor) conducted in 2011 for a photojournalism class, a series of photos and a video on a prominent iDollator named Davecat. The other authors became interested in the topic and began conducting a qualitative, sociological study of iDollators early in 2012. We have conducted interviews, entered chat rooms on the Internet, and have had conversations with those who possess dolls. We posted a questionnaire on several iDollator Websites, we interviewed participants, and we often reinterviewed them through email or via the telephone. We also conducted face-to-face interviews with three iDollators and met their dolls. The respondents live all over the United States, and overseas in Great Britain, Canada, Australia, Germany, Denmark, and the Netherlands. Participants range in age from 17 to their mid-70s. Their occupations range from blue-collar workers to upper-level management positions; some are retired. We were completely overt about our identities, the aims of the study, its voluntary participation, and the ability of the participants to end the interview at any time. At the time of writing this chapter, we have interviewed or otherwise talked with 48 participants. Thirty-one individuals have responded to our questionnaire. This chapter is based upon data from all of these sources. All the names have been replaced with pseudonyms to ensure the anonymity of the respondents. Although women do also possess synthetic dolls, and a few responded to our questionnaire, this chapter includes only information from men.

iDOLLATORS' MOTIVATIONS

Examination of the data indicates that the social worlds of iDollators are very complex. Some observers have dismissed these people as pathological, misogynistic, and potential rapists (Lasocky, 2005; Price, 2008). Psychologists have portrayed doll lovers as having a significant psychopathology. They argue that iDollators use dolls because of their "disturbed attachments," their inability to form normal attachments with humans; moreover, they contend that some iDollators may be suffering from Asperger's syndrome. Others argue that iDollators possess a "rubber fetish." However, on the basis of our extensive literature review, no systematic psychological research has been conducted to support these theoretical assertions of pathological behavior. On the contrary, our qualitative sociological study has found that iDollators present a radically different picture of themselves.

It is important to note that some men who purchase synthetic dolls are not iDollators. They have more of a "doll fetish" and use the dolls as mere sex toys.[2] Many consider the dolls as objects, store them in the closet, under the bed, or in the basement and pull them out when engaging in sexual pleasure. They subsequently hide them. These men are *not* the subject of this chapter. We were interested in finding, talking to, and understanding those men whom we refer to as iDollators, their social relationships with dolls, and their reasons for so doing.

Sexually Curious

About one-fifth of the men in our study indicated that they purchased a synthetic doll largely in response to sexual boredom. They were curious, wanted to seek out novel sexual experiences, or wanted to perform sexual acts that their partners were not interested in doing. Rico20, a 28-year-old graduate student, speaks to the issues of curiosity and novelty and the companionship that his doll provides him, when he states,

> I went on-line and was reading a lot about these high end sex dolls. I had to save up a long time before I could afford mine—she cost over $8,000. I had a lot of custom things done to her. The main reason I got involved with Maria [his doll] was first out of sheer curiosity. I wanted to know what it would feel like; I was surprised how realistic they are in every way.... I was also bored sexually with college girls and the same old, same old with them. So, I acquired my Maria and she has definitely not been a disappointment. We spend a lot of time together in my room—she is there when I am writing my thesis—I bounce things off her; she is my refuge and source of contentment as well as sexual excitement. She is many things to me.

Don, a 48-year old salesman and owner of three synthetic dolls says,

> I have always been sexually adventurous. I'm up for trying new things, especially when it comes to sex and my wife is pretty old-fashioned and a prude when it comes to these things. She is very religious. So, about three years ago, I purchased my first doll, Blanca. I could do a lot of things with her that my wife would definitely say is taboo. Then the next year, I bought Maria, my second doll. She was more customized— the eyes, the breasts, the butt, and even her clit. I find her very satisfying. Two months ago, I acquired Arabella. She is totally custom too. I had to work a lot of extra hours and make a lot of sales to afford her. All these girls allowed me to be very experimental—my sex life has been spiced up tremendously. My wife doesn't know about my girls, but

what she doesn't know won't hurt her....I have them out in my pole barn where she never goes....When I go out on the road for a week or two at a time selling [his product], I have, on occasion, taken Blanca with me. She is a whole lot of company. After a period of time, I have found that the girls mean much more to me than just sex dolls. Especially, Blanca and Maria, I would say that we have become partners in life. I talk to them, and they keep me company in ways that I never thought they would! They are a huge part of my life now.

In short, one of the major reasons that some men purchased and spent time with their dolls was initially based on their sexual intrigue and appeal, a curiosity that subsequently developed into a full-fledged relationship.

Physically Disqualified

About one quarter of the men in our study reported that a major reason that they turned to synthetic women with whom to have a relationship was a direct result of their own perceived physical unattractiveness. Following Goffman's (1963) seminal piece on stigma, such men possess *abominations of the body*. These men contend that aspects of their physical appearance, such as morbid obesity, a big nose, large ears, or an ugly face disqualified them in the eyes of others and prevented them from forming "normal" relationships with "organic women." As a result, they turned to what they perceived to be the next best thing: a synthetic form of woman. As Chuck, a 44-year-old single male, speaking on his dating past, or lack thereof, puts it,

> I was never a lady's man. Never could get a date. Kids always called me ugly and lard-ass. ... Hell, I was called ugly by pretty much everyone in my whole damn family. When I was a teenager, girls used to run from me or just ridicule me. I was picked on by everyone—the brunt of their cruel jokes. I was about 400 lbs. then. I never went to any proms or nothing. Sometimes I felt like they viewed me like I was the Elephant Man in that movie. About two years ago, after being alone all this time, I realized that I was never going to be able to get a woman like all my friends, so I looked into those Real Dolls and I *settled for her*. I had to save up a long while for her but she is worth every penny. She is my girlfriend, she's not organic, but *she is the next best thing* and we love each other. She doesn't care that I weigh 600 lbs. now. She loves me for who I am.

Similarly, Alberto, a 37-year-old single man, who never had a date in his entire life, adds,

> I have never been able to measure up to other guys. I am a failure at life when it comes to relationships and love. I was shy, wore big glasses and had a lot of acne. I am only 5 foot 7 and what girls want to date such a small guy? They're looking for big, football player types. Every time I'd get my nerve up to ask someone out, I would either be stood up or

they'd make some lame-ass excuse. Even when I was set up…the women would call at the last minute and cancel.…When I would call a woman, they would sometimes hang up on me.…That's when I turned to the Real Doll, my darling Vanessa. It is a step down in some ways—she is synthetic, but she never turns me down, or stands me up or makes fun of me. At least I don't feel like a fucking defect no more!

Following Merton's (1967) typology of deviants in which he distinguishes between the conformist and the ritualist, Chuck, Alberto, and some of the iDollators in our study once started out as conformists, seeking an attractive, organic woman, but gradually came to realize that this goal was unattainable to them; so, they deescalated their goals and settled for a synthetic woman with which to interact and develop relationships.

Rejected and Wary

A third reason that (one-fifth of) the men in our study turned to synthetic women was that they had experienced negative relationships with organic women. These men were either victims of infidelity, domestic abuse, deception, and financial and/or emotional exploitation, so they decided that it was easier and safer to have relationships with synthetic women who could "do them no wrong." For example, Vince, a 36-year-old fireman, turned to synthetic relationships after a long string of negative relationships with organic women:

I was tired of all the fighting and the drama involved with relationships with some women. Some are so high maintenance and some are just plain hard work. You never know what pleases them; you don't know what they want half the time. I was sick and tired of dealing with them; breaking up and making up—the lying, the cheating and all of the bull-crap that generally goes along with them. I went bankrupt twice because I was used! I got to the point where I was fucking fed up with it all. That's why, I found a solution that works for me: I have my Palamina [his doll] who doesn't talk back, she listens to me, she is always smiling; I can count on her 24/7.

Similarly, Big Ben, a thirtysomething-year-old who owns four dolls, speaks about his reasons for buying Real Dolls after a divorce from his unhappy marriage:

I was constantly depressed when married. Being around people in general stresses me out. I feel only love and peace with my dolls. They give me a refuge from the aggravations of the world. My relationship with my dolls is one of tranquility and joy. They comfort me and help me feel loved and accepted, which I have not found being in relationships with real women.

In sum, then, acquiring a synthetic doll and developing a social relationship with her helped these men to overcome overwhelming feelings of negativity and allowed them to be in control and have someone in their lives with whom they would feel comfortable, safe, and at peace.

Sad

Another major motivation why men purchased synthetic dolls was to fill a void after the loss of a partner. One-sixth of the men in our sample indicated that a wife or partner had died, and after a period of intense grief, they decided to purchase a synthetic doll to help in the grieving process. Chucky G., a 68-year-old retired policeman, tells of his decision to purchase a synthetic doll 3 months after the death of his wife:

> My wife and I had been married forty-two years before she passed. It was very sudden and I wasn't prepared for it. When she died, I thought my whole life was over; I lost everything, but a friend told me about these expensive dolls. He said I could cuddle with them. I bought myself one and it changed everything....at first I thought that Nina [his doll] would just help me get through the bad times, a temporary thing. But as time progressed, I started talking to her, being with her more, reading books with her, watching the television or watching a flick. She gradually took the place of my wife. Sure we have sex, but she is now my life partner. We are married to each other and I have a new lease on life.

In a similar vein, Ron, a 53-year-old teacher, adds,

> My live-in girlfriend got cancer very suddenly and in three months she was gone. I just couldn't process it. We were together over 20 years and I couldn't just go out and start dating again. Anyway, who would want to? After a few long months, I came over this article on the Internet about Real Dolls and I looked into them. They were very expensive and I need to save up to get one, but I thought that it would help me deal with my grief....When I first got Francesca, she was more of a sexual distraction—she took my mind off Dottie [his girlfriend]. But the oddest thing happened. She has turned into a girlfriend for me. We do everything together at home. I dress her, comb her hair; I read poetry to her. I talk about my day to her. I have even given her a ring. She has made life worth living again and I am thankful to have her in my life!

In short, then, for Chucky G., Ron, and others, their synthetic doll helped them to deal with the death of a loved one. Although they had initially thought to use the dolls as a therapeutic, transitional object, they gradually developed feelings for them, and these feelings burgeoned into full-fledged relationships.

Handicapped

Examination of the data indicated that another group of men (one eighth of the sample) who turned from organic to synthetic women were individuals who were either physically or mentally handicapped or who possessed a physical ill-ness that precluded dating and becoming involved with organic women. Myron, a 72-year-old wounded veteran who was partially disabled, speaks of the reasons that he became involved with a synthetic woman:

> I am an old veteran. I served my country by fighting in Korea. I can't walk. I am in a wheel chair. Who in the hell would date me or let alone, do anything else like in bed? I am missing one of my legs. It was blown off. Suzie Q [his doll] is the love of my life. She is beautiful, luscious; she is there for me. She has given me a new lease on life. I feel like a teenager with her around. She makes me feel like a man again!

Walter, a 42-year-old physically handicapped man adds,

> I know that I'm not normal. I have never been normal and I don't want to be normal. I've only wanted to find peace and joy in life. Due to my disability, I was often depressed and even suicidal. I felt a little better after my divorce, which I asked for so I would no longer feel guilty for preventing my wife at the time from having children....But then I felt lonely. I tried to date, but no one only could truly accept me and my disability...I grew tired of the rejection. I thought about hiring an escort to cuddle with me, but it seemed legally questionable and extremely expensive. I thought about body pillows and a few months later I pur-chased a doll, then another, and another....It may not be perfect, but for the first time in my life, I don't constantly hate my life...I do have some peace and joy thanks to my dolls.

Interestingly, we found that one mother acquired a doll for her adult son who was developmentally handicapped. Speaking about the significance that this doll played in her son's life, she says,

> I found out about these Real Dolls about five years ago. I ran across an article on the Internet. Then I talked to somebody about them. At first, I was grossed out and thought the idea was kind of perverted. But I knew my son needed someone, not just for a sexual outlet, but to truly have a bonding and kind of relationship with. It may sound kind of bizarre, but Leo [her son] doesn't just use Kitty [his doll] for sex; he watches TV with her, talks to her, carries her around, plays games with her. She is like a real girlfriend to him. And that makes me happy.

In short, then, synthetic dolls helped to make damaged men feel whole again. In turn, their self-images and identities were elevated.

Sexually Unfulfilled

The data also indicated that some men chose to purchase and develop relation-ships with dolls for other reasons. Specifically, one-fifth of the men in our study had wives or significant others who were ill and could not have sexual inter-course. Henry a 47-year-old, speaks to this issue when he states,

> My wife has MS [multiple sclerosis] and she is now in an advanced state. I have a healthy libido and I did not want to go to a prostitute or cheat on her in any way. It is devastating when one partner becomes ill....So, I decided to purchase Veronica [his doll] at the urging of a friend who already had a doll. She is not only my sex partner, but she also fills a void in my life for companionship. My wife can do less and less. She is becoming a shell of who she once was and I am scared. When my wife is asleep, I watch movies with Veronica. I talk to her; I confide in her all my fears and frustrations. I brush her hair. She is my escape from reality. She is like my shrink! I feel better after talking with her.

In a similar vein, Robert, a 62-year-old married man, commenting on his reasons for turning to iDollatry, says,

> My wife is no longer interested in having sex. She is over it. We have been married for over 35 years; we had kids and she and I raised them. We always had a healthy sex life; she says she still loves me and I love her, but she doesn't want to engage in the act anymore. I respect that. But I still have needs. So, I bought this synthetic doll, Maisy. Maisy is much more than just a sex toy. We have developed a very intense relationship, a loving relationship. I see her as my second wife. I have even given her a ring. In an odd way, we have become husband and wife.

It is interesting to note that a large number of men in our study stated that, although the dolls were first purchased primarily as a sex aid, the men later formed a relationship with them. Moreover, some of the wives had knowledge of the dolls and not only condoned their existence, but also established a relationship with them. Ron, a 57-year-old man, whose wife possesses a debilitating disease, remarked,

> My wife knows that I have Sandy [his doll]. She in fact gave her blessings to get her. Because of her paralysis, she can't have sex with me anymore, so it is a big relief for her to know that my urges are being met with Sandy. At first, Sandy was more or a sexual aid of sorts, but the funny thing happened, both my wife and I started talking to Sandy; my wife now dresses her up with me; and she has become a real person in both our lives. She's a companion for me and also for my wife too. When I am gone, Sandy sits with my wife and they keep each other company.

Fearful

Just as some men reported acquiring synthetic dolls, developing relationships with them, and even "marrying" them for the aforementioned reasons, others told us that they chose synthetic women over their organic counterparts because they were afraid of acquiring various sexually transmitted diseases, were afraid of getting a woman pregnant, or had already experienced these kinds of negative consequences. As SuperSid, a 49-year-old speaking on this matter, says,

> Out there in the world today, it is just too dangerous. It is not like it used to be. We don't live in a *Leave it to Beaver* world anymore. Everyone is hooking up with everyone else for one night stands. They're shacking up with anyone. My buddy got the clap and some other damn crap of a disease by fucking a whore. It's not worth it. I would rather have my silicone girl with me. I know that she is safe. She ain't going to give me nothing and I ain't going to pass anything on to her neither.

Similarly, Darwin the Great, a 41-year-old iDollator, speaking to the issue of pregnancy, adds,

> When I was younger, I got tricked into getting two girls pregnant. They both said they were on the pill or using some IUD or something, but they got knocked up anyway....I've learned the hard way, that it is much easier and safer to stay at home and be in a relationship with Vivienne [his synthetic doll]. No matter how many times we are intimate, I know that she is never going to get pregnant. I find that very satisfying and relieving.

In short, we found that some men who had had real or anticipated negative consequences from having sex with organic women alleviated their fears and anxieties by choosing a synthetic woman.

Ashamed

A final reason given by several of our respondents for turning to synthetic women with whom to have relationships centers on their strict religious dogma prohibiting homosexuality. Approximately one-fifth of the respondents in our study indicated that they had been socialized to believe that homosexuality was a "sin." Faced with the desire to have sex with other men, which would be highly stigmatizing in their minds, they turned instead to dolls. These men chose either male dolls or female dolls with penises; by so doing, they were able to maintain their heterosexual identities. According to Charlie69, a prominent deacon from a Catholic church,

> I grew up my entire life believing that homosexuality is a sin. I went to Catholic school, and it was drilled into me by the teachers and the priest. Now, I am in a position of authority within the Church; people respect me, and I have to go along with the Church's teachings. So I have abstained from having any relationship with a man....The congregation would shit if they knew I had this doll and what I did with her. But I do have my synthetic doll to satisfy my sexual wants and cravings. Kathleen has breasts and long brown hair, but the thing I adore about her is her 7 inch schlong. It is permanently attached. I didn't opt for the vagina thing; I do everything with her penis. It makes me hard

just thinking about it…And when I am making love to Kathleen, even though she has a tool, I still feel "normal" and I don't think I am going against the Church. She is a wonderful companion to me and I look forward to going home every night to be with her.

BigBarry, a 52-year-old lifelong Baptist, expresses similar sentiments when he states,

I am basically a Bible-Belt Christian, Baptist all of my living days. I have a Real Doll—George. He is 5 foot 3 and weighs 95 pounds. I ordered him with an 8 inch hard dong (penis) … He is actually very handsome. I would never fuck a guy—that's not my kind of thing and people around here don't think highly of gays and their so-called lifestyle. I have always liked guys and fantasized about it, but it is out of the question to start living a gay lifestyle—not around these parts, not in my family's eyes, and certainly not if I am going to stay a member of the Church…. So, for me, I am turned on by a male doll with a dong. I suck it and lick it. George also does the back-door thing on me too. He likes it and so do I. So in my mind, I am not going against the Church—I am still "normal"—I am not doing things with fags. I am just having sex with [a] male synthetic guy….I spend a lot of my free time hanging out with George at home—he's my best buddy.

In short, then, a number of the men in our study with heterosexual tendencies actively chose to purchase synthetic male and female dolls with penises giving them the ability to engage in guilt-free sex with what otherwise would be defined as sexually prohibited activities. By so doing, they were able to maintain "normal," heterosexual identities.

iDOLLATORS' REAL DOLL FUNCTIONS

Our study illustrates that synthetic dolls provided men with a number of benefits. In the majority of cases, we found that these men chose synthetic dolls as a way to combat, adapt to, or otherwise alleviate problems they were experiencing in their lives. First, for the vast majority, the acquisition of a synthetic life partner provided them with companionship to combat their loneliness, sadness, and/or grief. Second, having a synthetic doll in their lives allowed various men to deal with their past (or anticipated) anxiety and the difficulties that they experienced in some of their previous relationships with organic women. Third, iDollators who perceived themselves as either physically unattractive or disqualified in the eyes of others because of a physical disability, improved their self-images and identities via their relationship with their synthetic companions. Fourth, choosing to purchase and interact with synthetic dolls (male or female) allowed some men to engage in what they self-labeled as sexually prohibited behavior while, at the same time, not labeling themselves as deviant (homosexual) and avoiding the associated stigma. These men justified having sex and a relationship with a male doll or a female synthetic doll with a penis as different from engaging in a homosexual relationship with an organic being. They were thus able to maintain positive, heterosexual identities.

Although the preceding functions relate to these men turning to dolls to solve certain problems in their lives, we also found that dolls served an additional function by providing an outlet for owners' desires for sexual novelty and sexual satisfaction. Although some men initially used their dolls as a sexual toy, over time they formed meaningful social relationships with their dolls in a manner similar to that of the men who used the dolls to solve pressing life problems and challenges. A final function that we discovered, beyond the scope of this chapter, centers on how some iDollators used their synthetic dolls to engage in artistic expression, dressing them up, photographing and filming them, and posting their pictures and videos on Websites for others' views and comments.

DISCUSSION

In this chapter, we have focused on how a particular group of men find meaningful companionship. We have illustrated both how our iDollators correspond to Best and Luckenbill's (1982) ideal typical model of the individual deviant and the motivations and rationalizations of these men for pursuing this form of solitary deviance.

Close examination of solitary involvements in deviance suggests that people become involved in individual deviance in ways that are not so different from those characterizing group-based deviance. As Prus and Grills (2003:166) have aptly pointed out, "while some instances of individual deviance may come about as it was *imposed on them*; in some cases, people may also become involved in solitary endeavors through *instrumentalism, seekership* and *recruitment*." We have shown in our study that some iDollators came about their activities because it was *imposed* on them as a result of the loss of a partner, a personal illness, or a disability; in effect, they became involved by default. Others purchased synthetic dolls and pursued activities with them because they considered the dolls to be *instrumentally advantageous* (e.g., the dolls were safe from pregnancy and sexually transmitted diseases, some men avoided the pain of being turned down for dates with organic women, and others sought out dolls to have secret homosexual sex). Still others, through *seekership*, turned to an iDollatry lifestyle because they found it appealing and intriguing (e.g., these men sought sexual novelty and sexual satisfaction). Most recently, we have observed that some men are encouraged by others to pursue these interests through *recruitment* via iDollatry Websites.

Because we are dealing with issues of human agency with reference to instrumentalism, seekership, and recruitment, we need to point out that some men in our study did, at times, have reservations about having a synthetic doll in their lives (e.g., they fear discovery, they dislike the exorbitant costs involved, and they feel twinges of immorality). In contrast to those involved in deviant subcultures who have the benefit of others to help them deal with the problems associated with their deviant activity, those engaging in individual deviance have to develop more extensive rationales for engaging in their practices, have to deal with their reservations and misgivings, and must manage the anticipated discovery of their actions and concomitant stigma exclusively on their own (Herman, 2002; Prus and Grills, 2003).

In closing, the popular media that have often portrayed iDollators as creeps, perverts, or misogynists who want to have complete control over the "perfect woman" and/or physically mistreat her. In contrast, we have found that the iDollators in our study were experiencing some combination of being lonely, hurt, wary, unhappy, or depressed. They were individuals searching for companionship and unconditional love—needs that go far beyond their desire for sex. Rather than hating or mistreating women, the men who possessed these synthetic dolls cherished them. Like all of us in life, they used these dolls in their search for meaning and acceptance.

NOTES

1. An examination of the Web-based subculture of the iDollator community is beyond the scope of this chapter, which focuses on the individual, solitary nature of their deviance. The authors are currently investigating the subcultural world of iDollators.

2. There is some debate in the iDollator community that those with doll fetishes—that is, those who use the dolls primarily as sex toys and then store them away and do not develop relationships with them—are *not* "real iDollators." After consultation with long-term iDollators, for purposes of this chapter we have chosen to accept this distinction between those with doll fetishes and those who are full-fledged iDollators.

REFERENCES

Berger, Peter, and Thomas Luckmann. 1966. *The Social Construction of Reality.* New York: Doubleday.

Best, Joel, and David F. Luckenbill. 1980. "The Social Organization of Deviance." *Social Problems* 28(1):14–31.

Best, Joel, and David F. Luckenbill. 1982. *Organizing Deviance.* Englewood Cliffs, NJ: Prentice Hall.

Goffman, Erving. 1963. *Stigma: Notes on the Management of Spoiled Identities.* Englewood Cliffs, NJ: Prentice Hall.

Goode, Erich. 2005. *Drugs and American Society* (6th ed.). New York: McGraw Hill.

Herman, Nancy J. 1993. "Return to Sender: Reintegrative Stigma-Management Strategies of Ex-Psychiatric Patients." *Journal of Contemporary Ethnography* 22: 295–330.

Herman, Nancy J. 1994. "Former Crazies in the Community." pp. 25–42 in Mary Lorenz Dietz, Robert Prus, and William Shaffir (Eds.), *Doing Everyday Life: Ethnography as Human Lived Experience.* Toronto: Copp Clark Longman.

Herman, Nancy J. 2002. "'Mixed Nutters,' 'Loney Tuners,' and 'Daffy Ducks.'"

pp. 244–256 in Earl Rubington and Martin S. Weinberg (Eds.), *Deviance: The Interactionist Perspective* (8th ed.). Toronto: Allyn & Bacon.

Lasocky, Meghan. 2005. "Just like a woman." *Salon.com.* October 11, 2005. http://www.salon.com/2005/10/11/real_dolls/. Retrieved November 23, 2012.

Merton, Robert K. 1967. *Social Theory and Social Structure*. New York: Free Press.

Price, Catherine. 2008. "Your girlfriend seems so fake." *Salon.com.* April 1, 2008. http:/www.salon.com/2008/0401/robot_sex/. Retrieved November 11, 2012.

Prus, Robert, and Scott Grills. 2003. *The Deviant Mystique: Involvements, Realities, and Regulation*. Westport, CT: Praeger.

Snow, David A., and Leon Anderson. 1993. *Down on Their Luck: A Study of Homeless Street People*. Berkeley: University of California Press.

Thomas, W. I., and Dorothy Thomas. 1928. *The Child in America*. New York: Knopf.

41

Subculture and Community: Pain and Authenticity in SM Play

STACI NEWMAHR

Sexual deviance represents one of the large arenas for cooperative deviant interactions and relationships. Newmahr's study of a sadomasochism club, where people go to engage and watch others engaging in voluntary, cooperate public sex featuring dominance and submission, inflicting and receiving pain, and feeling fear and excitement presents an example of such an arena. Newmahr's research offers us a fascinating glimpse into this hidden subculture and the social meanings constructed by participants. As she describes her entry into, and growing membership in, this scene, we live with her through the subtle dynamics of the role interplay, the intensity of the sex and of the drama framing it, and the energy that fuels ultralate-night bouts of coffee and breakfasts at all-night diners afterward.

How does this portrayal compare with the impressions you have gathered about this scene from your everyday lives? What stereotypes has it reinforced or dispelled? In which role do you believe the power lies in these relationships? Which role appears to be the most demanding and sought after? How do these relationships compare with those of the iDollitors?

After a few minutes, Jesse asked me, "Do you like knives?"

"Sure," I replied.

"Close your eyes," Jesse said. She took my wrist. I felt a dull blade trail along the inside of my forearm. I opened my eyes and saw that it was not a blade at all, but a paper-thin plastic card. We marveled at how like a blade it felt.

Adam began to dig all of his sharps out of his bag. He held out his hand for mine. I gave it to him and watched as he placed a two-bladed finger cuff over his index finger. I had not seen a cuff like that before. It was a new toy

Reprinted with permission from the author Staci Newmahr.

for Adam also. He dragged it along the back of my hand. We discussed how to make them, how expensive they were, and where to find them.

Handing the cuff to Jesse, he said, "Here, try it on her neck."

"Do you mind?" Jesse asked me.

"No, it's okay," I replied, piling my hair atop my head with a hair band so that my hair wouldn't cause the blades to skip.

Jesse dragged the blades up and down my neck, softly at first. It gave me goose bumps. When I shivered, Adam wrapped my arms in his. Within a few seconds Jesse was no longer using the blade lightly enough to tickle, and I was no longer shivering. Adam reached into his pocket, removed his pocketknife and flipped it open. Taking my wrists in one hand, he stretched my arms across the table, palms up.

In soft voices, just above whispering, Jesse and Adam talked as they used the blades on my skin. I kept my eyes closed and focused on the feeling.

"She marks so nicely," Adam said.

"I know. And look at her face. It's like it's putting her to sleep," replied Jesse.

"Except when it hurts," Adam said as he pressed the knife into my skin.

The term "community" is contested in the social sciences. Its meanings vary widely, and criteria for its use are elusive. It is always, however, about boundaries. The notion of community is used to draw lines between insiders and outsiders. Participants in various subcultures or scenes find a bond created by shared life experiences, norms, values, and shared objectives.

The concept of community offers insight into the roles of identification and identity, community seeking and community building, and, ultimately, the intersection of community, identity, and interaction. In light of identities of marginality, it is not surprising that people conceive of their entrance into what I call the "Caeden" (a fictitious place-name) community, an urban sadomasochism (SM) community in the northeastern United States, as a metaphor for finding a home. Members of the scene readily share the perspective that they did not belong anywhere prior to finding the SM scene. They frequently assert that "people here get it," and the "it" refers not narrowly to SM interest, but to marginality more broadly, for the reasons and in the ways that are so common in Caeden. The sense of being understood, of being known, underlies the importance of community for many people in Caeden. By subsuming the marginalities under one overarching identity, the community offers understanding of the experiences of outsiderness that many have lived. The metaphor of home conveys not only like-minded people, but a belief in kindred spirits, acceptance, and connectedness.

Membership involves a sense of belonging and identification, a category in which those two terms include the feeling, belief, and expectation of fitting in, as well as a sense of acceptance within the group. As much as community member-ship is derived from identification, it is defined by drawing boundaries between people who belong in a group and those who do not. These lines are linked not only to familiarity, but to trust; the shift from outsider to insider begins with visibility and moves quickly to immersion in the public scene. For the people in Caeden, the most common criterion for inclusion is involvement; unknown SM participants may be "kinky," but they are not considered part of the com-munity. Players are able to, and many do, successfully arrange their lives around scene activities, avoiding "normals" (Goffman, 1963). During any given week, there are at least five SM-related events one can attend, of varying types: educa-tional, activist, or social and/or play oriented.

Acquiring membership in the scene includes socialization to the practices and the meanings those practices had for participants. In this chapter, I illustrate that notion by focusing on two concepts that are key to the SM subculture: pain and authenticity. Among sadomasochists, the use of pain is a complex issue; some members of the community believe that people have the power to reshape the experience of pain, while others view the unpleasantness of pain as central to its appeal. Regardless of the differences among community members in their rela-tionships to pain, it is always necessarily related to experiences of power and powerlessness. An understanding of pain-friendly behavior must therefore wrestle not only with the reasons people appear to seek pain, but whether it is "really" pain and what else it may be. To that end, I explore the ways in which partici-pants in sadomasochism construct and preserve their experiences of power imbal-ances through their framings of pain and its meanings for them.

METHOD

This analysis emerges out of a multisite ethnographic fieldwork project; I joined a well-established SM organization and attended informational lectures, demon-strations, and workshops; public and private SM parties; social lunches and dinners; organizational planning meetings; and activist fund-raising benefits. I was a member of this community for 4 years (2002–2006), normally participating in political and social activities, including SM interactions (called "scenes," or "play") several times a week. Very quickly, my involvement came to dominate my life in much the same way as it does for many people in this community.

While I was in the field, privately owned SM clubs functioned as the most important community space in Caeden. An SM "scene" is a social interaction that involves the mutually consensual and conscious use, among two or more people, of pain, power, perceptions about power, or any combination thereof, for psychological, emotional, or sensory pleasure. For most people, SM play is not feasible at home. Clubs provide adequate space, equipment, soundproofing, and privacy for SM play, as well as a place to socialize.

Often during my fieldwork, weekend nights at the primary SM club in Caeden began with dinner, followed by 6 hours at the club. After the club closed, community members normally went out to eat at an all-night diner. This get-together frequently resulted in several more hours of conversation and then breakfast. On several occasions, breakfast spilled over into lunch and the next night was another club night. There were thus weekends during which I was in the field and awake from Friday night until Sunday afternoon. Throughout the week, I maintained near-constant contact with community members via email, telephone, Web blogging, and instant messaging. I also attended multiday events in Caeden and other cities.

After approximately 6 months (and countless informal discussions), I began conducting formal ethnographic interviews. The interviews focused not only on entire life histories, but also explored SM-related questions. To ensure thematic uniformity, I employed an interview guide, but the format was flexible and dynamic in terms of structure, off-topic conversation, and sequence. In total, I conducted 20 ethnographic–thematic interviews, ranging from 4 to 11 hours. The average duration was 6½ hours.

SM AND PERFORMANCE

Outside the community, SM is often framed as "role play." In this image, consenting adults are free to suspend their individual lived realities for the sake of erotic enjoyment: The "teacher" spanks the misbehaving "student" in an eroticization of hierarchy. Pain is not central in these understandings of SM. It is either entirely absent or relegated to a less important role than the aesthetic of the interaction. More than any other mainstream image of SM, this view is "playful," innocent by way of the nonseriousness of pain. The role–play view of SM thus mitigates what might otherwise be understood as violence. It is, first and foremost, a game of "make-believe." Second, it does not really hurt.

This is *not* the prevailing discourse of SM within the community, in which role play occurs only occasionally. SM is not understood as either a pretense or a performance. When roles are adopted, pain is often a central aspect of the scene. For SM participants, there is no "show" for which to prepare on a conscious or discursive level. Nearly all the SM scenes begin without onlookers. There are no curtains to raise or lights to dim. Observers drift from scene to scene, moving through an SM club and sampling the goings-on, rather than witnessing a scene from the beginning to the end. Often, the most private play spaces in a venue are the most desirable, and at times players even enlist friends to help direct potential onlookers elsewhere.

Although the presence of onlookers certainly affects public play in numerous ways, SM participants are not playing to the audience. In fact, participants' reputations can be harmed if the participants appear aware of spectators beyond the extent necessary for safety. A "top" (a person who appears to be directing the action in an SM scene, in contrast to the "bottom") must be vigilant enough that she checks behind her before she throws a whip, but she will be sanctioned for appearing distracted, preoccupied, self-conscious, or otherwise inappropriately concerned with onlookers during a scene. SM is unlike other spontaneous performances, such as professional wrestling and improv, in which players generally attempt to affect the audience together. SM participants seek instead to affect each other in the presence of onlookers. The goal in SM experience is a successful performance, not for the sake of the audience, but for the sake of the players.

"POWER EXCHANGE": QUESTS FOR AUTHENTICITY

In the SM community nationwide, the term "power exchange" is used to describe both the objective and the dynamics of SM interactions. The term is used to describe a trade of sorts; generally, one person seeks to feel more powerful, the other less powerful. At its core, then, the link between SM participants is a quest for a sense of authenticity in experiences of power imbalance. To achieve this authenticity, participants must suspend belief in their own egalitarian relations for the duration of the scene. When they do, the sense of power imbalance *feels* real. This goal is what is sought, and what often occurs, in and through power exchange.

SM participants seek authenticity in emotional, physical, and psychological *experience*, rather than authenticity in their presentation to others. This achievement of authenticity is *beyond* that of what one might experience when playing a role. In other words, SM participants who, when they play, feel as if they are playing a role (as an actor might) do not achieve the authenticity of players who say that they *feel* afraid, helpless, evil, or invincible during their play. Unlike improv or other kinds of performance, SM provides authenticity to the extent that SM participants are able to convince themselves, and each other, of the realness of the experience.

SM is a carnal experience. It is enacted, performed, processed, lived, and experienced on and through the body. Bodily manifestations and consequences of SM, such as bruises, scratches, and scars, are deeply entwined in ideologies of power. For SM participants, "marks" are indicators of authenticity, as well as visible sites of its accomplishment. Similarly, the spilling of blood (less common in public play but not unusual) is a powerful symbol of authenticity:

> Phoebe stood in the brightly lit conference room beside a small
> steel table that held supplies for the demonstration. She unwrapped a
> cotton-looking scalpel pack and laid it beside a bottle of rubbing
> alcohol, a first aid kit, and a large box of cotton gauze. Aidan, shirtless
> and in jeans, was lying face down on the table with his arms at his sides.
> I moved farther into the room, closer to the top half of his body. . . .
>
> When Phoebe cut into the flesh of Aidan's shoulder, he hissed. He
> came up on his toes, feet flexed and his back muscles visibly rigid. She
> put her hand on his back and waited a second, while blood trickled
> from the wound. She continued her work, inserting the tip of the
> scalpel into his skin and making small slices. Slowly she cut a simple

pattern, angled and tribal looking. Every couple of minutes, she wiped the blood off of the scalpel on a swathe of gauze she kept on the small table. Once or twice she blotted his wound with a fresh piece of gauze (in order to see what she was doing, she later explained). Aidan was quiet throughout, punctuating the silence with only an occasional pained (sounding) moan—soft, deep and brief. (May 2003)

In the scene just described, whether Aidan feels that he was "hurting" or not, Phoebe is injuring his body. The blood testifies to her ability and willingness to wound him, and to his mortality. The power exchange—the suspension of belief in egalitarianism—here is assisted by the visibility of Aidan's blood.

In negotiating the tension between the aspirations for authentic experiences of power imbalance and the desire to play safely, SM participants must navigate conceptually muddy waters. Their experiences are constructed and interpreted through a complex, and sometimes competing, set of discursive and social-psychological strategies in the community. Pain, as a concept, is central to these strategies. In SM, pain may be experienced, disavowed, evidenced, sought, and avoided, but it plays a crucial role in the quest for authenticity.

FRAMING PAIN

My analysis identified four distinct discourses of pain, which I call "transformed pain," "sacrificial pain," "investment pain," and "autotelic pain." These discourses intersect with decisions to engage in or refrain from particular activities, motivations participants claim for engaging in SM, the SM identifications they adopt, and ideologies of power and powerlessness.

Discourses of pain in this SM community also intertwine with larger narratives of pain in interesting ways. Participants most commonly draw on the overarching cultural narrative of pain as fundamentally undesirable or a necessary evil. In three of the discourses—transformed pain, sacrificial pain, and investment pain—pain is framed as inherently unpleasant. Only one competing discourse challenges this assumption, and this challenge is met with resistance within the community.

In exploring these discourses, I have chosen to blur the distinctions between topping and bottoming (providing service or being "done to," respectively, in an SM scene), for several reasons. In part, the choice reflects my position that SM is best understood as a collaborative social interaction, rather than the site of interaction of two conceptually opposed objectives. In addition, many SM participants "switch," topping in some scenes and bottoming in others. Because of this practice, the attribution of particular discourses to either tops or bottoms would be misleading. Finally, the people in Caeden did not divide themselves socially along lines of "top" and "bottom." For my purposes here, I use "top" and "bottom" to refer to actions in moments in time, rather than as indicative of fixed identities.

Transformed Pain: Turning Pain into Pleasure

Among many members of this community, the belief that power differentials between participants in a scene are authentic (experienced as real) is fortified by the claim that the pain is *not* authentic. The transformed pain discourse centers on a disavowal of pain as such. SM participants who frame pain this way tend to engage in mild to moderate pain play, but when pain is experienced, it is understood as *not hurting*. Instead, pain is "transformed into pleasure." This transformation occurs almost instantly, usually in a process that is understood as conscious, though barely. Viewed in this manner, would-be painful situations are not experienced as hurt. This frame of mind relies on a conceptualization of pain as an objective stimulus, which may or may not result in the *feeling* of hurt. During a conversation at a restaurant one night, Faye captured the idea; she said that she "can convert pain to pleasure . . . make my body produce chemicals" by changing the context in her conscious experience.

This "processing" of pain sensations as pleasurable, within seconds or less, fuels a discourse in which pain can be *real* but not *bad*, without sacrificing perceptions of authentic power imbalances. For bottoms, this discourse reconciles masochism with rational thought: If pain does not "really" hurt, it is depathologized and therefore unproblematic—a thing to enjoy. Tops engage in the same discourse, potentially mitigating some of the struggles with guilt that often accompany topping, particularly for newer players. When I asked Seth about a scene I had watched in which it seemed to me he had caused Stephanie a good deal of (intended and desired) pain, I used the word "hurt." Seth was quick to correct me:

SETH: She doesn't want to be hurt. She wants to be given the sensation of pain. No. I want to provide the sensation of pleasure. If that pleasure is pain transmogrified into pleasure, I'm very happy to provide it.

ME: What if it's not?

SETH: I don't want to beat somebody who wants to be beaten so that they feel something. I'll beat somebody—I'll flog somebody or I'll cane somebody who is enjoying the sensation of being caned. The experience. It's having a good time. That's what I'm there for. . . . If they're going, "Fuck, that hurts!" Generally, my agreement is—what I say to people is, for me, if you say "Ow," in a way that indicates that you don't like it, I'm going to yellow. I'm gonna yellow on our scene and I'm going to slow down or do something else. I use "Ow" as a safeword. My default position is "Ow" is bad. Generally when someone says "Ow," it's something that they don't like.

Seth's sense was that his play partners' experience of pain is "I like pain; pain feels like pleasure," rather than "I like to be *hurt*." His definition of SM hinges on this distinction:

SM is the seeking of pleasure, I think, in a way, by people who can translate pain into pleasure, and by people who can translate the act of

giving pain . . . or seeing that the other person . . . is having pleasure.
I think a good sadist is somebody who is really empathic—somebody
who really can feel what the other person is feeling, and take joy in that.

By recasting pain as something other than hurt, Seth, like other participants
for whom this frame resonates, does not draw explicitly on discourses of violence
and victimization, relying on other aspects of SM play to construct an authentically imbalanced experience.

The recasting of pain as transformed frames the pain in accordance with the
hegemonic views of pain, but modifies the pain (and the narrative) by turning it
into something that is not pain. The participant who modifies pain is actively
changing the sensation, working to claim it and process it differently, toward
an eventual understanding of the pain as pleasure.

Sacrificial Pain: For a Greater Good

In a definitive contrast, sacrificial pain is framed as an undesirable sensation that
remains an undesirable sensation throughout (as in, for example, punishment and
discipline). In this conceptual move, pain does not transform into pleasurable
sensation. Pain is, and must remain, suffering, for the *suffering* is a sacrifice on
the part of the bottom. This sacrifice is conceptualized as being for the benefit
or desires of the top. Framed this way, pain is a primary tool used to reinforce
and construct a power imbalance between players. Leah said, of the first time
that she "got the concept of power exchange,"

> He's doing all this really horrible stuff to me. I'm going I don't like this.
> He's going, I do. And I want you to take five more, seven more, three
> more, whatever it was, depending on how miserable I looked at the
> time (laughs) I don't know where he was fishing his numbers from. But
> he was going, you know, it was just it was the first time it was like okay
> you have the power. Because you're doing all this nasty stuff. I'm trying
> to exert what little power I have, going "I don't like it." I'm not safe-
> wording, which might be stupid, or might not be, but I really didn't—
> I didn't want to, and I don't know why. . . . And it was like okay, well
> if you want to, and this is going to please you, then that's a good
> enough reason for me and I guess I can do this, and I'm going to just
> draw off on the fact that you want to and that's going to make you
> happy. And I'm just going to draw from that and that'll work for me at
> some level.

Actively constructing a narrative in which she is powerless to stop the activities in the scene, Leah uses the pain first to justify her acquiescence to this
"horrible" treatment and then to provide a sense of purpose: to make the top
happy. Understanding in this way her "miserable" experience as a sacrifice for
his happiness requires her to process the pain as suffering. If the pain were to
be pleasurable in its own right, she would be sacrificing nothing for him, and
thus the play would lose its value.

Sophie, who also draws on the discourse of "transformed pain," describes a situation here in which she actively resists transforming pain into pleasure, precisely for the sake of experiencing her pain as a sacrifice instead:

> What we had gotten into the practice of doing, is when he was using clamps on me, and it was almost impossible for me to take, he had gotten into the habit of having me sort of like focus on his eyes, almost like hypnotizing. So I would focus on his eyes, and he would say things like "You can take this because you know I want you to feel this, and you want to do it for me, and like, you know . . . and I would almost get into this trance sort of state, and then it would stop feeling—it wouldn't stop feeling like pain, but it would stop feeling like pain that I had to stop. Does that make sense? And it's not that it hurt less, necessarily. I think once, a long time ago, we were at a diner and you were saying things like you don't understand people who could transform pain into pleasure. It was important to me always to—like I've heard people ask "how do you cope with pain?" Well, I go into this space where I don't feel the pain anymore, or I move beyond the pain, or whatever. When playing with Joe, it was always really, really important to me to not do that. Because if the whole point of doing it was to let him exercise his sadism, which is what I also ultimately enjoyed, if I was not experiencing the pain, then he wasn't really getting to be sadistic. And it wasn't gonna be as much fun for him. So I would like almost deliberately stop myself from trying to do that defense thing that your mind might want to do, and be like, No! Concentrate on how bad it feels (laughs). So here I am, so I'm not not feeling the pain, but now I'm like "I'm feeling the pain and it feels horrible, but that's good because it's like this gift I'm giving you when I want to feel horrible for you." So I'd be doing that little mental gymnastic as he's looking at me.

Investment Pain: Pain Payoffs

In contrast, the investment pain discourse draws heavily on hypermasculine narratives of pain ("No pain, no gain"). This discourse frames pain as an unpleasant stimulus that promises future rewards. Not surprisingly, men, whether bottoming or topping, frame pain this way more often than women do.

Sociologists of sport find that pain is often framed as an investment toward a greater reward. Pain is understood not merely as an unfortunate by-product, but as a means to a particular end. While the hurting is not the goal in and of itself, it is rewarding both for what it evidences and for what it produces. Because this suffering is not for the sake of another, it is uniquely masculine. It is competitive—a challenge to the self—an investment given of free will and, more importantly, framed as such. Describing a hook-suspension scene, Kyle, for example, did not romanticize the pain itself, but wanted it for what it could provide him physiologically:

> After I got over the pain of it, and I was—you know, with any sort of play in the scene, there's a time early on where it just hurts. And then

after a while, the endorphins kinda build up and it doesn't hurt anymore. That's kinda how this was too. Once I got past the pain of it and
I could really pull back, and really pull, and have the hooks pull
forward.... at one point, early on, when that happened, I stopped caring
about the pain of it and just wanted the experience.

Similarly, when discussing a heavy flogging scene, Lawrence said:

It was a very intense buzz. My body was very light. I didn't feel the weight
of my body. I didn't lose awareness of where I was, but my head cleared up
completely, which was really wonderful, because I'm always thinking. I
have a very busy mind and sometimes that gets the better of me. And it was
wonderful just to be able to relax and not have to force yourself to relax.

ME: Did it hurt?

LAWRENCE: Oh, it hurt immensely.

Investment pain is often less personal than sacrificial pain, in which the experience of pain is wrapped up in the bottom's relationship to the person inflicting
the pain. Investment pain, in sharp contrast, is rewarded by the result of the pain,
regardless of the relationship to the inflictor.

Investment pain can also be a reward that comes from having withstood
pain, rather than from pain itself. The investment here is not in order to play
but for what the pain *itself* will yield. The pain is undesirable, and the experience
of pain is not for the sake of the sport (as it is in athletic contexts of pain), but
because it provides its own rewards. In this slant, investment pain remains relatively impersonal. It appears ideologically more selfish than sacrificial pain, but
nevertheless seeks to reconcile the experience of infliction with the notion of a
loving (rewarding) top, without sacrificing authenticity.

The investment pain discourse contains a few different variations on the
same theme. The overarching connection in this frame, however, is that there
are dividends to be earned as a result of the pain. Pain is thus inherently aversive,
but worth the endurance. Not surprisingly, this is a more common frame among
men who bottom, and the frame of sacrificial pain is more commonly used by
women who bottom.

Autotelic Pain: Liking the Hurting

The three discourses described thus far maintain and reproduce the conceptualization of pain as aversive. Pain is something to be withstood, endured, altered, or
conquered. To be able to do so provides rewards, but pain is still, in and of itself,
negative. Most people in the Caeden SM community draw on one or more of
these discourses. Generally, people who understand pain in these ways do not say
that pain itself feels good, do not claim to desire pain, and take care in the community to clarify that they are not pain seekers.

In contrast, the terms *sadist* and *masochist* are used to describe people who
frame their relationships to pain in positive terms. These identity labels are somewhat stigmatic in the community. In some instances, they are self-identifications.

They are also attributed to people who do not appear to rely on strategies to achieve authentic experiences of power imbalance. Participants who transform or provide pain, for example, distinguish themselves from masochists, who they believe "like the pain," and also from sadists, who "like to hurt people." Interestingly, the only discourse in the SM community in which pain appears as an (almost) unqualified "good" thing is, by far, the least common.

The foundation of this discourse is fairly simple for those who draw on it: The pain hurts, but the hurt also feels good. Participants who frame pain this way have an extraordinarily difficult time articulating their experience of pain. They generally distinguish between kinds of pain that they do like and kinds of pain that they do not like; the particular kind of pain, rather than the context, determines whether the response is favorable. In an interview, Laura (having already discussed the considerable extent to which pain hurts her) attempted to clarify for me what she liked about pain:

LAURA: Thuddy, deep pain. It feels good.

ME: While it hurts?

LAURA: Yes and no. It's a very difficult thing to explain. It registers as pain. But it also registers as good. Like, I like this feeling. Like flogging—it hurts but it doesn't. Spanking—it hurts but it doesn't. I don't like stingy pain all that much. A little bit, but not all that much. I like thuddy pain.

Laura's paradigm did not depend on the relationship, the rewards for her or for the top, or on the conceptualization of pain as not hurting. Instead, Laura articulated an intersection between pain and pleasure, a place where it hurts and it is enjoyable. People who frame pain this way struggle to express it in conversation, reluctant to choose between the seemingly dichotomous experiences of pain and pleasure. Usually, bottoms who view pain this way simply rely on the less stigmatized identity labels like "pain slut" and "heavy bottom"; these terms dismiss the question of pain experience and shut down conversation about the liking of the pain. Frank, for example, whom I had seen play with what is sometimes considered "heavy pain," used the phrase "processing pain," but had difficulty articulating this experience:

ME: How do you process pain?

FRANK: I used to breathe a lot and then I'd slump and I'd be mush. Now it's screaming, jumping up and down, lots of breathing.

ME: But how do you feel it—when it hits, does it hurt?

FRANK: Depends on the pain, depends on the instrument. . . . Like a flogging is going to be much more force . . . impact, hard, breath coming out of me, versus the singletail, you know, trying to resist the tearing sensation.

ME: Do you like the pain?

FRANK: I think so. It's not a like, like "oh yeah, yeah, give it to me." But I do, but it's not a hard-on thing, but, you know, it hurts, certainly. But not necessarily hurts. It's pain, I can identify it as pain . . .

ME: When someone says do you like pain, what's your answer?

FRANK: No. I guess no.

ME: But you . . . enjoy it, in the context of certain scenes?

FRANK: Yes. Yes.

Autotelic pain is experienced, valued, and appreciated as pain. Bottoms who frame pain this way say that it hurts and that they like it *anyway*. Unlike those who frame pain as transformed, those who view pain as autotelic do not feel that they engage in a conversion process; the hurting itself feels good, instantly and without work. For tops, this discourse casts them as villainous, drawing on a romantic concept of the Marquis de Sade, the seductive evildoer. Tops who frame pain this way are often desired as play partners precisely because of their sadism: The stronger the belief that the top enjoys the actual infliction of pain, the more authentic the scene becomes for bottoms.

Tops and bottoms who identify as wanting pain for its own sake and to its own ends are in the minority in the community. The autotelic pain discourse rejects conventional conceptualizations of pain as undesirable and, by extension, pain seeking as pathological. Most SM participants actively employ strategies to disavow, minimize, or rationalize their engagement with pain, perhaps precisely to avoid understanding their activities in the pathological terms of sadism and masochism.

Ultimately, this discourse appears to disentangle the enjoyment of pain from the understanding of pain as bad. Although the end result of transformed pain is pleasure, it becomes, posttransformation, pleasure *instead* of pain. Autotelic pain begins as pain, ends as pain, and is enjoyable nonetheless. The overarching context, however, must remain one of inflictor–inflicted. Sadists and masochists, either self-defined or other identified, do not appear to enjoy pain in other, solo contexts (such as medical pain, accidental harm, or self-injury). Nonetheless, they claim to enjoy pain in and of itself, extricated from contexts of power and control.

THE SOCIAL CONSTRUCTION OF PAIN

To understand pain, we must look at the situations in which people seek pain. In the SM community examined in this chapter, ideologies of power and discourses of pain are constructed in relation to one another. "*Power exchange*" is the attempt to achieve authentic experiences of power imbalances within social, legal, and ethical limitations. SM participants engage in the closest translation they can approximate within two sets of overlapping constraints: the community-policed mantra of "safe, sane, and consensual" and their own ethical and physical boundaries.

Discourses of pain assist in this translation process. All of these discourses blur the contradictions between otherwise egalitarian relations and embodied experiences of power differentials. In so doing, they each help construct SM experience in accordance with ideologies of powerfulness and powerlessness, without sacrificing authenticity.

These discourses of pain also lessen, or sometimes mitigate, ethical qualms for participants. Viewing the pain as not ultimately painful (transformed) or worth the cost of the pain (sacrificial or investment) allows participants to more comfortably understand their activities alongside moral proscriptions against hurting and being hurt. In the autotelic discourse, the pain hurts but can be simultaneously pleasurable, thereby justifying its appeal. More commonly, though, this discourse subsumes the claim "It hurts but I like it anyway." Here, the ethical "problem" of SM stands unconfronted, and people who make these claims are rare and stigmatized within the community. Nonetheless, by allowing pain to stand as nothing other than painful, authenticity is achieved more exclusively through carnal experience; when the idea is that one body is authentically hurting another body in a context that emphasizes the hurting, the belief in an egalitarian relationship can be fairly easily suspended.

Finally, these frames allow participants to carve out a range of metaphoric spaces in which to locate and understand their SM involvement. Community members can move along and between multiple dimensions of identity, pain, power, and gender relations. Hence, the same person may identify as a "masochist," a "service top," and a "slave," playing with shifting parameters of authentic experience of power imbalances in any given SM scene.

REFERENCE

Goffman, Erving. 1963. *Stigma: Notes on the Management of Spoiled Identity.* Englewood Cliffs, NJ: Prentice Hall.

42

Selling Excitement: Gender Roles at the Male Strip Show

MAREN T. SCULL

Scull takes us into another sexual arena with her portrayal of a sex show featuring male erotic dancers. In this setting, she has the opportunity to turn the gendered tables on the better known world of female strippers and ask the question whether the gendered dynamics of power and dominance play out similarly to the reversed gender roles of the performers and audience. Does an arena of heightened sexuality, in which the performers are selling sexual titillation and fantasy in exchange for a night out and a good time, translate into one in which the women are dominant?

Scull asserts, following the proclamations of the performers, that the men are in control, much like studies of female strippers which argue that these performers hold sway. In this study, we see women seizing the opportunity to get wild, to try to dominate the men, and inflicting bodily harm on the performers. But, ultimately, she argues, the men rise up with their hypermasculinity and put those women back in their traditional place.

How do you assess Scull's description of the balance of power in this arena? Does the money exchanged give the purchasers greater control? Does the repeated nature of the men's performances give the men greater control? How do these roles and power differentials translate outside of the strip club arena? How does the introduction of money reframe the sexuality compared with the SM partners in the previous chapter and the iDollators before that?

The exotic-dance literature is quite vast, a popular topic among academics since the 1960s. Given that, in the United States today, strip clubs featuring women are t far more common than strip clubs featuring men, much of this literature focuses on women who dance for men (WDM). There are also pockets of research regarding men who dance for men (MDM). These studies concentrate on strippers' personal characteristics: their feelings of objectification, issues of empowerment, deviant behaviors at the strip club, the importance of dramaturgy in their performances, and how the occupation shapes their identities. Scholars have also expanded their inquiries to other populations of exotic dancers, such

Reprinted with permission from the author, Maren T. Scull.

as transsexual strippers and female dancers who cater to Black same-sex-desiring women (BSSDW).

Given this extensive literature, it is surprising that so little attention has been paid to men who dance for women (MDW). The few studies that focus on this group explore reasons for entering the career, issues of social control, customer–dancer interactions, feelings of empowerment, and the motivations and opinions of female customers. In this research article, I focus exclusively on the perspectives and experiences of male dancers, concentrating on how they perform masculinity and the ways in which these performances promote and reinforce traditional gender roles. Although it is not always the case, masculine characteristics are usually intertwined with power, control, dominance, and aggression.

GENDER ROLES REVERSED OR REINFORCED?

Like almost every other space, strip clubs contain people who actively "do" and "perform" their gender. However, gendered performances are magnified and exaggerated at strip clubs because both dancers and customers rely on traditional, stereotypical ideas of masculinity and femininity while engaging in impression management. For instance, female strippers use hyperfeminine presentations of self to create superficial intimate relationships with their customers, while male patrons and bouncers use the space to enact and demonstrate their masculinity.

Some studies of MDW suggest that the behaviors exhibited by female audience members are an exception to this pattern. Specifically, the male strip show often emboldens women to act wild, assertive, and free to perform their gender differently than they do on a day-to-day basis. In other words, the male strip show encourages a reversal of gender roles, or what Petersen and Dressel (1982) refer to as "*gender role transcendence*." When women experience gender role transcendence, they behave in ways that mimic male stereotypes and act contrary to how they would in the presence of their husbands, partners, or boyfriends. The male strip show is a situation in which men become objects, rather than subjects, of the "gaze," with MDW exposed to the gaze.

Some male performers experience gender role transcendence, as some aspects of their job are not considered "manly" by conventional standards. For example, some MDW take a passive role by waiting for women to approach them rather than the other way around. In addition, male strippers accept money from women despite the fact that providing monetary support is usually defined as a man's responsibility. In fact, Dressel and Petersen (1982: 392) found that some men were "kept" by those of their customers who gave them gifts and large sums of money.

Others propose that the male strip show sustains and reinforces gender roles by promoting gender inequalities and men's domination over women. This is because the show is a space where women feel "forced" to interact with strippers who regularly attempt to humiliate and embarrass them. These performances also permit male dancers to exercise power and control over female patrons and coerce them into passive positions and roles. In this chapter, I will address this question.

THE STUDY

I used ethnographic methods and in-depth interviews for this research, spending over 18 months conducting fieldwork at a strip club that I call "Dandelion's," located in the western region of the United States. On Friday and Saturday nights, the management hired male strippers from a company called "Erotic Sensations" to perform from 9 p.m. to 1 a.m. Although female strippers performed completely naked in the nude room, male dancers were not permitted to dance naked. I attended Dandelion's almost every Friday and Saturday night from September of 2009 to March of 2011. Overall, I observed 42 male dancers and engaged in over 60 informal conversations with male and female strippers, patrons, bartenders, cocktail waitresses, bouncers, doormen, cashiers, bussers, and managers. Like many researchers who study exotic dance, I assumed the role of the "peripheral member" (Adler and Adler, 1987) and did not actively engage in stripping or tipping.

In addition to field observations, I conducted 22 semistructured, in-depth interviews with men who were employed as strippers at the time of the research. Interviews lasted from 45 minutes to 4 hours. They were conducted in a private room in a variety of locations, such as the respondent's home, a hotel room, a library, an office, or the dressing room. Respondents' ages ranged from 22 to 44 years, with a mean age of 32.5 years. There were many levels of experience among my participants. Some had been stripping for as long as 22 years, while others had been dancing for only 2 weeks at the time of our interview. There were many ethnicities represented in my sample. Ten respondents identified themselves as Caucasian and four as African American. The remaining participants described their ethnicity as Hawaiian, Hungarian, Laotian, Italian, Spanish, Latino, French–Native American, and Puerto Rican. All interviewees said they were heterosexual.

All interviews were tape-recorded and transcribed in full. Because writing field notes would have been inconsistent with the norms at Dandelion's, I used my cell phone to send myself messages via text. Sending extensive field notes was not possible, so I texted partial sentences and key words to trigger my memory about specific events. I also carried a tape recorder in my car to record verbal field notes as I drove home. I then used the combination of my recordings and texts to write detailed field notes on my computer, along with personal notes, methodological notes, and theoretical notes.

The Setting of Dandelion's

Dandelion's Strip Club was dark and *very* loud. The air was usually thick with cigarette smoke accompanied by the occasional scent of cologne or perfume. The space contained four octagonal stages, two bars, and a large dance floor. Disco balls, strobe lights, and multicolored spotlights hung from the ceiling, and almost all of the walls were lined with mirrors. In the back of the club was a giant screen featuring either videos that corresponded to the music, or sporting events such as football games, basketball games, and Ultimate Fighter

Championship matches. Dandelion's also hosted a variety of themed activities, competitions, and events. For example, during the summer, the club had a "Teeny Tiny Tan Line Contest," in which the female customer with the most attractive tan was awarded $100. Also featured was a "Mullet Mayhem Trailer Trash Bash," in which female patrons could win money for wearing the shortest "Daisy Duke" style shorts while men were encouraged to wear a mullet hairstyle. There were also events called "The Leather and Lace Fetish Party," "Bikes, Babes, and Beer Night," "Naked Dodgeball Night," and "Midget Wrestling."

Dandelion's closely resembled what Bradley-Engen and Ulmer (2009) categorize as a "hustle club," as employees regularly attempted to con customers into spending more than they had originally planned. In addition, as in other hustle clubs, some dancers used drugs, were subjected to sexual harassment, were offered sex for money, and/or stole from coworkers. However, Dandelion's also was similar to what Bradley-Engen and Ulmer (2009: 45) refer to as a "social club" because it was located in a blue-collar area and had a "good time" feeling. Like some strip clubs in other research, Dandelion's had an atmosphere that was like the television show *Cheers*, because, for many, it was not only a strip club, but also the local watering hole where people came to watch sports, socialize with the staff, and talk to other regulars.

MDW'S PHYSICAL INTERACTIONS
WITH CUSTOMERS

Strippers first illustrated the reinforcement of gender roles through the physically aggressive ways they handled female patrons.

Dominating

Although there were situations where dancers lost control over their interactions with women, the number of instances in which male performers physically dominated and controlled female audience members was overwhelming. My participants regularly touched women in ways that appeared to render them powerless both symbolically and literally. An excerpt from my field notes indicates this dominance:

> Many of the dancers wear big boots with large, silver, metal buckles that go all the way up the sides. It looks uncomfortable and dangerous when they wrap their legs around a woman's neck. This is interesting, as it does not appear to represent or mimic a sexual act. It is difficult to see the reaction of the women when they do this because their faces are almost completely buried. However, many of the women seem rattled and embarrassed when dancers let them go.

Male strippers actively encouraged women to touch the men's bodies by grabbing the women's hands and rubbing them against the men's chest, thighs, buttocks, and groin. It was also common for the male dancers to grab a patron's breasts, rub her vaginal area, or slap her rear end. In some instances, a stripper would grab a customer by the back of the head and press her face firmly against his crotch. Another variation of this move was to place a woman's head in the man's groin and either spank the woman with his hand or a belt or give the appearance that the stripper was humping the woman's face by gyrating his hips up and down. One stripper inserted a bottle partially filled with beer into the front of his G-string and instructed women to drink from the bottle.

Some strippers performed a move in which they sat a female patron on their lap with her back facing them. They would then jerk their hips up and down, causing the woman to bounce on their laps repeatedly. Although this was a fairly common gesture, I rarely saw women appear to enjoy it. One evening, I watched Luscious do this to a young, attractive female who was in the club with her friends. After he set her down, she turned to her female friend and waved her hand horizontally in front of her neck as if to say "cut." She then shook her head and made an exaggerated frightened face while her friends laughed.

Aggressive Touching

Other forms of aggressive touching included making customers grab themselves in sexual ways. Chicago regularly forced women to grab their own breasts during his performances. He would kneel down on the stage, take hold of a woman by her shoulders, and spin her around so that she was facing away from him. Holding her wrists tightly, he then placed her hands on each of her breasts. He would yell and scream as he moved his hands up and down, forcing the woman to jostle her breasts. In addition, Chicago frequently snatched women by the hand and made them rub their own crotch. Many customers seemed shocked when he did this, and some even exhibited obvious signs of fear. One evening, I witnessed a female patron run away from the stage after being subjected to one of his performances.

There were also instances when I observed audacious dancers use their body parts to "accidentally" hit women. One evening in particular, I saw Lover Boy smack a woman in the face with his G-string-covered penis. This contact appeared to be painful to the customer, as she winced when he did it. Instead of apologizing or checking to make sure that she was not hurt, he laughed loudly and then yelled to her friends "She just got bopped in the face!" Customers who were celebrating a special occasion were frequently the targets of physical mistreatment from overzealous strippers. These women were usually readily noticeable, as they wore lacy headbands, veils, crowns with blinking lights, sashes, or tinseled tiaras. One evening, I saw Ace be particularly pushy with a bachelorette:

> He grabbed her so hard that her bachelorette crown fell off. The combs
> in her hair came loose as well. She looked rattled by the time he let her
> go. She walked away trying to smooth out the top of her hair and
> reposition her crown. Her hair had been extensively styled, and it was
> too difficult for her to put her crown back in. She eventually gave up
> and set it on the table.

Humiliating

It was clear that many of my participants intentionally tried to humiliate female patrons for the purpose of entertaining the crowd. These women were expected to manage any negative emotions and endure the interaction without expressing discomfort. Scott, a 26-year-old Caucasian stripper, talked about how some women allowed unpleasant encounters to continue because they did not want to be seen as a "bitch":

> I think it's embarrassing for them. I think it's different if a girl were doing that to me because I wouldn't care. But there's this big burly man doing this to you, and you really don't want it to happen. But you can't really say, "Stop" because you'll look like a bitch. So you never say anything. Not everybody is willing to have that sort of attention. It happens that some girls love it and some girls...I can tell that they just don't like it at all.

Occasionally, dancers went beyond embarrassing women and violated them in more sexual ways. One evening in particular, Ace performed a lap dance for a bachelorette on stage and then lifted her entire body straight above his head. As she was suspended horizontally in the air, he inserted his thumbs inside her very, very short denim skirt. When he set her down, she looked surprised and confused. He then moved his hands down her back, lifted up her skirt, and rubbed her buttocks and vaginal area. Scott explained that some female customers complained about performers like Ace who were too aggressive:

> I've had many females come up to me in regard to one person in particular. They say, "Ya know, he's just a dirty slut. He's always grabbing my ass, he's grabbing my tits." So um, ya know, I'm speaking about Ace. They come up to me and say stuff about him like, "Tell him to stay the fuck away from me" and that kind of stuff. So when Ace is doing these things, he thinks they're enjoying it, which it seems like they are. But they're actually thinking, "This guy needs to get the fuck off of me. I'm gonna go tell Scott because he knows him." So, it's not like girls are up front with guys at all.

The physical control of women was also evident when strippers performed special tricks. As is the case with female dancers who use a "gimmick" while on stage, the majority of my respondents had at least one movement that was distinct from their coworkers. For example, Rico Suave, a 36-year-old Puerto Rican dancer, performed what was known as the "Rico Suave Special." To start, he would direct a woman to sit in a chair placed at the center of the stage. Once seated, he performed a lap dance for her. He moved his body seductively, swayed his hips from left to right, and removed articles of clothing while the audience screamed and cheered. He then sat in the patron's lap, pumped his hips up and down, and tugged on his boxer briefs to reveal his G-string. On several occasions, I observed him put the front of his boxer briefs over the woman's head. Once her head was securely wrapped in his underwear, he moved his hips back and forth to make it appear as though she was performing fellatio. After removing his boxer briefs to reveal his fringe G-string, he positioned himself in a handstand. Given the height of the chair in which the customer was seated, this position usually put his crotch at the same level as her head. He would then "pop" his crotch back and forth so that the fringe of his G-string brushed against her face. After he did this, he would sit in the chair while the patron performed a lap dance for him.

Although many of these actions were against club policies, dancers continued to engage in these behaviors without regard. In fact, I observed them touch women in sexually forceful ways so frequently that I was surprised to learn that these behaviors violated the club's rules. More commonly, however, the management at Dandelion's did not enforce club policies or mete out punishments when dancers violated them. The rules at private parties were even more lenient and, in some situations, nonexistent. This was one of the main reasons many dancers preferred performing at private residences. Sex Machine, a 27-year-old African-American dancer, compared the rules that operated at private parties versus those at Dandelion's:

> I prefer private parties. I can be more of who I am and my style of personality comes out. They allow me to have more freedom during the show. Dancing on stage is like, robotic. There's a big difference between dancing on the stage and dancing at a private party. At the private parties the girls can get really wild. At the parties, the girls are allowed to interact more with the dancers...way more compared to the club. At private shows we do all kinds of stuff. We let the women put whip cream on the dancers or we put it on the girls. Some of the dancers lick the whip cream off the girls, or vice versa. They get wild by trying to take off my G-string, smacking my butt, and putting whip cream on my butt and licking it off. When the parties get really wild, we go all nude and incorporate sex toys. If I'm dressed as a cop, I may put a jelly donut in between her breasts and eat it.

Overall, my participants were not passive beings while performing. Instead, they actively positioned themselves in ways that enabled them to physically control and manhandle female audience members. As Tewksbury (1993:174) notes, male strippers are not "merely objects available for the taking, as might be expected with female strippers."

DANCERS' HYPERMASCULINE
PRESENTATION OF SELF

Masculinity is not a given and is not inherent in a male body. Instead, it must be attempted, accomplished, performed, and defended. This was evident at Dandelion's, as dancers reinforced conventional gender roles through their hypermasculine presentation of self and their use of body technologies. Hypermasculinity is a type of masculinity that involves presenting oneself as possessing exaggerated "male" characteristics, such as being assertive, uncaring, forceful, strong, and dominant.

The body is central to exotic dance because performers must utilize their somatic selves to engage in impression management for the purpose of economic gain. Goffman (1959) explains that impression management is a process in which individuals, much like actors in a play, attempt to influence and sustain the perceptions others have of them through the use of props, gestures, and symbols. In the case of stripping, dancers must engage in "counterfeit intimacy," by conveying the impression that they are sexually interested in their customers, even though most are genuinely not. Using Goffman's language, the main "prop" strippers use is their body. By positioning their body in certain ways, MDW and WDM try to cater to the perceived sexual desires of club patrons.

In order to present themselves as manly and powerful, my respondents used what Wesely (2003) refers to as "***body technologies***." Body technologies are "the techniques we engage in to change or alter our physical appearance" (Wesely, 2003: 644). All of my respondents emphasized the importance of shaping their body and overall look, and engaged in many body technologies to create a "macho" appearance. These included wearing costumes, dieting, exercising, using steroids, maintaining the skin, and dealing with erections.

First, dancers wore a variety of ***outfits*** while performing, all of them conforming to gendered stereotypes. They wore clothing that resembled soldiers, businessmen, doctors, sailors, football players, firemen, police officers, gladiators, and cowboys. Hercules, a 42-year-old Caucasian dancer and the owner of Erotic Sensations, emphasized the importance of creating a convincing costume:

> In terms of getting costumes, you need to design them or put them together yourself. You need to have a theme to work off of. For example, your main costume could be a fireman, or a construction worker...whatever you think would be your normal costume. One of the guys, he's got like tall motorcycle boots, and so he dresses like a cop...to look like a motorcycle cop, and so he took it to another level. With my cop outfit, I'll have a flashlight and a gun. It's not a real gun. It looks real...it just has to *look* real.

In most cases, strippers' costumes also included what they referred to as "tear-aways." Tear-aways are pants that detach at the sides to allow strippers to rip them off with ease. Although some dancers buy these pants, most said that they were difficult to find and prohibitively expensive. Therefore, most participants made their own tear-aways by cutting the sides of a pair of jeans or dress pants, hemming the edges with a sewing machine, and inserting strips of Velcro® to hold the sides together. This arrangement enabled the dancers to remove and reassemble their clothing quickly during their performances. In addition, almost all of the dancers wore a G-string underneath their pants or tear-aways. Overall, these garments were used as props to create the impression that they were hypersexual beings and to maintain a lustful, romantic environment.

Dancers also used **diet and exercise** to create a hypermasculine appearance. Despite having different body types, *all* respondents emphasized the importance of exercising and taking care of their body. Rick, a 34-year-old Caucasian dancer, worked out 6 days a week and was vigilant about what, how, and when he ate, consuming six preweighed meals per day at specific intervals. Connor, a 29-year-old Caucasian dancer, also closely monitored what he ate:

> I diet a lot. When I wake up I eat 10 egg whites and five pieces of Ezekiel bread. And then for my second meal of the day I usually do a couple cans of low-sodium tuna and one cup of apple juice, and I blend that together and eat it. Then my third meal is tilapia with brown rice. And then there are a lot of protein shakes in there too. And then dinner is egg whites again with maybe a little avocado in there. Sometimes I'll cheat. Like every once in a while I'll have a piece of toast with peanut butter, or I'll go to McDonald's and order like five hamburger patties with nothing else. Then I go home and I just eat the patties.

My respondents not only adhered to a strict meal plan, but also talked about the importance of working out on a regular basis. On average, dancers reported exercising 15 hours per week. Some participants exercised immediately before their performances in order to make their muscles appear larger and more defined.

Many dancers incorporated anabolic-androgenic **steroids** into their diet. Given the physical demands associated with stripping and the expectations regarding physical appearance, this behavior is not surprising. Matt, a 33-year-old Caucasian stripper, talked about how the steroids enabled him to create the macho physique central to being an exotic dancer:

> All the guys take steroids, you know? And when I say all, I mean about 90% of them. Steroids put more testosterone in the body, and testosterone is the key component for every type of muscle building and protein synthesis. Basically, it helps make you stronger and faster. See, you won't last long in this industry if you don't use steroids. They all do steroids. You just can't look like this without doing steroids. It's physically impossible. Just building the muscle, getting the definition, getting the large size…it just goes with the territory.

There are several potential negative side effects associated with long-term steroid use, such as heightened blood pressure, acne, male-pattern baldness, a lower sperm count, a decrease in the size of the testicles, liver damage, changes in the libido, and gynecomastia (an increase in the amount of flesh in the pectoral region of men). Matt explained how the formation of gynecomastia, or what he called "gyno" or "bitch tits," develops:

> Over time [steroids] can cause gyno, which is bitch tits. It's sort of like extra fleshiness…it's not like boobs, but I mean I've got a little bit of it (he lifted up his shirt to reveal the pockets of flesh on his chest). I mean, every male has a certain amount of estrogen. It's a small amount. And then they have a certain amount of testosterone, and it has to be stable and in balance. When you take steroids, what happens is your

testosterone shoots up very high. The estrogen is naturally gonna shoot up too…to balance it. So, if you come off steroids too quick you can get it. That's what happens with a lot of people. And we're in an environment where we're not thinking about our futures. So what happens is, one day you don't have any steroids and you can't get more. So now it's like you're a chick because that testosterone has shot down and that estrogen just hasn't had a chance to catch up yet.

The benefit of having an exceptionally masculine physique through steroid use is often accompanied by the detriment of psychological and behavioral side effects, including mood swings, irritability, aggression, psychiatric problems, and, in extreme cases, psychotic episodes. Many of my respondents referred to this cluster of feelings as "'roid rage" and described its prevalence among strippers who used steroids. For the most part, these side effects were not problematic for dancers, because they were consistent with a hypermasculine presentation of self. Matt discussed the negative definition, noting that "The downside to steroids is that they enhance who you are. So if you're a dick, you'll be an even bigger dick when you're on steroids. They can also make you more aggressive."

My respondents also paid special attention to their **skin** when they were preparing themselves for an erotic performance. This attention came in the form of tanning, moisturizing, hair removal, and tattoos. Tanning was central to improving the look of their skin and attracting customers. Tanned skin made their bulging muscles look more defined, so most used self-tanners and tanning beds. Second, many strippers talked about moisturizing their skin before performances and the importance of warding off dry, flaky patches, or what some respondents disparagingly described as "ashy" skin. Most dancers accomplished this by coating their bodies with tanning lotion or baby oil before going on stage. Third, most of my participants shaved their arms, legs, back, stomach, and genital region to make their skin look more attractive. Preston, a 27-year-old Latino dancer, remarked that he liked to keep his body hair looking neat:

> With the shaving, you don't have to shave your legs and arms. Some guys shave everything…like their body is bald. I just trim myself, like my legs (he lifted up his pants to show me his leg). Obviously, I don't shave much. Like I don't have a lot of arm hair so I shave my arms. I have no chest hair, so it's just maintenance because you want to come out looking pretty so-to-speak. You don't want to look like a hairy dancer. You don't want to come out all grizzly. You want to turn women on, and they prefer someone who "manscapes" I guess the term would be.

The last body technology involved dealing with **erections**. There were no written rules forbidding male strippers from having, or appearing to have, an erection while performing at Dandelion's. On several occasions, I observed strippers who looked like they had a partially or fully erect penis while on stage. In fact, some participants appeared to have enormous penises bulging under their G-strings while other strippers' penises were hardly noticeable.

There was a wide range of viewpoints and behaviors regarding having an erection while on stage. Some dancers felt that it was important to have an erection while performing, and they actively engaged in techniques to produce and maintain one. One of these techniques was known as "tying off," which involved creating an erection through the use of visual materials and/or masturbation (also known as "fluffing"). They then placed a cock ring to act as a tourniquet to keep the penis engorged with blood. Rick talked about tying off:

> I think some people fluff themselves before getting on stage. They just get themselves hard in the back room. They may be looking at porn or pictures or something. But I don't think there's any standard or rule. There's tricks...there's cock rings that you can wear, which is really common. And a lot of guys get hard before a set with just masturbating. I think we all wear a cock ring. All it does is just push everything to the front of the thong. Also, there are certain thongs that are specifically made to insert the penis into. Sort of like a cock ring in and of itself, and you can tighten 'em and stuff like that. You can tell they're one of those if they have that elephant trunk–looking thing.

Other dancers talked about how they attempted to create the *illusion* of an erection without actually being fully stimulated. For example, Ryan, a 38-year-old Caucasian stripper, talked about how substances prescribed for erectile dysfunction helped him *appear* to have an erection without having to put considerable effort into sustaining one:

> There are some guys that will use Levitra or something before they go on their shift. When I first started here, not knowing the crowd, and not knowing the people, I would take half of a tab of Levitra. Really what it does is it relaxes the penis rather than inflates the penis. For me to have an erection, I have to have personal touch. I have to have stimulation to get an erection. Also, when you get nervous, the male body will retract and that's not good for business. So the Levitra allows you to loosen up, and it makes the blood flow to that area a little better. There are some guys that are very tactile and they may pop an erection in a minute but, you know, the professional dancers wear a cock ring. You don't necessarily wear it for an erection. You wear it because it pushes the testicles forward, you know? And it makes your package stand out in the front, so it's all up in the front. It just looks bigger and it just enhances the presentation of everything...a lot of guys do that.

Others defined erections as problematic, rather than desirable. For instance, Blade, a 42-year-old Hungarian stripper, preferred to have a flaccid penis while on stage. He made concentrated attempts to control his "friend," as he called it, by focusing on unattractive customers. Sex Machine expressed that, irrespective of one's choice to display an erection, having confidence in one's genitals was what was most important:

> Some guys try to put fake dongs on. They put a jelly dong on their dong, and you put it inside your G-string. Guys do it to look bigger. But I think that, overall, you just need to be secure about your package

when you go up there. There are some guys that believe in being natural. I like to be natural. At least I don't have to sit there and act like this and that. I don't want to see it in the dressing room. I think it is competiveness, but it's not spoken about. Each guy is thinking, "I'm better than you, I'm sexier than you, I'm hotter than you." Ace is a big dude and I don't get bothered by that at all. There are guys that are bigger...it's not about size; it's about how confident you are.

AGGRESSIVE WOMEN

Although my participants' actions reinforced the gender hierarchy in society through their physical control of women and their hypermasculine presentations of self, there were also examples of **gender role transcendence**. Specifically, there were times when women violated conventional gender expectations by **physically abusing and degrading** dancers. Hero, a 29-year-old Caucasian stripper, talked about a particular occurrence when he was scratched by a patron:

> There was a girl that was in here with a bachelorette party. She was really aggressive...really grabby. I put her arms around my sides and around my back. I kinda grabbed her hair and neck, and I was blowing on her neck and kissing her neck. And then she completely scratched me. She literally peeled the skin down my back. It was to the point where I walked back to the dressing room and I was like, "Oh my God." See, I didn't really feel it when I was up there. But then I got off [the stage] and I was like, "Alright you just need to stay away from that one." Then I go back in the dressing room and the guys are like, "You're bleeding." Hercules was pissed because he doesn't want us looking like that. He actually went up to her, but she didn't get kicked out. There have also been times when girls just grab your penis...and they grab hard.

Like DeMarco (2007:115), who reported that some MDM must deal with patrons who "try to put their fingers where they don't belong," some customers attempted to stick objects into a dancers' anus. Scott talked about a bachelorette who tried to violate him in this way:

> One girl attempted to put her finger up my butthole. It's just really odd to me that someone would just feel like it's okay to do that. I mean, regardless of them being inebriated, ya know? So I just grabbed her wrist and threw it back at her and said, "You can't do that." And when you do that, there's no playing around. It's serious. I've had a bachelorette try to do that to me in front of her whole bachelorette party. That to me is crazy. I'm like, "You're getting married within a week, and you're trying to put your finger up a guy's butthole? Does

that really make any sense?" It's like they think they can do anything they want.

Although these encounters were unpleasant and painful, like MDM, my respondents felt compelled to tolerate them in order to maximize their tips. Further, dancers reported that dealing with rowdy patrons was not particularly challenging, as women were easy to control and physically restrain. Strippers were so adept at dealing with forceful customers that there was only one occasion on which a performer, Hercules, walked out on a private party because a woman was overly forceful. He said,

> Some of the more aggressive ones…I don't think that they realize that they're hurting you. But when you have nail marks on your back, or when you get slapped, or when you are bit…that hurts. I've been bit. I was at a party and this girl bit my ass. Then she was jumping around and jumping on my back. It was like, "Okay we gotta tone this down a little bit." Finally when she bit me I said, "Ya know what? This is the first time I've ever had to shut down a party." I mean, I had teeth marks for two weeks. It's not common in that it doesn't happen all of the time, but it *does* happen.

Matt suggested that women acted so recklessly because they were unsatisfied with how their male partners treated them at home and therefore used dancers to vent and express their frustrations:

> I think a lot of women kinda get off on degrading us. You get all different kinds of people and some people think it's cool. I think they honestly just get a kick out of degrading us because their husbands treat them like shit, and this is their chance in their own little way to get even.

CONCLUSIONS

The performances of male dancers unveiled much about the gendered aspects of exotic dance and the reinforcement of gender roles. The male strip show both reflected and reproduced socially constructed notions of what it means to be a "man" in our society.

First, dancers demonstrated this notion by physically dominating and controlling female patrons. Although there were situations in which strippers were subjected to unruly customers, for the most part these occurrences were infrequent and inconsequential. Most importantly, my respondents rarely felt as though they lost control over those interactions. In fact, they said that even the most aggressive customers were easy to manage and manipulate. In addition, participants did not express any feelings of trauma or psychological stress as a result of physical mistreatment from overzealous customers. Instead, most respondents were simply bewildered or irritated by these occurrences. Overall, my findings are consistent with Pilcher's (2009: 230), that the way MDW interact with patrons is "not playful, fun, or 'liberating,' but rather violent, forceful and resembles the harassment of women more than it leaves room for women to be autonomous sexual adventurers."

Second, exploring the body technologies strippers used reveals much about how they performed their masculinity. Strippers were obsessed with creating a manly appearance, spending enormous efforts to prepare and maintain their bodies. By wearing costumes, dieting, exercising, using steroids, maintaining their skin, and dealing with erections, dancers were able to enact a specific *kind* of masculinity: one that was exaggerated, inflated, and overdone. In many ways, some dancers looked *so* masculine and *so* macho that they appeared to be caricatures of themselves. Enacting their masculinity in this way contributed to the reinforcement of traditional power differentials between genders, rather than to gender role transcendence. As Tewksbury (1993: 179) found in his research of MDM, my respondents adjusted the "traditional patriarchal privileges within the arena of sexual objectification and consumption," reconstructing a conventionally feminine occupation by introducing masculine elements into their performances. Thus, like female strip acts, the male dance revue at the Dandelion represented an environment in which masculinity was catered to and where it dominated.

Overall, the male strip show had the potential to both reinforce and overturn conventional, normative expectations associated with being either male or female. Although the male strippers were subjected to aggressive women, not only did they remain in control over their interactions with customers, but they dominated the customers. This domination contributed to the reinforcement of conventional, deeply ingrained gender norms and did little to generate feelings or behaviors associated with gender role transcendence.

REFERENCES

Adler, Patricia A., and Peter Adler. 1987. *Membership Roles in Field Research.* Newbury Park, CA: Sage Publications.

Bradley-Engen, Mindy S., and Jeffery Ulmer. 2009. "Social Worlds of Stripping: The Processual Orders of Exotic Dance." *Sociological Quarterly* 50: 29–60.

DeMarco, Joseph R. G. 2007. "Power and Control in Gay Strip Clubs." *Journal of Homosexuality* 53(1/2): 111–127.

Dressel, Paula L., and David M. Petersen. 1982. "Becoming a Male Stripper: Recruitment, Socialization, and Ideological Development." *Work and Occupations* 9(3): 387–406.

Goffman, Erving. 1959. *The Presentation of Self in Everyday Life.* Garden City, NY: Doubleday.

Petersen, David M., and Paula L. Dressel. 1982. "Equal Time for Women: Social Notes on the Male Strip Show." *Urban Life* 11(2): 185–208.

Pilcher, Katy Elizabeth Mary. 2009. "Empowering, Degrading or a 'Mutually Exploitative' Exchange for Women? Characterizing the Power Relations of the Strip Club." *Journal of*

International Women's Studies 10(3): 73–83.

Tewksbury, Richard. 1993. "Male Strippers: Men Objectifying Men." in Christine L. Williams (Ed.), *Doing Women's Work: Men in Nontraditional Occupations*: pp. 168–181.

Newbury Park, CA: Sage Publications.

Wesely, Jennifer K. 2003. "Exotic Dance and the Negotiation of Identity: The Multiple Uses of Body Technologies." *Journal of Contemporary Ethnography* 32(6): 643–669.

CONFLICT

43

Sexual Assault on Campus

ELIZABETH A. ARMSTRONG, LAURA HAMILTON, AND BRIAN SWEENEY

In the twenty-first century, American colleges and universities have been rocked by alcohol-related scandals leading to rape and death. School administrators have tried to clamp down on students by using punitive measures such as "strikes," probations, and expulsions. Students living in university housing have been particularly vulnerable to those measures, as they are under the watchful eye of resident advisors. The restrictions have pushed student partying—most prevalent among White, middle-class populations, according to Armstrong, Hamilton, and Sweeney—increasingly into the fraternity scene. There, women with these demographics are most at risk of rape, the authors find.

We have known for more than a decade that college women's risk of sexual assault by people they know far outweighs that of stranger rape. Armstrong, Hamilton, and Sweeney explore some of the popular explanations for the pervasiveness of fraternity rape, from individual bad boys, to the fraternity culture, to dangerous environments, and integrate them all by going beyond them. The authors locate the problem for women in a structural situation whereby they are forced out of their dorms by harsh sanctions against alcohol in the residence halls, leaving them to party elsewhere. The party scene they find is located in private environments where men control access to alcohol, transportation, and, more importantly, social status through their attentions to women. Here, traditional gender roles end up contributing to the victimization of women as men and women voluntarily engage in a "dance" where women seek flirtation to gain status and men offer flirtation and status to gain sex. Most women, being disempowered in these male-dominated settings, put themselves into situations where they are at high risk of victimization. When they are the victimized, they are likely to attribute blame to their girlfriends individually (remember Reinarman's comments about the vocabulary of attribution being individualistic

Source: Elizabeth A. Armstrong, Laura Hamilton, and Brian Sweeney, "Sexual Assault on Campus: A Multilevel, Integrative Approach to Party Rape." *Social Problems*, Vol. 53, No. 4, November 2006. © 2006 by the Society for the Study of Social Problems. Published by the University of California Press.

rather than structural?) rather than banding together to fight the unequal power structure in which men have the control and advantage. Older women and women of higher status look down on their less fortunate female schoolmates as "stupid," or "asking for it," despite the likelihood that they may have experienced these same troubles themselves at an earlier age.

Armstrong, Hamilton, and Sweeney offer disturbing insights into how women fight to protect their access to the party scene despite their risk of being taken advantage of and disempowered. We are reminded that fraternity culture, regardless of its deviant aspects, represents an enclave of the dominant culture where women are complicit in their own victimization and men use their gender advantage to callously foster their own ends.

What factors influence the relative status and opportunity of the women who are assaulted? What factors influence the relative status and opportunity of their assailants? What are the repercussions of the frat scene's location in the dominant majority culture for the position of women and men?

A 1997 National Institute of Justice study estimated that between one-fifth and one-quarter of women are the victims of completed or attempted rape while *in* college (Fisher, Cullen, and Turner 2000). College women "are at greater risk for rape and other forms of sexual assault than women in the general population or in a comparable age group" (Fisher et al. 2000: iii).[1] At least half and perhaps as many as three-quarters of the sexual assaults that occur on college campuses involve alcohol consumption on the part of the victim, the perpetrator, or both (Abbey et al. 1996; Sampson 2002). The tight link between alcohol and sexual assault suggests that many sexual assaults that occur on college campuses are "party rapes." A recent report by the U.S. Department of Justice defines party rape as a distinct form of rape, one that "occurs at an off-campus house or on- or off-campus fraternity and involves ... plying a woman with alcohol or targeting an intoxicated woman" (Sampson 2002: 6).[2] While party rape is classified as a form of acquaintance rape, it is not uncommon for the woman to have had no prior interaction with the assailant, that is, for the assailant to be an in-network stranger (Abbey et al. 1996).

Colleges and universities have been aware of the problem of sexual assault for at least 20 years, directing resources toward prevention and providing services to students who have been sexually assaulted. Programming has included education of various kinds, support for *Take Back the Night* events, distribution of rape whistles, development and staffing of hotlines, training of police and administrators, and other efforts. Rates of sexual assault, however, have not declined over the last five decades (Adams-Curtis and Forbes 2004: 95; Bachar and Koss 2001; Marine 2004; Sampson 2002: 1).

Why do colleges and universities remain dangerous places for women in spite of active efforts to prevent sexual assault? While some argue that "we know what the problems are and we know how to change them" (Adams-Curtis and Forbes 2004: 115), it is our contention that we do not have a complete explanation of the problem. To address this issue we use data from a study of college life at a large Midwestern university and draw on theoretical developments in the sociology of gender. Continued high rates of sexual assault can be viewed as a case of the repro-duction of gender inequality—a phenomenon of central concern in gender theory.

We demonstrate that sexual assault is a predictable outcome of a synergistic intersection of both gendered and seemingly gender neutral processes operating at individual, organizational, and interactional levels. The concentration of homogenous students with expectations of partying fosters the development of sexualized peer cultures organized around status. Residential arrangements inten-sify students' desires to party in male-controlled fraternities. Cultural expectations that partygoers drink heavily and trust party-mates become problematic when combined with expectations that women be nice and defer to men. Fulfilling the role of the partier produces vulnerability on the part of women, which some men exploit to extract nonconsensual sex. The party scene also produces fun, generating student investment in it. Rather than criticizing the party scene or men's behavior, students blame victims. By revealing mechanisms that lead to the persistence of sexual assault and outlining implications for policy, we hope to encourage colleges and universities to develop fresh approaches to sexual assault prevention.

APPROACHES TO COLLEGE SEXUAL ASSAULT

Explanations of high rates of sexual assault on college campuses fall into three broad categories. The first tradition, a ***psychological approach*** that we label the "individual determinants" approach, views college sexual assault as primarily a consequence of perpetrator or victim characteristics such as gender role attitudes, personality, family background, or sexual history. While "situational variables" are considered, the focus is on individual characteristics. For example, Antonia Abbey and associates (2001) find that hostility toward women, acceptance of ver-bal pressure as a way to obtain sex, and having many consensual sexual partners distinguish men who sexually assault from men who do not. Research suggests that victims appear quite similar to other college women (Kalof 2000), except that white women, prior victims, first-year college students, and more sexually active women are more vulnerable to sexual assault (Adams-Curtis and Forbes 2004; Humphrey and White 2000).

The second perspective, the ***"rape culture" approach***, grew out of second wave feminism. In this perspective, sexual assault is seen as a consequence of wide-spread belief in "rape myths," or ideas about the nature of men, women, sexuality, and consent that create an environment conducive to rape. For exam-ple, men's disrespectful treatment of women is normalized by the idea that men are naturally sexually aggressive. Similarly, the belief that women "ask for it" shifts responsibility from predators to victims. This perspective initiated an important shift away from individual beliefs toward the broader context. How-ever, rape supportive beliefs alone cannot explain the prevalence of sexual assault, which requires not only an inclination on the part of assailants but also physical proximity to victims.

A third approach moves beyond rape culture by identifying *particular contexts*—fraternities and bars—as sexually dangerous. Ayres Boswell and Joan Spade (1996) suggest that sexual assault is supported not only by "a generic culture surrounding and promoting rape," but also by characteristics of the "specific settings" in which men and women interact (p. 133). Mindy Stombler and Patricia Yancey Martin (1994) illustrate that gender inequality is institutionalized on campus by [a] "formal structure" that supports and intensifies an already "high-pressure heterosexual peer group" (p. 180). This perspective grounds sexual assault in organizations that provide opportunities and resources.

We extend this third approach by linking it to recent theoretical scholarship in the sociology of gender. Martin (2004), Barbara Risman (1998; 2004), Judith Lorber (1994), and others argue that gender is not only embedded in individual selves, but also in cultural rules, social interaction, and organizational arrangements. This integrative perspective identifies mechanisms at each level that contribute to the reproduction of gender inequality (Risman 2004). Socialization processes influence gendered selves, while cultural expectations reproduce gender inequality in interaction. At the institutional level, organizational practices, rules, resource distributions, and ideologies reproduce gender inequality. Applying this integrative perspective enabled us to identify gendered processes at individual, interactional, and organizational levels that contribute to college sexual assault....

METHOD

Data are from group and individual interviews, ethnographic observation, and publicly available information collected at a large Midwestern research university. Located in a small city, the school has strong academic and sports programs, a large Greek system, and is sought after by students seeking a quintessential college experience. Like other schools, this school has had legal problems as a result of deaths associated with drinking. In the last few years, students have attended a sexual assault workshop during first-year orientation. Health and sexuality educators conduct frequent workshops, student volunteers conduct rape awareness programs, and *Take Back the Night* marches occur annually.

The bulk of the data presented in this paper were collected as part of ethnographic observation during the 2004–2005 academic year in a residence hall identified by students and residence hall staff as a "party dorm." While little partying actually occurs in the hall, many students view this residence hall as one of several places to live in order to participate in the party scene on campus. This made it a good place to study the social worlds of students at high risk of sexual assault—women attending fraternity parties in their first year of college. The authors and a research team were assigned to a room on a floor occupied by 55 women students (51 first-year, 2 second-year, 1 senior, and 1 resident assistant [RA]). We observed on evenings and weekends throughout the entire academic school year. We collected in-depth background information via a detailed nine-page survey that 23 women completed[,] and [we] conducted interviews with

42 of the women (ranging from 1¼ to 2½ hours). All but seven of the women on the floor completed either a survey or an interview.

With at least one-third of first-year students on campus residing in "party dorms" and one-quarter of all undergraduates belonging to fraternities or sororities, this social world is the most visible on campus. As the most visible scene on campus, it also attracts students living in other residence halls and those not in the Greek system. Dense precollege ties among the many in-state students, class and race homogeneity, and a small city location also contribute to the dominance of this scene. Of course, not all students on this floor or at this university participate in the party scene. To participate, one must typically be heterosexual, at least middle class, white, American-born, unmarried, childless, [of] traditional college age, politically and socially mainstream, and interested in drinking. Over three-quarters of the women on the floor we observed fit this description.

There were no nonwhite students among the first and second year students on the floor we studied. This is a result of the homogeneity of this campus and racial segregation in social and residential life. African Americans (who make up 3 to 5 percent of undergraduates) generally live in living-learning communities in other residence halls and typically do not participate in the white Greek party scene. We argue that the party scene's homogeneity contributes to sexual risk for white women. We lack the space and the data to compare white and African-American party scenes on this campus, but in the discussion we offer ideas about what such a comparison might reveal....

Selves and Peer Culture in the Transition from High School to College

Student characteristics shape not only individual participation in dangerous party scenes and sexual risk within them but the development of these party scenes. We identify individual characteristics (other than gender) that generate interest in college partying and discuss the ways in which gendered sexual agendas generate a peer culture characterized by high-stakes competition over erotic status.

Nongendered Characteristics Motivate Participation in Party Scenes
Without individuals available for partying, the party scene would not exist. All the women on our floor were single and childless, as are the vast majority of undergraduates at this university; many, being upper-middle class, had few responsibilities other than their schoolwork. Abundant leisure time, however, is not enough to fuel the party scene. Media, siblings, peers, and parents all serve as sources of anticipatory socialization (Merton 1957). Both partiers and non-partiers agreed that one was "supposed" to party in college. This orientation was reflected in the popularity of a poster titled "What I Really Learned in School" that pictured mixed drinks with names associated with academic disciplines. As one focus group participant explained:

> You see these images of college that you're supposed to go out and
> have fun and drink, drink lots, party and meet guys. [You are] supposed

to hook up with guys, and both men and women try to live up to that. I think a lot of it is girls want to be accepted into their groups and guys want to be accepted into their groups.

Partying is seen as a way to feel a part of college life. Many of the women we observed participated in middle and high school peer cultures organized around status, belonging, and popularity (Eder 1985; Eder, Evans, and Parker 1995; Milner 2004). Assuming that college would be similar, they told us that they wanted to fit in, be popular, and have friends. Even on move-in day, they were supposed to already have friends. When we asked one of the outsiders, Ruth, about her first impression of her roommate, she replied that she found her:

> Extremely intimidating. Bethany already knew hundreds of people here. Her cell phone was going off from day one, like all the time. And I was too shy to ask anyone to go to dinner with me or lunch with me or anything. I ate while I did homework.

Peer Culture as Gendered and Sexualized   Partying was also the primary way to meet men on campus. The floor was locked to nonresidents, and even men living in the same residence hall had to be escorted on the floor. The women found it difficult to get to know men in their classes, which were mostly mass lectures. They explained to us that people "don't talk" in class. Some complained [that] they lacked casual friendly contact with men, particularly compared to the mixed-gender friendship groups they reported experiencing in high school.

Meeting men at parties was important to most of the women on our floor. The women found men's sexual interest at parties to be a source of self-esteem and status. They enjoyed dancing and kissing at parties, explaining to us that it proved men "liked" them. This attention was not automatic, but required the skillful deployment of physical and cultural assets. Most of the party-oriented women on the floor arrived with appropriate gender presentations and the money and know-how to preserve and refine them. While some more closely resembled the "ideal" college party girl (white, even features, thin but busty, tan, long straight hair, skillfully made-up, and well dressed in the latest youth styles), most worked hard to attain this presentation. They regularly straightened their hair, tanned, exercised, dieted, and purchased new clothes.

Women found that achieving high erotic status in the party scene required looking "hot" but not "slutty," a difficult and ongoing challenge. Mastering these distinctions allowed them to establish themselves as "classy" in contrast to other women. Although women judged other women's appearance, men were the most important audience. A "hot" outfit could earn attention from desirable men in the party scene. A failed outfit, as some of our women learned, could earn scorn from men. One woman reported showing up to a party dressed in a knee-length skirt and blouse only to find that she needed to show more skin. A male guest sarcastically told her "nice outfit," accompanied by a thumbs-up gesture....

Men also sought proof of their erotic appeal. As a woman complained, "Every man I have met here has wanted to have sex with me!" Another interviewee reported that: this guy that I was talking to for like ten/fifteen minutes says, "Could you, um, come to the bathroom with me and jerk me off?" And I'm like, "What!" I'm like, "Okay, like, I've known you for like, fifteen minutes, but no." The women found that men were more interested than they were in having sex. These clashes in sexual expectations are not surprising: men derived status from securing sex (from high-status women), while women derived status from getting attention (from high-status men). These agendas are both complementary and adversarial: men give attention to women en route to getting sex, and women are unlikely to become interested in sex without getting attention first.

University and Greek Rules, Resources, and Procedures

Simply by congregating similar individuals, universities make possible heterosexual peer cultures. The university, the Greek system, and other related organizations structure student life through rules, [the] distribution of resources, and procedures.

Sexual danger is an unintended consequence of many university practices intended to be gender neutral. The clustering of homogeneous students intensifies the dynamics of student peer cultures and heightens motivations to party. Characteristics of residence halls and how they are regulated push student partying into bars, off-campus residences, and fraternities. While factors that increase the risk of party rape are present in varying degrees in all party venues (Boswell and Spade 1996), we focus on fraternity parties because they were the typical party venue for the women we observed and have been identified as particularly unsafe (see also Martin and Hummer 1989; Sanday 1990). Fraternities offer the most reliable and private source of alcohol for first-year students excluded from bars and house parties because of age and social networks.

University Practices as Push Factors   The university has latitude in how it enforces state drinking laws. Enforcement is particularly rigorous in residence halls. We observed RAs and police officers (including gun-carrying peer police) patrolling the halls for alcohol violations. Women on our floor were "documented" within the first week of school for infractions they felt were minor. Sanctions are severe—a $300 fine, an 8-hour alcohol class, and probation for a year. As a consequence, students engaged in only minimal, clandestine alcohol consumption in their rooms. In comparison, alcohol flows freely at fraternities.

The lack of comfortable public space for informal socializing in the residence hall also serves as a push factor. A large central bathroom divided our floor. A sterile lounge was rarely used for socializing. There was no cafeteria, only a convenience store and a snack bar in a cavernous room furnished with big-screen televisions. Residence life sponsored alternatives to the party scene such as "movie night" and special dinners, but these typically occurred early in the evening. Students defined the few activities sponsored during party hours (e.g., a midnight trip to Walmart) as uncool....

Male Control of Fraternity Parties   The campus Greek system cannot operate without university consent. The university lists Greek organizations as student clubs, devotes professional staff to Greek-oriented programming, and disbands fraternities that violate university policy. Nonetheless, the university lacks full authority over fraternities; Greek houses are privately owned and chapters answer to national organizations and the Interfraternity Council (IFC) (i.e., a body governing the more than 20 predominantly white fraternities).

Fraternities control every aspect of parties at their houses: themes, music, transportation, admission, access to alcohol, and movement of guests. Party themes usually require women to wear scant, sexy clothing and place women in subordinate positions to men. During our observation period, women attended parties such as "Pimps and Hos," "Victoria's Secret," and "Playboy Mansion"—the last of which required fraternity members to escort two scantily clad dates. Other recent themes included: "CEO Secretary Ho," "School Teacher Sexy Student," and "Golf Pro/Tennis Ho."

Some fraternities require pledges to transport first-year students, primarily women, from the residence halls to the fraternity houses. From about 9 to 11 P.M. on weekend nights early in the year, the drive in front of the residence hall resembled a rowdy taxi-stand, as dressed-to-impress women waited to be car-pooled to parties in expensive late-model vehicles. By allowing party-oriented first-year women to cluster in particular residence halls, the university made them easy to find. One fraternity member told us this practice was referred to as "dorm-storming."

Transportation home was an uncertainty. Women sometimes called cabs, caught the "drunk bus," or trudged home in stilettos. Two women indignantly described a situation where fraternity men "wouldn't give us a ride home." The women said, "Well, let us call a cab." The men discouraged them from calling the cab and eventually found a designated driver. The women described the men as "just dicks" and as "rude."

Fraternities police the door of their parties, allowing in desirable guests (first-year women) and turning away others (unaffiliated men). Women told us of abandoning parties when male friends were not admitted. They explained that fraternity men also controlled the quality and quantity of alcohol. Brothers served themselves first, then personal guests, and then other women. Non-affiliated and unfamiliar men were served last, and generally had access to only the least desirable beverages. The promise of more or better alcohol was often used to lure women into private spaces of the fraternities.

Fraternities are constrained, though, by the necessity of attracting women to their parties. Fraternities with reputations for sexual disrespect have more success recruiting women to parties early in the year. One visit was enough for some of the women. A roommate duo told of a house they "liked at first" until they discovered that the men there were "really not nice."

The Production of Fun and Sexual Assault in Interaction

Peer culture and organizational arrangements set up risky partying conditions, but do not explain *how* student interactions at parties generate sexual assault. At the interactional level we see the mechanisms through which sexual assault is produced. As interactions necessarily involve individuals with particular characteristics and occur in specific organizational settings, all three levels meet when interactions take place. Here, gendered and gender neutral expectations and routines are intricately woven together to create party rape. Party rape is the result of fun situations that shift—either gradually or quite suddenly—into coercive situations. Demonstrating how the production of fun is connected with sexual assault requires describing the interactional routines and expectations that enable men to employ coercive sexual strategies with little risk of consequence.

College partying involves predictable activities in a predictable order (e.g., getting ready, pregaming, getting to the party, getting drunk, flirtation or sexual interaction, getting home, and sharing stories). It is characterized by "shared assumptions about what constitutes good or adequate participation"—what Nina Eliasoph and Paul Lichterman (2003) call "group style" (p. 737). A fun partier throws him or herself into the event, drinks, displays an upbeat mood, and evokes revelry in others. Partiers are expected to like and trust party-mates. Norms of civil interaction curtail displays of unhappiness or tension among party-goers. Michael Schwalbe and associates (2000) observed that groups engage in scripted events of this sort "to bring about an intended emotional result" (p. 438). Drinking assists people in transitioning from everyday life to a state of euphoria.

Cultural expectations of partying are gendered. Women are supposed to wear revealing outfits, while men typically are not. As guests, women cede control of turf, transportation, and liquor. Women are also expected to be grateful for men's hospitality, and as others have noted, to generally be "nice" in ways that men are not. The pressure to be deferential and gracious may be intensified by men's older age and fraternity membership. The quandary for women, however, is that fulfilling the gendered role of partier makes them vulnerable to sexual assault.

Women's vulnerability produces sexual assault only if men exploit it. Too many men are willing to do so. Many college men attend parties looking for casual sex. A student in one of our classes explained that "guys are willing to do damn near anything to get a piece of ass." A male student wrote the following description of parties at his (nonfraternity) house:

> Girls are continually fed drinks of alcohol. It's mainly to party but my roomies are also aware of the inhibition-lowering effects. I've seen an old roomie block doors when girls want to leave his room; and other times I've driven women home who can't remember much of an evening yet sex did occur. Rarely if ever has a night of drinking for my roommate ended without sex. I know it isn't necessarily and assuredly sexual assault, but with the amount of liquor in the house I question the amount of consent a lot.

Another student—after deactivating [his membership]—wrote about a fraternity brother "telling us all at the chapter meeting about how he took this girl home and she was obviously too drunk to function and he took her inside and had sex with her." Getting women drunk, blocking doors, and controlling transportation are common ways men try to prevent women from leaving sexual situations. Rape culture beliefs, such as the belief that men are "naturally" sexually aggressive, normalize these coercive strategies. Assigning women the role of sexual "gate-keeper" relieves men from responsibility for obtaining authentic consent, and enables them to view sex obtained by undermining women's ability to resist it as "consensual" (e.g., by getting women so drunk that they pass out).[2]

In a focus group with her sorority sisters, a junior sorority woman provided an example of a partying situation that devolved into a likely sexual assault.

ANNA: It kind of happened to me freshman year. I'm not positive about what happened, that's the worst part about it. I drank too much at a frat one night, I blacked out and I woke up the next morning with nothing on in their cold dorms, so I don't really know what happened and the guy wasn't in the bed anymore, I don't even think I could tell you who the hell he was, no I couldn't.

SARAH: Did you go to the hospital?

ANNA: No, I didn't know what happened. I was scared and wanted to get the hell out of there. I didn't know who it was, so how am I supposed to go to the hospital and say someone might've raped me? It could have been any one of the hundred guys that lived in the house.

SARAH: It happens to so many people, it would shock you. Three of my best friends in the whole world, people that you like would think it would never happen to, it happened to. It's just so hard because you don't know how to deal with it because you don't want to turn in a frat because all hundred of those brothers …

ANNA: I was also thinking like, you know, I just got to school, I don't want to start off on a bad note with anyone, and now it happened so long ago, it's just one of those things that I kind of have to live with.

This woman's confusion demonstrates the usefulness of alcohol as a weapon: her intoxication undermined her ability to resist sex, her clarity about what happened, and her feelings of entitlement to report it. We collected other narratives in which sexual assault or probable sexual assault occurred when the woman was asleep, comatose, drugged, or otherwise incapacitated.

Amanda, a woman [in] our hall, provides insight into how men take advantage of women's niceness, gender deference, and unequal control of party resources. Amanda reported meeting a "cute" older guy, Mike, also a student, at a local student bar. She explained that, "At the bar we were kind of making out a little bit and I told him just cause I'm sitting here making out doesn't mean that I want to go home with you, you know?" After Amanda found herself stranded by friends with no cell phone or cab fare, Mike promised that a sober friend of his would drive her home. Once they got in the car Mike's friend refused to take her home and instead dropped her at Mike's place. Amanda's concerns were heightened by the driver's disrespect. "He was like, so are you into ménage à trois?" Amanda reported staying awake all night. She woke Mike early in the morning to take her home. Despite her ordeal, she argued that Mike was "a really nice guy" and exchanged telephone numbers with him. These men took advantage of Amanda's unwillingness to make a scene. Amanda was one of the most assertive women on our floor. Indeed, her refusal to participate fully in the culture of feminine niceness led her to suffer in the social hierarchy of the floor and on campus. It is unlikely that other women we observed could have been more assertive in this situation. That she was nice to her captor in the morning suggests how much she wanted him to like her and what she was willing to tolerate in order to keep his interest.[3] This case also shows that it is not only fraternity parties that are dangerous; men can control party resources and work together to constrain women's behavior while partying in bars and at house parties. What distinguishes fraternity parties is that male dominance of partying there is organized, resourced, and implicitly endorsed by the university. Other party venues are also organized in ways that advantage men.

We heard many stories of negative experiences in the party scene, including at least one account of a sexual assault in every focus group that included heterosexual women. Most women who partied complained about men's efforts to control their movements or pressure them to drink. Two of the women on our floor were sexually assaulted at a fraternity party in the first week of school—one was raped. Later in the semester, another woman on the floor was raped by a friend. A fourth woman on the floor suspects she was drugged; she became disoriented at a fraternity party and was very ill for the next week.

Party rape is accomplished without the use of guns, knives, or fists. It is carried out through the combination of low level forms of coercion—a lot of liquor and persuasion, manipulation of situations so that women cannot leave, and sometimes force (e.g., by blocking a door, or using body weight to make it difficult for a woman to get up). These forms of coercion are made more effective by organizational arrangements that provide men with control over how partying happens and by expectations that women let loose and trust their party-mates. This systematic and effective method of extracting nonconsensual sex is largely invisible, which makes it difficult for victims to convince anyone—even themselves—that a crime occurred. Men engage in this behavior with little risk of consequences.

Student Responses and the Resiliency of the Party Scene

The frequency of women's negative experiences in the party scene poses a problem for those students most invested in it. Finding fault with the party scene potentially threatens meaningful identities and lifestyles. The vast majority of heterosexual encounters at parties are fun and consensual. Partying provides a chance to meet new people, experience and display belonging, and to enhance social position. Women on our floor told us that they loved to flirt and be admired, and they displayed pictures on walls, doors, and websites commemorating their fun nights out.

The most common way that students—both women and men—account for the harm that befalls women in the party scene is by blaming victims. By attributing bad experiences to women's "mistakes," students avoid criticizing the party scene or men's behavior within it. Such victim-blaming also allows women to feel that they can control what happens to them. The logic of victim-blaming suggests that sophisticated, smart, careful women are safe from sexual assault. Only "immature," "naïve," or "stupid" women get in trouble. When discussing the sexual assault of a friend, a floor resident explained that:

> She somehow got like sexually assaulted … by one of our friends' old roommates. All I know is that kid was like bad news to start off with. So, I feel sorry for her but it wasn't much of a surprise for us. He's a shady character.

Another floor resident relayed a sympathetic account of a woman raped at knife point by a stranger in the bushes, but later dismissed party rape as nothing to worry about "'cause I'm not stupid when I'm drunk." Even a feminist focus group participant explained that her friend who was raped "made every single mistake and almost all of them had to with alcohol.... She got ridiculed when she came out and said she was raped." These women contrast "true victims" who are deserving of support with "stupid" women who forfeit sympathy (Phillips 2000). Not only is this response devoid of empathy for other women, but it also leads women to blame themselves when they are victimized (Phillips 2000).

Sexual assault prevention strategies can perpetuate victim-blaming. Instructing women to watch their drinks, stay with friends, and limit alcohol consumption implies that it is women's responsibility to avoid "mistakes" and their fault if they fail. Emphasis on the precautions women should take—particularly if not accompanied by education about how men should change their behavior—may also suggest that it is natural for men to drug women and take advantage of them. Additionally, suggesting that women should watch what they drink, trust party-mates, or spend time alone with men asks them to forgo full engagement in the pleasures of the college party scene.

Victim-blaming also serves as a way for women to construct a sense of status within campus erotic hierarchies. As discussed earlier, women and men acquire erotic status based on how "hot" they are perceived to be. Another aspect of erotic status concerns the amount of sexual respect one receives from men. Women can tell themselves that they are safe from sexual assault not only because they are savvy, but also because men will recognize that they, unlike other women, are worthy of sexual respect. For example, a focus group of senior women explained that at a small fraternity gathering their friend Amy came out of the bathroom. She was crying and said that a guy "had her by her neck, holding her up, feeling her up from her crotch up to her neck and saying that I should rape you, you are a fucking whore." The woman's friends were appalled, saying, "no one deserves that." On other hand, they explained that: "Amy flaunts herself. She is a whore so, I mean …" They implied that if one is a whore, one gets treated like one.[4]

Men accord women varying levels of sexual respect, with lower status women seen as "fair game." On campus the youngest and most anonymous women are most vulnerable. High-status women (i.e., girlfriends of fraternity members) may be less likely victims of party rape.[5] Sorority women explained that fraternities discourage members from approaching the girlfriends (and ex-girlfriends) of other men in the house. Partiers on our floor learned that it was safer to party with men they knew as boyfriends, friends, or brothers of friends. One roommate pair partied exclusively at a fraternity where one of the women knew many men from high school. She explained that "we usually don't party with people we don't know that well." Over the course of the year, women on the floor winnowed their party venues to those fraternity houses where they "knew the guys" and could expect to be treated respectfully.

Opting Out While many students find the party scene fun, others are more ambivalent. Some attend a few fraternity parties to feel like they have participated in this college tradition. Others opt out of it altogether. On our floor, 44 out of the 51 first-year students (almost 90 percent) participated in the party scene. Those on the floor who opted out worried about sexual safety and the consequences of engaging in illegal behavior. For example, an interviewee who did not drink was appalled by the fraternity party transport system. She explained that:

> All those girls would stand out there and just like, no joke, get into these big black Suburbans driven by frat guys, wearing like seriously no clothes, piled on top of each other. This could be some kidnapper taking you all away to the woods and chopping you up and leaving you there. How dumb can you be?

In her view, drinking around fraternity men was "scary" rather than "fun."

Her position was unpopular. She, like others who did not party, was an outsider on the floor. Partiers came home loudly in the middle of the night, threw up in the bathrooms, and rollerbladed around the floor. Socially, the others simply did not exist. A few of our "misfits" successfully created social lives outside the floor. The most assertive of the "misfits" figured out the dynamics of the floor in the first weeks and transferred to other residence halls.

However, most students on our floor lacked the identities or network connections necessary for entry into alternative worlds. Life on a large university campus can be overwhelming for first-year students. Those who most needed an alternative to the social world of the party dorm were often ill-equipped to actively seek it out. They either integrated themselves into partying or found themselves alone in their rooms, microwaving frozen dinners and watching television. A Christian focus group participant described life in this residence hall: "When everyone is going out on a Thursday and you are in the room by yourself and there are only two or three other people on the floor, that's not fun, it's not the college life that you want."...

DISCUSSION AND IMPLICATIONS

Individual characteristics and institutional practices provide the actors and contexts in which interactional processes occur. We have to turn to the interactional level, however, to understand *how* sexual assault is generated. Gender neutral expectations to "have fun," lose control, and trust one's party-mates become problematic when combined with gendered interactional expectations. Women are expected to be "nice" and to defer to men in interaction. This expectation is intensified by men's position as hosts and women's as grateful guests. The heterosexual script, which directs men to pursue sex and women to play the role of gatekeeper, further disadvantages women, particularly when virtually *all* men's methods of extracting sex are defined as legitimate.

The mechanisms identified should help explain intra-campus, cross-campus, and overtime variation in the prevalence of sexual assault. Campuses with similar students and social organization are predicted to have similar rates of sexual assault. We would expect to see lower rates of sexual assault on campuses characterized by more aesthetically appealing public space, lower alcohol use, and the absence of a gender-adversarial party scene. Campuses with more racial diversity and more racial integration would also be expected to have lower rates of sexual assault because of the dilution of upper-middle class white peer groups. Researchers are beginning to conduct comparative research on the impact of university organization on aggregate rates of sexual assault. For example, Meichun Mohler-Kuo and associates (2004) found that women who attended schools with medium or high levels of heavy episodic drinking were more at risk of being raped while intoxicated than women attending other schools, even while controlling for individual-level characteristics. More comparative research is needed.

This perspective may also help explain why white college women are at higher risk of sexual assault than other racial groups. Existing research suggests that African American college social scenes are more gender egalitarian (Stombler and Padavic 1997). African American fraternities typically do not have houses, depriving men of a party resource. The missions, goals, and recruitment practices of African American fraternities and sororities discourage joining for exclusively social reasons (Berkowitz and Padavic 1999), and rates of alcohol consumption are lower among African American students (Journal of Blacks in Higher Education 2000; Weschsler and Kuo 2003). The role of party rape in the lives of white college women is substantiated by recent research that found that "white women were more likely [than nonwhite women] to have experienced rape while intoxicated and less likely to experience other rape" (Mohler-Kuo et al. 2004: 41). White women's overall higher rates of rape are accounted for by their high rates of rape while intoxicated. Studies of racial differences in the culture and organization of college partying and its consequences for sexual assault are needed.

Our analysis also provides a framework for analyzing the sources of sexual risk in nonuniversity partying situations. Situations where men have a home turf advantage, know each other better than the women present know each other, see the women as anonymous, and control desired resources (such as alcohol or drugs) are likely to be particularly dangerous. Social pressures to "have fun," prove one's social competency, or adhere to traditional gender expectations are also predicted to increase rates of sexual assault within a social scene.

This research has implications for policy. The interdependence of levels means that it is difficult to enact change at one level when the other levels remain unchanged. Programs to combat sexual assault currently focus primarily or even exclusively on education. But as Ann Swidler (2001) argued, culture develops in response to institutional arrangements. Without change in institutional arrangements, efforts to change cultural beliefs are undermined by the cultural commonsense generated by encounters with institutions. Efforts to educate about sexual assault will not succeed if the university continues to support organizational arrangements that facilitate and even legitimate men's coercive sexual strategies. Thus, our research implies that efforts to combat sexual assault on campus should target all levels, constituencies, and processes simultaneously. Efforts to educate both men and women should indeed be intensified, but they should be reinforced by changes in the social organization of student life.

Researchers focused on problem drinking on campus have found that reduction efforts focused on the social environment are successful (Berkowitz 2003: 21). Student body diversity has been found to decrease binge drinking on campus (Weschsler and Kuo 2003); it might also reduce rates of sexual assault. Existing student heterogeneity can be exploited by eliminating self-selection into age-segregated, white, upper-middle class, heterosexual enclaves and by working to make residence halls more appealing to upper-division students. Building more aesthetically appealing housing might allow students to interact outside of alcohol-fueled party scenes. Less expensive plans might involve creating more living-learning communities, coffee shops, and other student-run community spaces.

While heavy alcohol use is associated with sexual assault, not all efforts to regulate student alcohol use contribute to sexual safety. Punitive approaches sometimes heighten the symbolic significance of drinking, lead students to drink more hard liquor, and push alcohol consumption to more private and thus more dangerous spaces. Regulation inconsistently applied—that is, heavy policing of residence halls and light policing of fraternities—increases the power of those who can secure alcohol and host parties. More consistent regulation could decrease the value of alcohol as a commodity by equalizing access to it.

Sexual assault education should shift in emphasis from educating women on preventative measures to educating both men and women about the coercive behavior of men and the sources of victim-blaming. Mohler-Kuo and associates (2004) suggest, and we endorse, a focus on the role of alcohol in sexual assault. Education should begin before students arrive on campus and [should] continue throughout college. It may also be most effective if high-status peers are involved in disseminating knowledge and experience to younger college students....

NOTES

1. In ongoing research on college men and sexuality, Sweeney (2004) and Rosow and Ray (2006) have found wide variation in beliefs about acceptable ways to obtain sex even among men who belong to the same fraternities. Rosow and Ray found that fraternity men in the most elite

houses view sex with intoxicated women as low status and claim to avoid it.

2. On party rape as a distinct type of sexual assault, see also Ward and associates (1991). Ehrhart and Sandler (1987) use the term to refer to group rape. We use the term to refer to one-on-one assaults. We encountered no reports of group sexual assault.

3. Holland and Eisenhart (1990) and Stombler (1994) found that male attention is of such high value to some women that they are willing to suffer indignities to receive it.

4. Schwalbe and associates (2000) suggest that there are several psychological mechanisms that explain this behavior. Trading power for patronage occurs when a subordinate group accepts [its] status in exchange for compensatory benefits from the dominant group. Defensive ["]othering["] is a process by which some members of a subordinated group seek to maintain status by deflecting stigma [on]to others. Maneuvering to protect or improve [one's] individual position within hierarchical classification systems is common; however, these responses support the subordination that makes them necessary.

5. While "knowing" one's male partymates may offer some protection, this protection is not comprehensive. Sorority women, who typically have the closest ties with fraternity men, experience more sexual assault than other college women (Mohler-Kuo et al. 2004). Not only do sorority women typically spend more time in high-risk social situations than other women, but also arriving at a high-status position on campus may require [undergraduates] to begin their college social career as one of the anonymous young women who are frequently victimized.

REFERENCES

Abbey, Antonia, Pam McAuslan, Tina Zawacki, A. Monique Clinton, and Philip Buck. 2001. "Attitudinal, Experiential, and Situational Predictors of Sexual Assault Perpetration." *Journal of Interpersonal Violence* 16: 784–807.

Abbey, Antonia, Lisa Thomson Ross, Donna McDuffie, and Pam McAuslan. 1996. "Alcohol and Dating Risk Factors for Sexual Assault among College Women." *Psychology of Women Quarterly* 20: 147–69.

Adams-Curtis, Leah, and Gordon Forbes. 2004. "College Women's Experiences of Sexual Coercion: A Review of Cultural, Perpetrator, Victim, and Situational Variables." *Trauma, Violence, and Abuse: A Review Journal* 5: 91–122.

Bachar, Karen, and Mary Koss. 2001. "From Prevalence to Prevention: Closing the Gap between What We Know about Rape and What We Do." pp. 117–42 in *Sourcebook on Violence against Women*, edited by C. Renzetti, J. Edleson, and R. K. Bergen. Thousand Oaks, CA: Sage.

Berkowitz, Alan. 2003. "How Should We Talk about Student Drinking—And What Should We Do about It?" *About Campus* May/June: 16–22.

Berkowitz, Alexandra, and Irene Padavic. 1999. "Getting a Man or Getting Ahead: A Comparison of White and Black Sororities." *Journal of Contemporary Ethnography* 27: 530–57.

Boswell, A. Ayres, and Joan Z. Spade. 1996. "Fraternities and Collegiate Rape Culture: Why Are Some Fraternities More Dangerous Places for Women?" *Gender & Society* 10: 133–47.

Eder, Donna. 1985. "The Cycle of Popularity: Inter-personal Relations among Female Adolescents." *Sociology of Education* 58: 154–65.

Eder, Donna, Catherine Evans, and Stephen Parker. 1995. *School Talk: Gender and Adolescent Culture.* New Brunswick, NJ: Rutgers University Press.

Ehrhart, Julie, and Bernice Sandler. 1987. "Party Rape." *Response* 9: 205.

Eliasoph, Nina, and Paul Lichterman. 2003. "Culture in Interaction." *American Journal of Sociology* 108: 735–94.

Fisher, Bonnie, Francis Cullen, and Michael Turner. 2000. "The Sexual Victimization of College Women." Washington, DC: National Institute of Justice and the Bureau of Justice Statistics.

Holland, Dorothy, and Margaret Eisenhart. 1990. *Educated in Romance: Women, Achievement, and College Culture.* Chicago: University of Chicago Press.

Humphrey, John, and Jacquelyn White. 2000. "Women's Vulnerability to Sexual Assault from Adolescence to Young Adulthood." *Journal of Adolescent Health* 27: 419–24.

Journal of Blacks in Higher Education. 2000. "News and Views: Alcohol Abuse Remains High on College Campus, but Black Students Drink to Excess Far Less Often Than Whites." *The Journal of Blacks in Higher Education.* 28: 19–20.

Kalof Linda. 2000. "Vulnerability to Sexual Coercion among College Women: A Longitudinal Study." *Gender Issues* 18: 47–58.

Lorber, Judith. 1994. *Paradoxes of Gender.* New Haven, CT: Yale University Press.

Marine, Susan. 2004. "Waking Up from the Nightmare of Rape." *The Chronicle of Higher Education.* November 26, p. B5.

Martin, Patricia Yancey. 2004. "Gender as a Social Institution." *Social Forces* 82: 1249–73.

Martin, Patricia Yancey, and Robert A. Hummer. 1989. "Fraternities and Rape on Campus." *Gender & Society* 3: 457–73.

Merton, Robert. 1957. *Social Theory and Social Structure.* New York: Free Press.

Milner, Murray. 2004. *Freaks, Geeks, and Cool Kids: American Teenagers, Schools, and the Culture of Consumption.* New York: Routledge.

Mohler-Kuo, Meichun, George W. Dowdall, Mary P. Koss, and Henry Weschler. 2004. "Correlates of Rape While Intoxicated in a National Sample of College Women." *Journal of Studies on Alcohol* 65: 37–45.

Phillips, Lynn. 2000. *Flirting with Danger: Young Women's Reflections on Sexuality and Domination.* New York: New York University.

Risman, Barbara. 1998. *Gender Vertigo: American Families in Transition.* New Haven, CT: Yale University Press.

—————. 2004. "Gender as a Social Structure: Theory Wrestling with Activism." *Gender & Society* 18: 429–50.

Rosow, Jason, and Rashawn Ray. 2006. "Getting Off and Showing Off: The Romantic and Sexual Lives of High Status Black and White Status Men." *Department of Sociology*, Indiana University, Bloomington, IN. Unpublished manuscript.

Sampson, Rana. 2002. "Acquaintance Rape of College Students." *Problem-Oriented Guides for Police Series, No. 17.* Washington, DC: U.S. Department of Justice, Office of Community Oriented Policing Services.

Sanday, Peggy. 1990. *Fraternity Gang Rape: Sex, Brotherhood, and Privilege on Campus.* New York: New York University Press.

Schwalbe, Michael, Sandra Godwin, Daphne Holden, Douglas Schrock, Shealy Thompson, and Michele Wolkomir. 2000. "Generic Processes in the Reproduction of Inequality: An Interactionist Analysis." *Social Forces* 79: 419–52.

Stombler, Mindy. 1994. "'Buddies' or 'Slutties': The Collective Reputation of Fraternity Little Sisters." *Gender & Society* 8: 297–323.

Stombler, Mindy, and Patricia Yancey Martin. 1994. "Bringing Women In, Keeping Women Down: Fraternity 'Little Sister' Organizations." *Journal of Contemporary Ethnography* 23: 150–84.

Stombler, Mindy, and Irene Padavic. 1997. "Sister Acts: Resisting Men's Domination in Black and White Fraternity Little Sister Programs." *Social Problems* 44: 257–75.

Sweeney, Brian. 2004. "Good Guy on Campus: Gender, Peer Groups, and Sexuality among College Men." Presented at the American Sociological Association Annual Meetings, August 17, Philadelphia, PA.

Swidler, Ann. 2001. *Talk of Love: How Culture Matters*. Chicago: University of Chicago Press.

Ward, Sally, Kathy Chapman, Ellen Cohn, Susan White, and Kirk Williams. 1991. "Acquaintance Rape and The College Social Scene." *Family Relations* 40: 65–71.

Weschsler, Henry, and Meichun Kuo. 2003. "Watering Down the Drinks: The Moderating Effect of College Demographics on Alcohol Use of High-Risk Groups." *American Journal of Public Health* 93: 1929–33.

44

Opportunity Structures
for White-Collar Crime

OSKAR ENGDAHL

Returning once again to the crimes of the powerful, we see a smaller scale example of fraud and its relation to American culture than the BP Deepwater Horizon disaster. Our society has been rocked over the last decade by the unraveling of large numbers of white-collar crimes. New instances of fraud are revealed every month, with devastating financial consequences for millions. Not just the domain of the rich and well connected, fraud is alive and flourishing in our capitalist system from top to bottom.

Engdahl illustrates his analysis with the fraud perpetrated by a stockbroker who flew past the reach of regulators, buoyed beyond suspicion by a history of high financial returns. The author shows the way individuals with little expertise in the inner workings of the complex system of derivatives and other financial instruments put their trust in people to manage their hard-earned money. Firms with the responsibility for overseeing brokers' fiduciary responsibility gave these individuals enormous flexibility, as long as they generated profits for their organization. Regulators were likewise blinded, by their close ties to the financial industry and their political belief in the capitalist system. All parties to these transactions believed that the smart, well-connected segment of the population deserved to prosper at the expense of lesser folk. Engdahl depicts the loss of control that subsequently arose in the financial world.

How does this story compare with the events that led to the financial crash of 2008 and beyond? How does it compare with the giant Ponzi scheme perpetrated by Bernard Madoff or the real estate bubble that was buoyed by loose loans and their repackaging into complex packages of derivatives? Whose responsibility is it, or should it be, to regulate such behavior? Why has such regulation weakened, and why has it proved so difficult to reinstate? Has the public become protected from these kinds of financial frauds to any extent, and how so? What are the structural causes and motivations of this type of crime, compared with street crime? How is white-collar crime viewed as similar to or different from street crime? How is it viewed as similar to or different from infractions of social values, such as abortion or gay marriage?

Possibility or "opportunity" is a central component in many explanatory models for criminality. In research on white-collar crime, this [theme] emerges especially through an emphasis on the importance of social position for the possibility of carrying out crime. The emphasis has been a key aspect of economic-crime research since Sutherland, using the concept of white-collar crime, began to study criminality among upper social classes, which in his view was made possible by "positions in power" (Sutherland 1983: 7). By white-collar crime Sutherland meant [crime perpetrated] primarily [by] people who held positions in the sense that they were regarded as "respectable" and had "high social status." He connected such positions with the holders' "professional practice." Criminality then had to do with the fact that the position conveyed by the profession was misused in order to commit crime.

Sutherland's conceptualization has had great influence on how crime in the world of companies and businesses is understood and investigated in criminology and related disciplines. The discussion of which component is most central among those that Sutherland included in his concept has been extensive, as have attempts to refine the concept. Of the better-known notions that have defined the orientation of this research field, the most important are considered to be "corporate crime" and "occupational crime," which place the focus on crimes that occur through and on behalf of companies, or are committed by virtue of the position that accrues to a person in his or her professional role. Common to the diverse choices of concept is, however, a tendency to see it as essential to contextualize people and their activities by setting them in some form of social position with respect to each other. Equally often, the aim is seemingly to clarify that it is the *position* in the profession or company that create *opportunities* for crime. This [viewpoint] is plainly expressed by Vaughan (1983: 85) when, in a study of corporate misconduct, she says that "[p]osition is a key variable for understanding the response of members to the tensions organizations experience when competing for scarce resources. The skills used to commit a violation are those associated with a particular position in the organization."

In sharp contrast to the relatively great efforts made with other aspects of white-collar crime research, much less attention than might be expected has been devoted to explaining *how* social positions create opportunities for committing crime .... My approach to these issues will maintain the following:

1. The more or less explicit ideas advanced by research in the area can be systematized and classified according to different types of resources with whose help a position makes crime possible. These resources are understood in terms of access to authority, social contact networks, and technical–administrative systems.

2. Goffman's concepts of "barriers" and "back regions" can be used to clarify what it is in a social position that creates opportunities for crime, because [those concepts] have to do with how a concealed space for maneuvering is established by preventing others' insight ....

BARRIERS AS OPPORTUNITY STRUCTURES FOR WHITE-COLLAR CRIME: AN EMPIRICAL EXAMPLE

Background

In order to clarify both the theoretical content and the empirical relevance of the barrier concept, the argument that follows is illustrated with a case of white-collar crime. It involves a broker in a well-reputed brokerage firm who was found guilty of breach of trust (SNECB 2000; SDC 2001; SAC 2002; SC 2003). The case is treated as a typical one for strategic reasons so as to demonstrate and stress essential characteristics. At the core of the case is a losing deal that the broker perpetrated after having worked in the firm for a couple of years. The broker conducted a large number of deals on behalf of clients, among them the firm's most important private client. Both the client and the broker enjoyed high status at the firm. On a certain occasion, one of the broker's deals for this client did not proceed as planned, resulting in a loss corresponding to 1 million USD. The broker tried to hide the bad deal from the client and employees at the firm. He did so temporarily by transferring the deal to one of the firm's accounts for ongoing and unfinished deals, and permanently by conducting other deals secretly—through accounts belonging to clients or the firm—in order to generate profits that could cover the loss. The broker succeeded in concealing the loss for four years, but his secret deals failed to create profits that covered the initial loss. The total loss grew to 21 million USD by the time the irregularities were brought to light.

During those four years, no suspicions of the irregularities arose. The fact that the broker could hide extensive losses and deals for so long without being suspected can be explained: his relations with the client and the firm's other employees, chiefs, and accounts were permeated with barriers that obstructed suspicion and detection of crime. These [barriers] are referred to in the following analysis as financial self-interest, low priority of control, and interpretative primacy. [Their] great importance …[is] central to understanding the broker's case; it was not until the broker turned himself in that his irregularities came into light. During the four years of irregularities he had—at least [on] a couple of occasions—made so many good deals that he had been close to erasing his initial loss. But he never succeeded fully in erasing the loss, and when he later—once again—started making big losses he, in his own words, "felt that I was on the verge of what I could endure" (SNECB 2000: 88). During the four years he had, for periods of time, suffered from psychological stress, and the day following his confession he was hospitalized for a heart condition.

Financial Self-Interest

The relations between the broker and the client involved informal and oral agreements in person; no agreements were codified in writing. At first, the broker and client met "over a cup of coffee" to examine the deals, but this did not lead to any special basic orders. Instead, the client gave the broker general instructions of the type "do the deals you think are good" (SNECB 2000: 60, 211, 243; SDC 2001: 13f.). The client felt sure that the broker and his way of working were in line with reasonable business strategies. Moreover, the client believed that it was in his own interest to exploit the broker's knowledge, which he did by committing funds to the broker's care. The client not only regarded the broker as knowledgeable quite generally, but spoke of him as the country's "derivative king" (SNECB 2000: 62). It should be added that the client considered the broker to be very reliable. The client claimed to have wide experience of brokers who attempted to enrich themselves on others' deals, and according to him it was therefore normally necessary to constantly oversee brokers who were retained and to check their deals. However, he did not think such control necessary in the present case because he perceived this broker to be highly reliable. Hence, the client handed over the entire responsibility for all deals on one of his accounts to the broker, and handled the related papers himself only sporadically. The papers went "straight into the notebook" with no thorough control. The client described the broker as "the most serious person you can imagine" and commented further that "It's like I can rely on my own father" (SNECB 2000: 62). The broker was well aware that the client [neither] interfere[d] with whatever positions were taken on the account [nor] control[led] it afterward: "The only thing he looks at now and then is when he gets an account statement once a month and sees that the balance is more or less okay" (SNECB 2000: 177).

Thus, the client considered himself to have good knowledge of and trust in the broker. The business relationship functioned "extremely well" according to him, and was gradually complemented by the fact that, according to them both, they "got close to each other" (SNECB 2000: 62). This [closeness] was manifested by the broker being one of those who accompanied the client on trips, and by the client being best man at the broker's wedding. The client's perception of good knowledge about and trust in the broker contributed to his letting the broker manage his money. It should be noted, however, that good knowledge of how a broker functions can nonetheless include knowledge that leads one to refrain from committing assets to management by others. This knowledge, therefore, is not in itself decisive for the realization of an assignment of management. It is rather the perception that such commitment is in one's self-interest that is central for doing so. In the present case, the client judged the broker's skill, reliability, and loyalty to fulfill the requirements. His judgment was also confirmed in a manner clear to him: during the period of handling his deals, the broker was able to take out tens of million USD in profits.

From the broker's position, financial self-interest acted in two respects: besides the client's interest in letting the broker manage parts of his assets, the firm also showed this interest. The broker was known to be highly trusted among other employees in the firm, and quickly became operative chief of the options department. Initially this happened because he was one of the few who were acquainted with the business area of options, an area in which the firm wanted to expand its activities. The trust in him also increased when he conducted deals that satisfied clients, generated courtage for the firm, and built up the firm's stock trade with good results. Moreover, it was to him that other brokers in the firm left deals, clients, or problems of [a] relatively complicated nature. Not least due to these aspects, it lay in the firm's interest that he should take care of the present client. This client was very wealthy and constituted by far the biggest private client of the firm. For him, the firm managed large assets and carried out numerous deals that yielded sizable courtage income for it. The client was regarded as a "pro," strongly interested in "generating profits," and "demanding" (SNECB 2000: 60, 22, 243). By virtue of his position, the client had good insight into the firm as a whole and socialized with several of its employees. He helped the firm in various ways when it got into difficulties, and was also a landlord and something of a "royal purveyor" of residences for a number of the firm's employees. Thus, he was a key client in diverse respects. At the firm it was well known, too, that the client had great trust in the broker concerned, who was regarded as "smart at handling demanding clients" (SNECB 2000: 243). It therefore lay in the firm's interest that he took care of this client.

Low Priority of Control (or Preoccupation with Other Matters)

Another aspect that is central for understanding the broker's opportunity structure has its basis in the fact that the client was continually involved in many extensive deals, which he had no possibility of conducting and controlling by himself. The client was also more interested in and preoccupied with business expansion, than with administration and control of his own assets. Thus, he left it to the broker to carry out deals in certain areas and, as mentioned earlier, he was sparing with examination of the account statements and other papers that were sent to him. Satisfied with checking whether the account balance was plus or minus, he limited his information to the latest daily telephone talk that he had with the broker, or when he visited the firm. The client's neglect to control the conduct and administration of the deals was due partly to his scarce time for controlling all the deals, but mainly to his giving priority to the planning and initiation of deals. He himself thought that he was "almost unhealthily interested in deals," yet also that he found it "nice to avoid the constant supervision of deals" (SNECB 2000: 61, 38f.).

Possibly one can understand the client's actions in terms of ignorance, lack of interest, or the often-used term "absence of capable guardians" (Felson and Clarke 1998). However, in cases that can be said to concern absence of capable guardians, there is a strong tendency to point out primarily what does *not* happen, at the same time as one overlooks the relational components that allow this lack of control to arise. In the present case, the reason for neglect of control was that the client prioritized other matters than administration and control—a prioritization that created a barrier. It should thus be most plausible to conceptualize this aspect in terms of low priority of control, or perhaps of preoccupation with other matters. Here lies an explanation for why the control is deficient and creates opportunities for crime. In short, my view is that the problem does not essentially involve an "absence" of anything, but an activity's orientation toward something else so strongly that the control function is given low priority. It is the strength of this orientation that creates barriers that ultimately protect criminality, because people do not see what is happening "alongside" what they are preoccupied with. In the present brokerage firm, this barrier included a large number of people around the broker.

Just as the client was busy with other projects and therefore did not control his deals to any great extent, the firm was busy expanding in the years when the broker worked there. No great interest existed in administration and control. The firm made clear to the broker that he was employed to make deals and be income-generating. For this purpose he received responsibility and leeway for action as large as the concern was small in administration and control of him. He never got training or an introduction to relevant regulations in the area. The firm's attitude was expressed perhaps most plainly when he worked on new blank forms for options trade and one of his chiefs remarked, "Don't worry about the administrative stuff, you're employed to make deals" (SDC 2001: 25). As a result, the broker was able to create possibilities of concealing the bad deal. The clearest example was when he used the firm's "question-mark account." The broker knew that control of this account was deficient. Considering also that the firm had limited control over the security requirements when issuing new options, the broker began systematically to use the question-mark account for re-booking the loss so that nobody would notice it. He also used the account for conducting secret deals. Normally, options issued and bought by the firm were supposed to undergo an assessment so that the firm knew [that] the parties could fulfill their promises when the options expired. But this [assessment] was not done systematically and automatically—it required an agreement to be written. Only when this had been done were deals examined by the firm's lawyers. Therefore, the broker began to construct options by himself. These might be worthless, but because the broker wrote no agreement that he gave to the legal department, he avoided control. The constructed options were purchased by him for a small sum and placed on the client's account. Shortly afterward, he saw to it that the options were bought by the firm's question-mark account for a much higher price. The loss hole in the client's account was thereby covered. Obviously the firm's account had paid an over-price that would later be realized at a loss—but as long as the option had not expired, the loss was only potential and did not appear in the book-keeping as such for the uninitiated. The broker bought and sold options in such a way that the time it took for a loss to be realized was constantly "rolled" into the future. Worthless options passed to the firm, which neither had control over its accounts nor could judge the options' value. This [situation] brings us to another type of barrier.

Interpretative Primacy

That the type of deals that the broker conducted were notoriously complicated, and that his own knowledge in the area was matched by ignorance from clients, colleagues, chiefs, and other people related to his work, were further aspects that created opportunities for the broker to commit crime without being detected. Here it is less important that he was indisputably adept in the area, or that the deals he made were regarded as extremely complicated. It is not mainly in environments with such conditions that one should look to find opportunities for crime. Instead one should look in environments where the differences in knowledge give certain actors an *interpretative primacy* over others. In the broker's case, the knowledge gap meant that in practice he acquired an interpretative primacy regarding the content determination and evaluation of various transactions, accounts, and transfers. The broker also realized that the deals he made were hard to understand and check the contents of. This [difficulty] applied to his colleagues at the firm as well as to accounts and clients. Besides being aware that the client did not always check his deals very carefully, the broker thought that many of the deals were so complex that the client in question "never really understood what it was about" (SNECB 2000: 5) when he executed transactions on the client's account. The client knew this too, and thought that when it came to option deals a lot was "so complicated that I had no idea what was going in and out, shifting here and there...." (SNECB 2000: 62). Further, the broker had this advantage in relation to the firm's accountants:

> It's some women who have audited ... they have often sat and asked me about things they don't understand and then I've answered them about it. Certainly they haven't asked about [the account I myself managed for the client] ... as to whether this account is unbalanced or whether it has an actual loss. That has also been very hard to detect. I don't think their competence is up to seeing it. (SNECB 2000: 179)

SUMMARY

What have been designated here as financial self-interest, low priority of control, and interpretative primacy are three examples of barriers that obstruct other people (such as clients, colleagues, chiefs, etc.) from suspecting and discovering criminal events. Concretely they meant that the broker's relations with these other people were characterized *by* his being considered so able in the area that it was in their self-interest that he handled their deals; *by* them being more preoccupied with completing deals and expanding than with administration and control; and *by* the broker having an interpretative primacy in the area that led them to rely on his statements. These conditions functioned as barriers and allowed opportunities to accrue for his commission of acts whose real content was concealed, explaining why he could perform his hidden rolling program and keep it going for four years.

CONCLUSIONS

In this article, Goffman's (1959) concepts of "barriers" and "back regions" have been used to show more precisely how social positions strengthen the possibility of carrying out economic crime. The fundamental idea is that opportunities for crime arise in social situations where individuals, by virtue of their positions, build up barriers that hinder others from controlling a course of events. [These opportunities] can happen firstly through [those individuals] obstructing suspicion and detection, secondly through their impeding investigation and—once suspicion is established—assessment of an event as a crime, and thirdly through their preventing legal action and implementation of sanctions. With an empirical case study, the reasoning has been deepened as regards the type of barrier that creates opportunities for crime by hindering suspicion and detection. Here it was shown how three variants of this type of barrier—termed financial self-interest, low priority of control, and interpretative primacy—enabled a broker, by virtue of his position, to gather opportunities for operating without insight in back regions and committing crime.

The barrier concept offers, I think, a more viable explanation for how opportunities of white-collar crime are created than what has earlier been proposed in the field. It is undoubtedly true that the resources emphasized earlier, which in summary are access to authority, social contact networks, and technical–administrative systems, create possibilities for committing crime. As regards the previous explanatory models, however, the issue is rather of resources that improve the opportunity for realizing actions *in general*, not the opportunity of committing *criminal* acts. Through the concepts of barriers and back regions, the possibility of crime becomes clearer. These concepts should therefore, in my opinion, be incorporated and refined in the research on white-collar crime. They have good prospects of serving as an analytical apparatus that heightens the sensitivity to what occurs, or can occur, in certain environments and contributes to knowledge of decisive factors....

REFERENCES

Felson, M. and R. V. Clarke. 1998. "*Opportunity Makes the Thief*"—*Crime Detection and Prevention Series*. London, UK: Home Office.

Goffman, E. 1959. *The Presentation of Self in Everyday Life*. New York: Doubleday.

Sutherland, E. 1983. *White Collar Crime: The Uncut Version*. New Haven, CT: Yale University Press.

Vaughan, D. 1983. *Controlling Unlawful Organizational Behavior: Social Structure and Corporate Misconduct*. Chicago: The University of Chicago Press.

CASE MATERIAL CONSULTED

SAC (Svea Appeals Court). 2002. Decision of Svea Appeals Court 26 April 2002 in case B 5803-01.

SDC (Stockholm District Court). 2001. Decision of Stockholm District Court 12 July 2001 in case B 10030-94.

SC (Supreme Court). 2003. Decision of Supreme Court 14 October 2003 in case B 2100 02.

SNECB (Swedish National Economic Crime Bureau). 2000. Preliminary Investigative protocol in case B 10030-94. Stockholm: Swedish National Economic Crime Bureau, East Department.

PART VIII

Deviant Careers

One of the fascinating things about people's involvement in deviance is that it evolves, yielding a shifting and changing experience. Doing something for the first time is very different from doing it for the hundredth time. It is fruitful, then, to consider involvement in deviance from a career perspective, to see what the nature of deviance is and how it develops over the course of people's involvement with it. Sociologists have documented various stages of participation in such things as drug use, drug dealing, fencing, carrying out a professional hit, engaging in prostitution, and shoplifting. Although these activities are different in character, they have structural similarities in the way people experience them according to the stage of involvement in them. In fact, the career analogy has been applied fruitfully to the study of deviance because people go through the same cycle of entry, upward mobility, achieving career peaks, aging in the career, burning out, and getting out of deviance as they do in legitimate work.

Six themes tend to be most commonly addressed in the literature on deviant careers. **Entering deviance** attracts the greatest bulk of scholarly attention, for two reasons: Policy makers have great interest in finding out how and why people enter deviance so that they can prevent them from doing so, and entry represents fairly easy data for most scholars to gather, because every individual deviant or group of deviants can tell the story of how they got into the scene. Over the last couple of decades, sociologists have worked to discover and disseminate information that has been adopted by the public in making decisions aimed at influencing and deterring potential deviants. First, they have developed the concept of at-risk populations and articulated a range of risk factors associated with various forms of deviance, such as gang membership, dropping out of school, unwed pregnancy, suicide, depression, eating disorders, detainment, arrest, and incarceration (Loeber and Farrington, 1998; Werner and Smith, 1992).

Complementing these risk factors, sociologists have identified protective factors that, in the face of exposure to multiple risks, help prevent individuals from becoming involved in deviance (Werner and Smith, 1992).

Although some people venture into deviance on their own, the vast majority do it with the encouragement and assistance of others, often joining cooperative deviant enterprises. The turning points that mark significant phases in their transitions have been explored, as have their changing self-identities. Most commonly, people who become involved in deviance do so through a process of shifting their circle of friends. They drift into new peer groups as they are drifting into deviance, or their peer group as a whole drifts into deviance as the members enter a new phase of the life cycle.

Second, the career perspective incorporates an interest in the **training and socialization** of new deviants. Relatively little has been written about this area, for several reasons. One is that, although most deviants might be socialized to the norms and values of their activity through their contacts with fellow deviants, they get relatively little explicit training in how to perform their deviance, how to avoid detection, how to deal with the police, and other important concerns. Another reason is that real training generally occurs when deviants are working together, side by side, as a team in their enterprise. That leaves only crews and deviant formal organizations as the likely sources of training, and these forms of deviant organization are relatively rare. In their analysis of the career of a professional burglar and fence, Steffensmeier and Ulmer (2005) have suggested that the kind of professional criminals engaged in crew operations has diminished in number.

Third, focusing on careers in deviance enables people to study how individuals' involvement in their deviant worlds and with their deviant associates and activities **change over time**. This type of processual analysis represents a highly nuanced understanding of deviance that cannot easily be captured by the frozen-in-time snapshot of most survey research. Longitudinal studies of deviant careers are rare, but valuable, assets in the literature, as they distinguish for us some of the different motivations, rewards, conflicts, and problems that deviants encounter over the course of their participation in deviance. Because they illuminate motivations and deterrents which are more or less effective at different career stages, these studies are enormously helpful, not only to policy makers, but also to people who struggle to understand deviance and to help themselves, friends, and family members who are caught in its lure.

Over the course of their deviant careers, participants must face the various challenges involved in managing their deviance. They must navigate the changing dimensions of available opportunities to commit their deviance, evolving technologies that can catch them in committing their acts of deviance, their relationships within deviant communities, and their safety from agents of social control. They must also evolve a personal style for their deviance. They must balance their deviance with the nondeviant aspects of their lives, such as their relationships with family members, people in the community, and those on whom they rely to meet their legitimate needs. Yip's (1997) work on gay male Christian couples, for example, illustrates some of the creative ways that homosexuals find to maintain their relational commitment in a social environment in which their union is sanctioned by neither church nor state.

Wanting out, or **exiting deviance**, represents the fourth major area in this literature. As with entering deviance, there is a high political interest in the topic, with policy makers looking for ways to induce people to quit their deviance. Information on longer-term deviants, their attitudes toward the scene and the people in it, their satisfactions and dissatisfactions, and their hopes or dreams for the future is somewhat hard to get. People tend to feel most comfortable talking to others like themselves, and the most active researchers are young. Yet there are valuable studies of aging deviants.

A number of factors "push" people out of the deviant life and "pull" them back into the conventional world. People are pushed out by factors intrinsic to the deviant experience and lifestyle. They may burn out from the hours, the stress, the transience, and the drug use. Friends or associates who get arrested, jailed, injured, or killed may make them rethink their continuing involvement with deviance. Moreover, each time they get arrested, they face an increasing likelihood of doing a longer jail sentence. People who have spent some years in prison know that their next arrest is likely to lead to a more serious stint in prison. Loath to return, they may look for other things they can do to make a living. The longer people stay in deviance, the greater the likelihood is that there will be a change in the nature of the experience. What initially seemed daring and glamorous eventually becomes mundane, and the excitement turns to paranoia. People change during their involvement with deviance as well.

Pull factors are located outside of the deviant arena and entice people to leave that world behind and return to conventionality. Individuals such as friends, girlfriends, spouses, children, and other family members may encourage or intervene with deviants to entice or pressure them to quit their deviant ways. Legitimate recreational and occupational interests are key in helping to facilitate individuals' transition out of deviance. Yet returning to legitimate jobs in which their earning potential is reduced may involve restricting their spending patterns, something that people find difficult. They also become accustomed to the freewheeling lifestyle and open value system associated with a deviant community, in which conventional norms are disdained.

Reentering the straight world with its morality may chafe. Finding legitimate work may be difficult, especially if the participants were involved in occupational deviance, making money through illicit means. Former deviants often have difficulty putting together a résumé that accounts for their gap in legitimate employment and finding someone who will hire them. They may find adhering to the structure of the 9-to-5 straight world overly constraining. Yet most people do not want to spend their whole lives engaged in deviance.

Very little information is available on the **postdeviant** features of individuals' lives. These are the hardest data to get because, as we described in Part V, just as developing a deviant identity moves people out of their conventional friendships and social worlds into those populated by deviants, quitting deviance usually requires exiting from these same relationships and scenes. Once people decide to get out and actually make that move, they disperse and leave no forwarding address.

Finally, there is a literature on crime and deviance as work. This literature compares occupational deviance with legitimate jobs and examines **deviant versus legitimate careers**. Working in deviant fields holds many similarities to the skills, professionalism, connections, and attitudes needed for conventional jobs (Letkemann, 1973). Goods and services may be bought and sold, credit arranged and extended, costs and profits calculated, and business associates, suppliers, and customers assessed. Contracts cannot be legally enforced in deviant occupations, however, nor are associates as reliable or durable. Because of the high turnover of personnel and the greater likelihood of drug use involved in all facets of deviant work, people are less likely to have expertise in their trade or to deliver on promises made.

There are also limitations to the career analogy. Although legitimate work may have several structures (the compressed career, the bureaucratic career, the entrepreneurial career), the patterns for deviant careers are more generally entrepreneurial. Entry may take many shapes and lengths of time. Once one enters into deviance, behavioral shifts may be lateral and downward as well as upward, precipitous as well as gradual and controlled, repetitive as well as dissimilar, and they may involve continuity or a complete shift into other venues (Luckenbill and Best, 1981). Exits are problematic, varying in the degree to which the participant initiates the exit, in whether they are temporary or lasting, and in whether they involve anything from going out on top to slinking away in debt and disgrace. Perhaps the biggest contrast between deviant and legitimate careers (taking the bureaucratic organizational form for the latter) lies in the legitimate career's slower ascent at the beginning and the greater stability and security toward the end, compared with the deviant career's rapid upward mobility and earlier burnout (as we see in Chapter 47 on the pimp-controlled prostitute's career).

45

Deciding to Commit a Burglary

RICHARD T. WRIGHT AND SCOTT H. DECKER

In this classical occupational study of deviance, Wright and Decker share with us their insights into the motivations and behavior of residential burglars. Simply written and filled with rich quotes, this chapter affirms the view that most burglaries are committed spontaneously by semiskilled criminals. Crimes of opportunity, burglaries are fueled by perpetrators' desire for money. Although most attempt to diminish the stigma of their crimes by rationalizing that they steal to support their basic living expenses, Wright and Decker undercut these accounts as impression management, citing subjects' behavior and alternative explanations that they steal to gain money for drugs, for partying, to impress women, and to sustain an overall high lifestyle with the trappings of material success. Enmeshed in the world of instant gratification, these burglars give lip service to their desire for legitimate jobs, but have neither the patience nor the interest in developing the skills required for legitimate work or in working their way up the ladder of legitimate career success. For most of Wright and Decker's subjects, burglary is their "main line," although not the only line of deviant income. In addition to burglarizing for the financial yield, they are attracted to residential burglary by the excitement, the freedom, the adventure, the spontaneity, the identity it confers upon them, and the instant gratification. These people enact a criminal lifestyle that is reinforced by the norms and values of the deviant subculture in which they are ensconced.

The demographic characteristics of residential burglars have been well documented. As Shover (1991) has observed, such offenders are, among other things, disproportionately young, male, and poor. These characteristics serve to identify a segment of the population more prone than others to resort to breaking into dwellings, but they offer little insight into the actual causes of residential burglary. Many poor, young males, after all, never commit any sort of serious

From Richard T. Wright and Scott H. Decker, *Burglars on the Job: Streelife and Residential Break-ins*, pp. 35–61. © University Press of New England, Lebanon, NH. Reprinted with permission.

offense, let alone a burglary. And even those who carry out such crimes are not offending most of the time. This is not, by and large, a continually motivated group of criminals; the motivation for them to offend is closely tied to their assessment of current circumstances and prospects. The direct cause of residential burglary is a perceptual process through which the offense comes to be seen as a means of meeting an immediate need, that is, through which a motive for the crime is formed. Walker (1984: viii) has pointed out that, in order to develop a convincing explanation for criminal behavior, we must begin by "distinguishing the states of mind in which offenders commit, or contemplate the commission of, their offenses." Similarly, Katz (1988: 4), arguing for increased research into what he calls the foreground of criminality, has noted that all of the demographic information on criminals in the world cannot answer the following question: "Why are people who were not determined to commit a crime one moment determined to do so the next?" This is the question to which the present chapter is addressed. The aim is to explore the extent to which the decision to commit a residential burglary is the result of a process of careful calculation and deliberation.

In the overwhelming majority of cases, the decision to commit a residential burglary arises in the face of what offenders perceive to be a pressing need for cash. Previous research consistently has shown this to be so and the results of the present study bear out this point. More than nine out of ten of the offenders in our sample—95 of 102—reported that they broke into dwellings primarily when they needed money[:]

> Well, it's like, the way it clicks into your head is like, you'll be thinking about something and, you know, it's a problem. Then it, like, all relates. "Hey, I need some money! Then how am I going to get money? Well, how do you know how to get money quick and easy?" Then there it is. Next thing you know, you are watching [a house] or calling to see if [the occupants] are home. (Wild Will—No. 010) ...

These offenders were not motivated by a desire for money for its own sake. By and large, they were not accumulating the capital needed to achieve a long-range goal. Rather, they regarded money as providing them with the means to solve an immediate problem. In their view, burglary was matter of day-to-day survival[:]

> I didn't have the luxury of laying back in on damn pinstriped [suit]. I'm poor and I'm raggedy and I need some food and I need some shoes ...
> So I got to have some money some kind of way. If it's got to be the wrong way, then so be it. (Mark Smith—No. 030) ...

Given this view, it is unsurprising that the frequency with which the offenders committed burglaries was governed largely by the amount of money in their pockets. Many of them would not offend so long as they had sufficient cash to meet current expenses[:]

> Usually what I'll do is a burglary, maybe two or three if I have to, and then this will help me get over the rough spot until I can get my shit

straightened out. Once I get it straightened out, I just go with the flow until I hit that rough spot where I need the money again. And then I hit it … the only time I would go and commit a burglary is if I needed the money at that point in time. That would be strictly to pay light bill, gas bill, rent. (Dan Whiting—No. 102)

Long as I got some money, I'm cool. If I ain't got no money and I want to get high, then I go for it. (Janet Wilson—No. 060)

You know how they say stretch a dollar? I'll stretch it from here to the parking lot. But I can only stretch it so far and then it breaks. Then I say, "Well, I guess I got to go put on my black clothes. Go on out there like a thief in the night." (Ralph Jones—No. 018)

A few of the offenders sometimes committed a burglary even though they had sufficient cash for their immediate needs. These subjects were not purposely saving money, but they were unwilling to pass up opportunities to make more. They attributed their behavior to having become "greedy" or "addicted" to money[:]

I have done it out of greed, per se. Just to be doing it and to have more money, you know? Say, for instance, I have two hundred dollars in my pocket now. If I had two more hundreds, then that's four hundred dollars. Go out there and do a burglary. Then I say, "If I have four hundred dollars, then I can have a thousand." Go out there and do a burglary. (No. 018) …

Typically, the offenders did not save the money that they derived through burglary. Instead, they spent it for one or more of the following purposes: (1) to "keep the party going"; (2) to keep up appearances; or (3) to keep themselves and their families fed, clothed, and sheltered.

KEEPING THE PARTY GOING

Although the offenders often stated that they committed residential burglaries to "survive," there is a danger in taking this claim at face value. When asked how they spent the proceeds of their burglaries, nearly three-quarters of them—68 of 95—said they used the money for various forms of (for want of a better term) high-living. Most commonly, this involved the use of illicit drugs. Fifty-nine of the 68 offenders who spent the money obtained from burglary on pleasure-seeking pursuits specifically mentioned the purchase of drugs. For many of these respondents, the decision to break into a dwelling often arose as a result of a heavy session of drug use. The objective was to get the money to keep the party going.

The drug most frequently implicated in these situations was "crack" cocaine.

[Y]ou ever had an urge before? Maybe a cigarette urge or a food urge, where you eat that and you got to have more and more? That's how that crack is. You smoke it and it hits you [in the back of the throat] and you got to have more. I'll smoke that sixteenth up and get through, it's like I never had none. I got to have more. Therefore, I gots to go do another burglary and gets some more money. (Richard Jackson— No. 009) …

Lemert (1953: 304) has labelled situations like these "dialectical, self-enclosed systems of behavior" in that they have an internal logic or "false structure," which calls for more of the same. Once locked into such events, he asserts, participants experience considerable pressure to continue, even if this involves breaking the law[:]

> A man away from home who falls in with a group of persons who have embarked upon a two or three-day or even a week's period of drinking and carousing … tends to have the impetus to continue the pattern which gets mutually reinforced by [the] interaction of the participants, and [the pattern] tends to have an accelerated beginning, a climax and a terminus. If midway through a spree a participant runs out of money, the pressures immediately become critical to take such measures as are necessary to preserve the behavior sequence. A similar behavior sequence is [evident] in that of the alcoholic who reaches a "high point" in his drinking and runs out of money. He might go home and get clothes to pawn or go and borrow money from a friend or even apply for public relief, but these alternatives become irrelevant because of the immediacy of his need for alcohol. (Lemert, 1953: 303)

Implicit in this explanation is an image of actors who become involved in offending without significant calculation; having embarked voluntarily on one course of action (e.g., crack smoking), they suddenly find themselves being drawn into an unanticipated activity (e.g., residential burglary) as a means of sustaining that action. Their offending is not the result of a thoughtful, carefully reasoned process. Instead, it emerges as part of the natural flow of events, seemingly coming out of nowhere. In other words, it is not so much that these actors consciously choose to commit crimes as that they elect to get involved in situations that drive them toward lawbreaking.

Beyond the purchase of illicit drugs and, to a lesser extent, alcohol, 10 of the 68 offenders—15 percent—also used the proceeds from their residential burglaries to pursue sexual conquests. All of these offenders were male. Some liked to flash money about, believing that this was the way to attract women ….[:]

> [I commit burglaries to] splurge money with the women, you know, that's they kick, that's what they like to do. (Jon Monroe—No. 011) …

Like getting high, sexual conquest was a much-prized symbol of hipness through which the male subjects in our sample could accrue status among their peers on the street. The greatest prestige was accorded to those who were granted sexual favors solely on the basis of smooth talk and careful impression management. Nevertheless, a few of the offenders took a more direct approach to obtaining sex by paying a streetcorner prostitute (sometimes referred to as a "duck") for it. While this was regarded as less hip than the more subtle approach described above, it had the advantage of being easy and uncomplicated. As such, it appealed to offenders who were wrapped up in partying and therefore reluctant to devote more effort than was necessary to satisfy their immediate sexual desires[:]

> I spend [the money] on something to drink, ... then get me some [marijuana]. Then I'm gonna find me a duck. (Ricky Davis—No. 015)

It would be misleading to suggest that any of the offenders we spoke to committed burglaries *specifically* to get money for sex, but a number of them often directed a portion of their earnings toward this goal.

In short, among the major purposes for which the offenders used the money derived from burglary was the maintenance of a lifestyle centered on illicit drugs, but frequently incorporating alcohol and sexual conquests as well. This lifestyle reflects the values of the street culture, a culture characterized by an openness to "illicit action" (Katz, 1988: 209–15), to which most of our subjects were strongly committed. Viewed from the perspective of the offenders, then, the oft-heard claim that they broke into dwellings to survive does not seem quite so farfetched. The majority of them saw their fate as inextricably linked to an ability to fulfill the imperatives of life on the street.

KEEPING UP APPEARANCES

Of the 95 offenders who committed residential burglaries primarily for financial reasons, 43 reported that they used the cash to purchase various "status" items. The most popular item was clothing; 39 of the 43 said that they bought clothes with the proceeds of their crimes. At one level, of course, clothing must be regarded as necessary for survival. The responses of most of the offenders, however, left little doubt that they were not buying clothes to protect themselves from the elements, but rather to project a certain image; they were drawn to styles and brand names regarded as chic on the streets[:]

> See, I go steal money and go buy me some clothes. See, I likes to look good. I likes to dress. All I wear is Stacy Adams, that's all I wear. [I own] only one pair of blue jeans cause I likes to dress. (No. 011) ....

After clothes, cars and car accessories were the next most popular status items bought by the offenders. Seven of the 43 reported spending at least some of the money they got from burglaries on their cars[:]

> I spent [the money] on stuff for my car. Like I said, I put a lot of money into my car ... I had a '79 Grand Prix, you know, a nice car. (Matt Detteman—No. 072)

The attributes of a high-status vehicle varied. Not all of these offenders, for example, would have regarded a 1979 Grand Prix as conferring much prestige on its owner. Nevertheless, they were agreed that driving a fancy or customized car, like wearing fashionable clothing, was an effective way of enhancing one's street status ....[:]

> I don't know if you've ever thought about it, but I think every crook likes the life of thieving and then going and being somebody better. Really, you are deceiving people; letting them think that you are well off .... You've got a nice car, you can go about and do this and do that. It takes money to buy that kind of life.

Shover and Honaker (1990: 11) have suggested that the concern of offenders with outward appearances, as with their notorious high-living, grows out of what is typically a strong attachment to the values of street culture[:] values that place great emphasis on the "ostentatious enjoyment and display of luxury items." In a related vein, Katz (1988) has argued that for those who are committed to streetlife, the reckless spending of cash on luxury goods is an end in itself, demonstrating their disdain for the ordinary citizen's pursuit of financial security. Seen through the eyes of the offenders, therefore, money spent on such goods has not been "blown," but rather represents a cost of raising or maintaining one's status on the street.

KEEPING THINGS TOGETHER

While most of the offenders spent much of the money they earned through residential burglary on drugs and clothes, a substantial number also used some of it for daily living expenses. Of the 95 who committed burglaries to raise money, 50 claimed that they needed the cash for subsistence[:]

> I do [burglaries] to keep myself together, keep myself up.
>
> (James Brown   No. 025) ...

Quite a few of the offenders—13 of 50—said that they paid bills with the money derived from burglary. Here again, however, there is a danger of being misled by such claims. To be sure, these offenders did use some of their burglary money to take care of bills. Often, though, the bills were badly delinquent because the offenders avoided paying them for as long as possible—even when they had the cash—in favor of buying, most typically, drugs. It was not until the threat of serious repercussions created unbearable pressure for the offenders that they relented and settled their accounts[:]

> [Sometimes I commit burglaries when] things pressuring me, you know? I got to do somethin' about these bills. Bills. I might let it pass that mornin'. Then I start trippin' on it at night and, next thing you know, it's wakin' me up. Yeah, that's when I got to get out and go do a burglary. I *got* to pay this electric bill off, this gas bill, you know? (No. 009) ...

Spontaneity is a prominent feature of street culture (Shover and Honaker, 1992); it is not surprising that many of the offenders displayed a marked tendency to live for the moment. Often they would give every indication of intending to take care of their obligations, financial or otherwise, only to be distracted by more immediate temptations. For instance, a woman in our sample, after being paid for an interview, asked us to drive her to a place where she could buy a pizza for her children's lunch. On the way to the restaurant, however, she changed her mind and asked to be dropped off at a crack house instead....

Katz (1988: 216) has suggested that, through irresponsible spending, persistent offenders seek to construct "an environment of pressures that guide[s] them back toward crime." Whether offenders spend money in a conscious attempt to create such pressures is arguable; the subjects in our sample gave no indication of doing so, appearing simply to be financially irresponsible. One offender, for example, told us that he never hesitated to spend money, adding, "Why should I? I can always get some more." However, the inclination of offenders to free-spending leaves them with few alternatives but to continue committing crimes. Their next financial crisis is never far around the corner.

The high-living of the offenders, thus, calls into question the extent to which they are driven to crime by genuine financial hardship. At the same time, though, their spendthrift ways ensure that the crimes they commit will be economically motivated (Katz, 1988). The offenders perceive themselves as needing money, and their offenses typically are a response to this perception. Objectively, however, few are doing burglaries to escape impoverishment.

WHY BURGLARY?

The decision to commit a residential burglary, then, is usually prompted by a perceived need for cash. Burglary, however, is not the only means by which offenders could get some money. Why do they choose burglary over legitimate work? Why do they elect to carry out a burglary rather than borrow the money from a friend or relative? Additionally, why do they select burglary rather than some other crime?

Given the streetcorner context in which most burglary decisions were made, legitimate work did not represent a viable solution for most of the offenders in our sample. These subjects, with few exceptions, wanted money there and then and, in such circumstances, even day labor was irrelevant because it did not respond to the immediacy of their desire for cash (Lemert, 1953). Moreover, the jobs available to most of the offenders were poorly paid and could not sustain their desired lifestyles. It is notable that 17 of the 95 offenders who did burglaries primarily to raise money *were* legitimately employed[:]

> [I have a job, but] I got tired of waiting on that money. I can get money like that. I got talent, I can do me a burg, man, and get me five or six hundred dollars in less than a hour. Working eight hours a day and waiting for a whole week for a check and it ain't even about shit. (No. 022)....

Beyond this [attitude], a few of the offenders expressed a strong aversion to legitimate employment, saying that a job would impinge upon their way of life[:]

> I ain't workin' and too lazy to work and just all that. I like it to where I can just run around. I don't got to get up at no certain time, just whenever I wake up. I ain't gotta go to bed a certain time to get up at a certain time. Go to bed around one o'clock or when I want, get up when I want. Ain't got to go to work and work eight hours. Just go in and do a five minute job, get that money, that's just basically it. (Tony Scott—No. 085) ...

Nevertheless, a majority of the offenders reported that they wanted lawful employment; 43 of the 78 unemployed subjects who said that they did burglaries mostly for the money claimed they would stop committing offenses if someone gave them a "good" job[:]

> I'm definitely going to give it up as soon as I get me a good job. I don't mean making fifteen dollars an hour. Give me a job making five-fifty and I'm happy with it. I don't got to burglarize no more. I'm not doing it because I like doing it, I'm doing it because I need some [drugs]. (No. 079) ...

When faced with an immediate need for cash, then, the offenders in our sample perceived themselves as having little hope of getting money both quickly *and* legally. Many of the most efficient solutions to financial troubles are against the law (Lofland, 1969). However, this [fact] does not explain why the subjects decided specifically on the crime of residential burglary. After all, most of them admitted committing other sorts of offenses in the past, and some still were doing so. Why should they choose burglary?

For some subjects, this question held little relevance because they regarded residential burglary as their "main line" and alternative offenses were seldom considered when the need for money arose....[:]

> [I do burglary] because it's easy and because I know it. It's kind of getting a speciality or a career. If you're in one line, or one field, and you know it real well, then you don't have any qualms about doing it. But if you try something new, you could really mess up .... At this point, I've gotten away with so much [that] I just don't want to risk it—it's too much to risk at this point. I feel like I have a good pattern, clean; go in the house, come back out, under two minutes every time. (Darlene White—No. 100) ...

When these subjects did commit another kind of offense, it typically was triggered by the chance discovery of a vulnerable target ... most of the burglars we interviewed identified themselves as hustlers, people who were always looking to "get over" by making some fast cash; it would have been out of character for them to pass up any kind of presented opportunity to do so[:]

> If I see another hustle, then I'll do it, but burglary is my pet.
>
> (Larry Smith—No. 065) ...

THE SEDUCTIONS OF RESIDENTIAL BURGLARY

For some offenders, the perceived benefits of residential burglary may transcend the amelioration of financial need. A few of the subjects we interviewed—7 of 102—said that they did not typically commit burglaries as much for the money as for the psychic rewards. These offenders reported breaking into dwellings primarily because they enjoyed doing so. Most of them did not enjoy burglary per se, but rather the risks and challenges inherent in the crime[:]

> [I]t's really because I like [burglaries]. I know that if I get caught I'm a do more time than the average person, but still, it's the risk. I like doin' them. (No. 013)

> I think [burglary is] fun. It's a challenge. You don't know whether you're getting caught or not and I like challenges. If I can get a challenging [burglary, I] like that. It's more of the risk that you got to take, you know, to see how good you can really be. (No. 103)

These subjects seemingly viewed the successful completion of an offense as "a thrilling demonstration of personal competence" (Katz, 1988: 9). Given this [frame of mind], it is not surprising that the catalyst for their crimes often was a mixture of boredom and an acute sense of frustration born of failure at legitimate activities such as work or school[:]

> [Burglary] just be something to do. I might not be workin' or not going to school—not doing anything. So I just decide to do a burglary. (No. 017)

The offense provided these offenders with more than something exciting to do; it also offered them the chance to "be somebody" by successfully completing a dangerous act. Similarly, Shover and Honaker (1992: 288) have noted that, through crime, offenders seek to demonstrate a sense of control or mastery over their lives and thereby to gain "a measure of respect, if not from others, at least from [themselves]."…

While only a small number of the subjects in our sample said that they were motivated *primarily* by the psychic rewards of burglary, many of them perceived such rewards as a secondary benefit of the offense. Sixteen of the 95 offenders who did burglaries to raise cash also said that they found the crime to be "exciting" or "thrilling."…[:]

> [Beyond money], it's the thrill. If you get out [of the house], you smile and stand on it, breathe out. (No. 045)
>   It's just a thrill going in undetected and walking out with all they shit. Man, that shit fucks me up. (No. 022) …

Finally, one of the offenders who did burglaries chiefly for monetary reasons alluded to the fact that the crime also provided him with a valued identity.…

SUMMARY

Offenders typically decided to commit a residential burglary in response to a perceived need. In most cases, this need was financial, calling for the immediate acquisition of money. However, it sometimes involved what was interpreted as a need to repel an attack on the status, *identity*, or self-esteem of the offenders.

Whatever its character, the need almost invariably was regarded by the offenders as pressing, that is, as something that had to be dealt with immediately. Lofland (1969: 50) has observed that most people, when under pressure, have a tendency to become fixated on removing the perceived cause of that pressure "as quickly as possible." Those in our sample were no exception. In such a state, the offenders were not predisposed to consider unfamiliar, complicated, or long-term solutions (see Lofland, 1969: 50–54) and instead fell back on residential burglary, which they knew well. This [fallback] often seemed to happen almost automatically, the crime occurring with minimal calculation [and] as part of a more general path of action (e.g., partying). To the extent that the offense ameliorated their distress, it nurtured a tendency for them to view burglary as a reliable means of dealing with similar pressures in the future. In this way, a foundation was laid for the continuation of their present lifestyle which, by and large, revolved around the street culture. The self-indulgent activities supported by this culture, in turn, precipitated new pressures; and thus a vicious cycle developed.

That the offenders, at the time of actually contemplating offenses, typically perceived themselves to be in a situation of immediate need has at least two important implications. First, it suggests a mind-set in which they were seeking less to maximize their gains than to deal with a present crisis. Second, it indicates an element of desperation which might have weakened the influence of threatened sanctions and neutralized any misgivings about the morality of breaking into dwellings....

REFERENCES

Katz, J. 1988. *Seductions of Crime: Moral and Sensual Attractions in Doing Evil.* New York: Basic Books.

Lemert, E. 1953. "An Isolation and Closure Theory of Naive Check Forgery." *Journal of Criminal Law, Criminology, and Police Science* 44: 296–307.

Lofland, J. 1969. *Deviance and Identity.* Englewood Cliffs, NJ: Prentice Hall.

Shover, N. 1991. "Burglary." In Tonry, M., *Crime and Justice: A Review of Research.* vol. 14. pp. 73–113. Chicago: University of Chicago Press.

Shover, N., and Honaker, D. 1990. "The Criminal Calculus of Persistent Property Offenders: A Review of Evidence." Paper presented at the Forty-second Annual Meeting of the American Society of Criminology, Baltimore, November.

_____. 1992. "The Socially Bounded Decision Making of Persistent Property Offenders." *Howard Journal of Criminal Justice* 31. no. 4: 276–93.

Walker, N. 1984. "Foreword." In Bennett, T., and Wright, R. *Burglars on Burglary: Prevention and the Offender.* pp. viii–ix. Aldershot, UK: Gower.

46

Social Smoking: A Liminal Position

JASON WHITESEL AND AMY SHUMAN

Social smokers occupy a marginal status, belonging to neither the category of smokers nor that of nonsmokers. Although they self-identify more closely with the latter, society may place them more strongly in the camp of the former. All of the people in Whitesel and Shuman's study had previously been, at one point or another, regular smokers. But they intentionally removed themselves from this category to occupy a more difficult and ephemeral one falling somewhere in between; they smoke on social occasions only.

This chapter offers a fascinating discussion, that you are likely to find familiar, of some of the ways social smokers justify their status and somewhat "mooching" behavior. Techniques for bumming are discussed, as are the ways these individuals manage their position in relation to nonsmokers, regular smokers, and other social smokers. In reading this piece, pay particular attention to the social status and identity of social smokers as they relate to those three groups of people. How does their liminal status benefit them? Disadvantage them? How does this type of status compare with that of the bisexuals discussed by Weinberg, Williams, and Pryor in Chapter 24?

With increasing attention to health consequences, smoking in the United States has decreased in everyday practice, while, at the same time, continuing to be a means for claiming or denying a socially performed identity. Even in a climate of awareness regarding its health risks, social smoking helps negotiate complex roles of performance and exchange. In this paper, we discuss the availability and sustainability of the category of the "social smoker" in a world that requires a sharp distinction between the habitual [smoker] and the nonsmoker (both of which require identity work). The paradox for social smokers is that they believe that they are nonsmokers, but they tend to find their allies among habitual smokers, those who share a similar stigmatized and

From Jason Whitesel and Amy Shuman (2009). *Social Smoking: A Liminal Position.* *Sociological Focus*, 42(4), 330–349.

regulatory discourse. In the current climate of smoking bans in public places, the alienation of smokers is exacerbated. Social smokers move in and out of passing as either habitual smokers or as nonsmokers, even as passing in one category inevitably delegitimizes their claims to membership in the other. Social smokers do not belong to a coherent or fixed category; rather, social smoking destabilizes the rigidity of the habitual smoker/nonsmoker binary.

Social smokers eventually experience a disconfirming reality. This is an estrangement problem in which people lack social alliances in a society that does not recognize their experiences. Social smokers, who also experience a lack of category recognition, can suffer from a kind of hypervisibility in which olfactory and visible evidence forces them to question which version of reality is correct, or if they can sustain the reality they prefer. The alternative is not only for stigmatized individuals to pass as normal, but also—more importantly—to find a means to sustain the ambiguity of a contradictory position.

Conscious of the possibility that smoking produces real consequences for physical health and that smoking is stigmatized as a filthy habit, social smokers try to differentiate themselves from habitual smokers, and they convince themselves that they are effectively nonsmokers. Survey results representing the primary motivation for social smoking as "a dismissive attitude to the dangers of smoking" cannot account for its social dimension. Our research documents social smoking as a self-conscious performance involving elaborate, practiced, ongoing behavioral management. We argue that social smokers formulate intricate strategies that depend on reciprocity and mutual recognition to create a barely sustainable social identity associated with a stigmatized practice.

Social smokers use a variety of terms to distinguish between levels of commitment to smoking, including "antismoker," "situational smoker," and "stress smoker." Smokers categorize their practice according to either the contexts in which they smoke or the quantity of cigarettes consumed (i.e., "pack-a-day" smokers). Our research demonstrates how the particular category of social smoker is characterized by more than the superficial self-regulatory practice of smoking fewer cigarettes or smoking in limited circumstances. Consequently, cigarettes become an available resource for a slew of things, such as a way of working out social relationships or communicating social and economic positions. This [eventuality] leads to a relatively unknown dimension of the sociability of smoking: social smokers' shared understanding of the cigarette gift economy—that is, elaborate negotiations of exchange, even reciprocity, in which the decision not to purchase cigarettes helps create the self-definition of the nonsmoker who then borrows or bums cigarettes.

We explore how the contradictory and incompatible demands on social smokers' identity performance are negotiated. We investigate two broad research questions. *First, what self-presentation creates alignment with nonsmokers, habitual smokers, reformed smokers, and even antismokers? Second, as the number of existing social categories for smoking diminishes, how do social smokers adjust their identities?* In addressing these questions, we focus on the relationship between smoking practices and metacommentaries about those practices.

METHODS

After reviewing historical data and conducting participant observation research over a period of two years, we determined that the best means to identify and collect the performance strategies of self-identified social smokers was in-depth, semistructured interviews. We conducted seventeen interviews and ethnographic observations of women from Columbus, Ohio, and the surrounding Midwest area between the fall of 2001 and the summer of 2003, resulting in 442 single-spaced typed pages of data. To obtain as diverse a sample as possible, we recruited volunteers by posting and distributing project advertisements seeking social smokers at the local university, in an undergraduate newsletter, and in local bars. Purposive sampling was used to ensure that women who are social smokers would be included. We sampled women, based on prior research that suggests that female informants, unlike males, might be more attuned to external, social intangibles, such as holding or smelling a cigarette, or other factors that arouse the urge to smoke. Interviewees were mostly young women associated with a college, following recent research indicating the prevalence of social smoking among that group. This convenience sample offered us the opportunity to study smoking decisions among people who have the resources to consider alternatives to smoking.

In these confidential interviews, we inquired about informants' beneficial and negative experiences with smoking. One woman limited her smoking identity to a "closet smoker" and only four of the women had previously considered themselves habitual smokers. Though the extent of the women's participation in social smoking ranged from five months to 40 years, most fell within a three- to eight-year window of involvement, with five years being the median. Three types of potential informants were not interviewed, based on the purposive sampling design: irrefutable nonsmokers, current daily smokers, and exclusive cannabis smokers.

Informants described a wide range of smoking practices, including [the practices of] people who just "looked like" smokers but did not smoke. Some of the interviewees were antismokers, claimed to be nonsmokers for the purposes of medical records, [and] had previously considered themselves habitual smokers, and some were looking to become irrefutable nonsmokers. Despite these multiple, and, perhaps, contradictory self-labels, the respondents all shared the characteristics of nondaily smoking and [using smoking as] a social means for self-control (as opposed to addiction).

Of the women interviewed, fifteen were between 20 and 35 years old; one was 36, and one was 60. Of the seventeen interviewees, one was Asian, one African American, and the rest were white. The participants were invited to describe their relationship status. Nine of the respondents were single at the time of the interview, [and] the majority of them were heterosexual; eight of the respondents were in long-term relationships (three were married, three cohabited, two were in a lesbian partnership). Three young women identified as bisexual or "curious." Our research might support existing arguments for the correlation between already marginal (especially lower socioeconomic) people and smoking in the United States. However, we are less interested in whether already marginal people smoke than in how those in one category of smokers (social smokers) redefine themselves as outside the stigma. Our respondents utilize elaborate mechanisms to sustain their identities and disidentify with the stigmatized practice of smoking.

FINDINGS

Social smoking is a self-conscious category. We are interested not only in what social smokers do, but also in how they attach meaning to their practices and in their discourse about smoking as an identityshaping practice.

Smoking as a Social Activity

Smoking has always offered possibilities for camaraderie and social interaction. "Smoke breaks" were once a feature of many U.S. workplaces. In the new anti-smoking climate, especially in communities dominated by a nonsmoking ethos, smoking requires a more careful presentation of self, involving several paradoxes that challenge personal integrity. One veteran social smoker commented, "I think I'm the bisexual in the smoking world.... Bisexuals aren't really accepted by straights or gays—they're not accepted by either niche. [Likewise,] we're not accepted as either being a smoker or nonsmoker." We describe how social smokers characterize the integrity of their practice and how it is challenged by a complex gift economy in the exchange of cigarettes and lighters. Social smoking is a mode of social interaction characterized by self-deception, face-saving, and repositioning within a culturally unavailable category.

Our interviews with self-identified social smokers confirm smoking as a social activity. The interviewees recognized that even the "real ... 'tried and true' heavy smokers" often hype up their smoking in order to enjoy the company of others. One woman was bewildered when we asked her to try to identify any individuals with a smoking habit similar to hers. However, when she began to consider those individuals outside the imagined social smoking community, she readily acknowledged that those in her larger fellowship of bona fide smokers also tend to "increase the number of cigarettes they smoke and appear to use cigarettes in a more socially engaged manner." She elaborated that this sociability between all smokers occurs "in a way that is interactive with those around them," making smoking "a shared experience."

Among the interviewees, those who were students reported that stepping out for a cigarette provides the social intercourse that the solitary work of academic life often inhibits. One student reported that the irony of being a social smoker is that she became more conscious of how habitual smokers also employ cigarettes in order to fulfill an interpersonal need: "Regina, in my department, she'll always go around to the smokers, 'Are you going to smoke? Are you going to smoke?' She doesn't want to smoke by herself, even [when she's] a heavy smoker." According to a social smoker who works at a drugstore, social and habitual smokers smoke within group settings where cigarettes serve as the adhesive for "an atmosphere of conviviality." Unable to smoke in the company of nonsmokers, smokers seek each other out.

Bumming a Cigarette or a Light

In deciding not to purchase cigarettes, social smokers align themselves with non-smokers as opposed to habitual smokers, but then they must borrow or "bum" cigarettes. This situation requires occasional cigarette purchases, not just for themselves but for a group of nonpurchasing bumming social smokers. Not having cigarettes is a hallmark of the social smoker—thus, bumming takes on a different significance than it has for smokers generally, who also participate in the gift economy but who reciprocate more readily. One of the untenable dimensions of social smokers is that they should not have cigarettes; they prefer to bum them, but they seldom reciprocate. They occupy a contradictory, culturally unavailable position. Social smokers are beholden to and enabled by the very others with whom they do not want to be associated.

Many informants commented on the widespread phenomenon of bumming a cigarette, which involves engaging in elaborate negotiations of exchange, even reciprocity. According to the informants, "cadging a cigarette" not only restrains smoking involvement, but also sustains the self-identity of nonsmoker. They believe that if a pack of cigarettes seldom infiltrates their residence, there will probably be insufficient means to support routine smoking. One respondent stated, "I only bum cigarettes from others. I don't want to buy cigarettes for myself. ... Everyone I bum cigarettes from knows that I'm not interested in getting rehooked, so it's a casual thing. And everyone asks for my girlfriend's permission first." This quote clearly illustrates that for social smokers, the intent behind bumming is both to limit smoking to public occasions and to exclude cigarettes and paraphernalia from private domains.

In his foundational research on public gifting, Marcel Mauss demonstrated how every offering is actually embedded in a larger system of reciprocity. He argued [that] the relationship between giver and receiver is governed by the rule that the recipient should always give back more than what was received; there is no such thing as free alms ([1950] 1990). Consistent with this rule, the social smokers in this study discussed several ways in which the act of "bumming" enhanced their smoking objectives but potentially damaged their reputations. One informant insisted that with cigarettes costing "four dollars a package now," seeking and accepting them from others is an "intrusion." She asserted, "'neither a borrower nor a lender be' ... cigarettes are expensive.... I feel that it's socially incorrect [to bum], but it wouldn't stop me from doing it.... Nor would I resent it if anybody bummed a cigarette from me." These comments of a social smoker highlight that in contrast to market exchanges, gifts tend to be given "in the context of public drama," which makes them highly susceptible to public scrutiny, or, as the respondent suggested, to a judgment of fairness.

Cigarettes and "lights" have long been understood as public exchange commodities. Carol Brooks Gardner (1986) discusses exhaustible and renewable aid, the first corresponding to systems of giving, and the second to systems of borrowing. According to Gardner, a match or a cigarette is an exhaustible aid and a lighter is a renewable aid.

As part of their self-characterization as nonsmokers, social smokers also referred to "bumming a light." Though the majority of respondents at one time or another had made the customary request to bum a light, more than half managed to avoid purchasing lighters altogether: over a third just procured free matches at restaurants, bars, and weddings. One woman said, "I don't know when the last time I bought a lighter was.... [U]sually I'll bum someone else's if I don't have one." Some respondents remarked on how nondescript lighters, without prior claimants, are also cycled swiftly among members: "I'm a lighter clipper.... [S]ometimes you can't help it because sometimes you have the same lighter.... [B]ut, if I see a lighter lying around and nobody's got dibs on it, it's mine."

"Floating" lighters are symbolic of the social smoker's reluctance to join the category of smokers. Not having a lighter is about not smoking, although of course smokers, too, can participate in the exchange of lighters on the move. Respondents chronicled movement, meaning, and image-making as relevant to the social context of their behavior. Lewis Hyde identifies the central criterion of gifts as their movement: "*the gift must always move.* There are other forms of property that stand still, that mark a boundary or resist momentum, but the gift keeps going" ([1979] 1983:12 italics in original). One couple, both social smokers, reported the mysterious appearance of a lighter that prominently displayed a male stripper. The male striptease lighter, though appreciated as retro or kitsch, was detected as an anomaly in an economy in which lighters came and went without being accounted for. Here, the lighter seemed to be one of those objects that often escapes accountability and yet belongs to the exchange economy. In this case, identity is produced through circulation of rather than attachment to objects. The unbroken sequence of the lighter reiterates how social smokers' seemingly insignificant improvisational acts actively bring together a smoking community— if only assembling it by the evocations brought to mind through exchanged objects. Informants' reports about the fluidity of lighter ownership illustrate an alternative to the model of reciprocity described by Mauss[—that] is, that reciprocity is achieved not by giving in equal or greater amounts but by a shared knowledge that what goes around, comes around—one loses lighters, one finds them.

Four Characteristics of Bumming

The interviewees displayed four smoking behaviors associated with bumming: acquiring a cigarette, resolving how to ask, status differentiations, and inventing a reciprocal community.

Acquiring a Cigarette   The women often reported being offered a cigarette without having petitioned for it. These offerings frequently came as the result of a benefactor's knowledge about the informant's penchant for occasional tobacco use. Kitty said that it was common for her close friends to offer cigarettes spontaneously, but she also noted, "Even habitual smokers that are just kind of acquaintances, when they go to light up, they'll offer so you don't really have to ask." For one woman, knowing the donor determined whether she made a request or a demand: "I usually just say, 'Can I have one?' and that suffices.

If it's a cigarette from my boyfriend, I don't actually ask ... it's more of a command: 'Give me one.'" Most women described uncomfortable moments when they had not anticipated going out with less intimate compatriots, but the mood seemed appropriate for a cigarette. One young woman said:

> Occasionally, a group of friends will go to a bar after a meeting or something and I'll just kinda go to go along and they'll be like, "Well, I'm going to have a cigarette," and they'll be like, "Do you want one," and I'll be like, "Um, sure," or I'll say "No" and then I'll be like, "Uh, yeah, can I have a cigarette now? I changed my mind." Well, sometimes I feel awkward asking people like, if I'm at a bar, or party, or something, I'll be like, okay, who am I going to ask to bum a cigarette off of and how am I gonna say it?

This dilemma about whom to ask for a cigarette and how to ask for it caused the women to develop both styles of soliciting and more commonly, alternatives to conventional bumming. We can differentiate here between "demand sharing" within already existing social relationships (such as the woman telling her boyfriend to "give me one") and more subtle means for procuring a cigarette that requires added identity work.

Resolving How to Ask   About one-fourth of the women had at one time or another used flirtation or misrepresentation to resolve the quandary of how to ask for a cigarette. Informant tactics often depended on the perceived audience. One woman reiterated that demands can be made of people one knows, but, "if it's a stranger, I'll be just like uh, cute and be like, oh, 'Can I bum a cigarette?' in a sweet little voice." Similarly, another respondent remarked on the social advantage that can be garnered by altering one's intonation and mannerisms for dramatic effect. "I mean the way girls ask for [cigarettes] ... they have their little (especially if it's a guy but as a girl too), you kind of have the whole like, cock your head, like, voice raised, 'Can I have a cigarette?' Like that type of thing, it seems to kind of get girls pretty far." Another woman said that she opted for speech that was rather ingratiating as she cajoled potential gay male donors: "If I'm in a gay bar ... I'm like, 'Honey, can I have a cigarette?' I mean, it's very sort of gentle and friendly I think."

Blandishment to secure a cigarette from a stranger sometimes escalates to minor deception. One woman reported feigning either being out of cigarettes or claiming that she forgot them[:] "When I'm with one of my smoking friends, I don't even really ask, I just take it or point to the box and I'm like, 'Do you mind?'" However, she often felt compelled to offer a rationale for coaxing cigarettes from strangers: "'I left them in my car,' or you know, 'I didn't bring any with me,' or 'I ran out,' or whatever. Like if they look at you funny, that's when I supply the excuse. And they always say, 'No, it's fine. Go ahead.' So ... it's not that hard." As she explained, "You get free cigarettes that way and you don't have to pay for them yourself."

In Erving Goffman's terms, this woman's strategies are a means of face-saving ([1955] 1967). The excuse provides a retrospective warrant for the request and restores equality in a one-down situation when people "look at you funny". As Weiner (1976, 213) observed, "the untenable extremes of complete autonomy from others and total dependence on others are avoided through the mechanism of exchange. Exchange allows social space to be negotiable at the same time that personal space (one's autonomy) is inviolable." The problems of losing face or being onedown are compounded in the case of both bumming and smoking cigarettes. The concepts of face-saving and one-up/one-down relationships can be misleading if they are used to suggest the maintenance of existing social status. Instead, the use of flattery or excuses can create new social configurations.

Status Differentiation   Mauss demonstrated how the generalized norm of reciprocity operates at the level of public domain in which givers and receivers exchange roles somewhat equally. In social smoking, the roles are sometimes more stable, as for example when one person is a fixed moocher and another a fixed provider. Here our research on cigarette gifting provides an important corrective to current research on the moral dimension of smoking, perhaps the most misunderstood dimension. After conducting surveys about how much people smoke and when, researchers too often turn to moral explanations (the dismissal of health concerns) or the inability to quit (representing a weak will). Smokers know the moral arguments well, but their rehearsal of them does not indicate that they play a large role in social smoking. Instead, our research suggests that issues of morality figure more into social relationships, demonstrated by patterns of reciprocity.

Social actors begin to appraise bumming in terms of the constraint created by expense and the social opportunities available for bumming. Social smokers (and conceivably habitual smokers alike) realize that requesting a cigarette from strangers can be an imposition. One woman pointed out that "at the bars, [cigarettes are] not the cheapest thing. And I don't like approaching strangers and asking for them. Even though I would give them if somebody asked me for one, then I would give one, but I just don't like 'mooching.'" When we asked the young woman above, "What is mooching?" she plainly replied, "Going and getting cigarettes from people you don't know." In other words, a moocher imposes too much on other people's generosity by repeatedly trying to get something from them.

Most informants agreed that bumming etiquette dictated that only one cigarette from a stranger was permissible. A receptionist tried to avoid asking for cigarettes but when she did solicit, "it would be one and that would be the only time I would hit that person up for the night, and then, I would probably try to find a cigarette machine." She said she would reciprocate a single cigarette, "as long as they don't abuse it." Even social smokers who located cigarette machines or put a pack on their tab during an evening at the tavern, ultimately intending to return home without cigarettes, still endorsed the tenets set forth by the rule of one: "One's okay. I mean, I was at a bar one time and I gave the guy like ten cigarettes 'cause I didn't want the pack after that night. I was like, 'This is ... I'm drunk. Take these out of my hand.' But, yeah, I think I should have said, 'One is acceptable; then go find somebody else to ask or something.'"

Informants recognized that gifting and borrowing cigarettes could presumably entangle two different types of social actors: habitual [smokers] and social smokers. Habitual smokers by far were considered both susceptible and perhaps less at liberty to provide charity because of potential "nicotine fits." One respondent, in describing her time engaged in social smoking (generally at her favorite video bar), said she became wary of "bumming off habitual smokers." She explained that she did not "want to leave them with nothing because they might have it planned out." The informant simulated how this smoker's thoughts might be: "I have this one for before I go to bed. I'll have this one in the morning and I have this one for driving in the car on the way to work." Social smoking is less predictable; that is, social smokers do not predict their next smoke, but rather they see themselves as spontaneously responding to social occasions.

Inventing a Reciprocal Community   One solution designed to circumvent bumming... was to call on friends and family members to split a pack: "Usually my sister and I, or my friends and I, will buy a pack and split it, you know—so that I don't buy them that often." Another woman claimed that she and her friends anticipated their needs: "When I would go out with Matt and Candy, it would be like, 'Well, are you going to bring the cigarettes? Okay then, we don't have to stop and get any.' Like we'd be going out together and it'd be like, 'Oh, should we stop for cigarettes?'—'No, I brought some so don't worry about it.'" Calling on friends with a mutual understanding that "I'll have some of yours, but I will take a turn in providing them next time" was commonplace etiquette.

The women contrasted borrowing from friends with bumming from strangers. Vivian described a tacit code of social smokers: "I think at some point it's kosher just to buy, even if you buy for the table, it's just kosher to buy and contribute. I think there is something sort of [communal] about that, that even if you're buying to have one or two, you go ahead and buy because you buy to replace what you've bummed." Such social organization among cooperating friends with similar smoking interests was a widespread solution for effectively reducing cigarette consumption.

The community of social smokers is defined by the practice of borrowing and collectively purchasing cigarettes. Borrowing contributes to the mythology of the distinction between social smoking and habitual smoking, and sets up smoking in terms of social relations rather than in terms of health-based decision making. Social smokers restrict their tobacco use by using social controls based on an exchange economy, rather than on medical means of regulation.

Management Strategies

Social smokers attempt to manage their cigarette consumption in part through strategies for purchasing, not purchasing, gifting, borrowing, and "mooching" cigarettes. All of the respondents self-consciously monitored having cigarettes in their possession (a behavior also confirmed in our observations). Three categories of these tactics emerged, each corresponding to patterns of gifting and borrowing and each involving performed identities.

Self-regulation   In his study of the history of smoking habits, Jason Hughes (2003) argues that tobacco use has shifted from something one does to lose control, to something people characterize in terms of self-control. Overwhelmingly, the women attributed their ability to engage in limited tobacco use as the result of some form of self-regulation: "Basically, like if I don't bring cigarettes I won't want them, and then, maybe I'll only bring like maybe two…. [S]o there's different ways I try to regulate how much I smoke." One woman recounts using a stylized smoking accoutrement to personalize and creatively regulate her smoking[:] "Somebody bought me a little silver cigarette case and it used to be a ritual for me, because, it was like, you opened the case, you tapped the cigarette on the case and I would put in like maybe four, like three to five, to control my smoking." Similar to regulating the number of cigarettes kept on hand, some social smokers periodically tallied the quantity of cigarettes consumed at one time. One informant said that she, "just count[s] the cigarettes" and sets a maximum number to smoke…. She joked that she had become quite accustomed to smoking "a lot of stale cigarettes."

Accountability partners also helped some women achieve self-control. One informant's first year of college brought new freedoms and more frequent use of cigarettes, and, subsequently, the need to address a number of contradictions. Smoking was not something that Colleen ever wanted "to do all the time." She flatly tells her friends: "Jane, no matter what I do, don't let me smoke" or "Brooke, I'm not buying cigarettes." Unwaveringly, this respondent draws the line between habitual smoker and social smoker and asks that her close peers be involved in shaping her culpability during particular social occasions or time-frames. In short, a behavior plan aimed at a level of control and a contract with dependable supporters held respondents accountable to their self-selected guidelines and served to maintain their limited relationship with tobacco.

Reframing   Many of our respondents reframed their smoking, sometimes using incongruous frames. One woman reported: "I buy Marlboro Ultra Lights. I'm on a diet with my smoking." Typically, the word "diet" refers to efforts driven by health- and/or appearance-[conscious] individuals to reduce their consumption of food. Though there is evidence to suggest that women who smoke consider cigarettes to have an appetite-suppressing effect, the word "diet" in this context suggests that there is such a thing as a healthy consumption of cigarettes. It sets up an analogy between watching one's weight and abstaining from strong-tasting cigarettes. This informant followed up her slogan by stating [that] she would never smoke Marlboro "Reds" because they "scare [her], quite frankly" and they "just feel, like heavier" from the "heavier taste of the tobacco." Thus, the flavor is too heavy for a social smoker's tastes and she perceives having offset the undesired effects of tobacco by choosing a low tar and nicotine cigarette.

Although a nurse reported that she associates social smoking with indiscriminately enjoying the pleasures of the moment (with little regard for her future), most women reported that this notion of *carpe diem* was not necessarily their approach to occasional tobacco use, or, for that matter, to day-to-day life in general. Susie said that she was not exceedingly "scared about cancer or anything" and that her optimistic predisposition plays a role in limiting smoking and coun-terbalancing tobacco's effects through diet and exercise:

> Some people have the attitude, well just live for today and you know it's not going to do you any good to not smoke a cigarette today 'cause you might be dead tomorrow. I'm not like that because ... if that was the way [I thought, then] I wouldn't try to exercise or do anything else good for myself, I would just kind of give up. I guess that affect[s] my decision not to smoke all the time.

Several informants considered a lifestyle of "moderation" to stave off the ill effects of tobacco.

Most informants said that they had felt the need to avert potential harass-ment or manage possibly "discrediting information" brought on by the scrutiny of others (Goffman 1963). Their willingness to be open about their smoking sta-tus was, in part, based on anticipated criticism; many were secretive, though they realized that even without openly acknowledging smoking, the practice was potentially perceptible. The "perceptibility" of smoking is not only marked by sight, being spotted with cigarette wafting smoke, but also by smell (Goffman 1963:48). Informants reported being personally repelled by the smell of smoke; they used elaborate routines, including designated smoking garments and pro-ducts and folk remedies that remove odors lingering on their hair or clothing.

Negotiating "footing"   To maintain their footing as social smokers (which [invariably] meant nonsmoker), the women reported strategies of self-criticism and self-awareness. Goffman (1981, 128) observed, "A change in footing implies a change in the alignment we take up to ourselves and the others present." Though all the women could typically deliver "passable" impressions of either a smoker or nonsmoker, the respondents consistently reported being deliberately attentive to their cravings out of a fear of becoming addicted. Women who regularly reported being fearful also tested their doubts and ultimately implemen-ted new cognitive–behavioral systems such as avoiding having cigarettes in their possession. One respondent commented that she had renegotiated her smoking status:

> I didn't like what I saw. People look silly. They made themselves look silly. How would you like to see yourself stepping into a cigarette closet at my age? Silly really isn't the word for it. Perhaps "stupid."

As Goffman points out, performances offer the option for individuals to be "taken by [their] own acts or be cynical about them" (1959: 19). Here, the interviewee reflects somewhat cynically on her previous practice and performance and realigns herself with outside interpretations and negative evaluations of smoking.

Positioning oneself in opposition to tobacco companies, and yet smoking, represents at times unsustainableidentity construction. One respondent's social smoking served to minimize the guilt she feels for supporting tobacco companies, the unpleasant physical repercussions apparent after a smoking "streak," and the expense: she had been regionally conditioned by years spent in San Francisco, where purportedly cigarettes "jumped from like three dollars to, like, five or six." Alice said that when she first began experimenting with smoking that she would scale it back until she no longer craved a cigarette or at least had stopped thinking about them: "When I first started out, I would smoke a cigarette because I would think about it too much. I was afraid I was becoming addicted. So, I would smoke a cigarette and I said the next time I could smoke would be the time I forgot about it." Another woman tested out her physiological dependence based on an act of remembrance. A young Catholic woman framed her check as an issue of discipline and packaged it in an annual maintenance program that corresponded with her act of contrition: "I tried to give up smoking for Lent every year.... [F]or four or six weeks ... I would be like, 'I'm not going to smoke.... [I]t wasn't, like, because I was so religious, it was just kinda like, see if I could do it." Renouncing smoking for the sake of personal control was an integrative cognitive–behavioral approach used to manage and maintain an individual's limited use of tobacco.

The four- to six-week hiatus, devoted to nonsmoking, was a very common way for the women to test themselves. Richelle shared that there have been times when she genuinely longed for a cigarette; in those moments, management is crucial because those physical yearnings for a cigarette signify that she has been smoking too much:

> It's that thing of you never want to be addicted and you sometimes, you'll test yourself, like, well if I can go without it for a month that doesn't make me a smoker does it. One of my friends said, "Richelle, just admit it" (because I was smoking pretty heavily). "Just admit it: you're not just a social smoker, you smoke," because she was a smoker, and I was like, I don't want to say that I'm a smoker, so I cut back.

This reining in of one's smoking was associated with being able to maintain a foothold in a nonsmoking life world, an "untenable position." To illustrate further, a veteran smoker decided to engage in social smoking about a year after "quitting cold turkey." Recounting her experiences of withdrawal, she called attention to how the aggregate of the (1) "constant use of cigarettes"; (2) "nicotine habit"; and (3) "emotional investment" made her previous self-described addiction "very different from [her] smoking now." She explained that, "Now I rarely feel the urge to smoke, and it is easily overcome. I don't purchase cigarettes any longer; I don't include 'smoker' in my self-definition." By periodically giving up and routinely inhibiting smoking, the women were of the opinion that they were entitled to denounce publicly the identity label of smoker.

DISCUSSION

Historically, social smokers were considered marginal smoking figures, if they were characterized at all. Both academic and popular discussions have created a dominant narrative in which smoking is gradually disappearing among educated Western groups (who, if they smoke, know they ought to quit). Under a non-smoking *novus ordo seculorum*, the outlying social smoking characters ought to have become even less imaginable. However, the informant tactics reported here suggest a new smoking imagery, involving elaborate social frameworks that embrace and support smoking in contradiction of the existing medical and social discourses. Further, the strategies created by social smokers are entwined with other dimensions of identity production in which individuals engage in self-regulation to avoid the stigmatized category of smoker, including overt reciprocity and under-the-radar borrowing in their acquisition of cigarettes and lighters.

The women we studied occupy a marginalized status within a cultural context of discrimination, and they appear to have developed both an exonerating and oppositional orientation to smoking behavior and identity. They convince themselves that they occupy a social status distinct from habitual smokers. The interviewees expressed mixed motives that led them to practices such as bumming, self-regulatory measures, accountability partners, optimistic biases, compensatory measures, fumaphobia, frequent member checks, and smoking reprieves. These processes and rituals all serve to limit smoking and to verify that an individual is still outside the limits of habitual smoking. In this narrative, social smoking falls between ordered categories where behaviors that are more interesting can be observed. Our findings demonstrate the innovation of liminal beings who have found ways to participate in a highly stigmatized behavior, if only marginally.

To dismiss the category of social smoking by saying that all social smokers are smokers would miss the point. What is interesting here is how social smokers manage and sustain a culturally unavailable (stigmatized) category. Some self-defined habitual smokers practice the same behaviors our respondents reported and exhibited. However, our research suggests that self-defined social smokers differ in the fact that they self-consciously deploy particular strategies to differentiate themselves from habitual smokers. Moreover, they believe themselves to be successful in claiming and sustaining this role and do not consider either the discourses or practices they share with habitual smokers to undermine their claim.

To sustain untenable social categories, depending on the situation, individuals may pass, readjust their position, or strategically attempt to maintain ambiguous identities. The challenge is for individuals faced with a choice between a stigmatized and a culturally acceptable position. In the absence of a recognized category, individuals can create imaginary communities; in the case of social smokers these are sustained by tacit rules of reciprocity and exchange. Individuals who not only lack available categories, but who also are faced with disconfirming realities can self-consciously create seemingly unsustainable identities.

REFERENCES

Gardner, Carol Brooks. 1986. "Public Aid." *Urban Life* 15:37–69.

Goffman, Erving. [1955] 1967. "On Face-Work: An Analysis of Ritual Elements in Social Interaction." pp. 5–45 in *Interaction Ritual; Essays on Face-to-Face Behavior.* New York: Pantheon Books.

——. 1959. *The Presentation of Self in Everyday Life*. New York: Anchor Books Doubleday.

——. 1963. *Stigma: Notes on the Management of Spoiled Identity*. Englewood Cliffs, NJ: Prentice Hall.

——. 1981. "Footing." pp. 124–159 in *Forms of Talk*. Philadelphia: University of Pennsylvania Press.

Hughes, Jason. 2003. *Learning to Smoke: Tobacco Use in the West*. Chicago: The University of Chicago Press.

Hyde, Lewis. [1979] 1983. *The Gift: Imagination and the Erotic Life of Property*. New York: Vintage Books.

Mauss, Marcel. [1950] 1990. *The Gift: The Form and Reason for Exchange in Archaic Societies*. New York: W. W. Norton.

Weiner, Annette B. 1976. *Women of Value, Men of Renown: New Perspectives in Trobriand Exchange*. Austin: University of Texas Press.

47

Pimp-Controlled Prostitution

CELIA WILLIAMSON AND TERRY CLUSE-TOLAR

Although underrepresented in the sociological literature on prostitution, the relationship between pimps and prostitutes is legendary, yet often misunderstood, in popular culture. Based on qualitative interviews, Williamson and Cluse-Tolar's study offers an insightful analysis of pimp-controlled prostitutes' careers. Pimps often "turn prostitutes out," introducing them to the sex work industry. But while these women are seduced into accepting a pimp as their manager by the impression he gives that he is romantically interested in them, this behavior often turns out to be part of his "game," in which he does whatever it takes to lure them into an obligatory relationship that supports him in flashy style. The authors show how the power relationship between these prostitutes and their pimps shifts dramatically in a predictable fashion over the course of their deviant careers. At the beginning, the women have the greatest cachet, as the men must woo them into their service by "gaming" them with their charm and enticements of money and control over the customers. For a brief "honeymoon" period, the women have the freedom to move around from one pimp to another, but even here the men dominate these exchanges ("bros before hos"). The further the women get into the relationship, the less the pimps need to woo them or treat them well, and they eventually become subjected to severe forms of emotional and physical manipulation that is often violent in nature. Their careers in pimp-controlled prostitution peak early and go through a gradual, but early, decline, landing them worse off than they were before they entered the relationship. Only once they have hit bottom do they get the courage to exit, often fleeing with nothing more than their lives.

In what ways do the relationships here compare and contrast with those of married couples? How do they compare and contrast with the relationship between drug dealers and crack-addicted women? With that between male and female gang members? Can you think of any other deviant careers that follow a similar trajectory?

From Celia Williamson and Terry Cluse-Tolar, "Pimp-Controlled Prostitution," *Violence Against Women*, Vol. 8, No. 9. Copyright © 2002. Reprinted by permission of Sage Publications.

A pimp is one who controls the actions and lives off the proceeds of one or more women who work the streets. Pimps call themselves "players" and call their profession "the game." The context in which this subculture exists is called "the life" (Milner & Milner, 1972). Social scientists of the 1960s and 1970s devoted a significant amount of research energies toward exposing and understanding pimp-controlled prostitution within street-level prostitution (Goines, 1972; Heard, 1968; Milner & Milner, 1972; Slim, 1967, 1969). Street-level prostitution entails sexual acts for money or for barter that occur on and off the streets and include sexual activities in cars and motels, as dancers in gentlemen's clubs, massage parlor work, truck stops, and crack house work (Williamson, 2000). It represents that segment of the prostitution industry where there is the most violence.

... This study aims to examine pimp-related violence toward women involved in street-level prostitution within the context of pimp-controlled prostitution. To understand contemporary pimp-controlled prostitution and, more specifically, pimp-related violence, it is necessary to examine the type of relationships between pimps and prostitutes, the roles that each play in the business, and the social rules that accompany the lifestyle.

METHOD

... Information regarding the traditional pimp–prostitute phenomenon was obtained from a larger study that included both independent and pimp-controlled women. Criteria for inclusion in the study were women 18 years of age and older who were no longer involved in prostitution activities. Participants were selected through a process of purposive, or snowball, sampling by word of mouth. In total, 21 former street prostitutes from the Midwest were interviewed. Respondents ranged in age from 18 to 35. Of the total sample, 13 were Appalachian white women, 7 were African American women, and one was a Hispanic woman. The time spent in prostitution ranged from 3 months to 13 years. From this total sample, 6 of the women had pimps and 13 women worked independently. The small number of women found by the researcher who worked for pimps and were willing to be interviewed may underscore the limited access researchers have to this population and hence the importance of research in this area.

Of the six women who are the focus for this report, five were Appalachian white and one was Hispanic. They ranged in age from 18 to 28. For this subgroup, the time spent in prostitution ranged from four to eight years.

The researcher spent six months on the streets, three days per week, learning the culture, language, and geographic layout of the streets. The researcher learned where the dope houses were, who the pimps were, and how to identify a customer from a typical passerby.

Subsequently, in-depth, face-to-face interviews were conducted with six participants who were involved in pimp-controlled prostitution, and one interview was conducted with a pimp. Each interview lasted approximately two hours. Data were analyzed line by line. Codes were developed from the raw data. Codes were collapsed into larger themes. By connecting relevant themes, the researcher was able to develop the theoretical propositions that supported the subsequent theory of the lived experience of pimp-controlled prostitution.

In addition to these interviews, added interviews were conducted with some of the participants for the purpose of member checking, a process of clarification for qualitative methods, and to gather any additional missing data. Interviews were taped and transcribed verbatim. In addition to the member-checking techniques, the researcher engaged a group of social work experts in the area of street prostitution to critique the methodology and to provide guidance toward accurate interpretation of the data. This gathering is known as peer debriefing for the purpose of challenging the researcher's interpretations to increase the accuracy of the study findings. Both member checks and peer debriefing were used to enhance the credibility of the study.

FINDINGS

Pimping: Rules of the Game

Pimps involved in prostitution activities refer to this sector of the underground economy as "the game." Players, pimps, and macks are those at the top of the pimping game. To these men in power, it is a game in which they control and manipulate the actions of others subordinate to them. Monica, a prostituted woman for three years, explains,

> It's all about the game. Nothing in the game changes, but the name. It's all about getting that money. Some women have pimps that they give the money to, some are just out there on their own. (Monica)

A player or pimp has a particular manner or style of playing the game. The pimping game requires strict adherence to the rules. The idea of a game parallels the formal economy in that one can be said to be in a game; for example, he is in the real estate game. Pimps are also said to "have" game. To have game is to possess a certain amount of charisma and smooth-talking, persuasive conversation toward women.

There are several rules that one must be willing to follow to be a successful and professional pimp. Massi, a "bottom bitch"[1] to a pimp who boasts having six women in his stable,[2] outlines the rules for pimping. The most paramount rule in the pimping game is *"the pimp must get paid"* (Massi). This means [that] there cannot be any "shame in your game" (Massi); one must require and, if necessary, demand the money without shame. Second, any successful pimp will remember that the game is *"sold and not told."* This means that pimps are expected to sell it to a prospective prostitute that he wants to occupy his stable without revealing his entire game plan. To do this, he has to develop his game or "his rap." [This] consist[s] of a series of persuasive conversations similar to poetic and rhythmic scats that are philosophical in nature and ideological about life and making money. For Sonya, it was the combination of his rap and her need to feel loved by someone[:]

> For me, it was wanting to be loved and liked the words that was said. And you know, the nice things you got. I have two beautiful children that I wanted to take care of, and I guess that's the kind of hold they have on you. (Sonya)

... The third and final ingredient for successful pimping is that a pimp must have a woman or women that want to see him on top. He is looking for dedication. He is looking for someone who wants to see her man in fine clothes and driving fine cars. His success or lack of success is a reflection on her. If her man is not looking his best, then she is not a very successful ho, and this will make for an embarrassing impression. As a prostituted woman, she must work very hard to earn his respect and his love and to keep him achieving the best in material possessions. He in return invites her into his underground social network with the sense of belonging it brings and the promise of material possessions it provides[:]

> You just, you just take control of the tricks. You know what you gotta do to make your man happy.... Some prostitutes are out there for a man, for a pimp. We're out there bustin' our ass to get our money for a man. (Sonya)

> I worked for months to get my man into a new Cadillac. (Sandra)

The most well-respected pimps are called "macks." They are at the top of their game and employ many hardworking and successful prostitutes. Dominating the pimp scene are "players" who have an average stable of women, are well respected, and make a good living. Lowest in the hierarchy of pimping are tennis shoe pimps. These pimps may have one or possibly two prostitutes on the street. They are seen as least successful in the game, and unlike more successful pimps, they may do drugs and allow their women to do drugs. In this study, six of the women previously had pimps ranging from tennis shoe pimps to players.

Turning a Woman Out

A pimp's chance at gaining a woman's attention is by looking good, smelling good, flashing his possessions, and presenting himself as someone who can counter boredom with both adventure and excitement. This is rarely enough, however, to get a woman to prostitute herself. A pimp must be skilled at assessing a woman's needs and vulnerabilities. Understanding how to exploit those vulnerabilities and fulfill those unmet needs will enable him to prostitute her. Reese, a player in Toledo, Ohio, explains how he appeals to what women need:

> I tell her "Now, you need to leave them drugs alone" and I get her cleaned up. She may come here, on drugs or not on drugs, with nothing. I mean nothing. Dirty, strung out. Some of them don't even have a

social security card or state ID, nothing. I ask 'em if they want something better, you know, you can make some money. I'll set you up
right. Let you have a few things in your life. You wanna have nice
clothes, some good jewelry, be able to have your own place, maybe a
little car to drive around in? (Reese)

A pimp offers hope for the future, and women see this as an opportunity to
be financially successful. During the time a prostitute is entering the profession of
street-level prostitution, the pimp is said to be "turning her out" or has "turned
her out" on the streets to make a profit[:]

I knew this guy, and he brought me here and turned me out on the
streets. He was a pimp.... The first day, I was scared, but I got the
money. And once I seen the money, I mean, my first day I made $600
in a three-hour period. (Sonya)

Women involved with a pimp in this study were typically not engaged in
drug abuse. Pimps realize that crack is the competition and frown upon any drug
abuse in their stable. However, two women involved with tennis shoe pimps
indulged in drug use along with their pimp.

Pimp-controlled women in the study were told they were beautiful and that
men wanted them—that they desired them so much, they would pay hard-
earned money for them. In the words of Massi, the message is conveyed early
in the relationship that women are literally "sitting on a gold mine. If they
could work the game good enough, the game would work for them."

Although pimps never guaranteed emotional or financial security, the potential for success inspired women to test the waters in this new life. There was a
sense of belonging that women longed for, a sense of exciting hope for the
future, an adventure that would take them from their meager existence into a
life with a man who told them they had special skills, intelligence, and beauty.
In return for his attention, protection, and love, she would be required to work
to bring their dream into reality. Reese speaks in terms of "goals" as he works his
women:

I have them set little goals for themselves. Say they want to buy something they want; well, we would set a little goal. Say you make this
certain amount of money: Keep working and I'll set aside a little at a
time and you'll have what you want. Say if she wants a little piece a car,
she can work and I'll make sure that she gets that car. So I try to have
them set goals for themselves, something they're working for.

... Over time, as women learn the game and have become proficient in playing, they are known as thoroughbreds. Thoroughbreds are professionals in the
prostitution industry and are responsible for maintaining the market rates in the
profession. A thoroughbred is able to handle customers, command money, and
conduct business effectively and efficiently to maximize profits[:]

When you're turned out, you're just out there. You don't know what
you're doing. You're just being turned out for a new job. You're being

trained for it. And then once you get down the steps, you know, you become a thoroughbred. You don't let the guy take control of you, you take control of it. (Sonya)

Free Enterprise and Choosing Up

Pimps understand the meaning of business over personal ventures, that is, marketing a product and investing in your product first so your product can return profits. Thus, there is a honeymoon period or courting time between pimps and prostitutes. This is the time in which the pimp "runs his game." This may last one day or several months[:]

> He progressively led up to the fact that that's what he wanted. You know, he didn't come out that night when I met him and tell me, "This is what I am. This is what you need to do."... I think they really feel like they have to gain your trust before they can dump something like that on you. We spent a lot of time together. I mean ... we would go out to eat, go to the movies, and we did, you know, normal couple things. But ... in my head I'm just thinking it's just normal couple things, but he's thinking that he's winning ... that he's gonna win and I'm gonna end up doing what he wanted. And he was right. (Tracey)

Pimps understand the meaning of capitalism in that it is a pimp's prerogative to entice any woman away from another pimp. It is viewed as a component of free enterprise. Therefore, other pimps are free to attempt to seduce a woman away from her current pimp and into [their] stable for [their] financial gain. [They] may do this without retaliation from the current pimp, as the street rule is *"bros before ho's"* (Massi). The woman being approached is instructed not to respond to the seducing pimp's advances. She is never to make eye contact with another pimp. If she does, then she is "out of pockets," a term referring to a woman who puts a pimp's money at risk, and she is subject to "being broke," meaning physically reprimanded. On the other hand, the seducing pimp may also choose to "break her" and take all her money. These rules vary [from] situations where a woman is prohibited from making eye contact with a pimp to situations in which she is not allowed to make eye contact with any African American male[:]

> I mean, most of the time, if they're a true pimp, they're not gonna play like that. You know, they'll harass you and you mainly just turn away and look in the other direction or whatever and try not to come in contact with them, because if you do, then they do what you call "break you," they take your money.... He's allowed to harass you as much as he wants. But if I don't talk back to him, then I'm cool. But if I'm "out of pockets" that means you're doing something that you ain't suppose to be doing. You know, some pimps will beat you or you go through a lot of stuff.... They're in control. You do what they say. (Sonya)

In the event that a woman is dissatisfied with her current pimp, the appropriate way to switch pimps is to make a definite decision and "choose up":

> You choose up. And if you're with a pimp and you want to go with another pimp, you have to put the money in the other pimp's hand and let your man know, you know, you're leaving and going with somebody else.... I've been with three. (Elsie)

Reese explains the transaction between pimps when a woman chooses up:

> If he comes to me like a man and tells me "That's my ho now," and she done gave him the money, then that's cool. Leave your ho clothes here and go. You can take your regular clothes, the clothes I bought you to go see your family now and then, but you leave the ho clothes. But if she leaves here and is gone and then don't come back with my money and she been out there making money and giving it to him, "Nigga, that's *my* money you got."

Pimp and Prostitute Relationships in the Game

Each woman in the study had a pimp who set the rules, controlled her actions, and took her earnings. Most reported [that] they were infatuated with their pimp, but not always. Women involved with a tennis shoe pimp, a man who had only one or two women, were more likely to consider themselves in love and defined the involvement with their pimp as a relationship. The more corporate the pimp, for instance, a player possessing three or more women, the less likely it was for women to describe their feelings as love or to define their interaction as a relationship. Women's feelings were instead described as ones of infatuation, admiration, and loyalty. The more women involved with a pimp, the less probable it became for each woman to achieve a status that allowed for the comforts of his affections, time, and attention. It was more likely that each became a part of his pimp family or stable that was made up of many women. This type of arrangement between women is known as a "wife-in-law" situation, in which each prostituted woman is a member of the family that works for the benefit of the same pimp. They are known as wife-in-laws to each other. However, some women did not tolerate such arrangements and moved on, whereas others welcomed the prestige of being with a successful pimp and willingly took on the challenge and responsibilities as a prostituted woman under his direction[:]

> It's just like when a pimp goes out and gets another girl and she's in the family. She's whoring for him like you are. Like a wife-in-law is what they call it. Sometimes you just getting tired of it and you know cheating on me and you know the wife-in-law stuff where you know the wife-in-law is another girl that is working for him, and I just couldn't handle all that. (Tonya)

[Wives-in-law] may be responsible for ongoing training of recent inductees. However, the availability of wife-in-law training depends on how large the stable and how corporate the pimp. A bottom bitch, or number one lady, may also be required to work but may only use her mouth or hands when working and to save intercourse for her pimp. She may live with her pimp and may be required to train the new women joining his stable. Women may even drop off money to her after work in the event that their pimp is otherwise occupied[:]

> I know about the game because I was [his] bottom bitch. I knew everything about hoeing, tricking, or whatever. I was with [him] for eight years. He had women out here working their asses off. Wouldn't even ask him for money or nothing, not even $5, thinking that's making him respect them more. (Massi)

The true talents of a pimp, however, are in his ability to keep his women happy, command money, and portray a deep mysterious and somewhat mean demeanor about him, one that conveys the message that he is not to be crossed. He is then said to be "cold-blooded" or "icy," able to turn off any warm feelings and loving affection in exchange for certain emotional cruelty and physical harm. Two famous and successful pimps, Iceberg Slim and Ice Tea, were said to be so cold blooded they called themselves "Ice" to let everyone know their capacity for heartlessness[:]

> He would just snap. Like his whole expression would change. One day, he came to my motel room to beat my ass. And made it clear that he came over to beat on me. He said he had some extra time on his hands, that he didn't have anything to do, so he wanted me to know that he knew I was thinking about doing something stupid. And I was too. I was thinking about leaving him again. The last time I left him, I ended up in Cleveland .... He beat me until I blacked out .... But he was like that. He could be so much fun one time, silly and playing around, and the next minute, he could be something else, somebody you don't want to fuck with. (Massi)

A pimp's approach is never to cow down to his woman at any time. He cannot let love cloud his judgments concerning business. If he lets these weaknesses show, he will be left vulnerable and runs the risk of being less successful. Although pimps appear to be in control, in a sense every pimp becomes a whore to his prostitutes. The pimp rule is "purse first, ass last" (Massi). He may treat his [ho] in loving ways in return for the amount of money he requested she bring him. She must pay for his love with her sheer tenacity to work and bring him the money. She must in turn request little emotionally and financially. Because of his generosity, he gives her what he thinks she needs.

Pimp-Related Violence: Physical Control of Women

The extent to which women felt threatened by a pimp was, in part, a function of [their] evaluation of the likelihood that he [would] become violent. This threat had been realized by all of the women in the study. Pimp-related violence was sometimes unpredictable and took on many forms. However, the most revealing form of pimp-related violence was immediate attack following a violation of the rules. One such violation is leaving the "ho stroll" or designated work area early without making one's daily quota[:]

> Different pimps have different rules. I mean some of 'em set quotas with the girls, you have to make a certain amount of money before you can go home .... When I first started, I was bringing in $1,000 a day. (Tonya)

After it was found out that Debbie was holding back some money from her pimp, without hesitation she was quickly and brutally assaulted[:]

> He ended up getting mad at me one day and punched me in my chest and cracked my rib. That was cracked, and all I could remember is that I couldn't breathe. I mean, I passed out. I was knocked out all day. I was unconscious. (Debbie)

Some pimps use violence as a means of discouraging freelance work and coercing money from women who work within pimp territory but not for the benefit of any pimp[:]

> There was this guy that kept beatin' me in my head, telling me I was gonna pay him, and every time he'd see me, if I didn't give him some money, he'd punch me in my forehead. (Cara)

In the instance that a woman is found to be "out of pockets," she is subject to being broke. Carol told of an incident when she took an unauthorized leave without the permission of her pimp[:]

> When he caught me, he was like "I got you now," and he jumped out. We was in the projects. We were high as fuck off crack, me and Tony was. We were like "Oh, fuck." He was like "I got you now." He had a baseball bat, and Tony ran and left me. So, yes, I got the baseball bat, he beat me in my legs and told me "If you fall, bitch, I'm a hit you in your head and kill you." So I didn't fall, I just stood there and screamed and took it. The police came ... and they asked if I was going to press charges, and I said, "No." My face was swelling, I looked like a cabbage patch, I was horrible. (Carol)

Pimp-Related Violence: Emotional Control of Women

A pimp's success is dependent on arousing love and fear in his women. By giving his attention to more than one woman at the same time, he heightens both the love each woman desires all to herself and the fear that she may lose any part of it. However, the negative consequences of such arrangements may be jealousy and rivalry for his affections[:]

> He had got back with the girl that he had kids by, which she was already a seasoned hooker. She hated me from the jump start.

Since they were in that life, he made her deal with me. When she found out that I wanted him for me, she wanted to fight me every time she would see me and we did.... She hit me in the head with a beer glass, and I had stitches in my head. (Chris)

Relationships require a level of trust and a degree of vulnerability, and pimp–prostitute relationships are no different. Trust determines how vulnerable the person is willing to be. Without some degree of trust, interactions are limited to explicit contracts (Holmes, 1991), which is what prostitutes have with customers. "Trust involves coming to terms with the negative aspects of a partner, accepting or perhaps tolerating issues by buffering them in the broader context of the lifestyle" (Holmes, 1991, p. 79). Women take abuses from pimps in stride. They learn to cope with this relationship by not focusing on the abusive aspects for what they are but by instead encapsulating those aspects of their pimp that serve their needs for security and protection. Therefore, a pimp–prostitute relationship often lacks cognitive and behavioral consistency. What is believed and desired on the part of the prostitute and what actually happens in the relationship do not correspond and often require repeated leaps of faith on the part of the prostitute.

Leaving Pimp-Controlled Prostitution

Many factors prevent women from pursuing legal assistance. Often, women are fearful, intimidated by what may happen as a result of reporting, and may love their pimp despite his abuses. It is likely that [they] may buy into the rules of the pimping game and blame [themselves] for violating them. Women who have previously experienced the reluctance of law enforcement to take their claims about customer–related violence seriously are reluctant to take such risks where pimps are concerned[:]

I had this guy pull a gun on me, and he made me do things that I didn't want to do.... I ran to a gas station and called the police. The guy who worked at the gas station gave me his jacket to wear. All I had on was a shirt. I called the police and told them. They came. I described the guy to the police and showed them the spot. I told them what kind of car he had, what we was wearing, what he said, what he did, everything. They never even wrote anything down. They ran me for warrants, and when I didn't have any, they left. (Jerri)

Women who are fed up may choose to leave prostitution. The primary means for leaving pimp-controlled prostitution was escape[:]

I had gotten like four calls that day. And I hadn't seen him, so I had all the money on me and I just took it. I mean, none of my clothes, none of my nothing. I just took a cab to the bus station, and I went in to Amtrak police and told them what was going on. He had his own driver, and he knew that when I didn't get in the driver's car, that something was up. I took a regular cab, and I went to Amtrak police and I knew that he was coming, so I told 'em what was going on.... And, um, then the Amtrak police ... paid the cab driver to take me to the airport, and I caught a plane home. (Tracey)

DISCUSSION

Using the definition of pimping as controlling and living off the proceeds of one or more women, the findings suggest that pimp-controlled prostitution is still an integral part of street-level prostitution for some women and girls. Just how many is difficult to determine because pimp-controlled women and girls would be those most unlikely to be able to respond to requests for interviews. Pimp-controlled women in the study were reportedly subject to following the rules of the game. The old adage that "nothing in the game changes but the name" may be truer than not when viewing the dynamics of pimp-controlled prostitution. It is clear that many of the themes identified in this study have appeared in varied form in earlier studies of the 1960s and 1970s (Goines, 1972; Heard, 1968; Milner & Milner, 1972; Slim, 1967, 1969). The important point is that although pimps and prostitutes may differ [in] the extent to which they apply and adhere to the rules, even allowing for the wide range of situational differences, this code of conduct, termed *pimpology*, is a common practice in this underground society and still exists today (Hughes, Hughes, & Messick, 1999; Milner & Milner, 1972; Owens & Sheperd, 1998; Williamson, 2000).

On an interpersonal level, the power and control pimps maintain over women in their stable is akin to that used in abusive relationships. Just as pimps resemble batterers in intimate relationships (Giobbe, 1993), women working in pimp-controlled prostitution seem to be similar to those who are survivors of domestic violence. They often express feelings of love and admiration for the pimp, have their freedom and finances controlled, and may feel [that] they somehow deserve the violence they are dealt. However, there are differences in terms of the cycle of violence. Domestic violence survivors will often express that they knew when the violence was about to occur as evidenced by the building up of tension in their mate before an explosive episode. Beatings and other forms of violence occurring among pimp-controlled women may not follow a familiar pattern and may instead occur by surprise.

NOTES

1.   In a more corporate pimp family, the term *bottom bitch* refers to a woman who is the closest in rank to her pimp.

2.   A *stable* is what a pimp calls a group of women that prostitute for him.

REFERENCES

Giobbe, E. 1993. A Comparison of Pimps and Batterers. *Michigan Journal of Gender and Law*, 1(1): 33–57.

Goines, D. 1972. *Whoreson: The Story of a Ghetto Pimp*. Los Angeles: Holloway House.

Heard, N. C. 1968. *Howard Street*. New York: Signet.

Holmes, J. C. 1991. Trust and the Appraisal Process in Close Relationships. In W. H. Jones and D. Perlman (eds.), *Advances in Personal Relationships* (pp. 57–104). London: Jessica Kingsley.

Hughes, A., Hughes, A., and Messick, K. 1999. *American Pimp* [Film documentary]. United States: Metro Goldwyn Mayer Pictures.

Milner, C. A., and Milner, R. B. 1972. *Black Players*. Boston, MA: Little, Brown.

Owens, B. (Producer/Director), and Shepard, B. (Coproducer). 1998. *Pimps Up, Ho's Down* [Film documentary]. United States: Home Box Office.

Slim, I. 1967. *Trick Baby*. Los Angeles: Holloway House.

Slim, I. 1969. *Pimp: The Story of My Life*. Los Angeles: Holloway House.

Williamson, C. 2000. Entrance, Maintenance, and Exit: The Socioeconomic Influences and Cumulative Burdens of Female Street Prostitution. *Dissertation Abstracts International*, 61(02). (UMI No. 9962789).

48

Shifts and Oscillations in the Careers of Drug Traffickers

PATRICIA A. ADLER AND PETER ADLER

In this selection on exiting drug trafficking, Adler and Adler discuss the process by which people burn out of deviance. After spending several years in the upper echelons of the drug trade, many marijuana dealers and smugglers, who were initially attracted to drug trafficking by the same lure of the excitement, high life, and spontaneity that drew Wright and Deckers' burglars to their deviance, eventually find that the drawbacks of the lifestyle exceed the rewards. Their initial challenges and thrills turn to paranoia, people whom they know get busted all around them, and their risk of arrest grows. Years of excessive drug use take its toll on them physically, and they come to reevaluate the straight life they formerly rejected as boring. Yet they cannot easily quit dealing: They have developed a high-spending lifestyle that they are loath to abandon. One thing that they do is to shift around in the drug world, making changes in their involvement. When this approach doesn't bring them satisfaction, or when the factors pushing them out continue to grow, they try to retire from trafficking. Commonly, however, they quickly spend all their money and are drawn back into the business. Thus, their patterns of exiting often resemble a series of oscillations, or quittings and restartings, as they move out of deviance with great difficulty. It is interesting to note that this mode of oscillating in and out of deviance as a means of finally making an exit is often found in other forms of deviant careers, such as quitting cigarette smoking and leaving an abusive relationship.

Researchers and policy makers have a much better idea of the kinds of factors that draw people into deviance than of what happens to them once they enter these worlds. Unlike career trajectories in legitimate occupations, deviant pathways are less structured and more fluid. Many people working legitimate jobs are lodged in corporations or bureaucracies, where they have to work their way through a rigid hierarchy of rungs, progressing only as slots become available above them. Deviant work, in contrast, is entrepreneurial, lacking both the constraints and the security of conventional occupations.

Studies of career patterns in deviance are few, but there are two things we commonly do see. First, people do a lot of **shifting around**. They may rise up the ranks or slip downward; they may learn a set of skills or acquire new connections from people they know; they may hop around, combining part-time activities in one realm with those in another. These shifts are facilitated by the fact that most deviant work is unskilled or semiskilled and no credential is required. Second, deviant activities are the province of the young. As people age, they often burn out from the risk and the dangers; their friends leave the field, get arrested, or die; they start to think that a less glamorous, but more stable, lifestyle seems more appealing. So, at some point, most people think about exiting deviance. What they often find, however, is that getting out may not be as easy as getting in, especially if they have no legitimate skills to support themselves. Attempts to escape deviance, then, may be unsuccessful. What we often see is that people make several attempts to exit deviance, returning for a while and then trying to get out again. This pattern of **oscillating in and out** of deviance may characterize both occupational deviance as well as relational (battered women) or recreational (sex, cigarette smoking) forms.

We examine the case of upper-level drug dealers and smugglers to see one of the ways that this pattern is illustrated. The upper echelons of the marijuana and cocaine trade constitute a world that is rarely penetrated by sociologists. Importing and distributing tons of marijuana and kilos of cocaine at a time, successful operators can earn upward of a half million dollars per year. Their traffic in these so-called "soft" drugs constitutes a potentially lucrative occupation, yet few participants manage to accumulate any substantial sums of money, and most people envision their involvement in drug trafficking as only temporary. In this study, we focus on the career paths followed by members of one upper-level drug-dealing and drug-smuggling community. We discuss the various modes of entry into trafficking at these upper levels, contrasting them with entry into middle- and low-level trafficking. We then describe the pattern of shifts and oscillations that these dealers and smugglers experience. Once they reach the top rungs of their occupation, they begin periodically quitting and reentering the field, often changing the degree and type of their involvement upon their return. Their careers, therefore, offer insights into the problems involved in leaving deviance.

We begin by describing where our research took place, the people and activities we studied, and the methods we used. Second, we outline the process of becoming a drug trafficker, from initial recruitment through learning the trade. Third, we look at the different types of upward mobility displayed by dealers and smugglers. Fourth, we examine the career shifts and oscillations that veteran dealers and smugglers display, outlining the multiple conflicting forces that lure them both into and out of drug trafficking. We conclude by suggesting a variety of paths that dealers and smugglers pursue out of drug trafficking and discuss the problems inherent in leaving this deviant world.

SETTING AND METHOD

We based our study in Southwest County (a fictitious name), one section of a large metropolitan area in southwestern California near the Mexican border. Southwest County consisted of a handful of beach towns dotting the

Pacific Ocean, a location offering a strategic advantage for wholesale drug trafficking.

Southwest County smugglers obtained their marijuana in Mexico by the ton and their cocaine in Colombia, Bolivia, and Peru, purchasing between 10 and 40 kilos at a time. These drugs were imported into the United States along a variety of land, sea, and air routes by organized smuggling crews. Southwest County dealers then purchased the drugs and either "middled" them directly to another buyer for a small, but immediate, profit of approximately $2 to $5 per kilo of marijuana and $5,000 per kilo of cocaine or engaged in "straight dealing." As opposed to middling, straight dealing usually entailed adulterating the cocaine with such "cuts" as manitol, procaine, or inositol and then dividing the marijuana and cocaine into smaller quantities to sell them to the next-lower level of dealers. Although dealers frequently varied the amounts they bought and sold, a hierarchy of transaction levels could be roughly discerned. "Wholesale" marijuana dealers bought directly from the smugglers, purchasing anywhere from 300 to 1,000 "bricks" (averaging a kilo in weight) at a time and selling in lots of 100 to 300 bricks. "Multikilo" dealers, while not the smugglers' first connections, also engaged in upper-level trafficking, buying between 100 to 300 bricks and selling them in 25- to 100-brick quantities. These were then purchased by middle-level dealers, who filtered the marijuana through low-level and "ounce" dealers before it reached the ultimate consumer. Each time the marijuana changed hands, its price increase was dependent on a number of factors: the purchase cost; the distance it was transported (including such transportation costs as packaging, transportation equipment, and payments to employees); the amount of risk assumed; the quality of the marijuana; and the prevailing prices in each local drug market. Prices in the cocaine trade were much more predictable. After purchasing kilos of cocaine in South America for $10,000 each, smugglers sold them to Southwest County "pound" dealers in quantities of 1 to 10 kilos for $60,000 per kilo. These pound dealers usually cut the cocaine and sold pounds ($30,000) and half-pounds ($15,000) to "ounce" dealers, who in turn cut it again and sold ounces for $2,000 each to middle-level cocaine dealers known as "cut-ounce" dealers. In this fashion, the drug was middled, dealt, divided, and cut—sometimes as many as five or six times—until it was finally purchased by consumers as grams or half-grams.

Unlike low-level operators, the upper-level dealers and smugglers we studied pursued drug trafficking as a full-time occupation. If they were involved in other businesses, these were usually maintained to provide them with a legitimate front for security purposes. The profits to be made at the upper levels depended on an individual's style of operation, reliability, and security, as well as the amount of product he or she consumed. About half of the 65 smugglers and dealers we observed were successful, some earning up to three-quarters of a million dollars per year.[1] The other half continually struggled in the business, either breaking even or losing money.

Although dealers' and smugglers' business activities varied, the two kinds of drug traffickers clustered together for business and social relations, forming a moderately well-integrated community whose members pursued a "fast" lifestyle that emphasized intensive partying, casual sex, extensive travel, abundant drug consumption, and lavish spending on consumer goods. The exact size of South-west County's upper-level dealing and smuggling community was impossible to estimate because of the secrecy of its members. At these levels, the drug world was quite homogeneous. Participants were predominantly White, came from middle-class backgrounds, and had little previous criminal involvement. Although the dealers' and smugglers' social world contained both men and women, most of the serious business was conducted by the men, ranging in age from 25 to 40 years old.

We gained entry to Southwest County's upper-level drug community largely by accident. We had become friendly with a group of our neighbors who turned out be heavily involved in smuggling marijuana. Opportunistically, we seized the chance to gather data on this unexplored activity. Using key infor-mants who helped us gain the trust of other members of the community, we drew upon snowball sampling techniques and a combination of overt and covert roles to widen our network of contacts. We supplemented intensive participant observation between 1974 and 1980 with unstructured taped interviews and stayed in touch with our participants for many more years. Throughout, we employed extensive measures to cross-check the reliability of our data whenever possible. In all, we were able to closely observe 65 dealers and smugglers, as well as numerous other drug world members, including dealers' "old ladies" (girl-friends or wives), friends, and family members.

SHIFTS AND OSCILLATIONS

Despite the gratification that dealers and smugglers originally derived from the easy money, material comfort, freedom, prestige, and power associated with their careers, 90 percent of those we observed decided, at some point, to quit the business. This decision stemmed, in part, from their initial perceptions of the career as temporary ("Hell, nobody wants to be a drug dealer all their life"). Adding to these early intentions was a process of rapid aging in the career: Dealers and smugglers became increasingly aware of the restrictions and sacrifices their occupations required and got tired of living the fugitive life. They thought about, talked about, and in many cases took steps toward getting out of the drug business. But, as with entering, disengaging from drug trafficking was rarely an abrupt act. Instead, it more often resembled a series of transitions, or oscillations, out of and back into the business. For, once out of the drug world, dealers and smugglers were rarely successful in making it in the legitimate world, because they failed to cut down on their extravagant lifestyle and drug consumption. Many abandoned their efforts to reform and returned to deviance, sometimes picking up where they left off and other times shifting to a new mode of oper-ating. For example, some shifted from dealing cocaine to dealing marijuana, some dropped to a lower level of dealing, and others shifted their role within the same group of traffickers. This series of phaseouts and reentries, combined with career shifts, endured for years, dominating the pattern of their remaining involvement with the business. But it also represented the method by which many eventually broke away from drug trafficking, for each phaseout had the potential to be an individual's final departure.

Aging in the Career

Once recruited and established in the drug world, dealers and smugglers entered into a middle phase of aging in the career. This phase was characterized by a progressive loss of enchantment with their occupation. While novice dealers and smugglers found that participation in the drug world brought them thrills and status, the novelty gradually faded. Initial feelings of exhilaration and awe began to dull as individuals became increasingly jaded. This transformation was the result of an extended exposure both to the mundane, everyday business aspects of drug trafficking and to an exorbitant consumption of drugs (especially cocaine). One smuggler described how he eventually came to feel:

> It was fun, those three or four years. I never worried about money or anything. But after awhile it got real boring. There was no feeling or emotion or anything about it. I wasn't even hardly relating to my old lady anymore. Everything was just one big rush.

This frenzy of overstimulation and resulting exhaustion hastened the process of "burnout," which nearly all individuals experienced. As dealers and smugglers aged in the career, they became more sensitized to the extreme risks they faced. Cases of friends and associates who were arrested, imprisoned, or killed began to mount. Many individuals became convinced that continued drug trafficking would inevitably lead to arrest ("It's only a matter of time before you get caught"). Although dealers and smugglers generally repressed their awareness of danger, treating it as a taken-for-granted part of their daily existence, periodic crises shattered their casual attitudes, evoking strong feelings of fear. In reaction to such crises, they temporarily intensified security precautions and retreated into near isolation until they felt that the "heat" was off.

As a result of these accumulating "scares," dealers and smugglers increasingly integrated feelings of "paranoia" into their everyday lives. One dealer talked about his feelings of paranoia:

> You're always on the line. You don't lead a normal life. You're always looking over your shoulder, wondering who's at the door, having to hide everything. You learn to look behind you so well you could probably bend over and look up your ass. That's paranoia. It's a really scary, hard feeling. That's what makes you get out.

Drug world members also grew progressively weary of their exclusion from the legitimate world and the deceptions they had to manage to sustain that separation. Initially, the separation was surrounded by an alluring mystique. But as they aged in the career, this mystique became replaced by the reality of everyday boundary maintenance and the feeling of being an "expatriated citizen within one's own country." One smuggler who was contemplating quitting described the effects of the separation:

> I'm so sick of looking over my shoulder, having to sit in my house and worry about one of my non–drug world friends stopping in when I'm doing business. Do you know how awful that is? It's like leading a double life. It's ridiculous. That's what makes it not worth it. It'll be a lot less money [to quit], but a lot less pressure.

Thus, although the drug world was somewhat restricted, it was not an encapsulated community, and dealers' and smugglers' continuous involvement with the straight world made the temptation to adhere to normative standards and "go straight" omnipresent. With the occupation's novelty worn off and the "fast life" taken for granted, most dealers and smugglers felt that the occupation no longer resembled their early impressions of it. Once they reached the upper levels, their experience began to change. Eventually, the rewards of trafficking no longer seemed to justify the strain and risk involved. It was at this point that the straight world's formerly dull ambience became transformed (at least in theory) into a potential haven.

Phasing Out

Three factors inhibited dealers and smugglers from leaving the drug world. Primary among these factors were the hedonistic and materialistic satisfactions the drug world provided. Once accustomed to earning vast quantities of money quickly and easily, individuals found it exceedingly difficult to return to the income scale of the straight world. They were also reluctant to abandon the pleasure of the "fast life" and its accompanying drugs, casual sex, and power. Second, dealers and smugglers identified with, and developed a commitment to, the occupation of drug trafficking (Adler and Adler, 1982). Their self-images were tied to that role and could not be easily disengaged. The years invested in their careers (learning the trade, forming connections, building reputations) strengthened their involvement with both the occupation and the drug community. And since their relationships were social as well as business related, friendship ties bound individuals to dealing. As one dealer in the midst of struggling to phase out explained,

> The biggest threat to me is to get caught up sitting around the house with friends that are into dealing. I'm trying to stay away from them, change my habits.

Third, dealers and smugglers hesitated to quit the field voluntarily because of the difficulty involved in finding another way to earn a living. Their years spent in illicit activity made it unlikely for any legitimate organizations to hire them. This situation narrowed their occupational choices considerably, leaving self-employment as one of the few remaining avenues open.

Dealers and smugglers who tried to leave the drug world generally fell into one of four patterns.[2] The first and most frequent pattern was to postpone quitting until after they could execute one last "big deal." Although the intention was sincere, individuals who chose this route rarely succeeded: The "big deal" too often remained elusive. One marijuana smuggler offered a variation on this theme:

> My plan is to make a quarter of a million dollars in four months during the prime smuggling season and get the hell out of the business.

A second pattern we observed was individuals who planned to change immediately but never did. They announced that they were quitting, yet their outward actions never varied. One dealer described his involvement with this syndrome:

> When I wake up I'll say, "Hey, I'm going to quit this cycle and just run my other business." But when you're dealing you constantly have people dropping by ounces and asking, "Can you move this?" What's your first response? Always, "Sure, for a toot."

In the third pattern of phasing out, individuals actually suspended their dealing and smuggling activities but did not replace them with an alternative source of income. Such withdrawals were usually spontaneous and prompted by exhaustion, the influence of a person from outside the drug world, or problems with the police or other associates. These kinds of phaseouts usually lasted only until the individual's money ran out, as one dealer explained:

> I got into legal trouble with the FBI a while back and I was forced to quit dealing. Everybody just cut me off completely, and I saw the danger in continuing, myself. But my high-class tastes never dwindled. Before I knew it I was in hock over $30,000. Even though I was hot, I was forced to get back into dealing to relieve some of my debts.

In the fourth pattern of phasing out, dealers and smugglers tried to move into another line of work. Alternative occupations included (1) those they had previously pursued; (2) front businesses maintained on the side while dealing or smuggling; and (3) new occupations altogether. While some people accomplished this transition successfully, there were problems inherent in all three alternatives:

1. Most people who tried resuming their former occupations found that those occupations had changed too much while they were away. In addition, they themselves had changed: They enjoyed the self-directed freedom and spontaneity associated with dealing and smuggling, and were unwilling to relinquish it.

2. Those who turned to their legitimate front businesses often found that those businesses were unable to support them. Designed to launder, rather than earn, money, most of these ventures were retail outlets (restaurants, movie theaters, automobile dealerships, small stores) with a heavy cash flow. They had become accustomed to operating under a continuous subsidy from illegal funds. Once their drug funding was cut off, they could not survive for long.

3. Many dealers and smugglers utilized the skills and connections they had developed in the drug business to create a new occupation. They exchanged their illegal commodity for a legal one and went into import–export, manufacturing, wholesaling, or retailing other merchandise. For some, the decision to prepare a legitimate career for their future retirement from the drug world followed an unsuccessful attempt to phase out into a "front" business. One husband-and-wife dealing team explained how these legitimate side businesses differed from front businesses:

> We always had a little legitimate "scam" [scheme] going, like mail-order shirts, wallets, jewelry, and the kids were always involved in that. We made a little bit of money on them. Their main purpose was for a cover. But [this business] was different; right from the start this was going to be a legal thing to push us out of the drug business.

About 10 percent of the dealers and smugglers we observed began tapering off their drug world involvement gradually, transferring their time and money into a selected legitimate endeavor. They did not try to quit drug trafficking altogether until they felt confident that their legitimate business could support them. Like spontaneous phaseouts, many of these planned withdrawals into legitimate endeavors failed to generate enough money to keep individuals from being lured into reentering the drug world.

In addition to voluntary phaseouts caused by burnout, about 40 percent of the Southwest County dealers and smugglers we observed experienced a "bust-out" at some point in their careers. Forced withdrawals from dealing or smug-gling were usually sudden and motivated by external factors, either financial, legal, or reputational. Financial bustouts generally occurred when dealers or smugglers were either "burned" or "ripped off" by others, leaving them in too much debt to rebuild their base of operation. Legal bustouts followed arrest and, possibly, incarceration: Arrested individuals were so "hot" that few of their for-mer associates would deal with them. Reputational bustouts occurred when individuals "burned" or "ripped off" others (regardless of whether they intended to do so) and were banned from business by their former circle of associates. One smuggler gave his opinion on the pervasive nature of forced phaseouts:

> Some people are smart enough to get out of it because they realize, physically, they have to. Others realize, monetarily, that they want to get out of this world before this world gets them. Those are the lucky ones. Then there are the ones who have to get out because they're hot or someone else close to them is so hot that they'd better get out. But in the end when you get out of it, nobody gets out of it out of free choice; you do it because you have to.

Death, of course, was the ultimate bustout. Some pilots met this fate because of the dangerous routes they navigated (hugging mountains, treetops, or other air-craft) and the sometimes ill-maintained and overloaded planes they flew. Despite much talk of violence, few Southwest County drug traffickers died at the hands of fellow dealers.

Reentry

Phasing out of the drug world was temporary more often than not. For many dealers and smugglers, it represented but another stage of their drug careers (although this may not have been their original intention), to be followed by a period of reinvolvement. Depending on the individual's perspective, reentry into the drug world could be viewed as either a comeback (from a forced withdrawal) or a relapse (from a voluntary withdrawal).

Most people forced out of drug trafficking were anxious to return. The decision to phase out was never theirs, and the desire to get back into dealing or smuggling was based on many of the same reasons that drew them into the field originally. Coming back from financial, legal, and reputational bustouts was possible, but difficult, and was not always successful. The dealers or smugglers had to reestablish contacts, rebuild their organization and fronting arrangements, and raise the operating capital necessary to resume dealing. More difficult was the problem of overcoming the circumstances surrounding their departure. Once smugglers and dealers resumed operating, they often found their former colleagues suspicious of them. One frustrated dealer described the effects of his prison experience:

> When I first got out of the joint [jail], none of my old friends would have anything to do with me. Finally, one guy who had been my partner told me it was because everyone was suspicious of my getting out early and thought I made a deal [with police to inform on his colleagues].

Dealers and smugglers who returned from bustouts were thus informally subjected to a trial period in which they had to reestablish their trustworthiness and reliability before they could once again move in the drug world with ease.

Reentry from voluntary withdrawal involved a more difficult decision-making process, but was easier to implement. The factors enticing individuals to reenter the drug world were not the same as those which motivated their original entry. As we noted earlier, experienced dealers and smugglers often privately weighed their reasons for wanting to quit and wanting to stay in. Once they left, their images of, and hopes for, the straight world failed to materialize. They could not make the shift to the norms, values, and lifestyle of the straight society and could not earn a living within it. Thus, dealers and smugglers decided to reenter the drug business for basic reasons: the material perquisites, the hedonistic gratifications, the social ties, and the fact that they had nowhere else to go.

Once this decision was made, the actual process of reentry was relatively easy. One dealer described how the door back into dealing remained open for those who left voluntarily:

> I still see my dealer friends, I can still buy grams from them when I want to. It's the respect they have for me because I stepped out of it without being busted or burning someone. I'm coming out with a good reputation, and even though the scene is a whirlwind—people moving up, moving down, in, out—if I didn't see anybody for a year I could call them up and get right back in that day.

People who relapsed thus had not much of a problem obtaining fronts, reestablishing their reputations, or readjusting to the scene.

Career Shifts

Dealers and smugglers who reentered the drug world, whether from a voluntary or forced phaseout, did not always return to the same level of transaction or commodity that characterized their previous style of operation. Many individuals underwent a "career shift" and became involved in some new segment of the drug world. These shifts were sometimes lateral, as when a member of a smuggling crew took on a new specialization, switching from piloting to operating a stash house, for example. One dealer described how he utilized friendship networks upon his reentry to shift from cocaine to marijuana trafficking:

> Before, when I was dealing cocaine, I was too caught up in using the drug and people around me were starting to go under from getting into "base" [another form of cocaine]. That's why I got out. But now I think I've got myself together and even though I'm dealing again I'm staying away from coke. I've switched over to dealing grass. It's a whole different circle of people. I got into it through a close friend I used to know before, but I never did business with him because he did grass and I did coke.

Vertical shifts moved operators to different levels. For example, one former smuggler returned and began dealing; another, top-level, marijuana dealer came back to find that the smugglers he knew had disappeared and he was forced to buy in smaller quantities from other dealers.

Another type of shift relocated drug traffickers in different styles of operation. One dealer described how he tightened his security measures after being arrested:

> I just had to cut back after I went through those changes. Hell, I'm not getting any younger and the idea of going to prison bothers me a lot more than it did 10 years ago. The risks are no longer worth it when I can have a comfortable income with less risk. So I only sell to four people now. I don't care if they buy a pound or a gram.

A former smuggler who sold his operation and lost all his money during phaseout returned as a consultant to the industry, selling his expertise to those with new money and fresh manpower:

> What I've been doing lately is setting up deals for people. I've got foolproof plans for smuggling cocaine up here from Colombia; I tell them how to modify their airplanes to add on extra fuel tanks and to fit in more weed, coke, or whatever they bring up. Then I set them up with refueling points all up and down Central America, tell them how to bring it up here, what points to come in at, and what kind of receiving unit to use. Then they do it all and I get 10 percent of what they make.

Reentry did not always involve a shift to a new niche, however. Some dealers and smugglers returned to the same circle of associates, trafficking activity, and commodity they worked with prior to their departure. Thus, drug dealers' careers often peaked early and then displayed a variety of shifts, from lateral mobility, to decline, to holding fairly steady.

A final alternative involved neither completely leaving nor remaining within the deviant world. Many individuals straddled the deviant and respectable worlds forever by continuing to dabble in drug trafficking. As a result of their experiences in the drug world, they developed a deviant self-identity and a deviant *modus operandi*. They might not have wanted to bear the social and legal burden of full-time deviant work, but neither were they willing to assume the perceived confines and limitations of the straight world. They, therefore, moved into the entrepreneurial realm, where their daily activities involved some kind of hustling or "wheeling and dealing" in an assortment of legitimate, quasi-legitimate, and deviant ventures, and where they could be their own boss. This dual existence enabled them to retain certain elements of the deviant lifestyle and to socialize on the fringes of the drug community. For these individuals, drug dealing shifted from a primary occupation to a sideline.

LEAVING DRUG TRAFFICKING

This career pattern of oscillation into and out of active drug trafficking makes it difficult to speak of leaving drug trafficking in the sense of final retirement. Clearly, some people succeeded in voluntarily retiring. Of these, a few managed to prepare a postdeviant career for themselves by transferring their drug money into a legitimate enterprise. A larger group was forced out of dealing and either didn't or couldn't return: The bustouts were sufficiently damaging that these people never attempted reentry, or they abandoned efforts after a series of unsuccessful attempts. But there was no way of structurally determining in advance whether an exit from the business would be temporary or permanent. The vacillations in dealers' intentions were compounded by the complexity of operating successfully in the drug world. For many, then, no phaseout could ever be definitely assessed as permanent. As long as individuals had skills, knowledge, and connections to deal, they retained the potential to reenter the occupation at any time. Leaving drug trafficking may thus be a relative phenomenon, characterized by a trailing-off process in which spurts of involvement appear with decreasing frequency and intensity.

SUMMARY

Drug dealing and smuggling careers are temporary and fraught with multiple attempts at retirement. Veteran drug traffickers quit their occupation because of the ambivalent feelings they develop toward their deviant life. As they age in the career, their experience changes, shifting from a work life that is exhilarating and free to one that becomes increasingly dangerous and confining. But just as their deviant careers are temporary, so, too, are their retirements. Potential recruits are lured into the drug business by materialism, hedonism, glamor, and excitement. Established dealers are lured away from the deviant life and back into the mainstream by the attractions of security and social ease. Retired dealers and smugglers are lured back in by their expertise and by their ability to make money quickly and easily. People who have been exposed to the upper levels of drug trafficking therefore find it extremely difficult to quit their deviant occupation permanently. Their inability to quit stems, in part, from their difficulty in moving from the illegitimate to the legitimate business sector. Even more significant is the affinity they form for their deviant values and lifestyle. Thus, few, if any, of our subjects were successful in leaving deviance entirely. What dealers and smugglers intend, at the time, to be a permanent withdrawal from drug trafficking can be seen in retrospect as a pervasive occupational pattern of midcareer shifts and oscillations. More research is needed into the complex process of how people get out of deviance and enter the world of legitimate work.

NOTES

1. This is an idealized figure representing the profit a dealer or smuggler *could* earn and does not include deductions for such miscellaneous and hard-to-calculate costs as time or money spent in arranging deals (some of which never materialize); lost, stolen, or unrepaid money or drugs; and the personal drug consumption of a drug trafficker and his or her entourage. Of these costs, the single largest expense is the last one, accounting for the bulk of deductions from most Southwest County dealers' and smugglers' earnings.

2. At this point, a limitation on our data must be noted. Many of the dealers and smugglers we observed simply "disappeared" from the scene and were never heard from again. We therefore have no way of knowing if they phased out (voluntarily or involuntarily), shifted to another scene, or were killed in some remote place. We cannot, therefore, estimate the numbers of people who left the Southwest County drug scene via each of the routes discussed here. It is also impossible to determine the exact percentage of people falling into the different phaseout categories: Because of oscillation, people could experience several types of phaseout and thus appear in multiple categories.

REFERENCE

Adler, Patricia A., and Peter Adler. 1982. "Criminal Commitment Among Drug Dealers." *Deviant Behavior* 3: 117–135.

49

Obstacles to Exiting Emotional Disorder Identities

JENNA HOWARD

Howard's chapter on how people with emotional disorders grow to challenge and become discontented with their diagnoses shows us the identity dimensions of exiting deviance. In reading about fat and bisexual identities earlier, we saw the kinds of processes people underwent to change and enter into deviant identities. But Howard shows how it is equally difficult to exit them, reinforced by the tendency of comfort and friendships to hold the individuals to the status quo. The people with emotional disorders she describes have often bought into their psychological diagnoses with difficulty, suffering and in pain. Quitting them is likely to bring an equal amount of psychological trauma.

How does exiting emotional disorder identities compare with quitting drug trafficking? Are the stages of the process, the shifts around, or the oscillations in and out similar or different? What factors are pushing people out of each versus pulling them toward other activities or identities? How do Howard's existential, interactional, and cultural barriers to identity change compare with the active and passive status cues described by Degher and Hughes (Chapter 23) for the fat people?

Considering subjective identity career entrances and exits can be a rich source of social psychological insight into identity processes. At what point do particular self-definitions begin and at what point do they end? While some identity careers have "highly articulated" (Glaser and Strauss 1967) durations that are publicly marked by explicit symbols of entrance and exit (such as acquiring a uniform or receiving a diploma), the duration of other identity careers is more subjectively determined. One such identity career involves identities that are based on emotional disorder labels. Because the sensations and experiences that qualify as symptoms of these conditions tend to be internally located and lack visible boundaries, the determination of whether to consider oneself ill or not may be highly subjective. This study explores the subjective self-meanings of people who formerly identified with emotional disorder labels and no longer

From Jenna Howard. "Negotiating an Exit: Existential, Interactional, and Cultural Obstacles to Disorder Disidentification." *Social Psychology Quarterly*, Vol. 71(2). Copyright © 2008. Reprinted by permission of the American Psychological Association.

do (*delabelers*). My analysis of their narratives offers nuance to our understanding of identification and disidentification processes by highlighting the tension between the individual's decision to discard the disorder labels and the simultaneous reluctance to relinquish the associated identity. In particular, I focus on delabelers' descriptions of the difficulties involved in the disidentification process, classifying them as existential, interactional, and cultural obstacles to disidentification ....

This study is based on in-depth interviews with 40 individuals who claim to have formerly identified with emotional disorder labels and no longer use the label as a source of subjective identification. I refer to these individuals as *delabelers*. These individuals formerly identified with a wide range of emotional disorder labels as diverse as "anorexic," "codependent," "bipolar," or "agoraphobic"....[T]hese individuals, by definition, have subjectively disidentified with their emotional disorder labels, and their narratives offer insight into the changing subjective meanings of their disorder identities over time as well as into the exiting process itself. One of the outstanding themes common to a majority of delabelers' narratives depicts the disidentification process as difficult ....

This focus on the challenges to exiting experienced by this underresearched population offers insight into the experiences of delabelers specifically and it provides a sensitizing framework for broadening our inquiry into the processes and consequences of labeling within a mental health context.

DATA AND METHODS

The only criterion for being considered a delabeler is to have formerly identified oneself with a labeled emotional disorder. I do not distinguish between those who were professionally labeled or "self-labeled" (Thoits 1985), nor do I consider the current presence or absence of "symptoms" in my inclusion criterion. This means that being a delabeler is not necessarily synonymous with being "cured" (although in some cases it may be); it simply implies that one has chosen to no longer use the disorder label as a source of identity, for any reason. The scope of this work is limited to issues of self-identification; it does not address objective clinical outcomes ...

The unmarked nature of this population makes recruitment difficult because individuals who have disidentified from their labels generally cannot be found attending support group meetings or treatment centers. Therefore, to locate the forty delabelers that make up this sample, I depended on snowball sampling and advertising with flyers, including voluntary postings by two informants to online networks of therapists and social workers of which they are members. As a result, my informants came from ten different states, and all but four local respondents were interviewed in tape-recorded phone sessions.

It is a primarily female sample (31 women and nine men), ranging in age from 20 to 69. More than a third are over 50, and half are in social service, mental health, or other health-related professions. The gender skew may be attributed to the increasing feminization of psychotherapy and other care-giving professions as well to the socialized gender differences in the level of comfort with speaking about emotional problems. The occupational concentration may be due in part to the recruitment from the two online postings that target professionals in these fields, as well as to the "professional-ex phenomenon" (Brown 1991), which acknowledges that former sufferers are often attracted to helping professions. One consequence is that I may have inadvertently selected an especially self-reflective sample, both because reflective qualities can be expected to increase with age and because social service and mental health professions encourage sophistication in psychological matters. The prevalence of these occupations in this sample may have additionally influenced my findings because these individuals likely had greater access to professional discourses on mental health than the average person. Exposure to these discourses could have provided them with interpretive resources that may have informed their disidentification experiences.

While the range of disorder labels represented in this sample is wide, the distribution leans heavily toward the conditions that would be classified in the DSM-IV-TR (2000) as anxiety, eating, substance, and mood disorders. There are only eight cases of identification with psychotic or dissociative disorders. Nearly one quarter of the sample identified/disidentified with more than one label. A majority identified with their labels for less than ten years, while eleven delabelers report having identified with the labels for more than ten years ....

OBSTACLES TO DISIDENTIFICATION

All delabelers, by definition, ultimately decide to discard their disorder labels. For most, however, this identity exit is fraught with intra- and interpersonal conflicts that make the process emotionally difficult, thus posing obstacles to the disidentification process. In what follows, I use excerpts from delabelers' narratives to illustrate these obstacles on existential, interactional, and cultural levels. While these three types of obstacles are experientially interrelated, I discuss them separately for analytic clarity.

EXISTENTIAL OBSTACLES

Delabelers' narratives suggest that exiting disorder identities can be an existentially uncomfortable experience. Several delabelers describe having gone through a period of questioning, "Who am I now?" after deciding to disidentify with their labels, and some talk about having anticipated the identity transition with a great deal of anxiety about a potential lack of self-meaning. These delabelers describe the existential uncertainty that can accompany the decision to discard the disorder label as a profound obstacle in the disidentification process.

One of the reasons that discarding disorder identities can be so existentially unsettling is that it involves a transition from a culturally "marked" category to an "unmarked" (Brekhus 1998) one. Many identity transitions involve a transition from one identity status to a subsequent "ex-role" (Ebaugh 1988) (i.e., employee to retiree; spouse to divorcee; native to immigrant); however, disidentifying from emotional disorder labels does not involve adopting a new, labeled status. Rather, the transition moves the exiting individual from a highly marked, culturally recognized status (with associated support groups, national organizations, media attention, and self-help literature directed toward that status) to a completely unmarked, unrecognized *nonidentity*. Instead of replacing one identity with another (as when a transsexual trades one gender identity for a different one) this identity transition requires individuals to simply forfeit a source of identity ....

In addition to letting go of a marked identity in a cultural context that does not provide a marked ex-identity, delabelers face an additional existential challenge in their exiting processes due to their reason for discarding the disorder identity. Almost unanimously, delabelers explain their decision to discard the label as a result of becoming aware that the disorder identification had begun to have internally limiting consequences for their self-perceptions. More specifically, they commonly describe a realization that continuing to identify as "disordered" was narrowing the complexity of their self-concepts. In many cases, the disorder label had developed into a master status. As delabeler Gretchen remarks about her former "addict" identity, "I built who I was around it ... My entire life was recovery!"

Many delabelers explain that this recognition of the limiting consequences of the disorder identity initiated a more general questioning of the nature and consequences of all individual identity. For this reason, many conclude that simply replacing their attachment to the disorder identity with another categorical entity would not help them avoid the limitations that prompted them to discard the disorder identity in the first place. This creates an existential predicament for a number of delabelers after they decide to discard their disorder identities: they are faced with an *identity void*. Delabeler Eliza realized as she was discarding her disorder identity ("sex and love addict") that to replace it with another label would be "pointless": "It would be like sticking myself in another box that I would then have to climb out of!" From this perspective, the decision to discard the disorder labels is better characterized as walking into an identity void than as straddling two distinct identities.

For this reason many delabelers describe feeling a tremendous reluctance to turn in their illness identities. Anne's narrative offers some insight into the experience of the identity void. In her story about her exit from her identification as a "multiple" ([one with] multiple personality disorder) she explains why it can feel desirable to hold onto a label even after determining it has "out-served its usefulness." After having considered herself a "multiple" for many years she became aware that she "needed to be very cautious" about using the words "I am" in reference to her label. She ultimately came to believe, "By saying it over and over ('I am this; I am that') we start to define ourselves in a way that's hard to break because we really believe that's who we are, as opposed to being more than that." Despite this insight, however, the prospect of disidentifying with a label that had provided a way for her to understand her experience for so many years was unsettling. Articulating her reasoning for wanting to maintain her disorder identity, she poignantly explains, "It can be so comforting to know who you are, even if ifs a false self" ...

Naomi's narrative offers a striking example of an extended experience of such ambivalence. She was diagnosed with multiple personality disorder and was hospitalized in an intensive "MPD unit," where she was assumed to be a "textbook" case of the disorder The staff's assessment, she recalls, was that "if I wasn't MPD, nobody was." There was, however, much more conflict in her ability to understand her own condition. She describes a dual relationship with her disorder identity: on the one hand, she claims that she always doubted whether the diagnosis was accurate, and at the same time, she was deeply attached to the label. She describes feeling great anxiety at the very thought of not having that label with which to identify. Describing these contradictory feelings she explains:

> I knew this diagnosis was wrong on a level that was not allowed to be discussed .... But at the same time I had become terrified that if anyone contradicted the idea that I was a 'multiple,' then what in the hell had my life been about? ... If I wasn't 'multiple' then what the hell was going on? I was very invested in, 'You better believe me! Don't you dare tell me I'm not 'multiple'. And, on another level I knew I wasn't.

Naomi struggled with this dual understanding of her identification for several years. So much of her self-understanding depended on her disorder label, and yet she was also eager to let it go because she was simultaneously convinced that "this was not [her] story." She had, however, "invested" so much of her sense of self into this label that the prospect of discarding it forced her to face a "terrifying" identity void ....

The anxiety of this existential uncertainty can present a serious obstacle to the disorder identity exiting process. Identity transitions of any kind can be difficult, but when they involve giving up an identity without the cultural support of a recognized ex-identity, there can be a disconcerting sense of existential loss. This challenge, delabelers' narratives suggest, is compounded when the exiting process is actually prompted by a questioning of the nature and consequences of personal identity itself.

Interactional Obstacles

In addition to the existential anxieties about displaced identity, delabelers also describe interactional obstacles that make exiting disorder identities especially difficult and stressful. Issues of group solidarity are common in these narratives, particularly for individuals who participated in support groups or associated with other people who shared their disorder labels. Although the existential aspect of the exiting process tends to initiate feelings of anxiety, the interactional aspect of disassociating with the group identity can trigger feelings of guilt and fear. Both guilt and fear are closely associated with issues of loyalty to the group: one side includes individuals' feelings of indebtedness toward the group and the guilt that results from considering leaving (*the deserter complex*), whereas the other side involves the fear of being ostracized by the group for choosing to disidentify with the label (*reverse stigmatization*). Both are expressed in delabelers' narratives as consequences of considering the process of exiting the collective identity.

Deserter complex   Two similar examples of the deserter complex are expressed by Judy and Kasey who were participants in 12-step recovery groups, Co-dependents Anonymous (CoDA) and Alcoholics Anonymous (AA), respectively. After several years, they both felt that they had "outgrown" these groups and the disorder labels associated with them. Despite this recognition, deciding to exit was not straightforward for either of them, as they both suffered from guilt for deserting the group that they felt had, at one time, been so helpful to them. Judy explains that even though she was "ready to move on," she remembers:

> I felt like I was abandoning them … I thought I was supposed to stay. But, then I realized I can't rescue anyone. I had to do this for me. It's not that I didn't want to pay back what the group had done for me … [but] I came to realize that I had been giving all along by participating in the group, even in my need. The hard part, though, is letting go.

Kasey describes a similar sentiment in her story about deciding to disidentify as an "alcoholic" and leave AA after nine years:

> I wasn't drinking anymore, and I wasn't having cravings. So, some-where in my ninth year I stopped feeling the need to go to meetings … I didn't have as much time to spare to go to meetings since I started going to graduate school, but I was feeling guilty about not going.

Even after participation is no longer considered to be beneficial, loyalty to the group spawning the deserter complex can be understood by considering the group solidarity that is often generated in these small groups ….

Delabeler Reina describes having experienced feelings of solidarity guilt when she decided to minimize her involvement in the regional telephone sup-port network she had established for people suffering from anxiety and panic attacks. As the founder, Reina had developed a reputation within the network as being an especially supportive, sympathetic listener for "phobics" who called her in need. Similar to the 12-step philosophy, her network was founded on the principle that identifying with fellow sufferers and sharing similar experiences is mutually beneficial for both participants' recovery processes. As her own "re-covery" progressed, however, she found herself less interested in talking about anxiety and decided to remove herself from the network she had built. She experienced a great deal of guilt when someone would call, and she would have to tell them that she didn't have the time to talk that she used to have. In her guilt, she would question herself[:] "Am I still a good person? I don't want to be an insensitive, uncaring person by not being there to talk to every phobic … but I can't share that space with them anymore in the same way. I'm just not there."

This description of ambivalence and guilt for abandoning one's fellow sufferers parallels mobility experiences more generally because all mobility narratives, by definition, are formally similar in at least two ways: they involve both departure and progress. That is, one must not only leave (even if only symbolically) a former association but also surpass it for an association of a socially perceived higher status ....

Reverse Stigmatization Delabelers narratives express both sides of the interactional coin: group loyalty not only has the capacity to trigger feelings [of] guilt for "abandoning" the group but also fear of being ostracized by the group for choosing to leave. The prospect of losing close friends or being considered "in denial" is a difficult prospect for some individuals to face. This ostracism from the group (actual or anticipated) can be understood as a form of stigma for behaving as a deviant, from the perspective of the group of labeled individuals. I refer to this as reverse stigmatization because typical conceptions of stigma assume the culturally dominant "normals" to be the "reference group" according to which the stigmatized deviants are defined. In this case, the reference groups are reversed. Delabelers' narratives suggest that this reverse stigma (or even the perceived potential for it) can be quite distressing and can discourage disidentification.

Gretchen's narrative illustrates this dilemma when she discusses her fear of stigmatization by her former recovery community for choosing to discard her "addict" identity. As a college student, she had lived in a special dorm for recovering students, regularly attended AA and NA [Narcotics Anonymous] meetings, and interacted only with friends who were "in recovery." When the semester ended she moved out of the recovery dorm and chose to stop going to meetings. Despite her conviction that she had made the right decision for herself, she finds herself "deathly afraid of the judgments" of her "recovery friends." Anticipating their reactions she admits, "I know I'll look like a failure to them. Even though I don't see it like that, to them I am. I dread that." Although this fear did not ultimately prevent her from exiting the disorder identity, she admits that even now (several months later) she has remained fearful of encountering friends from recovery and facing the anticipated criticism ....

Both the deserter complex and reverse stigmatization demonstrate a deep degree of attachment to the recovery community. This attachment seems incongruous with delabelers' assertions about the disorder label having "outlived its purpose" or their having already "gotten what [they] could" out of participation in the recovery group. Considering the expressed desire for disidentification from the disorder identity, it seems surprising that delabelers would care so much about what these former group members would think about their decision to leave. Why should the opinions of the group members matter if delabelers have already decided to extricate themselves from these groups? One answer reflects the crucial issue involved in collective disorder identification: the recovery group is not only a treatment resource but also a source of community. Thus, to dissociate with such a community involves both forfeiting a sense of belonging as well as rejecting the microculture of the local group. From this perspective, it is understandable that individuals would experience attachment to their recovery communities and care about the opinions of the other participants, even after deciding to discard their disorder labels.

Cultural Obstacles

In addition to the existential and interactional obstacles to disidentifying with their labels, delabelers also describe cultural pressure to remain identified as "disordered" ...

Delabelers' narratives reflect this cultural trend of assuming that a label is needed in order to understand and cope with life's difficulties. For example, delabelers Keith and Vicky suggest that this belief influenced their identification with their respective disorder labels. When Keith tells of how he came to see himself as "disordered," he explains that because he was unhappy and having problems in his marriage, he felt he "had to go somewhere," implying that he should participate in some kind of therapeutic setting. He says that because he did not know where else to go, he went to an Alanon meeting "because it was available" and he was familiar with the name. When he eventually concluded that he had "gotten all that [he] could" from Alanon, he continued to believe that he "had to go somewhere," and so he began identifying as a "codependent" and attending CoDA meetings, upon the advice of his mother.

Similarly, Vicky explains that after five years of AA membership, she concluded that she was not an "alcoholic" but found herself assuming, "I should be in something ...." She explains:

> I went around and around and could never figure out which group to go to: Should I go to AA again? No, that doesn't fit anymore. Should I go to an Adult Child meeting? Alanon? Nothing really ever fit right so I ended up never going back, but many times I thought maybe I should.

Like Keith, Vicky felt that because she continued to have some emotional difficulties, ... she should find a new disorder label to classify her trouble.

The prevalent assumption that disorder labels are needed in order to understand and cope with difficult life experiences can be seen as symptomatic of the "triumph" of what Philip Rieff (1966) named the "therapeutic culture" in the 1960s ....

This increased psychologization of everyday life has borne an explosion in the number of therapeutic practitioners and therapeutic self-help groups; the interplay between the expert domain of psychological professionals[, on the one hand,] and popular self-help culture and media representations[, on the other,] further reinforces a cultural preoccupation with therapeutics. As Dana Becker (2005) points out "[T]he ideas that emanate from both the popular and professional cultures ... are often combined, transformed, recycled, and used for purposes other than the therapeutic." The result is a culture that increasingly views normal human emotion in pathological terms.

This transformation in our understanding of emotional disorder is not only a matter of increasing diagnostic categorization, [but] the very nature of emotional disturbance has also come to be seen more and more in biomedical terms. Karp (2006) cautions, "With the increasing acceptance of the biomedical model, we begin to believe that more and more of our feelings are illegitimate and abnormal and require biological intervention to correct." He asserts that the consequence of this belief, which is "unrelentingly pushed by pharmaceutical companies and some doctors," poses a threat to "personal autonomy and responsibility" (208).

This potential [threat] is suggested in several delabelers' narratives [which] reveal [that] an internalization of the disease concept of their emotional behavior made disidentifying with their disorder labels especially difficult. Eva, for example, was diagnosed as "clinically depressed" when she was in her early twenties. She explains:

> I was always told I had a "chemical imbalance" and that taking this pill every day would make everything wonderful. I was given various pills over a period of twenty years .... The idea that "depression" is an illness is so pervasive on the TV, and I used to read articles in magazines and books, and these people like [William] Styron come out with these books and say they suffered all their lives, and they had a chemical imbalance and now they take this one pill every day and everything is wonderful. I would read them, and I did kind of believe it was true ... I really wanted to believe because I was unhappy, and I wanted to believe that all we had to do was find the right pill and everything would be wonderful. And, I was like most people, wanting to put my trust and faith in somebody else, especially the medical profession and have them tell me what to do.

Eva's eventual inclination to disidentify with the disorder label posed a challenge for her because of her desire to trust the medical system, as well as the cultural encouragement to feel that she should trust it. In retrospect, she realizes that in the hands of the psychiatrists, she had become "very passive and dependent; dependent on somebody else to figure things out." She eventually began to internally question the psychiatric authorities, but she kept it a private struggle because she had "never heard of something that was not a biochemical analysis [of depression]." Deciding to discard the disorder label was, therefore, so difficult because it required her to take a stand against a cultural authority when she believed that she was alone in her views. She emphasizes, "I always thought it [doubting 'the system'] was just me." ....

DISCUSSION

Delabelers' narratives suggest that the decision to discard one's disorder identity is often just the beginning of a difficult process of disidentification, fraught with a variety of obstacles. It seems ... that the extent to which one is able to successfully negotiate the disorder identity exit would be related to the amount of "recovery capital" one possesses. Robert Granfield and William Cloud (1999) have coined this term to refer to the total number of one's resources that can be used to "promote and sustain a recovery experience" (179). They explain that such resources can include various forms of physical, social, and human capital such as one's financial status, friendship networks, and vocational skills ....

Individuals who have involvements in alternate jobs, relationships, and hobbies have a "far easier adjustment" than individuals who exited without such "bridges." As a general rule, Ebaugh (1998) suggests that "there seems to be a direct relationship between the number and quality of bridges and the degree of role adjustment and happiness after the exit."

In addition, delabeler Stacey suggests how financial resources may serve as a form of recovery capital. She started receiving Supplemental Security Income (SSI) after being diagnosed with Post-Traumatic Stress Disorder (PTSD) following what she describes as a traumatic loss. She explains that, until recently, her anxiety symptoms made holding a job impossible, and so she depended on the disability benefits for survival. She is now employed in her first part-time job since the diagnosis but does not feel quite ready for a full-time position. In the meantime, she continues to claim eligibility for half of her original benefits because her salary is only about $600 each month. She explains, although she no longer identifies with her diagnosis, using the disorder label to claim benefits makes a complete disidentification from the label difficult:

> One of the difficulties for me now is that, on the one hand, I don't want to be limited by a label or think I'm limited because of what I've gone through. But, at the same time, in terms of losing my benefits, sometimes I find myself having to argue for the limitations, and that's really frustrating for me. I don't want to be arguing it so convincingly that I start to convince myself!

Stacy's story suggests one way that financial resources can serve as recovery capital[:] if she had not been financially dependent on SSI, she likely would have felt free to disidentify from the label entirely.

To be clear, this discussion of disorder disidentification is not to suggest that everyone who identifies with an emotional disorder label *should* eventually disidentify with it. In fact, as Estroff (1981) suggests, for many chronically ill psychiatric patients, the diagnostic label may provide a sense of identity and resources that they may be unlikely to receive otherwise. This analysis of delabelers is thus not *recommending* disidentification. It is, however, highlighting, the intra- and interpersonal forces that may discourage disidentification, even in cases where it is desirable.

Several delabelers claim that their primary reason for participating in this study is to make it easier for others to discard their disorder labels than it was for them. Since the recovery discourse is so focused on acquiring disorder labels and coping with the symptoms of problematic emotional conditions, the possibility of exit is typically not emphasized. One important step in addressing the existential, interactional, and cultural obstacles to disidentification could be to incorporate a notion of "recovery from recovery" into the discourse so that disidentification can gain legitimacy as a possible course for some disorder identity careers.

REFERENCES

American Psychiatric Association. 2000. *Diagnostic and Statistical Manual of Mental Disorders*, 4th edition, text revision. Arlington, VA: author.

Becker, Dana. 2005. *The Myth of Empowerment: Women and the Therapeutic Culture in America*. New York and London: New York University Press.

Brekhus, Wayne. 1998. "A Sociology of the Unmarked: Redirecting Our Focus." *Sociological Theory* 16: 34–49.

Brown, David J. 1991. "The Professional Ex: An Alternative for Exiting the Deviant Career." *Sociological Quarterly* 32(2): 219–30.

Ebaugh, Helen Rose Fuchs. 1988. *Becoming an Ex: The Process of Role Exit*. Chicago and London: University of Chicago Press.

Estroff, Sue E. 1981. *Making It Crazy: An Ethnography of Psychiatric Clients in an American Community*. Berkeley and Los Angeles: University of California Press.

Glaser, Barney G., and Anselm L. Strauss. 1967. *The Discovery of Grounded Theory*. Chicago: Aldine.

Granfield, Robert and William Cloud. 1999. *Coming Clean: Overcoming Addiction Without treatment*. New York: New York University Press.

Karp, David. 2006. *Is It Me or My Meds? Living with Antidepressants*. Cambridge, MA: Harvard University Press.

Rieff, Philip. 1966. *The Triumph of the Therapeutic: Uses of Faith After Freud*. Chicago: University of Chicago Press.

Thoits, Peggy A. 1985. "Self-Labeling Processes in Mental Illness: The Role of Emotional Deviance." *American Journal of Sociology* 91: 149–221.

References for the General and Part Introductions

Adler, Patricia A., and Peter Adler. 1987. *Membership Roles in Field Research*. Newbury Park, CA: Sage.

_____. 2006. "Deviant Identity." In George Ritzer, *Encyclopedia of Sociology*. Malden, MA: Blackwell Publishing Ltd.

Bandura, Albert. 1973. *Aggression: A Social Learning Approach*. Englewood Cliffs, NJ: Prentice Hall.

Becker, Howard S. 1963. *Outsiders: Studies in the Sociology of Deviance*. New York: Free Press.

_____. 1973. "Labeling Theory Reconsidered." pp. 177–212 in *Outsiders*. New York: Free Press.

Best, Joel, and David F. Luckenbill. 1980. "The Social Organization of Deviants." *Social Problems* 28(1): 14–31.

_____. 1981. "The Social Organization of Deviance." *Deviant Behavior* 2: 231–59.

Cloward, Richard, and Lloyd Ohlin. 1960. *Delinquency and Opportunity*. Glencoe, IL: Free Press.

Cohen, Albert. 1955. *Delinquent Boys*. Glencoe, IL: Free Press.

Cohen, Stanley. 1972. *Folk Devils and Moral Panics: The Creation of the Mods and the Rockers*. London: MacGibbon and Kee Ltd.

Conrad, Peter, and Joseph W. Snyder. 1980. *Deviance and Medicalization*. St. Louis: Mosby.

Davis, Fred. 1961. "Deviance Disavowal: The Management of Strained Interaction by the Visibly Handicapped." *Social Problems* 9: 120–32.

Erikson, Kai T. 1966. *Wayward Puritans*. New York: Wiley.

Eysenck, Hans. 1977. *Crime and Personality*, Third Edition. London: Routledge and Kegan Paul.

Force, William Ryan. 2005. "There are no victims here: Determination versus disorder in pro-anorexia." Paper presented at the Couch-Stone Symposium of the Society for the Study for Symbolic Interaction, Boulder, CO, February.

Freud, Sigmund. 1925. *The Standard Edition of the Complete Psychological Works of Sigmund Freud.* London: Hogarth Press.

Goddard, Henry. 1979. *Feeblemindedness.* New York: Macmillan.

Godson, Roy, and William J. Olson. 1995. "International Organized Crime." *Society* 32: 18–29.

Goffman, Erving. 1963. *Stigma.* Englewood Cliffs, NJ: Prentice Hall.

Goode, Erich, and Nachman Ben-Yehuda. (1994). *Moral Panics.* Cambridge, MA: Blackwell.

Goring, Charles. 1913. *The English Convict.* London: His Majesty's Stationary Office.

Goffman, Erving. 1961. *Asylums.* New York: Doubleday.

Henry, Jules. 1964. *Jungle People.* New York: Vintage.

Hewitt, John P., and Randall Stokes. 1975. "Disclaimers." *American Sociological Review* 40: 1–11.

Hilgartner, Stephen, and Charles L. Bosk. 1988. "The Rise and Fall of Social Problems: A Public Arenas Model." *American Journal of Sociology* 94: 53–78.

Hooton, Earnest. 1939. *The American Criminal.* Westport, CT: Greenwood Press.

Hughes, Everett. 1945. "Dilemmas and Contradictions of Status." *American Journal of Sociology* (March): 353–59.

_____. 1980. "Coming Out All Over: Deviants and the Politics of Social Problems." *Social Problems* 28: 1–13.

Krauthammer, Charles. 1993. "Defining Deviancy Up." *The New Republic* (November 22): 20–25.

Lemert, Edwin. 1951. *Social Pathology.* New York: McGraw-Hill.

_____. 1967. *Human Deviance, Social Problems, and Social Control.* New York: Prentice Hall.

Letkemann, Peter. 1973. *Crime as Work.* Englewood Cliffs, NJ: Prentice Hall.

Loeber, Rolf, and David P. Farrington. 1998. *Serious and Violent Juvenile Offenders.* Thousand Oaks, CA: Sage.

Lombroso, Cesare. 1876. *On Criminal Man.* Milan, Italy: Hoepli.

_____. 1920. *The Female Offender.* New York: Appleton.

Luckenbill, David F., and Joel Best. 1981. "Careers in Deviance and Respectability: The Analogy's Limitations." *Social Problems* 29(2): 197–206.

Lyman, Stanford M. 1970. *The Asian in the West.* Reno/Las Vegas, Nevada: Western Studies Center, Desert Research Institute.

Matza, David. 1964. *Delinquency and Drift.* New York: Wiley.

Merton, Robert. 1938. "Social Structure and Anomie." *American Sociological Review* 3 (October): 672–82.

Miller, Walter. 1958. "Lower Class Culture as a Generating Milieu of Gang Delinquency." *Journal of Social Issues* 14(3): 5–19.

Mills, C. Wright. "Situated Actions and Vocabularies of Motive," *American Sociological Review,* V (December): 904–913.

Moynihan, Daniel Patrick. 1993. "Defining Deviancy Down." *The American Scholar* 62(1): 17–30.

Polsky, N. 1967. *Hustlers, Beats, and Others.* Chicago: Aldine.

Quinney, Richard. 1970. *The Social Reality of Crime*. Boston: Little, Brown.

Schur, Edwin. 1971. *Labeling Deviant Behavior*. New York: Harper & Row.

_____. 1979. *Interpreting Deviance*. New York: Harper and Row.

Scott, Marvin, and Stanford Lyman. 1968. "Accounts." *American Sociological Review* 33(1): 46–62.

Sellin, Thorsten. 1938. "Culture Conflict and Crime." A Report of the Subcommittee on Delinquency of the Committee on Personality and Culture, *Social Science Research Council Bulletin* 41. New York.

Sheldon, William. 1949. *Varieties of Delinquent Youth*. New York: Harper & Row.

Skinner, B. F. 1953. *Science and Human Behavior*. New York: Macmillan.

Smith, Alexander B., and Harriet Pollack. 1976. "Deviance as a Method of Coping." *Crime and Delinquency* 22: 3–16.

Spector, Malcolm and John Kitsuse. 1977. *Constructing Social Problems*. Menlo Park, CA: Cummings.

Steffensmeier, Darrell, and Jefferey Ulmer. 2005. *Confessions of a Dying Thief*. New Brunswick, NJ: Aldine Transaction.

Sumner, William. 1906. *Folkways*. New York: Vintage.

Sutherland, Edwin. 1934. *Principles of Criminology*. Philadelphia: J. B. Lippincott.

Sykes, Gresham, and David Matza. 1957. "Techniques of Neutralization: A Theory of Delinquency." *American Sociological Review* 22: 664–70.

Tannenbaum, Frank. 1938. *Crime and the Community*. Boston: Ginn.

Turner, Ralph H. 1972. "Deviance Avowal as Neutralization of Commitment." *Social Problems* 19 (Winter): 308–21.

Weatherford, Jack. 1986. *Porn Row*. New York: Arbor House.

Werner, E., and R. Smith. 1992. *Overcoming the Odds: High-Risk Children from Birth to Adulthood*. New York: Cornell University Press.

Whitt, Hugh P. 2002. "Inventing Sociology: André-Michel Guerry and the *Essai sur la statistique morale de la France*." pp. ix–xxxvi in Hugh P. Whitt and Victor W. Reinking (trans.), *André-Michel Guerry's Essay on the Moral Statistics of France: An 1833 Report to the French Royal Academy of Science*. Lewiston, NY: Edwin Mellen Press.

Yip, Andrew K. T. 1997. "Gay Male Christian Couples and Sexual Exclusivity." *Sociology* 31(2): 289–306.

Yochelson, Samuel, and Stanton Samenow. 1976. *The Criminal Personality*, Vols 1 & 2. New York: Jason Aronson.